APPLIED LINEAR ALGEBRA with *APL*

APPLIED LINEAR ALGEBRA with *APL*

Garry Helzer
UNIVERSITY OF MARYLAND

LITTLE, BROWN AND COMPANY
Boston Toronto

Library of Congress Cataloging in Publication Data

Helzer, Garry.
Applied linear algebra with APL.

1. Algebras, Linear—Data processing.
2. APL (Computer program language) I. Title.
QA184.H463 1982 512′.5 82-82210
ISBN 0-316-35526-7

Library of Congress Catalog Card No. 82-82210

ISBN 0-316-35526-7

9 8 7 6 5 4 3 2 1

MV

Published simultaneously in Canada
by Little, Brown and Company (Canada) Limited

Printed in the United States of America

TO SUE

Preface

My purpose in this book is to provide material of immediate practical use while building upon and strengthening the mathematical skills the reader brings to it. The book exhibits some continuity in both style and content with previous work while providing the practical details necessary for application.

The book's readers will come from varying backgrounds — for linear algebra is important not only in the physical sciences and engineering but also in economics, biology, and, through multivariate statistics, in any field where large amounts of data are gathered and analyzed. These readers share the problem of processing their data in computers. Chapter 1 begins on this common ground, introducing vectors and matrices as data structures suitable for storing and manipulating sets of data. This gets us directly into $\mathbf{R}^n$ without any reference to higher-dimensional geometry. Connections are made with previous mathematics via evaluation of sums and approximation of integrals.

Chapter 2 develops matrix algebra in the manipulative spirit of secondary school algebra. This concrete and familiar approach permits greater depth: one-sided inverses and least-squares solutions are discussed. Deeper proofs are postponed until Chapter 3.

Linear dependence is first mentioned in Chapter 2 in connection with the existence of left inverses. Thereafter the concept occurs from time to time throughout the text. There is no abstract chapter on linear dependence and dimension theory.

Restricting the discussion to $\mathbf{R}^n$ and emphasizing matrix algebra allows one to deal with a class of functions that is both wider and less abstract than the set of linear transformations. Affine and quadratic functions are introduced in Chapter 2 and dealt with throughout the book. This makes the applications more natural and accessible. Newton's method in higher dimensions and the linearization of a nonlinear differential equation, for example, are quite natural in the affine context, whereas phrasing these procedures in purely linear terms raises barriers to understanding.

By the end of Section 2.4 the reader should be able to quickly produce solutions to well-behaved systems of, say, thirty equations in thirty unknowns or to fit a cubic polynomial to a couple of hundred data points.

General (singular) systems of linear equations are taken up in Chapter 3. The properties of the row-reduced echelon form of a matrix developed here are the basis of the development of flats (affine subspaces), coordinate systems, and subspaces given in Chapter 4.

Chapter 4 begins with the analytic geometry of lines, planes, and conics in $\mathbf{R}^2$ and $\mathbf{R}^3$. The discussion of conics, which is restricted to $\mathbf{R}^2$, is needed for the diagonalization of symmetric matrices and the second-derivative test for max-min in Chapter 5. The lines and planes are generalized to their higher-dimensional versions and connected with sets of solutions of systems of linear equations. Matrix algebra is used to define coordinate systems on subspaces of $\mathbf{R}^n$ and derive coordinate-change formulas for affine and quadratic functions. Only then are the more abstract notions of subspace and basis introduced.

Distance and angle are introduced in Chapter 5 and orthonormal coordinate systems are defined. Then symmetric matrices are diagonalized via the Jacobi algorithm, and some eigenvalue theory is developed. This allows readers interested mainly in statistical applications to avoid the general eigenvalue theory of Chapter 7.

Perpendicular projection is emphasized and the least-squares computations of Chapter 2 are justified.

In some texts, linear programming is presented as a strange, out-of-the-blue trick (the simplex algorithm) that can be used to solve dog-food-mixing problems. Chapter 6 emphasizes the geometric aspects of the problem and offers a broader view of the possible applications and methods of solution.

Chapter 7 emphasizes the geometric aspect of determinants, develops the standard facts about eigenvalues, and closes with some reasonably efficient functions for computing eigenvalues. A reader finishing Section 7.7 should have little trouble analyzing the eigenvalues of matrices of small order (up to 20 by 20, say).

The ability to get answers to nontrivial problems is what makes practical application possible. This ability is provided by the APL system. It would be wrong, however, to conclude that this is the sole function of the APL notation or that "programming" is being taught at the expense of "mathematics." The "programming" is here to reinforce and aid the mathematics. It is a teaching tool. It is not something apart from the course's mathematical content.

This reinforcement is possible because, in spite of its name,† APL is not really a programming language at all. It is a system of mathematical notation developed for the purpose of expressing algorithms symbolically.

If you cannot express an algorithm precisely — that is, symbolically — then your understanding of it is in some measure incomplete. "You claim you understand the mathematics? Show me the output."

In this text the use of APL aids abstract understanding in several ways. The APL language is based on the idea of a function. In APL one does not write programs, one defines functions. It can be used to teach the idea that a function is a rule — one that cannot necessarily be expressed as an algebraic formula.

Also, APL is used in this book to teach mathematical induction. There are few loops in this text; instead, functions are written recursively. "Do you understand induction? Why doesn't your function work?"

†APL stands for A Programming Language.

Proofs lose some of their abstract character when we see that they often describe algorithms that may be coded more or less directly into functions. The proof that Gaussian reduction can be used to produce a row echelon form, for example, is coded directly into an APL function that takes a matrix and computes its row echelon form.

This approach has certain implications for both "programming style" and the treatment of numerical analysis.

The programming in this book is completely functional. All the APL functions take an explicit input and produce an explicit output. They do not alter their environment. In particular, messages are not printed and global variables are neither used nor changed. Any of the functions may be used in any expression where they make mathematical sense. This allows the construction, from chapter to chapter, of extremely powerful and flexible tools for solving linear algebra problems.

Since the computation is designed to help the learning process and since a large portion of the theory is based upon the properties of the row echelon form of a matrix, our Gaussian reduction function computes the row-reduced echelon form of a matrix instead of, say, the LU factorization. This is in spite of the theoretical savings in computation time offered by such methods. For the same reason we are not particularly concerned about operation counts. We are interested in avoiding codings that cause the computational expense to grow exponentially, but this is about the extent of our concern.

The style of computation envisioned here is personal computing.† Operation counts are less vital when letting your microcomputer run all weekend costs a negligible fraction of your electricity bill. For the most part, FORTRAN-based cost estimates are meaningful for higher-level processors, such as APL, only when applied to large problems. The coding of the QR algorithm used in Chapter 7 runs much faster than would APL implementations of current state-of-the-art FORTRAN code for matrices of the size considered in this text. (Further, as of this writing at least one APL vendor has implemented an eigenvalue solver as a primitive APL function and extended domino to arbitrary matrices.)

Accuracy is a different matter. Error estimates are beyond the scope of this text. As a consequence, we tend to answer the question, "How accurate is your answer?" with a second question, "What do you mean by an 'answer'?" Given a purported answer $\bar{x}$ to a linear system $Ax = b$, we accept it as an "answer" if $b - A\bar{x}$ is sufficiently small. This spot checking is perfectly feasible in the interactive APL environment. The more subtle question of how close such an $\bar{x}$ need be to the "real" answer must be left for more advanced work. The practice with actual computation, however, is excellent experience for more advanced work.

I am indebted to P. E. Hagerty for advice and criticism on the use of APL, to Ellen Correl, Hsin Chu, Jerome Dancis, and Avron Douglis who taught from preliminary versions of this book, and especially to the University of Maryland Mathematics Department technical typist staff, Berta Casanova, Cindy Black, June Slack, and Linda Fiori, who typed so many drafts of the manuscript.

† Several popular microcomputers may be equipped with APL interpreters for a few hundred dollars.

DEPENDENCY DIAGRAM

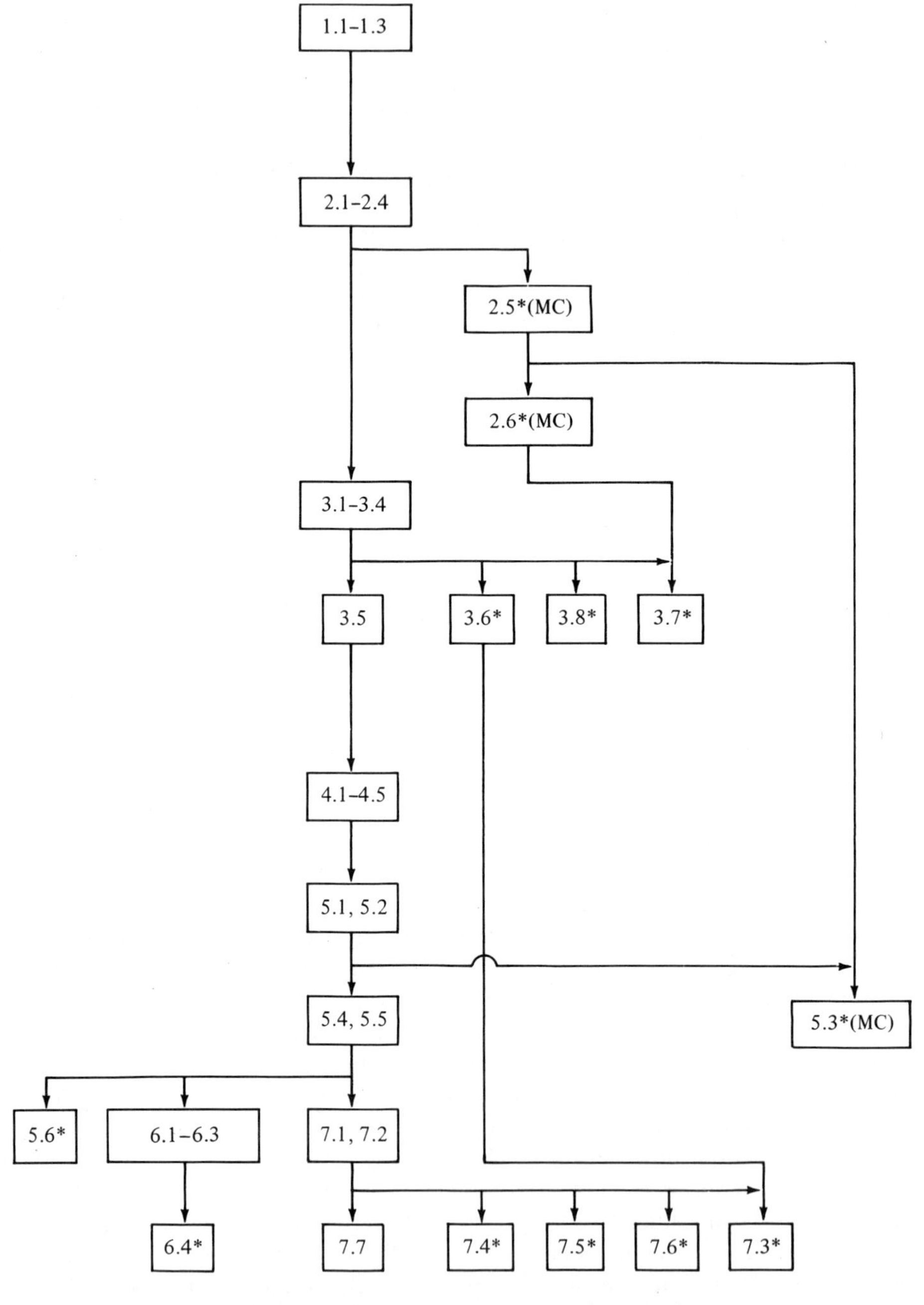

Sections within boxes to be covered consecutively

MC = multivariate calculus

* = optional section

Instructor's guide

This text may be used for a variety of linear algebra courses at a variety of levels. The sections marked by asterisks are optional. No later sections not marked by asterisks depend upon them. The sections labeled "multivariate calculus" assume some familiarity with partial differentiation. A dependency diagram is provided.

A sophomore-level course. This course proceeds straight down the dependency diagram from Section 1.1 to Section 7.7, omitting Chapter 6 and the optional sections. Class-time discussion of Sections 3.6 and 5.5 may be omitted and the results of these sections provided as library functions. If time permits, optional Sections 3.6* and 7.3* or 7.4* may be added.

A junior/senior-level course emphasizing physical applications. This course consists of the sophomore-level course plus the optional multivariate calculus Sections 2.5*, 2.6*, and 5.3*. Since such courses typically have fewer class meetings per semester than a sophomore-level course, this material is often all that can be covered comfortably. If time permits, however, Sections 7.6*, 3.7*, and possibly 3.8* may be added.

A junior/senior-level course emphasizing nonphysical applications. This course consists of the sophomore course with Chapter 6 replacing Chapter 7. This scheme avoids complex numbers but covers eigenvalues for symmetric matrices—the main statistical application. If time permits, Section 7.1 (determinants) should be added.

A second-semester course on applications or mathematical models. This course assumes a previous sophomore-level course in linear algebra and consists of Chapter 6 and the optional sections other than 7.4*. If the sophomore course does not use this text, then Sections 1.1, 3.1, and 3.4 are necessary for APL background. Additional APL operations such as matrix multiplication and inversion (domino) may be quickly explained as needed. If the students' previous course emphasized linear transformations, then a rapid run through Sections 1.2, 1.3, 2.2, and 2.3 will improve their grasp of matrix algebra.

Contents

CHAPTER SEVEN

Eigenvalues and Eigenvectors 448

APPLIED LINEAR ALGEBRA with *APL*

CHAPTER ONE

Vectors, Matrices, and APL

Because modern research methods deal with collections of numbers, some knowledge of vector and matrix concepts is almost indispensable for research in most fields. The numbers dealt with may be, for example, the tabulated results of a questionnaire or of many repetitions of a laboratory measurement. Sometimes the numbers do not come from measurements but are generated in a calculation, as during computer solutions of differential equations. Matrix algebra provides a convenient formalism for handling groups of numbers, and this formalism is commonly used to process such data by digital computers.

In fact, the most powerful scientific computers built since about 1980 are often referred to as vector machines. They are designed to perform a group of tasks simultaneously (*parallel processing*) instead of taking up each individual task of the group in order (*sequential processing*).

In this chapter vectors and matrices are introduced as *data structures* — devices for organizing and manipulating collections of numbers. The chapter also introduces the APL notational system.

Although APL is usually described as a programming language (in fact, the acronym APL is derived from the phrase "A Programming Language"), it originated as a formal notation for describing algorithms.† It is a mathematical notation that can be directly executed by a computer equipped with an APL processor. In effect an APL processor turns a computer into a (very powerful) calculator. As a result many difficult calculations may be carried out on such a machine without resort to programming. Programming — or, more properly, *function definition* — is taken up in a later chapter.

As we shall see, the APL notation is particularly apt when parallel or vector processing is involved.

Section 1.1 is devoted to elementary arithmetic in APL notation. In Section 1.2 vectors (singly subscripted arrays) are introduced and the APL operations are extended to vectors. In Section 1.3 matrices (doubly subscripted arrays) are defined and the APL operations are further extended to this case.

†K. E. Iverson, *A Programming Language* (New York: Wiley, 1962).

1.1 Some APL Notation

This section introduces the APL system of mathematical notation. This system has been designed to execute directly on computers, avoiding the usual intermediate step of first translating the mathematics into a "program" and then running the program on a computer.

Performing calculations in APL is similar to using a hand calculator. In fact, an APL processor turns a computer into a very effective calculator. This machine orientation gives rise to a notation that differs in several respects from the conventional. For example, one writes `X*2` for x^2 and `3×A` for $3A$. Other differences will become apparent as the work proceeds; some of these are as difficult to assimilate as the change from miles to kilometers. The advantages, however, outweigh the disadvantages.

In the text proper we will consider the use of the APL symbolism but not the practical details of connecting with a computer, controlling printout format, and so on. Some hints are given in Appendix C, however, and the various APL function symbols are listed in Appendix B.

The basic operations of arithmetic — addition, subtraction, multiplication, and division — are denoted by `+`, `-`, `×`, `÷`, respectively. In APL an expression is regarded as an instruction to perform the indicated operations. For example, if you type `2+3` (followed by a carriage return), the computer types `5` at the left margin and then moves the typehead in seven spaces on the next line to signal that it is ready for the next command.

```
      2+3
5
```

Here are more examples.

```
      3-2          you type 3-2, the computer types 1
1
      2×3          you type 2×3, the computer types 6
6
      10÷5         you type 10÷5, the computer types 2
2
```

In the representation of negative numbers the APL notation for basic arithmetic differs from conventional notation. In APL negative numbers are indicated not by the subtraction sign but by a separate symbol — the "high minus."

```
      2-3          3 subtracted from 2 is negative 1
¯1
      2+¯3         2 plus negative 3 is negative 1
¯1
      2-¯3         negative 3 subtracted from 2 is 5.
5
```

The subtraction symbol, -, is at the upper right of the APL keyboard (with +, ×, ÷), whereas the high minus ¯ is at the upper left (shift-2). The reason for the high minus will become apparent when vectors are discussed in Section 1.2.

Machine computations are rarely exact. In fact, most numbers cannot even be stored in the machine exactly. The number of digits a computer can handle varies with the manufacturer. Standard APL systems will display ten digits unless told otherwise.

```
      2÷3
0.6666666667
```

This is too many digits for most purposes, and so most examples are displayed to three digits.

```
      )DIGITS 3
WAS 10
      2÷3
0.667
```

The)DIGITS is an example of an APL system command. It sets the number of digits in the displayed answer. It does not affect the accuracy of the calculation in any way.† System commands such as)DIGITS are dealt with in Appendix C.

The APL language has many arithmetic functions in addition to +, -, ×, and ÷. A complete list of APL functions is in Appendix B. Here are a few that we shall use in subsequent examples.

```
      2*2
4
      2*.5
1.414213562
```

$a*b$ is APL notation for a^b. Notice that $\sqrt{a}$ is $a^{1/2}$.

* is called *exponentiation.*

```
      !3
6
      !6
720
```

! is the factorial function. Notice that in APL one writes "!3" rather than "3!." (The exclamation point does not appear on the APL keyboard. It is typed as quote-back-space-period.)

```
      ⌊3.141
3
      ⌊¯3.141
¯4
```

⌊ is called *Floor.* ⌊a is the greatest integer that is less than or equal to a (sometimes called "the greatest integer function").

```
      |3
3
      |¯3
3
```

|a is the absolute value of a.

†The examples in this text were all computed on a UNIVAC 1100/42 using the APL-1100 processor developed at the University of Maryland by P. E. Hagerty. This processor does most calculations in double-precision floating-point arithmetic, which for UNIVAC machines is about eighteen decimal digits. Other machines and processors may produce slightly different answers.

VARIABLES

The phrase "Let A denote the number 3" is written in APL symbols as A←3. One may then use the letter A anywhere that the number 3 would be appropriate.

```
      A←3
      1+A
4
      A*A
27
      A        (Display A)
3
```

A is called a *variable*. To display the value of a variable, type its name. This has been done in the last two lines above. Variable names are not restricted to single letters. They may be almost any combination of letters, digits, and the symbol ∆.

```
      SAM←A
      A×SAM
9
```

Here are more examples.

```
      B←1+A
      B
4
      A←B+B
      A
8

      A←1+A
      A
9
```

Look at this example carefully. It reads from right to left. "Take A, add 1 to it, and call the result A" or "Increment A by 1."

MONADIC AND DYADIC FUNCTIONS

In the APL notation most symbols denote two different (usually related) functions. Here is an example.

```
      !6
720
      !0
```

!n is $n(n - 1)(n - 2) \cdot ... \cdot 2 \cdot 1$, the factorial function.† k!n is the binomial coefficient $\binom{n}{k}$ (the number of subsets of size k that may be chosen from a set of size n).

†In fact, !n is $\Gamma(n + 1)$ where Γ denotes the gamma function. This means k need not be an integer. The function k!n is closely related to $\Gamma(n + 1)/\Gamma(k + 1)\Gamma(n - k + 1)$.

```
1
      2!3
3
      3!2
0
```

A *monadic* function has a single argument, which is written to the right of the function. If f is a monadic function, we write fa in APL for the more conventional $f(a)$. One can write $f(a)$ if one wishes. The parentheses are not wrong; they are simply unnecessary.

```
      !(6)
720
```

A *dyadic* function has two arguments. One is written on the left and the other on the right. If f is a dyadic function, we write afb in APL for the conventional $f(a, b)$. Unlike the monadic case, the conventional notation $f(a, b)$ cannot be used instead of afb. The symbols $f(a, b)$ mean something quite different in APL.

Dyadic functions seem a bit strange until we realize that the basic arithmetic functions +, −, ×, ÷, * are all dyadic. We conventionally write $a+b$, $a×b$, $a÷b$ rather than the "$f(a, b)$" forms $+(a, b)$, $×(a, b)$, $÷(a, b)$. One could keep the dyadic form for the functions +, ×, ÷, −, say, and use the $f(a, b)$ for all other functions of two variables. But the existence of special cases would complicate the automatic machine execution of formulas.

The need to treat all functions in the same way is also the reason one writes `!6` rather than `6!`, `|-3` rather than `|-3|`, and so on.

Here are some more monadic-dyadic forms.

```
      2*3
8
      2*.5
1.414213562
      2*¯1
0.5
```

$a*b$ is a^b

```
      *1
2.718281828
      *.5
1.648721271
      *¯1
0.3678794412
```

$*a$ is e^a, where e is the base of natural logarithms. Since $e^1 = e$, the number e may be obtained as `*1`.

```
      ⌊3.56
3
      ⌊¯4.67
¯5
```

$⌊a$ is the floor of a, defined previously.

```
      3⌊4
3
      2⌊(2*¯1)
0.5
```

$a \lfloor b$ is the minimum of the two numbers a and b. Since `2*¯1` is $\frac{1}{2}$, the second computation gives the minimum of 2 and $\frac{1}{2}$.

The symbols +, −, ×, ÷ denote monadic functions as well as the usual arithmetic functions. Two important ones are monadic ÷ and monadic −.

```
      ÷2
0.5
```

÷a is $1/a$

```
      A←3
      −A
¯3
      ¯A
LEXICAL ERR
```

−a is the negative of a. Notice that the high minus ¯ does not denote a function. It is a part of a number — like a digit or decimal point.

EXAMPLE 1.1 Write the expression $\sqrt{1+2}$ in APL notation.

Solution `(1+2)*.5` or `(1+2)*(÷2)` ■

EXAMPLE 1.2 Write the expression $|-3|^3$ in APL notation.

Solution `(|¯3)*3` or `(|−3)*3` ■

In the first expression the high minus is used to indicate a negative 3. In the second expression the monadic − is used to change `3` to `¯3`.

EXAMPLE 1.3 Write the expression $1 + e^{\sqrt{5}}$ in APL notation.

Solution `1+*(5*.5)` or `1+*(5*(÷2))` ■

EXAMPLE 1.4 Write the expression $\sin^3 4$ in APL, where 4 means 4 radians.

Solution From Appendix B we see that `1○`x is the sine of x radians. Thus the answer is

`(1○4)*3` ■

EXAMPLE 1.5 Write the number π in APL notation.

Solution From Appendix A we see that ○a is π times a. Thus the solution is

`○1` ■

Example 1.6 Translate the statement

"Let X be 7" or "let $X = 7$"

into APL notation.

Solution These are standard mathematical phrases that correspond to the APL expression

```
X←7
```

■

Example 1.7 Translate into APL:

Set A equal to 7
and set B equal to e^{-A}

Solution

```
A←7
B←*(-A)
```

Notice that one must use the monadic -. The high minus ¯ cannot be used.

The solution can be written on one line:

```
B←*(-(A←7))
```

■

Example 1.8 Translate into APL:

Set D equal to $\sqrt{B^2 - 4AC}$

Solution

```
D←((B*2)-4×A×C)*.5
```

Note: Typing this expression on the computer will result in an error message unless A, B, and C have already been assigned values. ■

ORDER OF EVALUATION

The parentheses about `B*2` in the last example are necessary, although they are not necessary in conventional notation. Furthermore, many of the parentheses in the examples above are not necessary but are simply there for clarity. For example, the solution to Example 1.3 could be written

```
1+*5*÷2
```

This is because the *assumed order of evaluation* in APL expressions is different from the conventional order.

```
      1+2×3
7
```

```
      2×3+1
8
      (2×3)+1
7
      2×(3+1)
8
```

Evaluation Rule

Start at the right-hand end of the line and work to the left, evaluating each function as you encounter it.

In the example above

```
1 + 2 × 3
    └────────── First compute 2×3 obtaining 6
7
└────────────── then compute 1+6 obtaining 7
2 × 3 + 1
    └────────── First compute 3+1 obtaining 4
8
└────────────── then compute 2×4 obtaining 8
```

Notice that you can force any order of evaluation you prefer by using parentheses. Expressions in parentheses are evaluated first. When in doubt use parentheses!

EXAMPLE 1.9 Evaluate the APL expression

```
1-2-3
```

and translate it into conventional notation.

Solution Working from the right, we first evaluate 2-3 and then we subtract the result from 1. Thus we have

```
1-(2-3)
```

or 2. The conventional expression is $1 - (2 - 3)$ or $1 - 2 + 3$. ■

EXAMPLE 1.10 Translate the APL expression

```
2**5
```

into conventional notation.

Solution Starting at the right, we first encounter *5. Notice that this is the monadic * because there is a function symbol rather than a number or variable to the left of this *. Thus the expression is equivalent to

```
2*(*5)
```

or

$$2^{e^5}$$ ■

EXAMPLE 1.11 Translate the APL expression

```
2×3+4×5
```

into conventional notation. Evaluate.

Solution Working from the right, we have

```
2×(3+(4×5))
```

or

```
46
```

■

EXAMPLE 1.12 Evaluate the APL expression

```
⌊3⌈7⌊¯2⌈12
```

Solution Working from the right, we obtain

```
⌊(3⌈(7⌊(¯2⌈12)))
```

To evaluate the expression, we note from Appendix B that dyadic ⌈ is the maximum function, dyadic ⌊ is the minimum function, and monadic ⌊ is floor. Thus

```
      ⌊(3⌈(7⌊(¯2⌈12)))
is    ⌊(3⌈(7⌊12))
is    ⌊(3⌈7)
is    ⌊7
or    7
```

■

EXAMPLE 1.13 Translate the APL expression

```
-A-B
```

into conventional notation.

Solution Starting from the right, we first encounter $A - B$; thus the expression is $-(A - B)$ or $B - A$. ■

The rules for evaluating complex APL expressions may be formally stated as follows:

RULE 1 If a function symbol has number, variable, or expression in parentheses to its left, then it represents a dyadic function. Otherwise it represents a monadic function.

RULE 2 A monadic function operates on everything to its right.

RULE 3 A dyadic function operates on everything to its right and the number, variable, or expression in parentheses to its left.

RELATIONAL AND LOGICAL FUNCTIONS

(=, ≠, <, >, ≤, ≥, ∧, ∨, ~, ⍲, ⍱)

A mathematical notation that is both powerful and machine readable must allow us to write statements normally expressed in English sentences as "formulas" — that is, as formal expressions.

For example, the statement

Let A be the set of prime numbers between 1 and 20

can be written symbolically in APL† and processed by a computer equipped with an APL processor.

The writing of such expressions is made possible in part by the APL meanings of the symbols =, ≠, <, >, ≤, ≥. In APL these symbols denote dyadic functions, called the *relational* functions. Here is how they work:

```
      3<2
0
      2<3
1
```

The expression 3<2 is considered to be a question that requires a yes or no answer: 0 means no and 1 means yes.

Again:

```
      1=3     Is 1 equal to 3?
0             No
      1≠3     Is 1 not equal to 3?
1             Yes
```

The relational functions may be used in expressions in the same manner as any dyadic function.

†A←(2=+⌿0=(⍳20)∘.|⍳20)/⍳20. This is a formal description of the sieve of Eratosthenes; see exercise 46 in Exercises 1.3.

Example 1.14 Evaluate the APL expression

```
4=3+1
```

Solution As in any APL expression, one may begin at the right and work to the left, evaluating each function as it is encountered. Doing this (or applying Rule 3 directly to the dyadic function =), we see that the expression is equivalent to

```
4=(3+1)
```

or `4=4`. The answer is `1`. ■

Example 1.15 Evaluate the APL expression

```
3+1=4
```

Solution Again either by working from the right or by applying Rule 3, we see that the expression is equivalent to

```
3+(1=4)
```

This is `3+0` or `3`. ■

Example 1.16 Evaluate the APL expression

```
4=⍟*4
```

Solution The expression is equivalent to `4=(⍟(*4))`. From Appendix B we see that `*4` is e^4 and monadic `⍟` is the natural logarithm. Thus `⍟*4` is $\ln(e^4)$ or 4. The answer is 1. ■

The *logical* functions ∧, ∨, ~, ⍲, ⍱ are called *and, or, not, Nand,* and *Nor*. They operate exclusively with the "true/false" values 1 and 0. The operation tables for *AND* and *OR* are

∧	0	1
0	0	0
1	0	1

∨	0	1
0	0	1
1	1	1

The function *NOT* is monadic and changes 1 to 0 and 0 to 1. The table for ∧ can be read "false *AND* false is false," "false *AND* true is false," "true *AND* false is false," and "true *AND* true is true." Similarly for *OR*. The operation of ~ can be read "*NOT* true is false" and "*NOT* false is true."

EXAMPLE 1.17 For what values of X does the APL expression

```
(X≤2)∧(X≥¯1)
```

result in a `1`? a `0`?

Solution For the expression to result in a `1`, both `X≤2` and `X≥¯1` must result in a `1`. Thus `X` must be less than or equal to 2 *and* `X` must be greater than or equal to -1. The result of the expression is a `1` if `X` lies in the interval $[-1, 2]$. Otherwise the result is a `0`. ∎

Notice that in such expressions `∧` really means "and." The next example shows why the function `∧` is necessary in the expression above.

EXAMPLE 1.18 What is the result of the APL expressions?

```
X←3
2≤X≤4
```

Solution Using the rules for evaluating APL expressions, we see that the second expression is equivalent to

```
2≤(X≤4)
```

since `X` is `3`, `X≤4` is `1` and the expression reduces to `2≤1` or `0`. Thus, although `X` lies in the interval $[2, 4]$, the expression `2≤X≤4` results in a `0` or "false." To test if `X` lies in the interval $[2, 4]$, one must use such expressions as

`(2≤X)∧X≤4` or `(4≥X)∧2≤X` ∎

EXAMPLE 1.19 For what values of the variables A and B does the APL expression

```
(~A∨B)=(~A)∧(~B)
```

result in a 1?

Solution In order for the expression to make sense, the variables A and B must have values `0` or `1`. Thus we are immediately reduced to four possible cases `A←0, B←0; A←0, B←1; A←1, B←0; A←1, B←1.`

Suppose `A←0` and `B←0`. Then the expression is

```
   (~0∨0)=(~0)∨(~0)
or (~0)=1∨1
or 1=1
or 1
```

Checking the other three cases in this way shows that the expression always returns a 1. This means that the expression is an identity. In formal logic this identity is known as one of DeMorgan's Laws. ■

EXERCISES 1.1

Write the following arithmetic expressions in APL notation. All angles are in radians. A list of APL functions may be found in Appendix B.

1. $\sqrt{2}$ 2. $3^2 + 1$ 3. 3^{2+1} 4. $1 + \frac{2}{3}$
5. $7^{-3/2}$ 6. $\ln 3 + 2$ 7. e^4 8. e^{2^2}
9. $(e^2)^2$ 10. $\sqrt{3+7}$ 11. $\sqrt{3} + 7$ 12. $1/\sqrt{2}$
13. $\pi/180$ 14. $\tan 7$ 15. $\sin(3\pi/2)$ 16. $\cot 7$
17. $\sec^2 3 + 1$ 18. $|-\frac{3}{2}|$ 19. $\sinh^2 4 + \cosh^3 5$ 20. $\mathrm{Tan}^{-1}(\cot(3\pi))$
21. $\log_{10}(\ln 3)$ 22. e^{π} 23. $e^{\pi} - \pi^e$

Write the following algebraic expressions in APL notation. All angles are in radians. A list of APL functions may be found in Appendix B.

24. $\dfrac{x+y}{x-y}$ 25. e^{xy} 26. $\sin 3x$ 27. $\sqrt{x^2 + y^2}$
28. $\sqrt{b^2 - 4ac}$ 29. $\frac{1}{2}bh$ 30. πr^2 31. $\frac{1}{3}\pi r^2 h$
32. $\dfrac{\sqrt{a+h} - \sqrt{a}}{h}$ 33. $\dfrac{1}{\sqrt{1+x^2}}$ 34. $\ln|u|$ 35. $\dfrac{1+\sqrt{x}}{1-\sqrt{x}}$

Evaluate the APL expressions. Answers should be worked by hand and checked on a computer, if one is available. A list of APL functions may be found in Appendix B.

36. `1+2×3` 37. `1+2÷3` 38. `1-2-3` 39. `2×3-4×7`
40. `¯3*2` 41. `-3*2` 42. `3*-2` 43. `3*¯2`
44. `⍟*1` 45. `*⍟6` 46. `1○¯1○÷2` 47. `1○¯1○○1`
48. `(-1-3)*÷2` 49. `(12!20)-(!20)÷(!12)×!20-12`
50. `(2×6○7)-(*7)+*-7` 51. `⌊5÷3` 52. `⌈5÷3`
53. `(5÷3)-⌊5÷3` 54. `5-(3×⌊5÷3)` 55. `⌈1○○÷4`
56. `⌊.5+100÷3` 57. `1⌈2⌈3⌈4⌈5⌈6` 58. `1⌈2⌊3⌊4⌈5⌈6`
59. `1⌊2⌊3⌊4⌊5⌊6` 60. `(A+|A)-2×0⌈A` where `A` is any number.

Translate the APL expressions into conventional notation.

61. `A÷B÷C` 62. `1σ○X` 63. `Z-B-C+D` 64. `A×B-C`
65. `X*Y*2` 66. `X*2+Y*2` 67. `3○¯3○X` 68. `1--*X`
69. `1++*X` 70. `○R*2` 71. Which is larger, e^{π} or π^e?

What is the result of the following APL expression(s)? Answers should be worked by hand and checked on a computer, if one is available. A list of APL functions may be found in Appendix B.

72. `1=4-3`

73. `4-3=1`

74. `4-3≠1`

75. `1≠4-3`

76. `X←3`
    `1≤X`

77. `X←3`
    `(1≤X)∧X≤12`

78. `X←3`
    `(4≤X)∨X≤12`

79. `K←44`
    `1*0≠2|K`

80. `0=1=2=3=4=5`

1.2 Vectors

In applying mathematics to real-world situations, it is often necessary to process large quantities of data. The vector concept, although originating in the physics of three-dimensional space, has become indispensable for the symbolic manipulation of large quantities of data. The related concept of a matrix will be discussed in the next section.

We begin with an important example.

THE LEAST-SQUARES STRAIGHT LINE

Suppose that fifteen adult males are chosen at random and their height (in inches) and their weight (in pounds) recorded. The resulting data are given in Table 1.1 and plotted in Figure 1.1.

TABLE 1.1

Height (x)	60	61	62	63	64	65	66	68	69	70	71	72	74	75	76
Weight (y)	120	120	135	135	130	135	150	140	170	145	160	160	160	160	175

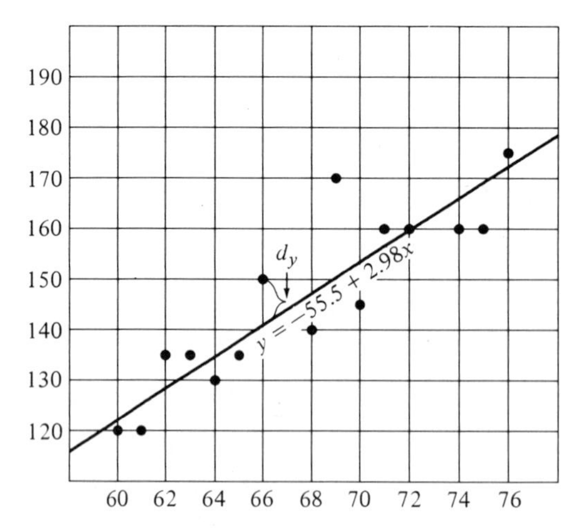

FIGURE 1.1

Assuming that weight, denoted by x, is a straight-line function of height, denoted by y, what is the best estimate of this function?†

That is, we are assuming a relationship of the form

$$y = a_0 + a_1 x \tag{1.1}$$

and we wish to estimate a_0 and a_1 from our data.

The estimate of the line (1.1) most often used in this situation is the *least-squares* straight line. This line is characterized by the fact that a_0 and a_1 are chosen to minimize the sum

$$\sum_{i=1}^{n} (a_0 + a_1 x_i - y_i)^2 \tag{1.2}$$

where the measured data points are $(x_1, y_1), (x_2, y_2), \ldots, (x_n, y_n)$. Since $a_0 + a_1 x_i$ is the point on the line (1.1) with x coordinate equal to x_i, one sees that (1.2) is the sum Σd_i^2 where d_i is the vertical distance from the point (x_i, y_i) to the line (1.1).

The coefficients a_0, a_1 that minimize the expression (1.2) can be shown to be the solutions of the *normal equations*

$$\begin{aligned} n a_0 + \left(\sum x_i\right) a_1 &= \sum y_i \\ \left(\sum x_i\right) a_0 + \left(\sum x_i^2\right) a_1 &= \sum x_i y_i \end{aligned} \tag{1.3}$$

where all sums are taken as i ranges from 1 to n. These equations will be proved several times later in increasing generality (see Exercise 61 and Section 5.4).

The solution of (1.3) is given by

$$\begin{aligned} a_0 &= \frac{\left(\sum y_i\right)\left(\sum x_i^2\right) - \left(\sum x_i\right)\left(\sum x_i y_i\right)}{\Delta} \\ a_1 &= \frac{n\left(\sum x_i y_i\right) - \left(\sum x_i\right)\left(\sum y_i\right)}{\Delta} \\ \Delta &= n\left(\sum x_i^2\right) - \left(\sum x_i\right)^2 \end{aligned} \tag{1.4}$$

† One would expect weight to be proportional to volume and volume to be proportional to $(\text{height})^3$. In the limited range of heights of adult males in a culturally homogeneous population, however, a straight line might be good enough.

TABLE 1.2

X	Y	XY	X^2
60	120	7,200	3,600
61	120	7,320	3,721
62	135	8,370	3,844
63	135	8,505	3,969
64	130	8,320	4,096
65	135	8,775	4,225
66	150	9,900	4,356
68	140	9,520	4,624
69	170	11,730	4,761
70	145	10,150	4,900
71	160	11,360	5,041
72	160	11,520	5,184
74	160	11,840	5,476
75	160	12,000	5,625
76	175	13,300	5,776
$\Sigma X = 1016$	$\Sigma Y = 2195$	$\Sigma XY = 149{,}810$	$\Sigma X^2 = 69{,}198$

$a_0 = [(2195)(69{,}198) - (1016)(149{,}810)]/[(15)(6918) - (1016)^2] = -55.5$
$a_1 = [(15)(149{,}810) - (1016)(2195)]/[(15)(6918) - (1016)^2] = 2.98$

To facilitate hand computation, the data may be arranged as in Table 1.2. The first two columns X, Y contain the original data. The columns XY and X^2 are calculated line by line from X and Y. The results are then summed and a_0, a_1 calculated. The line $y = a_0 + a_1x$ is plotted in Figure 1.1.

We can denote the first column of the table by the single variable X and the second column by the single variable Y. Then X and Y denote *vectors.*

DEFINITION 1.1 A *vector* is an ordered list of numbers.† The individual numbers in the list are the *components* of the vector.

When we say that the list is ordered, we mean that there is a first component, a second component, and so on. In the example above the first component of X is 60, the second component is 61, and the tenth is 70. In Y the first and second components are both 120, the seventh is 150, and the eighth is 140.

In APL notation the vectors X and Y of the table are defined by the expressions

```
X←60 61 62 63 64 65 66 68 69 70 71 72 74 75 76
Y←120 120 135 135 130 135 150 140 170 145 160 160 160 160 175
```

We display the value of a vector by listing the components separated by

†This is probably the most widely accepted definition of the term "vector." It comes from computer science, where a vector generally means any singly indexed list of data — numeric or not. The original geometric meaning of the term can be found in Chapter 4.

blanks. Other examples of vectors in APL notation are

```
.1 2.7 3          three components
1 ¯2 4 7          four components; the high minus required on the second com-
                  ponent is explained below.
```

The number of components in a vector `X` is the called the *size* of `X` or the *shape* of `X` and is denoted by `⍴X`. Using the above *X* and *Y*,

```
      ⍴X
15
      ⍴Y
15
      ⍴1 2 3
3
```

VECTOR ARITHMETIC

In APL notation the usual arithmetic operations have been extended to vectors. If *A* and *B* are vectors of the same size — that is, if `(⍴A)=⍴B` is true — then `A+B` is the result of adding corresponding components of `A` and `B`.

```
      1 .3 6 + 8 ¯7 2
9 ¯6.7 8
      1 2 3 4 + 5 6 7 8
6 8 10 12
      1 2 3 + 1 2
LENGTH ERROR
      1 2 3+1 2
          ∧
```

The same is true of the other operations −, ×, ÷, *.

```
      1 2 3−4 5 6
¯3 ¯3 ¯3
      1 2 3 × 4 5 6
4 10 18
      1 2 3 ÷ 4 5 6
0.25 0.4 0.5
      1 2 3 *4 5 6
1 32 729
```

Now let us return to the least-squares example above. The first two columns of Table 1.2 have been stored in the variables X and Y. The next two columns were computed as

```
      XY←X×Y
      XY
7200 7320 8370 8505 8320 8775 9900 9520 11730 10150 11360 11520 11840 12000 13300
      X2←X×X
      X2
3600 3721 3844 3969 4096 4225 4356 4624 4761 4900 5041 5184 5476 5625 5776
```

These vector multiplications are examples of *parallel processing*. In this case fifteen multiplications are to be carried out and may be done in any order or simultaneously. Operations such as X×Y may execute quite quickly on vector-oriented machines even when X and Y are quite large. On any type of computer, however, APL expressions will operate most efficiently when parallelism is exploited. More important, the use of vector operations tends to simplify formulas, making them easier to understand. Notice that here and below we have eliminated the subscripts from Equations (1.4).

The next task in computing the least-squares straight line is summing the components of the vectors X, Y, XY, and X2. In APL notation ΣX is written +/X and read "plus over X." If X is any vector, then +/X is the sum of the components of X. Thus the sums of the columns of Table 1.2 are given by the following calculations (the monadic + — see Appendix B — has been used to define and display a variable on a single line).

```
      +SX←+/X
1016
      +SY←+/Y
2195
      +SXY←+/XY
149810
      +SX2←+/X2
69198
```

Now we can compute a_0, a_1 from Equations (1.4).

```
      +Δ←(15×SX2)-SX*2
5714
      +A0←((SY×SX2)-SX×SXY)÷Δ
¯55.53902695
      +A1←((15×SXY)-SX×SY)÷Δ
2.98039902
```

We have given the least-squares calculation as an example of how vectors are used to handle calculations involving large quantities of data. The procedure above is not the most efficient way of making a least-squares calculation. As more linear algebra and APL is introduced, the least-squares calculation will be made more general and more efficient.

Incidentally, the extension of the usual arithmetic operations to vectors is what makes the "high minus" necessary. Consider the two calculations below.

```
      A←1 2 ¯3 4
      A
1 2 ¯3 4
      A←1 2 -3 4
      A
¯2 ¯2
```

In the first calculation we have a vector whose third component is negative. In the second calculation the vector `3 4` is subtracted from the vector `1 2`.

The arithmetic functions `+`, `-`, `×`, `÷`, `*` are not the only dyadic functions that operate componentwise on vectors. Any of the scalar dyadic or monadic functions operate componentwise on vectors. Here are some examples. The definitions of the functions may be found in Appendix B.

```
      !1 2 3 4 5
1 2 6 24 120
      A←1 2 ¯3 4
      +A
1 2 ¯3 4
      -A
¯1 ¯2 3 ¯4
      ÷A
1 0.5 ¯0.3333333333 0.25
      ×A
1 1 ¯1 1
      *A
2.718281828 7.289056099 0.04978706937 54.59815003
      |A
1 2 3 4
    1 2 3⌈4 5 6
4 5 6
    1 2 3⌊4 5 6
1 2 3
      2 3 4⍟4 5 6
2 1.464973521 1.29248125
      1 2 3○4 5 6
¯0.7568024953 0.2836621855 ¯0.2910061914
      2 4 8*÷1 2 3
2 2 2
```

EXAMPLE 1.20 What is the end result of the APL expressions:

```
X←1 3 2 7
Y←¯1 2 7 1
X−Y
```

Solution `1 3 2 7 − ¯1 2 7 1` or `2 1 ¯5 6` ■

EXAMPLE 1.21 What is the result of the APL expressions:

```
A←2 ¯2 7 ¯12 14
A+|A
```

Solution From Appendix A we see that monadic `|` is absolute value. Thus `A+|A` is `2 ¯2 7 ¯12 14+2 2 7 12 14` or `4 0 14 0 28.` ■

REDUCTION

The expression `+/X` used above to sum the components of a vector is an example of the use of the APL *reduction operator* "`/`". The function "+" may be replaced by any of the dyadic functions from Appendix B. For example, to multiply the components of a vector together use "×" instead of "+":

```
      X←5 4 3 2 1
      ×/X
120
```

If α is a dyadic function from Apppendix B, then the expression `α/X` produces the same result as putting an α between each pair of components of X. Thus, with the above X, `+/X` produces the same result as `5+4+3+2+1` and `×/X` produces the same result as `5×4×3×2×1`. Here are three other useful reductions:

1. *The maximum component of X.* The expression `⌈/X` gives the maximum component of X. For example, with the above X, `⌈/X` is `5⌈4⌈3⌈2⌈1` or `5⌈(4⌈(3⌈(2⌈1)))`.

```
  ⌈/X
5
```

2. *The minimum component of X.* Similarly, `⌊/X` picks out the smallest component of the vector X.

```
  ⌊/X
1
```

3. *The alternating sum.* The expression `-/X` computes the alternating sum of the components of *X*. For the above *X*, `-/X` is `5-4-3-2-1`.

In conventional notation this is

$$5 - (4 - (3 - (2 - 1))) \qquad \text{or} \qquad 5 - 4 + 3 - 2 + 1$$

```
-/X
3
```

VECTORS AND SCALARS

A vector `X` is a list of numbers and `⍴X` gives the size of the list. An assignment of the form `A←3`, however, defines `A` to be a *scalar*. A scalar is simply a number that is not considered to be part of any list. The expression `⍴A←3` or `⍴3` will cause a blank line to be printed. We will return to this point shortly.

We have seen that `A×B`, for example, is defined if `A` and `B` are both scalars or if `A` and `B` are vectors of the same size. The expression `A×B` is also defined if only one of the variables `A` or `B` is a scalar.

```
      2×3  4  5
6  8  10
      3  4  5×6
18  24  30
```

The scalar is first extended to a vector of the proper length and then the multiplication is carried out. Thus the first calculation is the same as `2 2 2×3 4 5` and the second is the same as `3 4 5×6 6 6`. Similar remarks apply to the other scalar dyadic functions.

```
      2+3  4  5
5  6  7
      3  4  5*6
729   4096   15624
```

The utility of this special action for scalars is illustrated by the next example.

POLYNOMIALS

The conventional expression $3x^2 - 2x + 7$ may be written in APL as

`+/3 ¯2 7×X*2 1 0,`   `X` a scalar

To see this we work from the right, evaluating each function as we encounter it. This procedure shows that the APL expression is equivalent to

```
+/(3 ¯2 7×(X*2 1 0))
```

If x is a scalar, then `X*2 1 0` creates the vector $x^2 \quad x^1 \quad x^0$. The multiplication then results in the vector $3x^2 \quad -2x \quad 7$ and the `+/` sums the components of this vector.

EXAMPLE 1.22 Compute the value of the polynomial $x^2 - 2x + 1$ when $x = 12$.

Solution

```
    +/1 ¯2 1×12*2 1 0
121
```

■

EXAMPLE 1.23 Compute the value of the polynomial $3x^5 - 6x + 2$ for $x = 6$, -2, $\sqrt{7}$, e.

Solution

```
        P←3 ¯6 2
        E← 5 1 0
        +/P×6*E
23294
        +/P×¯2*E
¯82
        X←7*.5
        +/P×X*E
375.0509349
```

Since several evaluations are requested, we store the coefficient vector and exponent vector.

Alternatively, `+/P×(7*.5)*E`. But the latter expression is more prone to typing error.

```
        X←*1
        +/P×X*E
430.9297863
```

Alternately, `+/P×(*1)×E`. ■

LINEAR COMBINATIONS

Two of the vector operations defined above are fundamental in the study of linear algebra. These operators are

1. The addition of two vectors.
2. The multiplication of a vector and a scalar.

Let v be a vector and α a scalar. In the conventional mathematical notation `α×V` is written αv. Further, vectors are usually set off in parentheses with their components separated by commas. Thus the vector `1 ¯3 7` is conventionally written $(1, -3, 7)$. There is some conflict between APL notation and conventional notation here, but both systems should be mastered. To avoid confusion, we shall always display APL expressions in a special typeface.

In conventional notation the definitions of vector addition and vector-scalar multiplication become

1. $(\alpha_1, \alpha_2, \alpha_3, \ldots, \alpha_n) + (\beta_1, \beta_2, \beta_3, \ldots, \beta_n) = (\alpha_1 + \beta_1, \alpha_2 + \beta_2, \ldots, \alpha_n + \beta_n)$.
2. $\alpha(\beta_1, \beta_2, \beta_3, \ldots, \beta_n) = (\alpha\beta_1, \alpha\beta_2, \alpha\beta_3, \ldots, \alpha\beta_n)$.

Combining the two operations in a single expression gives a linear combination.

DEFINITION 1.2 Let $v_1, v_2, \ldots, v_n$ be vectors of the same size. Let $\alpha_1, \alpha_2, \ldots, \alpha_n$ be scalars. The vector

$$v = \alpha_1 v_1 + \alpha_2 v_2 + \cdots + \alpha_n v_n$$

is a *linear combination* of the vectors $v_1, v_2, \ldots, v_n$.

EXAMPLE 1.24 Let $v_1 = (1, 3, 7)$, $v_2 = (-1, 2, 12.3)$, $v_3 = (0, 2, -4)$, $\alpha_1 = 6$, $\alpha_2 = .3$, $\alpha_3 = 7$. Compute the linear combination

$$v = \alpha_1 v_1 + \alpha_2 v_2 + \alpha_3 v_3$$

Solution

```
      +V←(6×1 3 7)+(.3×¯1 2 12.3)+7×0 2 ¯4
5.7  32.6  17.69
```

Notice that the APL order of evaluation makes the parentheses necessary. ∎

CATENATION

So far we have not explained how to express a vector such as $(2, \sqrt{3}, 4)$ in APL. If we try the expression

```
2 3*.5 4
```

we obtain

```
2 3*(.5 4)
```

which is $(\sqrt{2}, 3^4)$ in conventional notation.

This problem is overcome by introducing the *catenation* function. This function is denoted in APL by the comma (,). Catenation sticks vectors and scalars together to make larger vectors.

```
      2,3
2 3
      2,3 4 5
2 3 4 5
      2 3 , 4 5 6 7
2 3 4 5 6 7
      2,(3*.5),4
2  1.732050808  4
```

Catenation is a dyadic function and is treated like any other dyadic function. Thus, working in from the right on the last expression, we see that it is equivalent to

```
2,((3*.5),4)
```

whereas `2,3*.5,4` is equivalent to

```
2,(3*(.5,4))
```

which, in conventional notation, is $(2, \sqrt{3}, 3^4)$.

EXAMPLE 1.25 Write the vector $(\sqrt{2}, e^3, -7, -14)$ in APL notation.

Solution `(2*÷2),(*3),¯7 ¯14` or `(2*.5),(*3),−7 14` ■

EXAMPLE 1.26 Write the APL expression

```
2,*3,−2 1
```

in conventional notation.

Solution Working in from the right, we have

```
2,(*(3,−(2 1)))
2,(*3 ¯2 ¯1)
```

In conventional notation this is the vector $(2, e^3, e^{-2}, e^{-1})$. ■

INDEX GENERATOR

Consider the problem of evaluating the sum

$$\sum_{k=1}^{400} (k)^{\sqrt{k}}$$

If K is the vector of indices (1, 2, 3, . . . , 400), then the APL notation for this sum is

```
+/K*K*÷2
```

This observation reduces the problem of computing the sum to the problem of generating the vector of indices K. The function that accomplishes this in APL is the *index generator,* whose symbol is the monadic ⍳. If n is a nonnegative integer, then ⍳n is the vector $(1, 2, 3, \ldots, n)$.

```
      ⍳5
1 2 3 4 5
      ⍳2
1 2
      ⍳23
1 2 3 4 5 6 7 8 9 10 11 12 13 14 15 16 17 18 19 20 21 22 23
```

Now we can compute $\sum_{k=1}^{400} (k)^{\sqrt{k}}$.

```
      K←⍳400
      +/K*K*÷2
6.045595546E52
```

The E indicates scientific notation. For example, `6.2E¯2` means 6.2×10^{-2} or .062.

Example 1.27 Compute $\sum_{k=1}^{100} (k^2 + 4k)$.

Solution

```
      K←⍳100
      +/(4×K)+K*2
358550
      +/(4×⍳100)+(⍳100)*2
358550
```

■

Example 1.28 Compute $\sum_{k=-12}^{20} k^2 + k$.

Solution

```
      K←¯13+⍳33
      +/K×K+1
3652
```

■

The result of the index generator is always a *vector*. Thus `ι1` is not the scalar `1` but a vector of size 1.

```
      ι1
1
      ρι1
1
```

Further, `ι0` is defined. It is a vector without components, and it prints as a blank line.

```
      ι0
              ⟵ blank line
      ρι0
0
```

The vector `ι0` is really quite useful because it is an identity for catenation, just as 0 is an identity for addition and 1 is an identity for multiplication.†

We are now in a position to describe what `ρ` does with a scalar — it returns `ι0`.

`ι0.`

```
      ρ3
              ⟵ blank line
```

The result of monadic `ρ` is always a vector.

INDEXING

Often we must deal with individual components of vectors. If `V` is a vector, then the third component of `V` is written `V[3]`

```
      V←1 2 ¯7 12 9
      V[3]
¯7
      V[4]
12
      V[3 4]
¯7 12
      V[4 3]
12 ¯7
```

Notice that we are not restricted to single indices. A vector of indices will produce a vector result. Indexing may also be used to change individual components of a vector.

†That is, `V,ι0` is `V` just as `V+0` and `V×1` are both `V`.

```
      V
1 2 ¯7 12 9
      V[2]←6
      V
1 6 ¯7 12 9
      V[4 5]←-2 3
1 6 ¯7 ¯2 ¯3
```

EXAMPLE 1.29 What is the result of the APL expressions:

```
V←1 3 5 7
V[1 3]←V[3 1]
V
```

Solution `V[3 1]` is the vector `5 1`; thus the second line is `V[1 3]←5 1`. Thus `V` becomes `5 3 1 7`. ■

EXERCISES 1.2

What is the result of the following APL expressions? Calculate by hand and check your answers at a terminal if one is available.

1. `1 2 -3 4`
2. `1 2÷3 4`
3. `1 2÷3 4 -5 6`
4. `1 2×3 4÷1 2`
5. `A←1 ¯2 3 ¯4 5 ¯6`
   `(A+|A)÷2`
6. `A←1 ¯2 3 ¯4 5 ¯6`
   `(A-|A)÷2`
7. `(!3 2 1)-3 2 1×2 1 1×1 1 1`
8. `⌊5 6 7 8÷1 2 3 4`
9. `⌊5 6 7 8÷1 ¯2 3 ¯4`
10. `2|1 2 3 4 5 6 7`
11. `1 2 3 4 5 5 6 7|10`
12. `1 2 3○○÷4`
13. `2 4 6 8⍟2 4 6 8*¯49.832`
14. `(-1 2 3)○1 2 3○¯49.832`

Write APL expressions to evaluate the polynomials at the points indicated. Compute the results at a terminal if available.

15. $3x^2 - 2x + 1$ at $x = 1, 2$
16. $5x^4 - x + 2$ at $x = 3, \sqrt{2}, \sin(\pi/13)$
17. $x^{11} - x^{10} - x^2 + 3x - 2$ at $x = .98, 1, 1.01$
18. $x^{12} - 2x^{11} + x^{10} - x^3 + 4x^2 - 5x + 2$ at $x = .98, 1, 1.01$

For the following problems, write APL expressions to compute the linear combinations $\alpha_1 v_1 + \alpha_2 v_2 + \alpha_3 v_3 + \cdots + \alpha_n v_n$. Evaluate the expressions at a terminal if available.

19. $v_2 = (1, 0, 1)$, $v_2 = (0, 1, 0)$, $\alpha_1 = 6$, $\alpha_2 = 3$
20. $v_1 = (1, 0, 1)$, $v_2 = (0, 1, 0)$, $\alpha_1 = -12$, $\alpha_2 = 16$
21. $v_1 = (1, \frac{2}{3}, 1)$, $v_2 = (1, 2, 3)$, $\alpha_1 = 6$, $\alpha_2 = \frac{1}{2}$
22. $v_1 = (1, \sqrt{2}, 3, \sqrt{5}, 7)$, $v_2 = (1, 2, 3, 4, 5)$, $v_3 = (2, -7, 3, -6, \frac{1}{2})$ $\alpha_1 = 2$, $\alpha_2 = -1$, $\alpha_3 = \sqrt{7}$

Write APL expressions to evaluate the following sums. Evaluate the expressions at a terminal if available.

23. $\sum_{k=1}^{30} k$

24. $\sum_{k=1}^{40} (k^2 + k)$

25. $\sum_{k=1}^{100} (k^2 + 1)$

26. $\sum_{k=0}^{25} \frac{1}{k!}$

27. $\sum_{k=0}^{20} \frac{(-1)^k 3^{2k+1}}{(2k+1)!}$

28. $\sum_{k=0}^{99} \sin \frac{2\pi k}{100} \cos \frac{2\pi k}{100}$

29. Recall that

$$e^x = \sum_{k=0}^{\infty} \frac{x^k}{k!}$$

Thus for each n, $\Sigma_{k=0}^{n} (x^k/k!)$ is an approximation to e^x. Taking $x = -10$, what value of n provides the closest approximation to `*¯10` on your machine?

Let f be a continuous function on the interval $[a, b]$. We can approximate

$$\int_a^b f(t)\, dt$$

by the sum

$$\sum_{k=1}^{n} f(x_k)\, \Delta x; \quad \text{where} \quad x_k = A + \frac{b-a}{n} k \quad \text{and} \quad \Delta x = \frac{b-a}{n}.$$

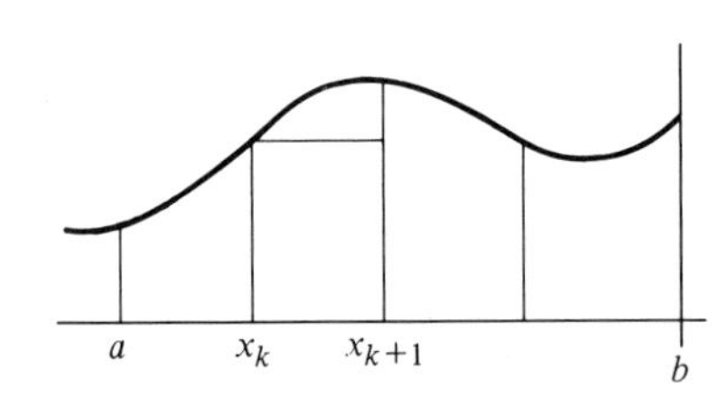

This approximates the area under the curve by a sum of areas of rectangles — as in the definition of Riemann integral.

In the following problems compare the approximation $\Delta x\, \Sigma_{k=1}^{n} f(x_n)$ with the exact value of $\int_a^b f(t)\, dt$ for the given f, a, and b. Take $n = 100$.

30. $f(x) = x$, $a = 0$, $b = 1$
31. $f(x) = x^2$, $a = 0$, $b = 1$
32. $f(x) = \cos x$, $a = 0$, $b = \pi/2$
33. $f(x) = \sin x$, $a = 0$, $b = \pi$
34. $f(x) = (\ln x)/x$, $a = 1$, $b = 10$
35. $f(x) = \cos 3x \sin 5x$, $a = 0$, $b = 2\pi$

What is the value of the vector V? Answers should be worked by hand and checked at the terminal if one is available.

36.
```
V←⍳3
V[2]←2*.5
```

37.
```
V←⍳ 9
V[1 3 7]←2 4 6
```

38.
```
V←⍳7
V[2 3]←V[3 2]
```

39.
```
W←¯4+⍳7
V←W[1 5 1 5 1 5]
```

40.
```
V←⍳5
V←V[6-V]
```

41.
```
V←⍳7
V[1 3 7]←¯6
```

42.
```
V←⍳8
V[1 1]←¯6
```

43.
```
V←⍳3
V[1 1]←V[1 1]
```

44. Let V be the vector denoted by $(\alpha_1, \alpha_2, \ldots, \alpha_n)$ in conventional mathematical notation. By translating the APL expression

```
AVE←+/V÷⍴V
```

to conventional notation show that AVE is the average of the components of V.

Let V and W be the vectors denoted by $(\alpha_1, \alpha_2, \ldots, \alpha_n)$, $(\beta_1, \beta_2, \ldots, \beta_m)$ in conventional notation. By translating to conventional mathematical notation, show that the following expressions are true in the sense that the result contains no zeros.

45. `(⍴V,W)=(⍴V)+⍴W`

46. `(+/V+W)=+/V,W` (Assume `(⍴V)=⍴W`)

47. `((+/V)++/W)=+/V,W`

48. `V=V[⍳⍴V]`

49. The result of the monadic function ⍴ is always a vector. Using this fact, what is ⍴⍴V for a vector V? A scalar V?

50. Show that if the identity of exercise 47 is to hold for *all* vectors V, then `0=+/⍳0`.

Compute the following expressions and check your answers at a terminal if one is available.

51. `×/2 4 6`

52. `×/⍳6`

53. `(!17)-×/⍳17`

54. `-/⍳3`

55. `⌈/6 2 5 1 ¯7`

56. `⌊/6 2 5 1 ¯7`

57. `*/2 2 2`

58. `|/1 2 5 9`

59. `○/2 1,○2`

60. The following table gives the production of steel from 1946 to 1956.

Year	1946	1947	1948	1949	1950	1951	1952	1953	1954	1955	1956
Tons of steel	66.6	84.9	88.6	78.0	96.8	195.2	93.2	111.6	88.3	117.0	115.2

(a) Fit a least-squares straight line to these data, writing steel production as a function of time.

(b) Using your equation from part (a), solve for time as a function of steel production.

(c) Compute the least-squares straight line, giving time as a function of steel production. Is this the equation obtained in part (b)?

61. Let $(x_1, y_1), (x_2, y_2), \ldots, (x_n, y_n)$ be given and set

$$s = \sum_{i=1}^{n} (y_i - b - ax_i)^2$$

Solve the equations

$$\frac{\delta s}{\delta a} = 0 \quad \text{or} \quad (\Sigma\, x_i^2)a + (\Sigma\, x_i)b = \Sigma\, x_i\, y_i$$

$$\frac{\delta s}{\delta b} = 0 \quad \text{or} \quad (\Sigma\, x_i)a + nb = \Sigma\, y_i$$

to find the values of a and b that minimize s.

The following table summarizes information from eight gasoline credit card receipts. (The car, a 1976 Dodge Aspen station wagon, had its engine tuned just before the first gasoline purchase recorded here.)

Odometer reading	39922	40125	40327	40526	40762	40937	41097	41243
Date	12/20/78	12/26/78	12/28/78	12/30/78	1/3/79	1/11/79	1/23/79	1/30/79
Gallons purchased	14	14.6	13	13.4	14.9	13.8	14.5	13.6
Cost in dollars	10.90	11.36	10.41	10.45	11.73	10.92	11.56	10.48

Define the following vectors.

M = miles = first row of table
T = time = second row of table with the dates rewritten as days from some arbitrary point in time. Say, day 1 is 12/1/78, so that 12/20/78 becomes 20, 12/26/78 becomes 26, and so on.
G = gallons = third row of table
C = cost = fourth row of table.

Use these vectors to compute the following quantities:

62. The price per gallon, in cents, for each of the eight purchases.
63. The average price per gallon, in cents, for the time covered by the table.
64. The times, in days, between purchases.
65. The distances traveled, in miles, between purchases.
66. The average distance traveled per day between each pair of purchases.
67. Compute the gasoline mileage, in miles per gallon:
    (a) Between purchases
    (b) Overall
68. Let y be gasoline mileage and x be miles per day traveled. Fit a least-squares straight line to the data from exercises 66 and 67a to estimate y as a function of x. Plot the line and data on the same graph.

1.3 Matrices

A matrix is a rectangular array or "table" of numbers. An example of a matrix is the body of Table 1.2 of Section 1.2, which we will denote by M (the row of totals has been omitted).

```
  M
60     120    7200    3600
61     120    7320    3721
62     135    8370    3844
63     135    8505    3969
64     130    8320    4096
65     135    8775    4225
66     150    9900    4356
68     140    9520    4624
69     170   11730    4761
70     145   10150    4900
71     160   11360    5041
72     160   11520    5184
74     160   11840    5476
75     160   12000    5625
76     175   13300    5776
```

This matrix has 15 rows and 4 columns. Any entry in the matrix is uniquely determined by giving its row and column. For example, the element on the ninth row and third column is 11730, the entry in the first row and first column is 60.

Matrices are useful whenever one has a collection of numbers that can be naturally arranged in a row-column format.

For example, consider the system of linear equations

$$\begin{aligned} 3x - 7y + 12z &= 4 \\ 8x - 9y \qquad &= 7 \\ 2x + y - 16z &= -1 \end{aligned}$$

The coefficients of the unknowns form a matrix:

$$\begin{matrix} 3 & -7 & 12 \\ 8 & -9 & 0 \\ 2 & 1 & -16 \end{matrix}$$

which we will use to solve such systems.

In APL notation the entry in the ninth row and third column of a matrix `M` is denoted by `M[9;3]`, the entry in the first row and first column is `M[1;1]`, and so on. Using the matrix `M` defined above, for example, we have

```
      M[9;3]
11730
      M[1;1]
60
      M[7;4]
4356
```

The individual rows and columns of a matrix are vectors. Often we shall regard a matrix as a convenient way of manipulating a set of vectors. In APL notation, we refer to a row vector by omitting the column reference, and we refer to a column by omitting the row reference

```
      M[1;]
60 120 7200 3600
      M[;3]
7200 7320 8370 8505 8320 8775 9900 9520 11730 10150 11360 11520 11840 12000 13300
```

This notation is also used to replace entries of a matrix.

```
    M[10;3]←0
    M
60     120     7200     3600
61     120     7320     3721
62     135     8370     3844
63     135     8505     3969
64     130     8320     4096
65     135     8775     4225
66     150     9900     4356
68     140     9520     4624
69     170    11730     4761
70     145        0     4900
71     160    11360     5041
72     160    11520     5184
74     160    11840     5476
75     160    12000     5625
76     175    13300     5776
```

The 10; 3 entry has been changed from 10150 to 0.

```
    M[12;]←4  3  2  1
    M
60        120       7200       3600
61        120       7320       3721
62        135       8370       3844
63        135       8505       3969
64        130       8320       4096
65        135       8775       4225
66        150       9900       4356
68        140       9520       4624
69        170      11730       4761
70        145          0       4900
71        160      11360       5041
 4          3          2          1
74        160      11840       5476
```

```
75      160    12000     5625
76      175    13300     5776
```

In this case the vector M[12;] has been replaced by the vector 4 3 2 1.

ρ FOR MATRICES

The monadic function ρ is called *size*. The action of ρ on vectors and scalars has been discussed in Section 1.2. If V is a vector, then ρV gives the number of components of V. If A is a scalar, then ρA is the empty vector (ι0). In both cases the result of ρ is a vector. The vector ρX has one component if X is a vector and zero components of X is a scalar.

If M is a matrix, then ρM is a vector with two components. The first component is the number of rows of M and the second component is the number of columns of M.

```
      ρM
15 4
```

The symbol ρ also represents a dyadic function called *reshape*. The dyadic ρ is used to define matrices. Here is how it works.

```
      3 2 ρ 1 3 7 9 4 6
1 3
7 9
4 6
      2 2 ρ 1 3 7 9 4 6
1 3
7 9
      3 1 ρ 1 3 7 9 4 6
1
3
7
```

The dyadic function ρ takes a vector left argument and any variable right argument. It reshapes the right argument to the shape specified by the left argument.

The dyadic ρ uses as many data from the right-hand entry as are needed. If there are not enough data, it returns to the start and reuses the data.

```
      3 3ρ1 2 3 4
1 2 3
4 1 2
3 4 1
```

```
        3 2⍴1 2 3
1  2
3  1
2  3
        3 3⍴5
5  5  5
5  5  5
5  5  5
```

If the left argument is a scalar, it is taken to be a single component vector. In this case the result is a vector.

```
        6⍴5
5 5 5 5 5 5
```

To enter a matrix into a machine, we first enter the data as a vector and then reshape the vector.

EXAMPLE 1.30 Enter the matrix

$$\begin{matrix} 1 & 2 & 7 & 9 \\ 6 & 3 & 8 & 11 \\ 12 & 1 & 6 & 14 \\ 4 & 7 & 9 & 3 \\ 1 & 8 & 2 & 4 \end{matrix}$$

from the terminal.

Solution

```
        X←1 2 7 9 6 3 8 11 12 1 6 14 4 7 9 3 1 8 2 4
        X←5 4⍴X
        X

 1   2   7   9
 6   3   8  11
12   1   6  14
 4   7   9   3
 1   8   2   4
```
■

EXAMPLE 1.31 What is *M* if

```
M←1 3 ¯5 7 2
M←2 3⍴M
```

Solution The second line reshapes M to a 2-by-3 matrix. This requires six entries. After using the five numbers 1, 3, −5, 7, 2, we return to the beginning. Thus the 1 is used twice. M is

```
1 3 ¯5
7 2  1
```
■

EXAMPLE 1.32 What is M if

```
M←2 3⍴7 12 ¯1 4 ¯9 8 6
M[1;3]←0
M[;2]←¯6 5
```

Solution The first line defines M to be the 2-by-3 matrix

```
7  12  ¯1
4  ¯9   8
```

The 6 is not used. The second line replaces the 1;3 entry of M by zero. Thus M becomes

```
7  12  0
4  ¯9  8
```

The last line replaces the second column of M by the vector ¯6 ¯5. Thus M is finally

```
7  ¯6  0
4  ¯5  8
```
■

The APL indexing scheme for matrices is somewhat more elaborate than has been indicated so far. Suppose, for example, that M is the matrix

```
 1  ¯6  5  12
 6   3  0  ¯1
¯1  12  4   2
¯6   4  3   1
```

An expression such as M[2 4;1 3] specifies the submatrix of M at the intersection of rows 2 and 4 with columns 1 and 3

```
           1   ¯6   5   12
row 2 →   (6)   3  (0)  ¯1
          ¯1   12   4    2
row 4 →  (¯6)   4  (3)   1
           ↑        ↑
       column 1  column 3
```

Hence `M[2 4;1 3]` is the matrix

```
 6  0
¯6  3
```

In fact, an expression such as `M[1;]`, which specifies the first row of `M`, is just an abbreviation for `M[1;⍳N]`, where `N` is the number of columns of `M`. Similarly, `M[;6]` means `M[⍳K;6]`, where `K` is the number of rows of `M`. (`M[1;]` is a vector rather than a 1-by-`N` matrix, however. This somewhat subtle point is developed further in Exercise 16.)

The next example is important for later use.

EXAMPLE 1.33 Assume that `M` has at least four rows. What is the effect of the expression:

```
M[2 4;]←M[4 2;]
```

Solution `M[4 2;]` is an abbreviation for `M[4 2;⍳N]`, where `N` is the number of columns of `M`. Thus `M[4 2;]` is a 2-by-`N` matrix whose first row is the fourth row of `M` and whose second row is the second row of `M`. This matrix is to replace the submatrix of `M` denoted by `M[2 4;]` — that is, the submatrix consisting of the second and fourth rows of `M`. The end effect of the expression is thus to interchange the second and fourth rows of `M`. The expression

```
M[4 2;]←M[2 4;]
```

has precisely the same effect. ■

Notice in passing that just as a vector may have size 0, (`⍳0`) matrices may have 0 rows or columns.

```
      M←10 0⍴0
      M
                    ←M is displayed as a blank line
      ⍴M
10 0
```

SCALAR FUNCTIONS FOR MATRICES

The primitive scalar functions (Appendix B) extend to matrices as well as to vectors (parallel processing).

Monadic functions simply operate on each entry.

```
      +M←3 3⍴⍳9
1  2  3
4  5  6
7  8  9
```

```
      -M
¯1 ¯2 ¯3
¯4 ¯5 ¯6
¯7 ¯8 ¯9

      *M
    2.720       7.390      20.100
   54.600     148.000     403.000
 1100.000    2980.000    8100.000

      ÷M
 1.000   0.500   0.333
 0.250   0.200   0.167
 0.143   0.125   0.111
```

If two matrices are the same shape, then dyadic scalar functions operate on corresponding entries.

```
      N←M[3 2 1;2 1 3]
      N
 8  7  9
 5  4  6
 2  1  3
      N-M
 7  5  6
 1 ¯1  0
¯5 ¯7 ¯6

      N÷M
 8.000   3.500   3.000
 1.250   0.800   1.000
 0.286   0.125   0.333
      N*M
      8.000      49.000      729.000
    725.000    1020.000    46700.000
    128.000       1.000    19700.000
      N≠M
 1  1  1
 1  1  0
 1  1  1
```

If `A`, `B` are variables (scalars, vectors, or matrices) and α is a scalar dyadic function, then `AαB` is defined if `(ρA)=ρB`.

However, `AαB` is also defined if one of `A` or `B` is a scalar or vector of size 1. In this case, the scalar or vector of size 1 is first reshaped to match the other variable. Then the computation is carried out.

```
      3×M
  3   6   9
 12  15  18
 21  24  27
      M-3
¯2 ¯1  0
 1  2  3
 4  5  6
```

```
      +V←1ρ3
3
      ρV
1
      V×M
  3   6   9
 12  15  18
 21  24  27

      M-V
¯2 ¯1  0
 1  2  3
 4  5  6
```

The dyadic ρ is used here to reshape the scalar 3 to a vector with 1 component. The function Ravel, defined below, could also be used.

Two operations on matrices are of special importance for linear algebra: addition and scalar multiplication. If α is a scalar and A is a matrix, we will abbreviate $\alpha \times A$ to αA or $A\alpha$. We give formal definitions of these two operations.

DEFINITION 1.3 Let A and B be matrices of the same shape. The *sum* $A + B$ of A and B is defined by

$$(A + B)[i;j] = A[i;j] + B[i;j]$$

for all valid indices i, j.

DEFINITION 1.4 Let A be a matrix and α a scalar. The product αA (or $A\alpha$) is defined by

$$(\alpha A)[i;j] = \alpha A[i;j]$$

for all valid indices i, j.

RAVEL

Ravel is a monadic function denoted by the comma ",". For any variable X — scalar, vector or matrix — ,X is a vector.

For example ,3 is the vector of size 1, whose single component is 3.

```
      +V←,3
3
      ρV
1
```

If X is a vector, then ,X is simply X. If X is a matrix, then ,X strings the components out into a vector. In this case Ravel reverses the effect of the reshape function.

```
      +M←2 3ρ4 6 3 0 2 4
4 6 3
0 2 4
      ,M
4 6 3 0 2 4
```

CATENATION FOR MATRICES

The catenation function permits matrix arguments. The form we will find most useful is the simplest. If P and Q have the same number of rows, then P, Q is the matrix obtained by catenating corresponding rows. If P is m by n and Q is m by r, then P, Q is m by $(n + r)$.

```
      +P←2 3ρι6
1 2 3
4 5 6
      +Q←2 3ρ6+ι6
 7  8  9
10 11 12
      P,Q
 1  2  3  7  8  9
 4  5  6 10 11 12
```

The columns may also be catenated. We accomplish this by indexing the symbol ",". The expression ",[1]" means catenate along the *first* or row index. The "P,Q" above is actually an abbreviation of "P,[2]Q", which means catenate along the second or column index.

As a general rule in APL, if the index is omitted, the last index (the column index for matrices) is the one that is affected.

```
      P,[1]Q
 1  2  3
 4  5  6
 7  8  9
10 11 12
```

```
        P,[2]Q
  1   2   3   7   8   9
  4   5   6  10  11  12
        ρP
2 3
        ρQ
2 3
        ρP,[1]Q
4 3
        ρP,[2]Q
2 6
```

P and *Q* need not both be matrices. One may be a scalar (which will then be reshaped first) or a vector of the proper size.

REDUCTION

The summation "+/" operates on matrices in a fashion similar to catenation. One can write "+/[1]" to sum along the row *index* (that is, to sum *vertically*) or "+/[2]" to sum along the column *index* (that is, to sum horizontally). "+/[2]" is assumed if "+/" is written. In this case, however, there is an alternate notation for "+/[1]." The symbol "+⌿" (plus, /, backspace, -) means "sum along the first index." The result is a vector.

```
        P
  1  2  3
  4  5  6
        +/P
6  15
        +⌿P
5  7  9
```

All the remarks above apply to general reductions α/P, where α is a scalar dyadic function (Appendix B).

```
        ⌈/P
3  6                Pick the maximum from each row.
        ⌊⌿P
1  2  3             Pick the minimum from each column.
```

EXAMPLE 1.34 Let M be the matrix

```
60      120     7200    3600
61      120     7320    3721
62      135     8370    3844
63      135     8505    3969
```

```
64    130    8320    4096
65    135    8775    4225
66    150    9900    4356
68    140    9520    4624
69    170   11730    4761
70    145   10150    4900
71    160   11360    5041
72    160   11520    5184
74    160   11840    5476
75    160   12000    5625
76    175   13300    5776
```

This is the body of Table 1.2 without the last row of column totals. Write an APL expression to redefine *M* to be Table 1.2 with the last row included.

Solution

```
      +M←M,[1]+⌿M
  60          120          7200          3600
  61          120          7320          3721
  62          135          8370          3844
  63          135          8505          3969
  64          130          8320          4096
  65          135          8775          4225
  66          150          9900          4356
  68          140          9520          4624
  69          170         11730          4761
  70          145         10150          4900
  71          160         11360          5041
  72          160         11520          5184
  74          160         11840          5476
  75          160         12000          5625
  76          175         13300          5776
1016         2195        149810         69198
```
■

STANDARD SCORES (Z-SCORES)

Statistical measurements are often made in quite arbitrary units, so that the raw data from different experiments are difficult to compare. To give a simple example, if the heights of the individuals in a population are measured by one experimenter in English units (feet, inches) and the same population is measured by another experimenter in metric units, then, although the raw numbers are quite different, the distribution of the two sets of data is in a sense the same. To bring this out, the first step in many statistical analyses is to transform the raw data into *standard scores.*

To do this, one uses the formula

$$z = \frac{x - \overline{x}}{\sigma}$$

Here x is a measurement, $\overline{x}$ is the mean of the measurements, and σ is the standard derivation of the measurements.

As an illustration of data manipulation using matrices, we will turn a matrix of raw data into a matrix of standard scores.

Let X be the vector of measured heights from Table 1.1 and Y the vector of measured weights. Let A be a matrix with two columns: X and Y.

```
      A
60   120
61   120
62   135
63   135
64   130
65   135
66   150
68   140
69   170
70   145
71   160
72   160
74   160
75   160
76   175
```

First compute the X and Y means.

```
      +M←(+⌿A)÷15
67.73   146.3
```

Next subtract the means from the corresponding raw scores and call the result B.

```
      +B←A-(ρA)ρM
¯7.7330  ¯26.3300
¯6.7330  ¯26.3300
¯5.7330  ¯11.3300
¯4.7330  ¯11.3300
¯3.7330  ¯16.3300
¯2.7330  ¯11.3300
¯1.7330    3.6670
 0.2667   ¯6.3330
 1.2670   23.6700
```

```
2.2670  ¯1.3330
3.2670  13.6700
4.2670  13.6700
6.2670  13.6700
7.2670  13.6700
8.2670  28.6700
```

Notice that the expression `(⍴A)⍴M` reshapes `M` to the shape of `A`.

```
      (⍴A)⍴M
67.730   146.300
67.730   146.300
67.730   146.300
67.730   146.300
67.730   146.300
67.730   146.300
67.730   146.300
67.730   146.300
67.730   146.300
67.730   146.300
67.730   146.300
67.730   146.300
67.730   146.300
67.730   146.300
67.730   146.300
```

Now the standard deviation σ of a set of numbers $x_1, \ldots, x_n$ with mean $\overline{x} = (\Sigma\, x_i)/n$ is defined by

$$\sigma^2 = \frac{1}{n} \sum (x_i - \overline{x})^2.$$

Thus the standard deviations are

```
      +SD←((+⌿B*2)÷15)*÷2
5.039   16.78
```

and the standard scores are

```
      +Z←B÷(⍴B)⍴SD
¯1.5350 ¯1.5690
¯1.3360 ¯1.5690
¯1.1380 ¯0.6754
¯0.9393 ¯0.6754
¯0.7408 ¯0.9734
¯0.5424 ¯0.6754
```

```
¯0.3440  0.2185
 0.0529 ¯0.3774
 0.2514  1.4100
 0.4498 ¯0.0794
 0.6482  0.8145
 0.8467  0.8145
 1.2440  0.8145
 1.4420  0.8145
 1.6400  1.7080
```

THE OUTER PRODUCT

The outer product is an APL function used to create matrices whose entries are given by simple formulas. It is the dyadic operation denoted by the pair of symbols "∘." (often referred to as "jot-dot"). The outer product is used in conjunction with a scalar dyadic function α. The form in which we will use it is

```
M←V∘.αW
```

where V and W are vectors.

If V has m components and W has n components, then M is an m-by-n matrix (i.e., ⍴M is (⍴V),⍴W) whose i, j component is given by the formula

```
M[I;J]=V[I]αW[J]
```

EXAMPLE 1.35 (*Kronecker product of two vectors*) What is the result of the expression:

```
4 ¯7 3 2∘.×2 1 2
```

Solution This is of the form V∘.αW with V the vector 4 ¯7 3 2, W the vector 2 1 2, and α the function ×. V is size 4 and W is size 3. Thus the result is 4 by 3. The I;J component of the result is V[I]×W[J]. Thus the result is

```
 4×2   4×1   4×2
¯7×2  ¯7×1  ¯7×2
 3×2   3×1   3×2
 2×2   2×1   2×1
```

or

```
  8   4   8
¯14  ¯7 ¯14
  6   3   6
  4   2   4
```
■

DEFINITION 1.5 If V, W are vectors, the *Kronecker product* of V and W is the matrix V∘.×W.

We shall have occasion to use the Kronecker product in subsequent sections.

EXAMPLE 1.36 (*Hilbert matrix*) The n-by-n Hilbert matrix is the matrix

$$\begin{matrix} 1 & \frac{1}{2} & \frac{1}{3} & \cdots & 1/n \\ \frac{1}{2} & \frac{1}{3} & \frac{1}{4} & \cdots & 1/(n+1) \\ \frac{1}{3} & \frac{1}{4} & \frac{1}{5} & \cdots & 1/(n+2) \\ \vdots & & & & \vdots \\ 1/n & 1/(n+1) & 1/(n+2) & \cdots & 1/(2n-1) \end{matrix}$$

write an APL expression to create a 5-by-5 Hilbert matrix.

Solution The $i;j$ entry of the matrix is $1/(i+j-1)$, and the indices i and j both run from 1 to 5. First set up vectors of indices

```
I←J←⍳5
```

Then a 5-by-5 matrix H with entries $H[i;j] = 1/(i+j-1)$ is just

```
      ⎕←H←÷I∘.+J−1
1.0000  0.5000  0.3333  0.2500  0.2000
0.5000  0.3333  0.2500  0.2000  0.1667
0.3333  0.2500  0.2000  0.1667  0.1429
0.2500  0.2000  0.1667  0.1429  0.1250
0.2000  0.1667  0.1429  0.1250  0.1111
```
■

If A is an m-by-n matrix, then +/+/A is $\Sigma_{i=1}^{m}\Sigma_{j=1}^{n} A[i;j]$ and +/+⌿A is $\Sigma_{j=1}^{n}\Sigma_{i=1}^{m} A[i;j]$. This observation can be used in conjunction with the outer product to evaluate such double sums.†

EXAMPLE 1.37 Evaluate

$$\sum_{i=1}^{20}\sum_{j=1}^{30} 2i + j^3$$

Solution If A is a 20-by-30 matrix with $A[i;j] = 2i + j^3$, then +/+/A is the sum sought. First, set up vectors of indices

```
I←⍳20
J←⍳30
```

† Both these sums are, of course, equal to +/,A as well. But +/+/ emphasizes the double-sum nature of the computation.

Then the sum sought is

```
      +/+/(2×I)∘.+J*3
4337100
```

■

EXAMPLE 1.38 Evaluate

$$\sum_{i=1}^{20}\sum_{j=1}^{10}(i+j)(i-j)$$

Solution The easiest way to do this is to set up two 20-by-10 matrices A and B with $A[i;j] = i + j$ and $B[i;j] = i - j$ and compute +/+/A×B.

```
      I←⍳20
      J←⍳10

      +/+/(I∘.+J)×I∘.-J
21000
```

■

EXAMPLE 1.39 Evaluate

$$\sum_{i=1}^{20}\sum_{j=1}^{20}(-1)^{i+j}ij$$

Solution

```
      I←J←⍳20
      +/+/(I∘.×J)×¯1*I∘.+J
100
```

■

EQUALITY CONVENTION

There is a conflict between the use of the equality sign "=" in conventional mathematical discourse and the use of the dyadic APL function "=". If A and B are matrices, then in conventional mathematical discourse "$A = B$" means that A and B are the same matrix — that is, have the same entries. The APL expression "A=B" results in a matrix of zeros and ones. A one means that the corresponding entries of A and B are equal, and a zero indicates that the corresponding entries of A and B are not equal. The conventional expression "$A = B$" corresponds to the case where the APL result "A=B" contains no zero entries.

To reconcile the two, we use the following convention: An APL expression, such as A=B, which results in a matrix of zeros and ones is assumed to contain no zeros. This convention is adopted to allow us to write phrases such as "A×B is defined whenever (⍴A)=⍴B." Exceptions to the convention, such as "What is the result of the expression 3=3×¯1 0 1?" should be clear from context.

EXAMPLE 1.40 (*Identity matrix*) The n-by-n identity matrix ID has $ID[i; j] = 0$ when $i \neq j$ and $ID[i; i] = 1$. Write an APL expression to create the N-by-N identity matrix.

Solution The APL expression `I=J` results in a `1` when `I` and `J` are equal and a `0` when they are not. Thus the identity matrix can be represented as

```
1=1   1=2   1=3   ...
2=1   2=2   2=3   ...
3=1   3=2   3=3   ...
 ⋮     ⋮     ⋮
```

Hence a solution is

```
ID←(ιN)∘.=ιN
```

An entirely different solution is

```
ID←(N,N)ρ1,Nρ0
```
■

EXERCISES 1.3

In exercises 1 through 15 a matrix M is defined. What is M? Answers should be worked out by hand and checked at a terminal if one is available.

1. `M←2 2ρι4`
2. `M←3 2ρι4`
3. `M←1 2ρι4`
4. `M←3 2ρι4`
   `M[3;2]←6`
5. `M←3 1ρ1`
   `M[2;1]←0`
6. `M←5 4ρι5`
   `M[3;]←4ρ0`
   `M[;3]←5ρ¯1`
7. `M←5 4ρι5`
   `M[;3]←5ρ¯1`
   `M[3;]←4ρ0`
8. `M←5 5ρι6`
   `M[3;]←M[;3]`
9. `M←5 5ρι6`
   `M[3;]←M[3;6−ι5]`
10. `M←5 5ρι6`
    `M[3;]←M[6−ι5;3]`
11. `M←3 3ρ(3ρ1),(3ρ2),3ρ3`
12. `M←3 3ρ(3ρ1),(3ρ2),3ρ3`
    `M[1 3;]←M[3 1;]`
13. `M←3 3ρι9`
    `M[1 3;1 3]←2 2ρ9 7 3 1`
14. `M←3 3ρι9`
    `M[1 3;1 3]←M[3 1;3 1]`
15. `M←3 3ρι9`
    `M[3;]←M[1;3ρ1]`

16. Let `M` be a matrix and let `I`, `J` be arrays of valid indices for M. The rule determining the shape of `M[I;J]` is `(ρM[I;J])=(ρI),ρJ` — for example, `(ρM[1 2;1 2 3])=(ρ1 2),ρ1 2 3` or 2-by-3 — and, since the shape of a scalar is `ι0`, `ρM[2;3]` is `(ρ2),ρ3` or `(ι0),ι0` or `ι0`. Thus `M[2;3]` is a scalar.

Use this rule to determine the shape of the arrays defined below. The answers may be checked at a terminal. Assume `10 10=ρM`.

(a) `M[1 2;1 2]` (b) `M[1;1 2]` (c) `M[1 2;1]`

(d) `M[1 2;,1]` (e) `M[,1;1 2]` (f) `M[,3;]`

(g) `M[;,6]` (h) `M[;]` (i) `M[,3;,4]`

(j) `M[,3;4]` (k) `M[3;,4]` (l) `M[2 2ρι4;1]`

(m) `M[1;2 2ρι4]`

In exercises 17 through 30 a matrix *M* is defined. What is *M*?

17.
```
A←2 3ρι4
B←2 3ρ2
M←A+B
```

18.
```
A←2 3ρι4
B←A[2 1;3 2 1]
M←A−B
```

19.
```
A←2 3ρι4
B←A[2 1;3 2 1]
M←A×B
```

20.
```
A←2 3ρι4
B←A[2 1;3 2 1]
M←B÷A
```

21.
```
A←2 3ρι4
B←A[2 1;3 2 1]
M←B⌈A
```

22.
```
A←2 3ρι4
B←A[2 1;3 2 1]
M←⌊A○B
```

23.
```
A←2 3ρι4
B←A[2 1;3 2 1]
M←A,B
```

24.
```
A←2 3ρι4
B←A[2 1;3 2 1]
M←A,[2]B
```

25.
```
A←2 3ρι4
B←A[2 1;3 2 1]
M←A,[1]B
```

26.
```
A←2 3ρι4
M←5 6,A
```

27.
```
A←2 3ρι4
M←A, ¯1
```

28.
```
A←2 3ρι4
M←1,[1]A
```

29.
```
A←2 3ρι4
M←2,1,A
```

30.
```
A←2 3ρι4
M←0,[1]0,(A,[1]0),0
```

31. This exercise is based on Table 1.3. Measurements were taken from two groups consisting of boys aged 9 to 11. The measurements were repeated after six months. One group, the *control*, was a sample from the general population. The other group, the *experimental*, consisted of boys engaged in competitive swimming.

The measurements were height (in inches), weight (in kilograms), and fitness (in liters

TABLE 1.3

Height		*Weight*		*Fitness*	
Pre	*Post*	*Pre*	*Post*	*Pre*	*Post*
53.00	54.70	31.59	33.35	1.28	1.33
54.10	55.10	27.27	26.80	1.66	1.49
53.10	54.30	29.55	30.50	1.28	1.37
60.50	61.90	42.59	44.25	1.97	2.23
55.20	56.10	29.55	30.30	1.64	1.70
56.30	58.00	34.77	35.00	1.41	1.83
57.40	58.80	35.68	36.50	1.43	1.61
54.80	55.10	27.05	27.25	1.43	1.41
57.50	58.90	36.14	37.20	1.61	1.55
59.00	61.50	39.55	39.10	1.79	1.90
55.80	56.90	41.73	41.90	1.74	1.79
57.50	59.80	30.68	33.35	1.33	1.88
53.60	54.70	32.73	35.30	1.53	1.65
51.60	52.70	26.00	25.20	1.48	1.54
56.80	57.90	31.59	33.10	1.36	1.44
Control					

Height		*Weight*		*Fitness*	
Pre	*Post*	*Pre*	*Post*	*Pre*	*Post*
50.50	51.50	25.50	26.82	1.34	1.65
56.75	57.60	35.40	37.00	1.80	1.82
56.50	57.60	33.60	32.72	1.88	1.93
54.50	55.80	36.90	40.30	.95	1.84
54.25	55.20	28.60	29.54	1.52	1.89
53.00	53.70	30.40	32.10	1.60	2.10
53.00	54.30	29.10	32.00	1.24	1.65
55.25	56.70	30.90	32.30	1.23	1.50
49.50	50.70	24.10	25.20	1.32	1.50
56.00	56.90	30.90	31.00	1.32	1.73
54.80	58.00	26.40	27.30	1.42	1.59
58.75	60.70	37.30	39.25	1.92	2.05
51.20	52.70	26.80	27.60	1.05	1.76
57.00	54.00	47.30	51.10	2.10	2.27
50.60	51.40	26.48	28.30	1.33	1.54
Experimental					

SOURCE: M. Karpman, personal communication.

of oxygen per minute). The "fitness" measure is oxygen consumption while walking a treadmill.

(a) Store these tables as two matrices, *CON* and *EXP*.

(b) "Fitness" is an absolute measure that does not take body size into account. Add two new columns to *CON* and *EXP* consisting of fitness divided by weight expressed in milliliters/kilogram-minute.

(c) Reduce *CON* and *EXP* to standard scores.

Write APL expressions to compute the double sums in exercises 32 through 40. Evaluate the expressions at a terminal if one is available.

32. $\sum_{i=1}^{10}\sum_{j=1}^{20} i+j$

33. $\sum_{i=1}^{10}\sum_{j=1}^{20} i-j$

34. $\sum_{i=1}^{10}\sum_{j=1}^{10} i-j$

35. $\sum_{i=1}^{10}\sum_{j=1}^{20} (i^2+j^2)$

36. $\sum_{i=1}^{20}\sum_{j=1}^{10} (i^2+j^2)^{-1}$

37. $\sum_{i=1}^{10}\sum_{j=1}^{10} (\sin i\pi/6)(\cos j\pi/6)$

38. $\sum_{i=1}^{10}\sum_{j=1}^{10} \max(\sin i, \cos j)$

39. $\sum_{i=0}^{10}\sum_{j=0}^{10} \frac{i}{j+1}$

40. $\sum_{i=-6}^{14}\sum_{j=-12}^{-2} \frac{3+i}{(1+j)^3}$

Let R be the rectangle in the xy plane bounded by the coordinate axes, the line $x=a$ and the line $y=b$. Divide the interval $[0,a]$ into n subintervals $x_0=0<x_1<\cdots<x_n=a$ and divide the interval $[0,b]$ into m subintervals $y_0=0<y_1<\cdots<y_m=b$. Let A_{ij} be the area of the rectangle: $x_{i-1}\le x\le x_i$, $y_{j-1}\le y\le y_j$. Then an approximation to the double integral $\iint_R F(x,y)\,dA$ is the double sum $\sum_{j=1}^{m}\sum_{i=1}^{n} f(x_i,y_j)\,A_{ij}$. If $x_i-x_{i-1}=a/n$ and $y_j-y_{j-1}=b/m$, then $A_{ij}=ab/nm$, and the approximation becomes $ab/nm\,\sum_{i=1}^{n}\sum_{j=1}^{m} f(x_i,y_j)$.

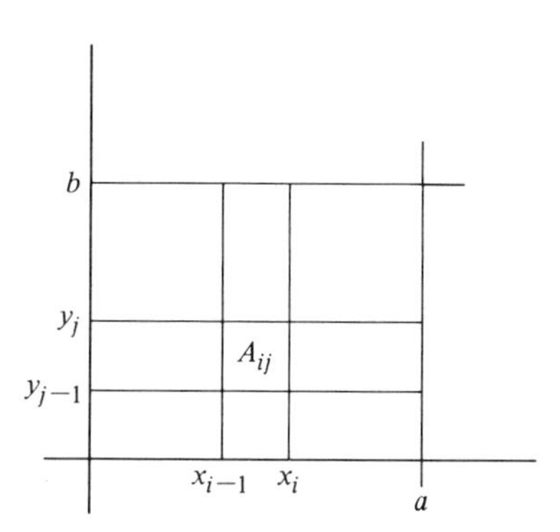

For example, if $a=b=1$ and $n=m=25$, then an approximation to $\iint_R 4xy\,dA=1$ is given by `I←J←(⍳25)÷25` and

```
      (+/+/4×I∘.×J)÷25*2
1.08
```

In exercises 41 through 45 use this approximation to estimate values of the given double integral using the given values of m and n.

41. $\iint_R x \sin y \, dA,$ R: $\begin{cases} 0 \le x \le \pi, \\ 0 \le y \le \pi, \end{cases}$ $m = n = 30$

42. $\iint_R e^{x+y} \, dA,$ R: $\begin{cases} 0 \le x \le \ln 7, \\ 0 \le y \le \ln 8, \end{cases}$ $m = n = 30$

43. $\iint_R \cos x \sin y \, dA,$ R: $\begin{cases} 0 \le x \le \pi, \\ 0 \le y \le \pi, \end{cases}$ $m = n = 30$

44. $\iint_R \sqrt{1 - x^2 + y^2} \, dA,$ R: $\begin{cases} 0 \le x \le 1, \\ 0 \le y \le 1, \end{cases}$ $m = n = 40$

45. $\iint_R \tan xy \, dA,$ R: $\begin{cases} 0 \le x \le \pi/2, \\ 0 \le y \le \pi/2, \end{cases}$ $m = n = 25$

46. Sieve of Eratosthenes

(a) By hand calculation show that

```
V←2=+⌿0=(⍳10)∘.|⍳10
```

is a vector of zeros and ones with $V[i] = 1$ if and only if i is a prime number less than 10.

(b) Let N be a positive integer and define A by `A←(⍳N)∘.|⍳N`. Show that for any positive integer i less than or equal to N the number of zero components of the vector $A[;i]$ is the number of distinct positive divisors of i.

(c) Using part (b), show that the vector V of zeros and ones defined by

```
V←2=+⌿0=(⍳N)∘.|⍳N
```

has $V[i] = 1$ if and only if i is a prime less than or equal to N.

Note: To produce the primes themselves from the vector V one may use the expression `V/⍳N`, where "`/`" denotes the compression function defined in Section 3.4.

CHAPTER TWO

Matrix Algebra

In this chapter we begin the study of linear algebra proper with the algebra of matrices.

A multiplication for matrices is defined in Section 2.1. The set of matrices endowed with this multiplication and the addition defined in Chapter 1 (Definition 1.3) form an algebraic system extending the familiar algebra of scalar quantities.

Using this new algebra, we may write systems of scalar equations as a single matrix equation and then proceed to solve the matrix equation in a manner analogous to that for solving a single scalar equation.

This matrix algebra is developed in Sections 2.1, 2.2, and 2.3. The other sections of this chapter, except for Section 2.4, contain common applications of matrix algebra.

Sections 2.5* and 2.6* apply matrix algebra to the calculus of functions of several variables. These sections will be most useful to readers who have some acquaintance with partial differentiation. They are called *multivariate calculus* sections.

Section 2.4 discusses the class of functions central to matrix algebra — affine and quadratic functions. The content of Section 2.4 is not necessary for Chapter 3 but is needed for Chapter 4 and the multivariate calculus Sections 2.5* and 2.6*.

2.1 Matrix Multiplication

First a word about the differences between matrices and vectors. Matrices are *doubly* indexed arrays. If M is a matrix, then it takes *two* indices to specify a component of M — for example, $M[2; 4]$. Vectors are *singly* indexed arrays. If V is a vector, then it takes *one* index to specify a component of V — for example, $V[3]$.

Because of the way in which vectors are displayed, it is tempting to think of a vector as a matrix with but a single row. We will never do this.

In matrix algebra, however, it is often very convenient to blur the distinction between vectors and *matrices with a single column*. In formal situations and in writing expressions where indexing is involved (for example, `M[3;2]←V[4]`), it is necessary to carefully preserve the matrix-vector distinctions. In other situations, however — especially when writing out the matrix-vector products defined below in Definition 2.1 — it is convenient to write vectors vertically and ignore the details involved in indexing components. The reason is that the vectors we are concerned with are often the columns of a given matrix.

Vectors written in a vertical format will be referred to as *column vectors.*

In conventional mathematical notation, matrices are usually set off typographically by brackets, parentheses, or braces. For example,

$$\begin{bmatrix} 1 & 2 \\ 3 & 4 \end{bmatrix} \text{ or } \begin{pmatrix} 1 & 2 \\ 3 & 4 \end{pmatrix} \text{ or } \begin{Bmatrix} 1 & 2 \\ 3 & 4 \end{Bmatrix}$$

Examples of this notation are:

$$\begin{bmatrix} 1 & 2 \\ 3 & 4 \end{bmatrix} + \begin{bmatrix} 2 & 1 \\ 7 & 2 \end{bmatrix} = \begin{bmatrix} 3 & 3 \\ 10 & 6 \end{bmatrix}$$

$$3\begin{bmatrix} 1 & 2 \\ 7 & 6 \\ 0 & 1 \end{bmatrix} = \begin{bmatrix} 3 & 6 \\ 21 & 18 \\ 0 & 3 \end{bmatrix}$$

MATRIX TIMES VECTOR

Let A be a matrix with n columns. The columns of A are vectors. Let $v_i = A[;i]$ for $i = 1, \ldots, n$. Let $v = (\alpha_1, \ldots, \alpha_n)$ be a vector with n components. The *matrix-vector product* Av is the vector

$$w = \alpha_1 v_1 + \alpha_2 v_2 + \cdots + \alpha_n v_n$$

That is, Av is that linear combination of the columns of A with scalars equal to the components of v.

EXAMPLE 2.1

$$\begin{bmatrix} 1 & 3 \\ 2 & 0 \\ -1 & 5 \end{bmatrix}\begin{bmatrix} 2 \\ -2 \end{bmatrix} = 2\begin{bmatrix} 1 \\ 2 \\ -1 \end{bmatrix} + (-2)\begin{bmatrix} 3 \\ 0 \\ 5 \end{bmatrix} = \begin{bmatrix} 2\cdot 1 + (-2)\cdot 3 \\ 2\cdot 2 + (-2)\cdot 0 \\ 2\cdot(-1) + (-2)\cdot 5 \end{bmatrix} = \begin{bmatrix} -4 \\ 4 \\ -12 \end{bmatrix}$$

Here we have written the vectors $(2, -2)$ and $(-4, 4, -12)$ as column vectors. ∎

A more formal definition will be useful later for checking identities.

DEFINITION 2.1 Let A be a matrix and V a vector. Suppose that `(ρA)[2]=ρV`. Then AV is the vector with `(ρA)[1]=ρAV` and

$$(AV)[i] = \sum_{k=1}^{\rho V} A[i;k]V[k]$$

for all valid indices i.

SYSTEMS OF LINEAR EQUATIONS

Matrix-vector multiplication enables one to write a system of linear equations in compact form. Consider, for example, the system of linear equations

$$\begin{aligned} 3x + 2y &= 7 \\ 9x - 5y &= 12 \\ 6x - 7y &= 5 \end{aligned}$$

These three scalar equations may be written as the single vector equation

$$\begin{bmatrix} 3 \\ 9 \\ 6 \end{bmatrix} x + \begin{bmatrix} 2 \\ -5 \\ -7 \end{bmatrix} y = \begin{bmatrix} 7 \\ 12 \\ 5 \end{bmatrix}$$

or, using matrix-vector multiplication,

$$\begin{bmatrix} 3 & 2 \\ 9 & -5 \\ 6 & -7 \end{bmatrix} \begin{bmatrix} x \\ y \end{bmatrix} = \begin{bmatrix} 7 \\ 12 \\ 5 \end{bmatrix}$$

More generally, given a system of m equations in the n unknowns $x_1, \ldots, x_n$:

$$\begin{aligned} a_{11}x_1 + a_{12}x_2 + a_{13}x_3 + \cdots + a_{1n}x_n &= b_1 \\ a_{21}x_1 + a_{22}x_2 + a_{23}x_3 + \cdots + a_{2n}x_n &= b_2 \\ a_{31}x_1 + a_{32}x_2 + a_{33}x_3 + \cdots + a_{3n}x_n &= b_3 \\ &\vdots \\ a_{m1}x_1 + a_{m2}x_2 + a_{m3}x_3 + \cdots + a_{mn}x_n &= b_m \end{aligned}$$

The m scalar equations may be rewritten as the vector equation

$$\begin{bmatrix} a_{11} \\ a_{21} \\ \vdots \\ a_{m1} \end{bmatrix} x_1 + \begin{bmatrix} a_{12} \\ a_{22} \\ \vdots \\ a_{m2} \end{bmatrix} x_2 + \cdots + \begin{bmatrix} a_{1n} \\ a_{2n} \\ \vdots \\ a_{mn} \end{bmatrix} x_n = \begin{bmatrix} b_1 \\ b_2 \\ \vdots \\ b_m \end{bmatrix}$$

or

$$\begin{bmatrix} a_{11} & a_{12} & a_{1n} \\ a_{21} & a_{22} & a_{2n} \\ \vdots & \vdots & \vdots \\ a_{m1} & a_{m2} & a_{mn} \end{bmatrix} \begin{bmatrix} x_1 \\ \\ \vdots \\ x_n \end{bmatrix} = \begin{bmatrix} b_1 \\ \\ \vdots \\ b_m \end{bmatrix}$$

or, finally,

$$AX = B$$

where $A[i;j] = a_{ij}$, $X[j] = x_j$, $B[i] = b_i$.

In this chapter and the next an algebra of matrices will be developed to solve such equations.

APL NOTATION

If A is a matrix, V a vector, and `(ρA)[2]=ρV`, then the matrix-vector product is written in APL notation as

```
A+.×V
```

The construction "+ ×" (read "sum of products") is a particular example of the APL *generalized inner product* discussed later in this section. A formal definition will be useful later.

DEFINITION 2.2 Let V, W be vectors with `(ρV)=ρW`. Then the scalar `V+.×W` is defined by the identity

```
(V+.×W)=+/V×W
```

That is, `V+.×W` is, in conventional notation,

$$\sum_{k=1}^{\rho V} V[k]W[k].$$

DEFINITION 2.3 Let A be a matrix, V and W vectors with `(ρV)=(ρA)[2]` and `(ρW)=(ρA)[1]`. Then the vectors `A+.×V` and `W+.×A` are defined by the identities

```
(A+.×V)[i]=A[i;]+.×V
(W+.×A)[i]=W+.×A[;i]
```

for all valid indices i.

Notice that since $A[i;]$ and $A[;i]$ are vectors, the right-hand sides of the identities in Definition 2.3 are defined in Definition 2.2. Using the conventional notation from Definition 2.3 above, `(A+.×V)[i]` is

$$\sum_{k=1}^{\rho V} A[i;\,k]V[k]$$

the definition of matrix-vector multiplication given above.

EXAMPLE 2.2 Perform the computation of Example 2.1 in APL.

```
      A←3 2⍴1 3 2 0 ¯1 5
      V←2 ¯2
      +W←A+.×V
¯4  4  ¯12
      ⍴A
3 2
      ⍴V
2
      ⍴W
3
```

Notice that V is a vector, not a 2-by-1 matrix, and W is also a vector. ■

MATRIX TIMES MATRIX

If A and B are matrices and the number of columns of A equals the number of rows of B, then the matrix-matrix product AB is defined as follows. Each column of B is a vector. The ith column of AB is A times the ith column of B:

$$(AB)[;i] = AB[;i]$$

EXAMPLE 2.3

$$A = \begin{bmatrix} 1 & 3 \\ 2 & 0 \\ -1 & 5 \end{bmatrix}, \qquad B = \begin{bmatrix} -1 & 3 \\ 1 & 1 \end{bmatrix}$$

$$\begin{bmatrix} 1 & 3 \\ 2 & 0 \\ -1 & 5 \end{bmatrix}\begin{bmatrix} -1 \\ 1 \end{bmatrix} = (-1)\begin{bmatrix} 1 \\ 2 \\ -1 \end{bmatrix} + 1\begin{bmatrix} 3 \\ 0 \\ 5 \end{bmatrix} = \begin{bmatrix} 2 \\ -2 \\ 6 \end{bmatrix}$$

$$\begin{bmatrix} 1 & 3 \\ 2 & 0 \\ -1 & 5 \end{bmatrix}\begin{bmatrix} 3 \\ 1 \end{bmatrix} = 3\begin{bmatrix} 1 \\ 2 \\ -1 \end{bmatrix} + 1\begin{bmatrix} 3 \\ 0 \\ 5 \end{bmatrix} = \begin{bmatrix} 6 \\ 6 \\ 2 \end{bmatrix}$$

Thus,

$$\begin{bmatrix} 1 & 3 \\ 2 & 0 \\ -1 & 5 \end{bmatrix} \begin{bmatrix} -1 & 3 \\ 1 & 1 \end{bmatrix} = \begin{bmatrix} 2 & 6 \\ -2 & 6 \\ 6 & 2 \end{bmatrix} \quad \blacksquare$$

The procedure of writing out the linear combination of the columns each time two matrices are multiplied is not practical. One works through the columns of B in order, calculating the corresponding columns of the product, skipping the intermediate step of writing out the linear combinations.

$$\begin{bmatrix} 1 & 3 \\ 2 & 0 \\ -1 & 5 \end{bmatrix} \begin{bmatrix} -1 & 3 \\ 1 & 1 \end{bmatrix} = \begin{bmatrix} 1 \cdot (-1) + 3 \cdot 1 & 1 \cdot 3 + 3 \cdot 1 \\ 2 \cdot (-1) + 0 \cdot 1 & 2 \cdot 3 + 0 \cdot 1 \\ (-1) \cdot (-1) + 5 \cdot 1 & (-1) \cdot 3 + 5 \cdot 1 \end{bmatrix}$$

$$= \begin{bmatrix} 2 & 6 \\ -2 & 6 \\ 6 & 2 \end{bmatrix}$$

APL *Notation*

If A and B are matrices, the matrix product is again given in APL by the expression `A+.×B`. In this case the definition is by means of the identity

```
(A+.×B)[;i]   =   A+.×B[;i]
```

The right-hand side of this equation is a matrix-vector product.

EXAMPLE 2.4 Perform the computation of Example 2.3 in APL.

```
      A←3 2ρ1 3 2 0 ¯1 5
      B←2 2ρ¯1 3 1 1
      A+.×B
 2.000  6.000
¯2.000  6.000
 6.000  2.000
```

■

The following formula is sometimes useful.

PROPOSITION 2.1 Let A and B be matrices with AB defined:

$$(A +.\times B)[i;j] = A[i;] +.\times B[;j]$$

Alternately,

$$(AB)[i;j] = \sum_{k=1}^{(\rho A)[2]} A[i;k]B[k;j].$$

Proof $(A +.\times B)[i; j]$ is the ith component of the vector $(A +.\times B)[;j]$. The ith component of $V = (A +.\times B)[;j]$ is written $V[i] = ((A +.\times B)[;j])[i]$. This gives us the identity

$$(A +.\times B)[i; j] = ((A +.\times B)[;j])[i]$$

but $(A +.\times B)[;j] = A +.\times B[;j]$ by definition. The product $A +.\times B[;j]$ is the product of the matrix A and the vector $B[;j]$, and hence by the definition of matrix-vector product

$$(A +.\times B[;j])[i] = A[i;] +.\times B[;j].$$

In summary,

$$\begin{aligned}(A +.\times B)[i; j] &= ((A +.\times B)[;j])[i] \\ &= (A +.\times B[;j])[i] \\ &= A[i;] +.\times B[;j] \qquad \blacksquare\end{aligned}$$

If A has m rows and n columns and B has p rows and q columns, then AB is not defined unless $n = p$. If $n = p$, then $AB = C$ is defined and C has n rows and q columns:

$$m\ \overset{n}{\boxed{A}}\ n\ \overset{q}{\boxed{B}} = m\ \overset{q}{\boxed{C}}$$

Even if AB is defined, BA need not be (for example, `3 2=⍴A` and `2 5=⍴B`). If both AB and BA are defined, they need not be of the same shape, as the next example shows.

EXAMPLE 2.5

$$A = \begin{bmatrix} 1 & 0 \\ 0 & 0 \\ 0 & 1 \end{bmatrix} \qquad B = \begin{bmatrix} 1 & 0 & 0 \\ 0 & 0 & 1 \end{bmatrix}$$

$$AB = \begin{bmatrix} 1 & 0 \\ 0 & 0 \\ 0 & 1 \end{bmatrix}\begin{bmatrix} 1 & 0 & 0 \\ 0 & 0 & 1 \end{bmatrix} = \begin{bmatrix} 1 & 0 & 0 \\ 0 & 0 & 0 \\ 0 & 0 & 1 \end{bmatrix}$$

$$BA = \begin{bmatrix} 1 & 0 & 0 \\ 0 & 0 & 1 \end{bmatrix}\begin{bmatrix} 1 & 0 \\ 0 & 0 \\ 0 & 1 \end{bmatrix} = \begin{bmatrix} 1 & 0 \\ 0 & 1 \end{bmatrix} \qquad \blacksquare$$

A matrix is called *square* if the number of rows equals the number of columns. If A and B are square and AB is defined, then BA is defined. Further, AB and BA are both square and have the same shape. Still, $AB \neq BA$ in general.

EXAMPLE 2.6

$$A = \begin{bmatrix} 1 & 2 \\ 3 & 4 \end{bmatrix}, \quad B = \begin{bmatrix} 5 & 6 \\ 7 & 8 \end{bmatrix}$$

$$AB = \begin{bmatrix} 1 & 2 \\ 3 & 4 \end{bmatrix}\begin{bmatrix} 5 & 6 \\ 7 & 8 \end{bmatrix} = \begin{bmatrix} 19 & 22 \\ 43 & 50 \end{bmatrix}$$

$$BA = \begin{bmatrix} 5 & 6 \\ 7 & 8 \end{bmatrix}\begin{bmatrix} 1 & 3 \\ 2 & 4 \end{bmatrix} = \begin{bmatrix} 17 & 39 \\ 23 & 53 \end{bmatrix} \quad \blacksquare$$

In algebra the identity $ab = ba$ is called the *commutative law* of multiplication. Matrix multiplication does not obey the commutative law. Matrix multiplication is *noncommutative.*

Transpose

Let A be a matrix. The transpose of A, written ⍉A† in APL and A^T otherwise, is the matrix whose rows are the columns of A. Thus, if A is m-by-n, then A^T is n-by-m.

EXAMPLE 2.7

$$\begin{bmatrix} 1 & 2 & 3 \\ 4 & 5 & 6 \\ 7 & 8 & 9 \end{bmatrix}^T = \begin{bmatrix} 1 & 4 & 7 \\ 2 & 5 & 8 \\ 3 & 6 & 9 \end{bmatrix}$$

$$\begin{bmatrix} 1 & 4 & 3 \\ -2 & 7 & 1 \end{bmatrix}^T = \begin{bmatrix} 1 & -2 \\ 4 & 7 \\ 3 & 1 \end{bmatrix}$$

$$\begin{bmatrix} 1 \\ 4 \\ 6 \\ 9 \end{bmatrix}^T = [1 \quad 4 \quad 6 \quad 9] \quad \blacksquare$$

The *main diagonal* of a matrix A is the vector with components $A[1;\ 1]$, $A[2;\ 2]$, $A[3;\ 3]$, . . . , $A[n;\ n]$:

$$\begin{bmatrix} a_{11} & a_{12} & a_{13} \\ a_{21} & a_{22} & a_{23} \\ a_{31} & a_{32} & a_{33} \end{bmatrix}$$

The matrix A^T is obtained by "flipping" A about its main diagonal — hence the APL notation ⍉. Incidentally, the main diagonal of A is denoted in APL by 1 1 ⍉ A.

† ⍉ is typed ○-backspace-\.

EXAMPLE 2.8

```
      +A←3 5⍴⍳25
 1   2   3   4   5
 6   7   8   9  10
11  12  13  14  15
      ⍉A
 1   6  11
 2   7  12
 3   8  13
 4   9  14
 5  10  15

      1 1⍉A
1 7 13
```

■

A formal definition of A^T is useful in checking identities.

DEFINITION 2.4. Let A be a matrix. The matrix ⍉A is defined by

1. (ρA)[2 1] = ρ ⍉A
2. (⍉A)[i; j] = A[j; i] for every pair of indices i; j.

If A is a vector or scalar, then A = ⍉A.

The next proposition gives two useful identities. The first identity emphasizes the noncommutative nature of matrix multiplication.

PROPOSITION 2.2. Let A, B be matrices with AB defined. Then

$$(AB)^T = B^T A^T$$
$$(A + B)^T = A^T + B^T$$

Proof Let A be m-by-n and B n-by-q. Then AB is m-by-q, so that $(AB)^T$ is q-by-m. Since B^T is q-by-n and A^T is n-by-m, $B^T A^T$ is defined and is q-by-m. Thus the shapes are correct. The equality of the individual components is most easily checked using APL notation.

Since (⍉A)[i; j] = A[j; i], it follows that (⍉A)[i;] = A[;i] and (⍉A)[;j] = A[j;]. Thus,

(⍉A +.× B)[i; j] = (A +.× B)[j; i]   (definition of ⍉)
= A[j;] +.× B[;i]
= (⍉A)[;j] +.× (⍉B)[i;]   (Proposition 2.1)
= +/(⍉A)[;j]×(⍉B)[i;]

$$= +/(\boxtimes B)[i;] \times (\boxtimes A)[;j] \qquad \text{(these arrays are vectors)}$$
$$= (\boxtimes B)[i;] +.\times (\boxtimes A)[;j]$$
$$= ((\boxtimes B) +.\times \boxtimes A)[i;j]$$

for all indices $i; j$. ■

A matrix is called *symmetric* if $A = A^T$. The symmetric matrices form an important class of matrices that will be studied in some detail in later chapters. It follows from the above proposition that matrices of the form A^TA or AA^T are always symmetric (exercise 6).

GENERALIZED INNER PRODUCT

The "sum of products" operation "+.×" used above is a specific example of a more general APL construction that we will sometimes use. The construction generalizes in two ways.

First the "+" and "×" may be replaced by any of the scalar dyadic functions from Appendix B. Thus, given vectors A and B, one could form the "sum of quotients" `A+.÷B` (= `+/A÷B`) or the product of differences `A×.-B` (= `×/A-B`). If either A or B is a matrix, then the obvious changes in the identities used to define $A +.\times B$ give the appropriate definitions.

Second, the operation is always defined if one of the arguments is a scalar or single-component vector. The scalar is first reshaped to a vector of the proper size.

EXAMPLE 2.9 (*Sum of squares*) One often wishes to compute the sum of squares of the components of a vector V. This is just `V+.*2` (recall that `*` is exponentiation). For example,

`1 ¯1 3+.*2` is $1^2 + (-1)^2 + 3^2$ or 11. ■

EXAMPLE 2.10 (*Averages*) One often wishes to compute the average of a set of numbers. Let the numbers be stored as the components of a vector V. Then the average is just `V+.÷ρV`.

Suppose that `V ← 1 2 3`. Then `ρV` is `3` (the single-element vector) and `V+.÷ρV` is `1 2 3 +.÷3`, which is

$$\frac{1}{3} + \frac{2}{3} + \frac{3}{3} = \frac{1+2+3}{3} = 2. \qquad ■$$

EXAMPLE 2.11 (*Equality of vectors*) In later chapters we will occasionally need an APL expression that will return a 1 if two vectors are equal and a 0 if they are unequal. If V and W are scalars, then the expression `V=W` will do. If V and W are vectors, then the scalar dyadic function `=` operates component by component. Thus `1 2 3 = 1 0 3` is `1 0 1`.

To check the equality of the vectors V and W use `V∧.=W`. For example, `1 2 3 ∧.= 1 0 3` is `(1=1)∧(2=0)∧(3=3)`, which is `1∧(0∧1)` which is `0`.

In particular, the expression `V∧.=0` is `1` if V is the zero vector and is `0` otherwise. ∎

EXERCISES 2.1

1. Compute, by hand and by machine if available, the matrix-vector product $w = Av$.

(a) $A = \begin{bmatrix} 2 & -2 \\ 1 & 3 \end{bmatrix}$ $v = (1, 1)$

(b) $A = \begin{bmatrix} 1 & 4 \\ 0 & 0 \\ 3 & 2 \end{bmatrix}$ $v = (4, -1)$

(c) $A = \begin{bmatrix} 1 & 3 & -1 \\ -2 & -4 & 6 \end{bmatrix}$ $v = (1, 1, 4)$

(d) $A = \begin{bmatrix} \frac{1}{3} & \frac{2}{3} & 0 \\ \frac{1}{2} & 1 & -\frac{1}{2} \\ 600 & -598 & -1 \end{bmatrix}$ $v = (1, 1, 1)$

(e) $A = \begin{bmatrix} 0 & 1 \\ 1 & 0 \end{bmatrix}$ $v = (a, b)$

(Omit machine computation for this problem.)

(f) $A = [1 \quad 3 \quad -4 \quad 2]$
$v = (1, 1, 1, \frac{1}{2})$

(g) $A = \begin{bmatrix} 1 \\ 3 \\ 2 \end{bmatrix}$ $v = (-2)$

(h) $A = [3]$ $v = (4)$

2. Compute, by hand and by machine if available, the matrix-matrix product indicated, if possible.

(a) $\begin{bmatrix} 1 & 3 \\ -3 & 1 \end{bmatrix}\begin{bmatrix} 1 & 2 \\ 2 & 1 \end{bmatrix}$

(b) $\begin{bmatrix} 1 & 2 \\ 2 & 1 \end{bmatrix}\begin{bmatrix} 1 & 3 \\ -3 & 1 \end{bmatrix}$

(c) $\begin{bmatrix} 1 & 3 \\ 2 & 4 \\ 7 & -7 \end{bmatrix}\begin{bmatrix} 2 & 1 & 1 & 0 \\ 1 & 2 & 1 & 0 \end{bmatrix}$

(d) $\begin{bmatrix} 1 & 1 \\ 0 & 1 \\ -2 & -4 \end{bmatrix}\begin{bmatrix} -3 & 7 & 5 & 3 \\ 0 & 0 & 1 & 0 \\ 7 & 2 & 2 & 1 \end{bmatrix}$

(e) $\begin{bmatrix} 7 & 5 & 3 \\ 0 & 1 & 0 \\ 2 & 2 & 1 \end{bmatrix}\begin{bmatrix} 1 & 1 & -3 \\ 0 & 1 & 0 \\ -2 & -4 & 7 \end{bmatrix}$

(f) $\begin{bmatrix} 1 & 3 & 0 \\ 3 & 0 & 1 \end{bmatrix}\begin{bmatrix} 1 & 4 \\ 2 & 0 \end{bmatrix}$

(g) $\begin{bmatrix} 6 & 1 & -1 & 2 \\ 2 & -1 & -2 & 2 \end{bmatrix}\begin{bmatrix} 1 & -2 \\ -2 & 5 \\ 1 & -3 \\ -1 & 2 \end{bmatrix}$

(h) $\begin{bmatrix} 1 & -2 \\ -2 & 5 \\ 1 & -3 \\ -1 & 2 \end{bmatrix}\begin{bmatrix} 6 & 1 & -1 & 2 \\ 2 & -1 & -2 & 2 \end{bmatrix}$

(i) $[1 \quad 3 \quad 2]\begin{bmatrix} 7 \\ -1 \\ 1 \end{bmatrix}$

(j) $\begin{bmatrix} 1 \\ 4 \\ 3 \\ -1 \end{bmatrix}[2]$

(k) $\begin{bmatrix} 1 \\ 4 \\ 3 \end{bmatrix}[2 \quad 6 \quad 1]$

(l) AB where $A = \begin{bmatrix} 1 & -1 & 1 & 0 & 3 & 4 & -1 & 9 & -5 & 6 \\ 0 & 1 & 0 & -1 & 0 & -1 & 0 & -3 & 1 & -2 \end{bmatrix}$

$$B^T = \begin{bmatrix} -4 & 4 & 1 & 4 & 4 & -1 & 2 & 0 & 1 & 1 \\ 3 & -2 & -2 & -4 & -4 & 0 & -1 & 1 & -1 & -1 \end{bmatrix}$$

(m) $\begin{bmatrix} 0 & 1 \\ 1 & 0 \end{bmatrix}\begin{bmatrix} 1 & 2 \\ 3 & 4 \end{bmatrix}$

(n) $\begin{bmatrix} 1 & 2 \\ 3 & 4 \end{bmatrix}\begin{bmatrix} 0 & 1 \\ 1 & 0 \end{bmatrix}$

(o) $\left(\begin{bmatrix} 0 & 1 \\ 1 & 0 \end{bmatrix}\begin{bmatrix} 1 & 2 \\ 3 & 4 \end{bmatrix}\right)\begin{bmatrix} 1 & 0 \\ 0 & 1 \end{bmatrix}$

(p) $\begin{bmatrix} 0 & 1 \\ 1 & 0 \end{bmatrix}\left(\begin{bmatrix} 1 & 2 \\ 3 & 4 \end{bmatrix}\begin{bmatrix} 0 & 1 \\ 1 & 0 \end{bmatrix}\right)$

(q) $\begin{bmatrix} 1 & 2 & 3 \\ 4 & 5 & 6 \\ 7 & 8 & 9 \end{bmatrix}\begin{bmatrix} 1 & 0 & 0 \\ 0 & 0 & 0 \\ 0 & 0 & 1 \end{bmatrix}$

(r) $\begin{bmatrix} 1 & 0 & 0 \\ 0 & 0 & 0 \\ 0 & 0 & 1 \end{bmatrix}\begin{bmatrix} 1 & 2 & 3 \\ 4 & 5 & 6 \\ 7 & 8 & 9 \end{bmatrix}$

3. Let $A = \begin{bmatrix} a & -b \\ b & a \end{bmatrix}$, $B = \begin{bmatrix} c & -d \\ d & c \end{bmatrix}$. Show that $AB = BA$ in this case.

4. Let $A = \begin{bmatrix} a & 0 \\ 0 & a \end{bmatrix}$ and let B be any 2-by-2 matrix. Show that $AB = BA$.

5. Let A be a 2-by-2 matrix. Suppose that $AB = BA$ for *every* 2-by-2 matrix B. Show that there is some number a such that

$$A = \begin{bmatrix} a & 0 \\ 0 & a \end{bmatrix}$$

6. Use Proposition 2.2 to check that any matrix of the form AA^T or A^TA is symmetric.

7. Rewrite the following systems of equations as a single matrix equation.

(a) $\begin{aligned} 3x + 4y &= 2 \\ 7x - 6y &= 1 \end{aligned}$

(b) $\begin{aligned} 4y &= 3 \\ 2x + y &= 0 \end{aligned}$

(c) $\begin{aligned} 3x_1 + x_2 - 4x_3 &= 7 \\ x_2 + x_3 &= 1 \\ 4x_1 \quad\quad &= 0 \end{aligned}$

(d) $\begin{aligned} y_1 &= 3x_1 + 12x_2 - 7x_3 \\ y_2 &= 4x_1 + x_2 \end{aligned}$

(e) $\begin{aligned} x &= 4t \\ y &= 3t \end{aligned}$

(f) $\begin{aligned} x &= 4t + 7 \\ y &= -6t + 1 \end{aligned}$

(g) $\begin{aligned} 3x + 12y &= 7u - v \\ 12x - y &= u + 6v \\ y &= 1 + 2t \end{aligned}$

8. Matrices may have no rows or columns. That is, a matrix might be 0-by-4 or 5-by-0. To create such a matrix in APL use `A←0 4⍴0` or `A←5 0⍴⍳0`.

(a) If A is 0-by-10 and B is 10-by-0, then $C = AB$ should be 0-by-0. Check this on a machine if one is available.

(b) If A is n-by-0 and V is a vector without components (`0=⍴V`), then `W←A+.×V` is a vector with n components. Use the identity `(A+.×V)[I]=A[I;]+.×V`, exercise 50 of Exercises 1.2, and the fact that `(V+.×W)=+/V×W` if V and W are vectors to show that the components of W are zeros. This exercise shows that a linear combination of an empty set of vectors gives a zero vector.

(c) Let A be n-by-0 and B be 0-by-m. Show that AB is an n-by-m matrix of zeros. [Hint: See part (b).] Verify this on a machine if one is available.

9. What is the result of the APL computations?

(a) `1 2 3 +.×2`

(b) `1 2 3 -.×2`

(c) `1 2 3 +.*2`

(d) `1 4 9 +.÷.5`

(e) `2*.×1 2 3`

(f) `1 12 6 ⌈.⌊ 3 11 7`

(g) `1 12 6 ⌊.⌈ 3 11 7`

(h) `1 2 3∧.=1 3 3`

(i) `1 2 3 ∨.=1 3 3`

(j) `1 2 3 ∧.≠1 3 3`

(k) `1 2 3 ∧.≤1 3 3`

(l) `1 2 3 ∧.<1 3 3`

(m) `1 2 3 ∨.≠1 3 3`

(n)
```
A←3 2ρι6
A∧.=3 4
```

(o)
```
A←3 2ρι6
A∨.≠3 4
```

(p)
```
A←4 2ρ1 3 0 0
+/A∧.=0
```

(q)
```
A←4 2ρ1 3 0 0
+/A∨.≠0
```

(r)
```
A←⍉4 2ρ1 3 0 0
+/0∧.=A
```

(s)
```
A←⍉4 2ρ1 3 0 0
+/0∨.≠A
```

10. Write APL expressions to:
    (a) Count the number of zero rows in the matrix A.
    (b) Count the number of nonzero rows in the matrix A.
    (c) Count the number of zero columns in the matrix A.
    (d) Count the number of nonzero columns in the matrix A.
11. Use Proposition 2.1 and Definition 2.4 to show that if A^TA is a matrix of zeros, then A is a matrix of zeros.

2.2 Inverse Matrices

There is an algebra of matrices similar in many respects to the familiar algebra of real numbers. There are, however, some fundamental differences. One of these differences has already been encountered. Given matrices A and B, it is *not* true that $AB = BA$. Matrix algebra is noncommutative.

We shall see shortly that cancellation also fails. If $AB = AC$, we cannot "cancel the A's" and deduce that $B = C$. In general $B \neq C$.

One important property of scalar multiplication that does hold for matrix multiplication is the *associative law*.

PROPOSITION 2.3 Let A, B, C be matrices with AB and BC defined. Then $A(BC) = (AB)C$.

Proof Let A be m-by-n. Since AB is defined, B must have n rows. Say B is n-by-p. Since BC is defined, C must have p rows. Say C is p-by-q.

Then BC is n-by-q and $A(BC)$ is m-by-q. On the other hand, AB is m-by-p, and so $(AB)C$ is m-by-q. Thus $A(BC)$ and $(AB)C$ have the same shape. Next we show that corresponding entries are equal.

Consider the $i;j$ entry of $A(BC)$ — that is, $(A(BC))[i;j]$.

$$\begin{aligned}(A(BC))[i;j] &= \sum_{h=1}^{n} A[i;h](BC)[h;j] && \text{by Proposition 2.1}\\ &= \sum_{h=1}^{n} A[i;h]\left(\sum_{l=1}^{p} B[h;l]C[l;j]\right) && \text{by Proposition 2.1}\\ &= \sum_{h=1}^{n}\sum_{l=1}^{p} A[i;h]B[h;l]C[l;j]\end{aligned}$$

Similarly,

$$\begin{aligned}((AB)C)[i;j] &= \sum_{l=1}^{p} (AB)[i;l]C[l;j]\\ &= \sum_{l=1}^{p}\left(\sum_{h=1}^{n} A[i;h]B[h;l]\right)C[l;j]\\ &= \sum_{l=1}^{p}\sum_{h=1}^{n} A[i;h]B[h;l]C[l;j] \qquad \blacksquare\end{aligned}$$

EXAMPLE 2.12 It occasionally happens that a calculation is made easier by invoking the associative law. As an extreme example consider the following calculation. Here A is 4 by 1, B is 1 by 5, and C is 5 by 1.

$$\left(\begin{bmatrix}1\\-1\\2\\3\end{bmatrix}\begin{bmatrix}1 & 3 & -2 & 5 & 7\end{bmatrix}\right)\begin{bmatrix}1\\-1\\2\\-2\\1\end{bmatrix} = \begin{bmatrix}1 & 3 & -2 & 5 & 7\\-1 & -3 & 2 & -5 & -7\\2 & 6 & -4 & 10 & 14\\3 & 9 & -6 & 15 & 21\end{bmatrix}\begin{bmatrix}1\\-1\\2\\-2\\1\end{bmatrix}$$

$$= \begin{bmatrix}-9\\9\\-18\\-27\end{bmatrix}$$

$$\begin{bmatrix}1\\-1\\2\\3\end{bmatrix}\left(\begin{bmatrix}1 & 3 & -2 & 5 & 7\end{bmatrix}\begin{bmatrix}1\\-1\\2\\-2\\1\end{bmatrix}\right) = \begin{bmatrix}1\\-1\\2\\3\end{bmatrix}\begin{bmatrix}-9\end{bmatrix} = \begin{bmatrix}-9\\9\\-18\\-27\end{bmatrix} \qquad \blacksquare$$

The trick illustrated in Example 2.12 is important in machine implementation of some important linear algebra algorithms. If all three vectors were of length n, for example, then the first calculation involves performing $2n^2$ multiplications and keeping track of up to $n^2 + n$ numbers, whereas the second calculation involves $2n$ multiplications and keeping track of $3n$ numbers. For $n = 100$, say, this is a significant difference.

Next we take up the problem of the cancellation law. We begin with an example to show that it does not work. We need three matrices A, B, C such that $AB = AC$ but $B \neq C$.

EXAMPLE 2.13

$$A = \begin{bmatrix} 1 & -2 & -1 & 1 \\ -2 & 5 & 2 & -4 \end{bmatrix}, \quad B = \begin{bmatrix} 6 & 1 \\ 2 & -1 \\ 1 & 2 \\ 0 & -1 \end{bmatrix}, \quad C = \begin{bmatrix} 5 & 3 \\ 0 & 1 \\ 3 & 1 \\ -1 & 0 \end{bmatrix}$$

$$AB = \begin{bmatrix} 1 & -2 & -1 & 1 \\ -2 & 5 & 2 & -4 \end{bmatrix} \begin{bmatrix} 6 & 1 \\ 2 & -1 \\ 1 & 2 \\ 0 & -1 \end{bmatrix} = \begin{bmatrix} 1 & 0 \\ 0 & 1 \end{bmatrix}$$

$$= \begin{bmatrix} 1 & -2 & -1 & 1 \\ -2 & 5 & 2 & -4 \end{bmatrix} \begin{bmatrix} 5 & 3 \\ 0 & 1 \\ 3 & 1 \\ -1 & 0 \end{bmatrix} = AC$$

But $B \neq C$. ■

In the system of real numbers cancellation is related to division. Cancellation for the real numbers can be justified by the following argument.

Suppose that $ab = ac$ and $a \neq 0$ (if $a = 0$, we cannot cancel). Since $a \neq 0$, we may divide both sides of the equation by a to get $b = c$. Let us look at this division in another way. Since $a \neq 0$, we can form $1/a = a^{-1}$. Now multiply both sides of the equation by a^{-1}:

$$\frac{1}{a}(ab) = \frac{1}{a}(ac)$$

Now use the associative law of multiplication.

$$\left(\frac{1}{a} \cdot a\right)b = \left(\frac{1}{a} \cdot a\right)c$$

Since $a^{-1}a = 1$ this becomes

$$1 \cdot b = 1 \cdot c$$

Since $1 \cdot b = b$ and $1 \cdot c = c$, we have the result.

Since cancellation does not hold for matrices, this argument must fail for matrices. By Proposition 2.3 the associative law holds, and so the problem is the equation $a^{-1}a = 1$. There are two problems here. What is "a^{-1}" and what is "1" for matrices? The second question is easy to answer.

DEFINITION 2.5 The *identity matrix of order n* is the n-by-n matrix with ones on the main diagonal and zeros elsewhere [i.e., `(ιn)∘.=ιn`].

The identity matrices of orders 1, 2, 3 are

$$[1] \qquad \begin{bmatrix} 1 & 0 \\ 0 & 1 \end{bmatrix} \qquad \begin{bmatrix} 1 & 0 & 0 \\ 0 & 1 & 0 \\ 0 & 0 & 1 \end{bmatrix}$$

The zero-by-zero matrix `(0 0 ρ 0)` is an identity matrix.

PROPOSITION 2.4 Let A be an m-by-n matrix.

1. If I is the m-by-m identity matrix, then $IA = A$.
2. If I is the n-by-n identity matrix, then $AI = A$.

Proof By Proposition 2.1,

$$(IA)[i; j] = \sum_{h=1}^{m} I[i; h]A[h; j].$$

Now $I[i; h]$ is 1 if $i = k$ and 0 if $i \neq k$. So the only nonzero term of this sum is $I[i; i]A[i; j] = A[i; j]$. Similarly, $AI = A$. ■

The question of "a^{-1}" for matrices is considerably more complicated.

DEFINITIONS 2.6

(i) The matrix L is a *left inverse* for A if $LA = I$, an identity matrix.

(ii) The matrix R is a *right inverse* for A if $AR = I$, an identity matrix.

(iii) The matrix B is an *inverse* for A if it is both a left inverse and a right inverse for A. In this case A is called *invertible* or *nonsingular* and B is denoted by A^{-1}.

EXAMPLE 2.14 Consider the matrices of Example 2.13. We have $AB = I$, so that A is a left inverse of B and B is a right inverse of A. A is also a left inverse of C, and C is also a right inverse of A.

This shows that a matrix may have more than one right inverse. Taking transposes and invoking Proposition 2.2, we have $I = I^T = (AB)^T = B^TA^T$. Similarly $I = C^TA^T$. And so a matrix may have more than one left inverse. ■

Example 2.14 shows that a matrix may have several left or right inverses but they must all have the same shape. Suppose that $LA = I$ and A is m-by-n. Then I has n columns and L has m columns. Since I is square, L has n rows. Thus if A is m-by-n, then L is n-by-m. Similarly a right inverse would be n-by-m also.

The next example shows that a matrix may have no inverses at all.

EXAMPLE 2.15 Let

$$A = \begin{bmatrix} 1 & 0 \\ 0 & 0 \end{bmatrix}$$

Suppose that A has a left inverse L. Then L must be 2 by 2 also. Say

$$L = \begin{bmatrix} a & b \\ c & d \end{bmatrix}$$

Since $LA = I$, we have

$$\begin{bmatrix} a & b \\ c & d \end{bmatrix}\begin{bmatrix} 1 & 0 \\ 0 & 0 \end{bmatrix} = \begin{bmatrix} 1 & 0 \\ 0 & 1 \end{bmatrix}$$

$$\begin{bmatrix} a & 0 \\ c & 0 \end{bmatrix} = \begin{bmatrix} 1 & 0 \\ 0 & 1 \end{bmatrix}$$

Since $0 \neq 1$, this is impossible. Thus A does not have a left inverse. A similar argument shows that A does not have a right inverse. ■

If a matrix is invertible, then there is only one inverse. For suppose that $LA = I_n$ and $AR = I_m$, where I_n, I_m are identity matrices of orders n and m. Then A is m-by-n, and L, R are both n-by-m. Multiply through the first equation by R on the right:

$$\begin{aligned} (LA)R &= I_n R & \\ (LA)R &= R & \text{by Proposition 2.4} \\ L(AR) &= R & \text{by Proposition 2.3} \\ LI_m &= R & \\ L &= R & \text{by Proposition 2.4} \end{aligned}$$

If $\bar{L}$ is another left inverse for A, then the argument shows $\bar{L} = R$ also. Hence $L = \bar{L}$. Similarly all right inverses are equal. Hence all left and right inverses are equal.

This justifies using the symbol A^{-1} when A is invertible, since there is only one inverse.

The following result will be proved in Chapter 3.

PROPOSITION 2.5 Let A be a matrix.

1. If A is invertible, then A is square.
2. If A is square and A has a left inverse, then A is invertible.
3. If A is square and A has a right inverse, then A is invertible. ∎

The import of this theorem is that if A is square and $BA = I$, then $B = A^{-1}$ and so $AB = I$ automatically. Conversely if $AB = I$, then $B = A^{-1}$ and $AB = I$ automatically. Thus only one of the conditions, $AB = I$ or $BA = I$, need be checked if B is a candidate for A^{-1}.

Returning to the question of cancellation, if $AB = AC$ and A has a left inverse L, then $B = C$. The argument is as follows:

$$\begin{aligned} AB &= AC \\ L(AB) &= L(AC) \\ (LA)B &= (LA)C \qquad \text{by Proposition 2.3} \\ IB &= IC \\ B &= C \qquad \text{by Proposition 2.4} \end{aligned}$$

Solving Linear Equations

We saw in Section 2.1 that any system of linear equations could be rewritten as a single matrix equation of the form

$$AX = B$$

Now if L is any left inverse of A, then

$$\begin{aligned} L(AX) &= LB \\ (LA)X &= LB \\ X &= LB \end{aligned}$$

So *if a solution X exists,* it must be LB. Thus we have proved the next proposition.

PROPOSITION 2.6 Let A, B be matrices and assume that A has a left inverse L. Then the equation

$$AX = B$$

has the solution $X = LB$ or it has no solution at all. ∎

If we are given L, we set $X = LB$ and then check to see if this is a solution.

The problem, of course, is finding L. In Chapter 3 we will develop a procedure to do this. Another procedure will be mentioned in Chapter 5. These procedures can be used to give X directly, so that often L is not explicitly calculated.

The procedures will also give X when L does not exist. For the moment, however, we give a formula for L when A is 2 by 2 and a "canned" procedure for more general A.

PROPOSITION 2.7 Let

$$A = \begin{bmatrix} a & b \\ c & d \end{bmatrix}$$

Then A is invertible if and only if $\Delta = ad - bc \neq 0$, and if $\Delta \neq 0$, then

$$A^{-1} = \frac{1}{\Delta}\begin{bmatrix} d & -b \\ -c & a \end{bmatrix}$$

Proof If $\Delta \neq 0$, then

$$\frac{1}{\Delta}\begin{bmatrix} a & -b \\ -c & a \end{bmatrix}\begin{bmatrix} a & b \\ c & d \end{bmatrix} = \frac{1}{\Delta}\begin{bmatrix} \Delta & 0 \\ 0 & \Delta \end{bmatrix} = \begin{bmatrix} 1 & 0 \\ 0 & 1 \end{bmatrix}$$

by the rule for multiplying scalars and matrices. Thus A^{-1} has the stated form by Proposition 2.5.

Suppose, however, that $\Delta = 0$. Then if $a = b = c = d = 0$ we have

$$A = \begin{bmatrix} 0 & 0 \\ 0 & 0 \end{bmatrix}$$

which is not invertible, since

$$LA = \begin{bmatrix} 0 & 0 \\ 0 & 0 \end{bmatrix} \neq I.$$

If A has a nonzero entry then let

$$B = \begin{bmatrix} d & -b \\ -c & a \end{bmatrix}$$

Now

$$AB = \begin{bmatrix} \Delta & 0 \\ 0 & \Delta \end{bmatrix} = \begin{bmatrix} 0 & 0 \\ 0 & 0 \end{bmatrix} = A\begin{bmatrix} 0 & 0 \\ 0 & 0 \end{bmatrix}$$

If A were invertible, it could be canceled, giving

$$B = \begin{bmatrix} 0 & 0 \\ 0 & 0 \end{bmatrix}$$

which is not true. ■

Writing out matrices consisting of all zeros, as in the proof above, is tedious.

DEFINITION 2.7 A *zero matrix* is a matrix of any shape consisting entirely of zeros. Zero matrices will be denoted by 0. The shape can be deduced from the context.

EXAMPLE 2.16 Solve the linear system.

$$\begin{aligned} x + 2y &= 3 \\ 3x + 4y &= 5 \end{aligned}$$

Solution Rewriting as a matrix equation gives

$$\begin{bmatrix} 1 & 2 \\ 3 & 4 \end{bmatrix} \begin{bmatrix} x \\ y \end{bmatrix} = \begin{bmatrix} 3 \\ 5 \end{bmatrix}$$

So

$$\begin{bmatrix} x \\ y \end{bmatrix} = \begin{bmatrix} 1 & 2 \\ 3 & 4 \end{bmatrix}^{-1} \begin{bmatrix} 3 \\ 5 \end{bmatrix} = \frac{1}{4 - 6} \begin{bmatrix} 4 & -2 \\ -3 & 1 \end{bmatrix} \begin{bmatrix} 3 \\ 5 \end{bmatrix} = -\frac{1}{2} \begin{bmatrix} 2 \\ -4 \end{bmatrix} = \begin{bmatrix} -1 \\ 2 \end{bmatrix}$$

is the only possible solution. To check if it is a solution, substitute back into the original equation.

$$\begin{bmatrix} 1 & 2 \\ 3 & 4 \end{bmatrix} \begin{bmatrix} -1 \\ 2 \end{bmatrix} = \begin{bmatrix} -1 & +4 \\ -3 & +8 \end{bmatrix} = \begin{bmatrix} 3 \\ 5 \end{bmatrix}$$

Thus the system has the unique solution $x = -1$, $y = 2$. ■

The next example shows that the equation $AX = B$ need not have a solution even if A has a left inverse.

EXAMPLE 2.17 Let

$$A = \begin{bmatrix} 11 & -1 \\ -2 & 3 \\ -3 & 3 \end{bmatrix}, \qquad B = \begin{bmatrix} 1 \\ 2 \\ 3 \end{bmatrix}$$

and consider the system $AX = B$. By checking directly, we may verify that

$$L = \begin{bmatrix} 6 & 1 & 1 \\ 8 & 1 & 2 \end{bmatrix}$$

is a left inverse for A. Thus, multiplying through by L gives

$$X = LB = \begin{bmatrix} 6 & 1 & 1 \\ 8 & 1 & 2 \end{bmatrix} \begin{bmatrix} 1 \\ 2 \\ 3 \end{bmatrix} = \begin{bmatrix} 11 \\ 16 \end{bmatrix}$$

Substituting in the original equation, we have

$$A\begin{bmatrix}11\\16\end{bmatrix} = \begin{bmatrix}1 & -1\\-2 & 3\\-3 & 3\end{bmatrix}\begin{bmatrix}11\\16\end{bmatrix} = \begin{bmatrix}-5\\24\\15\end{bmatrix} \neq B$$

Thus the system has no solution. ∎

Proposition 2.7 gives an easily checked condition for a 2-by-2 matrix to be invertible ($\Delta \neq 0$). Next we get a condition for a general matrix to have a left inverse. The full proof will have to wait until Chapter 3. First we state some useful identities. The proofs are left as exercises.

PROPOSITION 2.8 Let A, B, C, D be vectors or matrices such that AB, $B + C$, and CD are defined. Let α and β be scalars. Then

1. $A(B + C) = AB + AC$
2. $(B + C)D = BD + CD$
3. $A(\alpha B) = \alpha AB = AB\alpha$
4. $(\alpha + \beta)A = \alpha A + \beta A$ ∎

Now consider a matrix A in which, say, the third column is the sum of the first column and twice the second column: $A[;3] = A[;1] + 2A[;2]$. Then A cannot have a left inverse L. To see this, suppose that L were a left inverse of A. Then $LA = I$. Now by the definition of matrix multiplication: $(LA)[;3] = LA[;3]$. So

$$\begin{aligned}
I[;3] &= (LA)[;3] \\
&= LA[;3] \\
&= L(A[;1] + 2A[;2]) \\
&= LA[;1] + L(2A[;2]] && \text{by Proposition 2.8 Statement 1} \\
&= LA[;1] + 2LA[;2] && \text{by Proposition 2.8 Statement 2} \\
&= I[;1] + 2I[;2]
\end{aligned}$$

But for an identity matrix I,

$$I[;3] \neq I[;1] + 2I[;2]$$

since

$$I[;3] = (0, 0, 1, 0, \ldots) \neq (1, 2, 0, \ldots) = I[;1] + 2I[;2]$$

In fact, no column of I is a linear combination of the other columns. It follows by an argument similar to the above that if A has a left inverse, then no column of A can be a linear combination of the other columns.

DEFINITION 2.8 A set of vectors is called *linearly independent* if no one of them is a linear combination of the others.

The proof of the next result must be postponed until Chapter 3.

PROPOSITION 2.9 A matrix has a left inverse if and only if the columns of A are linearly independent. ■

We have been concentrating on left inverses and will continue to do so. Statements about right inverses can be derived from corresponding statements about left inverses via the next proposition. For example, a matrix will have a right inverse if and only if the rows of A are linearly independent.

PROPOSITION 2.10 The matrix R is a right inverse for the matrix A if and only if R^T is a left inverse for A^T.

Proof $AR = I$ if and only if $I = I^T = (AR)^T = R^T A^T$. ■

The above definition of linear independence is somewhat clumsy to apply in most circumstances. The next proposition gives a more usable formulation.

PROPOSITION 2.11 Let A be a matrix and X an unknown vector. The columns of A are linearly independent if and only if the equation

$$AX = 0$$

has only the solution $X = 0$.

Proof The zeros in the statement of the proposition are, of course, zero vectors. Let $v_i = A[;i]$ and let $x_i = X[i]$. $X \neq 0$ if $x_i \neq 0$ for some i.

First assume that $AX = 0$ has a solution $X \neq 0$. Then by definition of matrix-vector multiplication:

$$x_1 v_1 + x_2 v_2 + \cdots + x_i v_i + \cdots + x_n v_n = 0$$

with, say, $x_i \neq 0$. Then v_i is a linear combination of the other v's. In fact

$$v_i = -\frac{x_1}{x_i} v_1 - \frac{x_2}{x_i} v_2 - \cdots - \frac{x_n}{x_i} v_n$$

The argument is reversible. If

$$v_i = \alpha_1 v_1 + \alpha_2 v_2 + \cdots + \alpha_{i-1} v_{i-1} + \alpha_{i+1} v_{i+1} + \cdots + \alpha_n v_n,$$

then $x = (\alpha_1, \alpha_2, \ldots, \alpha_{i-1}, -1, \alpha_{i+1}, \ldots, \alpha_n)$ is a nonzero solution. ■

Domino

A matrix A may have several left inverses if it is not square (see Exercise 4). One of these, the pseudo-inverse defined in Section 2.4, is denoted in APL by `⌹A`. The symbol ⌹ (⎕-backspace-÷) is called *domino*.

Thus if A is a matrix with linearly independent columns, then `L←⌹A` produces a left inverse L for A.†

Example 2.18

```
      A
 ¯38     6    49   ¯86
 ¯21   ¯50    49     1
  17  ¯245   ¯97   844
  12   ¯80    39   ¯22
  45   ¯59    29   ¯86
      +L←⌹A
¯0.003 0.034 0.003 ¯0.071 0.053
0.003 0.030 0.001 ¯0.061 0.032
0.010 0.058 0.004 ¯0.093 0.059
0.002 0.014 0.002 ¯0.027 0.015

      L+.×A
 1.00E0     ¯1.03E¯17    3.10E¯17   ¯3.17E¯17
¯9.87E¯18    1.00E0      1.93E¯17   ¯1.65E¯17
¯1.73E¯17   ¯1.33E¯17    1.00E0     ¯3.30E¯17
¯3.69E¯18   ¯3.00E¯18    9.22E¯18    1.00E0
```
■

The example above illustrates a serious problem with machine computation. *The calculations are not exact.* In fact, since only a finite number of digits can be stored in a machine, a number such as $\frac{1}{3}$ cannot even be stored exactly.

As the numbers are stored in the machine, errors occur. As calculations are made and numbers are rounded off, more errors occur. In bad situations these errors can accumulate and render the computed answer meaningless.

We will guard against this problem by always *checking the answer*. In the example above we computed $L +.\times A$ to see if L was an acceptable left inverse. The answer was not exactly an identity matrix but was close enough. The computation was done on a UNIVAC machine, which carries about eighteen decimal places. The deviations of the entries of $L +.\times A$ from I are in the last few places. This illustrates our working assumption for machine calculations.

Working Assumption: The last few decimals are *always wrong*.

There is no absolute rule for accepting or rejecting the result of a machine calculation. The amount of accuracy needed for the particular application is all that is required.

† Provided A is not "ill-conditioned." See Section 3.5.

Further, if the numbers in a problem are the result of laboratory or statistical measurements, then one cannot expect an answer more accurate than the original measurements.

For the examples in this book, any number that is fifteen orders of magnitude less than the original data will usually be considered to be zero.

EXAMPLE 2.19

```
      A
¯3.80E13     6.00E8     4.90E5   ¯8.60E1
¯2.10E13    ¯5.00E9     4.90E5    1.00E0
 1.70E13    ¯2.45E10   ¯9.70E5    8.44E2
 1.20E13    ¯8.00E9     3.90E5   ¯2.20E1
 4.50E13    ¯5.90E9     2.90E5   ¯8.60E1

      +L←⌹A
¯3.55E¯15    3.48E¯14   3.17E¯15  ¯7.10E¯14   5.32E¯14
 3.12E¯11    3.04E¯10   1.99E¯11  ¯6.12E¯10   3.24E¯10
 1.05E¯6     5.83E¯6    4.61E¯7   ¯9.36E¯6    5.93E¯6
 2.10E¯3     1.49E¯2    2.22E¯3   ¯2.71E¯2    1.51E¯2

      L+.×A
 1.00E0      4.28E¯21  ¯3.55E¯26  ¯4.50E¯29
¯8.35E¯14    1.00E0     5.29E¯22  ¯2.42E¯25
¯2.21E¯9     7.21E¯13   1.00E0    ¯4.87E¯21
¯4.65E¯6     1.04E¯9    3.20E¯14   1.00E0
```

We accept the new L because the size of the entries of $L +.\times A$ is less than 10^{-15} times the *original entries* of A. ■

The factor 10^{-15} that we are using here is peculiar to the machine used to calculate the examples. Other factors must be used for other machines. The factor is called the *comparison tolerance* or *fuzz*. (To obtain the fuzz for a particular machine, type ⎕CT. The symbol ⎕CT will be discussed in detail in Section 3.4 of Chapter 3.)

EXAMPLE 2.20 Attempt to solve the system of equations

$$
\begin{aligned}
-88x + 30y - 2z &= 1 \\
-41x + 38y + 16z &= 2 \\
69x + 90y + 35z &= 3
\end{aligned}
$$

Solution Let A be the matrix of coefficients

```
      A←3 3⍴¯88 30 ¯2 ¯41 38 16 69 90 35
      A
```

```
¯88  30  ¯2
¯41  38  16
 69  90  35
```

Let L be the pseudo-inverse.

```
      +L←⌹A
¯0.001 ¯0.012 0.005
0.025 ¯0.029 0.015
¯0.064 0.101 ¯0.021
```

Then the candidate for a solution is

```
      +X←L+.×1 2 3
¯0.00916  0.0114  0.0744
```

Check the solution by looking at $B - AX$:

```
      1 2 3−A+.×X
1.04E¯17  3.47E¯18  3.47E¯18
```

These numbers, called *residuals,* are negligible compared to the input data (for the machine used to compute the example), and so we accept X as a solution; that is, to three significant figures, $x = -.00916$, $y = .0114$, $z = .0744$. ■

EXAMPLE 2.21 Attempt to solve the system of equations

$$\begin{aligned} 57x_1 + 94x_2 - 22x_3 &= 1 \\ -40x_1 - 88x_2 + 48x_3 &= 2 \\ -19x_1 - 8x_2 + 50x_3 &= 3 \\ -563x_1 - 1220x_2 + 114x_3 &= 4 \end{aligned}$$

Solution Let A be the matrix of coefficients, L the pseudo- (left) inverse, and X the possible solution defined by L.

```
      +A←4 3⍴57 94 ¯22 ¯40 ¯88 48 ¯19 ¯8 50 ¯563 ¯1220 114
  57     94    ¯22
 ¯40    ¯88     48
 ¯19     ¯8     50
¯563  ¯1220    114
      +L←⌹A
0.048 0.041 ¯0.020 0.000
¯0.021 ¯0.017 0.010 ¯0.001
0.007 0.018 0.007 ¯0.000
      +X←L+.×1 2 3 4
0.0729  ¯0.0309  0.0632
```

Checking the residuals:

```
      1 2 3 4-A+.×X
1.14  ¯0.835  0.98  0.142
```

These residuals are too large to be acceptable. We conclude that the system has *no solution.* ■

EXAMPLE 2.22 Attempt to solve the system

$$x + 2y + 3z = 6$$
$$4x + 5y + 6z = 15$$
$$7x + 8y + 9z = 24$$

Solution Proceeding as in Examples 2.20 and 2.21,

```
      +A←3 3⍴⍳9
1  2  3
4  5  6
7  8  9
      +L←⌹A
DOMAIN ERROR
      +L←⌹A
        ∧
```

A has no left inverse and our procedure breaks down: *no conclusion.* ■

Notice that the last system *does* have solutions: $x = y = z = 1$ is a solution, as is $x = 0$, $y = 3$, $z = 0$. Systems such as this will be solved in Chapter 3.

Dyadic ⌹

The operator ⌹ has a dyadic form. If A is a matrix and B is a matrix or vector, then `X←B⌹A` directly produces a candidate for a solution to $AX = B$. This form of domino uses less machine time than `X←(⌹A)+.×B`.

EXAMPLE 2.23 Attempt to solve the linear system

$$8u - v + 5w = -4$$
$$12v - 7w = 0$$
$$3u - 5v + 3w = 1$$

Solution

```
      +A←3 3ρ8 ¯1 5 0 12 ¯7 3 ¯5 3
8  ¯1   5
0  12  ¯7
3  ¯5   3

      +X←¯4 0 1⌹A
0.377  ¯0.927  ¯1.59

      ¯4  0 1-A+.×X
2.08E¯17  ¯5.55E¯17  2.95E¯17
```

The residuals are sufficiently small. The solution is, to three significant figures, $u = .377$, $v = -.927$, $w = -1.59$. ■

So far we have no general criterion for the equation $AX = B$ to have a solution. The following proposition is obvious from the definition of matrix-vector multiplication but is worth stating explicitly nonetheless.

PROPOSITION 2.12 Let A be a matrix and B a vector. The equation $AX = B$ has solutions if and only if B is a linear combination of the columns of A. ■

EXERCISES 2.2

1. Show that $\begin{bmatrix} 1 & 0 \\ 0 & 0 \end{bmatrix}$ does not have a right inverse.

2. (a) Show that $L = A^T$ is a left inverse for $A = \begin{bmatrix} 1 & 0 \\ 0 & 0 \\ 0 & 1 \end{bmatrix}$

   (b) Show that A does not have a right inverse.

3. Let L_1, L_2 both be left inverses for A. Check that $\alpha L_1 + \beta L_2 = L$ is a left inverse for A whenever $\alpha + \beta = 1$.

   Hint: Use Proposition 2.8. This exercise shows that if A has two distinct left inverses, then it has an infinity of left inverses.

4. $L_1 = \begin{bmatrix} 1 & -1 & 0 & -3 & 6 \\ -2 & 3 & 0 & 7 & -15 \\ 0 & 1 & 1 & 1 & -3 \end{bmatrix}, \quad L_2 = \begin{bmatrix} -1 & 1 & 1 & 4 & -7 \\ -1 & 3 & 0 & 6 & -12 \\ -1 & 3 & 2 & -5 & -13 \end{bmatrix}$

$$A^T = \begin{bmatrix} 18 & -16 & -2 & -3 & -7 \\ 5 & -4 & -1 & -1 & -2 \\ -3 & 3 & 1 & 0 & 1 \end{bmatrix}$$

   (a) Show that L_1, L_2 are left inverses for A.

   (b) Find three more left inverses for A.

   Hint: See Exercise 3.

5. (a) Show that if A, B are invertible and AB is defined, then $B^{-1}A^{-1}$ is an inverse for AB. Hence AB is also invertible and $(AB)^{-1} = B^{-1}A^{-1}$.

   (b) Show by example that $(AB)^{-1} \neq A^{-1}B^{-1}$ in general.

6. Show that if A is invertible, then $(A^{-1})^T$ is an inverse for A^T and hence A^T is also invertible with $(A^T)^{-1} = (A^{-1})^T$. Further, the inverse of a symmetric matrix, when it exists, is symmetric.

7. Show that if A is invertible, $(A^{-1})^{-1} = A$ and $(A^n)^{-1} = (A^{-1})^n(A^n = A \cdot A \cdot \cdots \cdot A$, n times).

8. Show that if A, B, $A + B$ are all invertible, $(A^{-1} + B^{-1})^{-1} = A(A + B)^{-1}B = B(A + B)^{-1}A$.

9. Show that if R_1 and R_2 are right inverses for A and $B = R_1 - R_2$, then $AB = 0$.

   Hint: See Proposition 2.8.

10. Show that $\begin{bmatrix} 3 & 0 & 0 \\ 0 & 2 & 0 \\ 0 & 0 & -1 \end{bmatrix}^{-1} = \begin{bmatrix} \frac{1}{3} & 0 & 0 \\ 0 & \frac{1}{2} & 0 \\ 0 & 0 & -1 \end{bmatrix}$

11. Let $D = \begin{bmatrix} d_1 & & & 0 \\ & d_2 & & \\ & & \ddots & \\ 0 & & & d_n \end{bmatrix}$ be a diagonal matrix ($D[i; j] = 0$ if $i \neq j$).

    (a) Check that if $d_i \neq 0$ for all i, then $D^{-1} = \begin{bmatrix} d_1^{-1} & & 0 \\ & \ddots & \\ 0 & & d_n^{-1} \end{bmatrix}$

    (b) Check that if $d_i = 0$ for some i, this D is not invertible.

    Hint: Imitate Example 2.15. Or, by inspection, find an $X \neq 0$ such that $AX = 0$ and invoke Propositions 2.9 and 2.11.

12. Solve the following linear systems by the method of Example 2.16.

    (a) $\begin{aligned} x + y &= 1 \\ x - y &= 1 \end{aligned}$ (b) $\begin{aligned} 2x - y &= 0 \\ x + 2y &= 1 \end{aligned}$ (c) $\begin{aligned} 3x + 5y &= 1 \\ 2x - 7y &= 0 \end{aligned}$

    (d) $\begin{aligned} 6x - 24y &= 0 \\ 18x + 12y &= 0 \end{aligned}$ (e) $\begin{aligned} x + y &= 3 \\ x - y &= 3 \end{aligned}$

13. In the problems below L is a left inverse for A. Check this fact and use it to solve for the unknown matrix X, if it exists, in the equation $AX = B$.

    (a) $A = \begin{bmatrix} 1 & -1 & 1 \\ -2 & 3 & -4 \\ -2 & -1 & 5 \end{bmatrix}$, $L = A^{-1} = \begin{bmatrix} 11 & 4 & 1 \\ 18 & 7 & 2 \\ 8 & 3 & 1 \end{bmatrix}$, $B = \begin{bmatrix} 2 \\ -8 \\ 11 \end{bmatrix}$

    (b) $A = \begin{bmatrix} 1 & -2 & -1 \\ -1 & 3 & 1 \\ 1 & -4 & 0 \end{bmatrix}$, $L = A^{-1} = \begin{bmatrix} 4 & 4 & 1 \\ 1 & 1 & 0 \\ 1 & 2 & 1 \end{bmatrix}$, $B = \begin{bmatrix} -6 \\ 8 \\ -7 \end{bmatrix}$

    (c) $A = \begin{bmatrix} 1 & 1 & -2 & -3 \\ 0 & 1 & 0 & -2 \\ -3 & -4 & 7 & 11 \\ -3 & -3 & 5 & 10 \end{bmatrix}$, $L = A^{-1} = \begin{bmatrix} 13 & 2 & 3 & 1 \\ 12 & 3 & 2 & 2 \\ 3 & 1 & 1 & 0 \\ 6 & 1 & 1 & 1 \end{bmatrix}$, $B = \begin{bmatrix} -15 \\ -6 \\ 54 \\ 46 \end{bmatrix}$

(d) $A = \begin{bmatrix} 1 & 1 \\ 0 & 1 \\ -3 & -4 \\ -3 & -3 \end{bmatrix}, \quad L = \begin{bmatrix} 13 & 2 & 3 & 1 \\ 12 & 3 & 2 & 2 \end{bmatrix}, \quad B = \begin{bmatrix} -2 \\ 0 \\ 7 \\ 5 \end{bmatrix}$

(e) $A = \begin{bmatrix} 1 & -1 \\ -3 & 4 \\ -3 & 1 \end{bmatrix}, \quad L = \begin{bmatrix} -5 & -1 & -1 \\ -6 & -1 & -1 \end{bmatrix}, \quad B = \begin{bmatrix} -1 \\ 5 \\ -1 \end{bmatrix}$

(f) $A = \begin{bmatrix} 1 & -3 & 1 \\ -2 & 7 & -2 \\ -2 & 4 & -1 \\ -1 & 3 & -4 \end{bmatrix}, \quad L = \begin{bmatrix} -18 & -5 & -4 & -1 \\ 2 & 1 & 0 & 0 \\ 6 & 2 & 1 & 0 \end{bmatrix}, \quad B = \begin{bmatrix} -2 \\ 6 \\ 3 \\ -7 \end{bmatrix}$

14. Use Propositions 2.9 and 2.12 and the results of exercise 13(c) to show, without computation, that exercise 13(d) cannot have a solution.

15. (Computer assignment) Use ⌹ to determine whether the matrices below have left inverses. (If not, you will get a DOMAIN ERROR.)

(a) $\begin{bmatrix} -98 & -49 & -86 \\ -63 & 602 & 136 \\ -79 & 51 & -42 \end{bmatrix}$ (b) $\begin{bmatrix} 61 & 53 & 16 \\ 2 & -18 & 35 \\ -21 & -57 & -81 \end{bmatrix}$

(c) $\begin{bmatrix} -94 & 9 & 42 \\ 28 & 1 & -26 \\ 51 & -26 & 76 \\ -132 & 88 & -260 \end{bmatrix}$ (d) $\begin{bmatrix} 26 & -71 & 99 \\ -41 & -82 & -8 \\ 0 & -29 & -2 \end{bmatrix}$

(e) $\begin{bmatrix} -40 & 62 & 48 \\ -80 & 124 & 96 \\ -200 & 310 & 240 \\ -80 & 124 & 96 \end{bmatrix}$ (f) $\begin{bmatrix} 37 & 99 & 2 & 5 & -29 \\ -13 & -24 & -96 & -27 & 82 \\ 1 & -65 & 40 & 7 & -88 \\ 46 & 49 & -52 & -40 & 56 \\ 80 & -21 & -12 & -57 & -35 \end{bmatrix}$

16. (Computer assignment) Attempt to solve the given systems of linear equations using the *dyadic* function ⌹.

(a)
$$\begin{aligned} 58x + 22y + 73z &= 58 \\ 97x - 38y + 31z &= -35 \\ -42x + 86y - 51z &= 36 \end{aligned}$$

(b)
$$\begin{aligned} -472u + 454v - 50w &= -22 \\ -97u + 88v + 16w &= 2 \\ -69u + 54v + 54w &= 12 \end{aligned}$$

(c)
$$\begin{aligned} 680x_1 + 271x_2 &= -397 \\ -44x_1 - 96x_2 &= -89 \\ -40x_1 + 38x_2 &= 44 \\ -28x_1 - 19x_2 &= 98 \end{aligned}$$

(d)
$$\begin{aligned} -2x - 82y - 43z + 27t &= -64 \\ 22x + 72y + 42z + 70t &= -80 \\ -74x - 99y - 4z + 59t &= -69 \\ -57x - 53y - 56z - 48t &= -48 \end{aligned}$$

(e)
$$\begin{aligned} -38x_1 - 46x_2 - 93x_3 &= -7 \\ -23x_1 + 67x_2 + 47x_3 &= -53 \\ -95x_1 \qquad\quad + 36x_3 &= -68 \\ 133x_1 - 25x_2 + 423x_3 &= 50 \\ 42x_1 + 18x_2 - 33x_3 &= 48 \end{aligned}$$

(f)
$$\begin{aligned} 35t_1 - 92t_2 - 48t_3 &= -6 \\ -19t_1 - 21t_2 - 87t_3 &= -38 \\ 479t_1 - 606t_2 - 244t_3 &= 142 \\ 44t_1 + 69t_2 - 80t_3 &= 22 \\ 755t_1 + 103t_2 - 534t_3 &= 384 \end{aligned}$$

17. (Computer assignment) Use [computer icon] to determine if the given matrices have right inverses.

(a) $\begin{bmatrix} 81 & -93 & 8 \\ -53 & -68 & 1 \end{bmatrix}$ (b) $\begin{bmatrix} -7 & 53 & -31 & 78 \\ -26 & 40 & -91 & -46 \\ 26 & 74 & -92 & 22 \end{bmatrix}$

(c) $\begin{bmatrix} -96 & -41 & 99 & 82 \\ 52 & -82 & -56 & 50 \\ -66 & 62 & -16 & -83 \end{bmatrix}$ (d) $\begin{bmatrix} 32 & 33 & -97 & -20 \\ -12 & -329 & 105 & 296 \\ -5 & 74 & -2 & -69 \end{bmatrix}$

(e) $\begin{bmatrix} -80 & -30 & 21 & 72 & -54 \\ -5 & -82 & -1 & -27 & 92 \\ -27 & -58 & -27 & -69 & 60 \\ -49 & -89 & 53 & -88 & -38 \end{bmatrix}$

18. Verify statements 1 through 4 of Proposition 2.8.

2.3 Matrix Algebra

The identities of Propositions 2.3 and 2.8 allow us to manipulate matrix equations in a manner similar to the way we manipulate scalar equations. There are differences, however. For matrices, $AB \neq BA$. This means that one cannot simply multiply an equation by a matrix. The equation must be multiplied either "on the left" or "on the right."

EXAMPLE 2.24 Assume that A is invertible and solve the matrix equations for X.

(a) $AX = B$ (b) $XA = B$ (c) $AXA = B$ (d) $A^2X = B$

Solutions

(a) Multiply the equation *on the left* by A^{-1}.

$$\begin{aligned} A^{-1}(AX) &= A^{-1}B \\ (A^{-1}A)X &= A^{-1}B \\ X &= A^{-1}B, \qquad \text{if } X \text{ exists at all} \end{aligned}$$

(b) Multiply the equation *on the right* by A^{-1}.

$$\begin{aligned} (XA)A^{-1} &= BA^{-1} \\ X(AA^{-1}) &= BA^{-1} \\ X &= BA^{-1}, \qquad \text{if } X \text{ exists at all} \end{aligned}$$

(c) First multiply the equation on the left by A^{-1}, then on the right by A^{-1}.

$$
\begin{aligned}
A^{-1}(AXA) &= A^{-1}B \\
XA &= A^{-1}B \\
(XA)A^{-1} &= A^{-1}BA^{-1} \\
X &= A^{-1}BA^{-1}, \qquad \text{if } X \text{ exists at all}
\end{aligned}
$$

(d) Here $A^2 = AA$. Multiply on the left by $(A^{-1})^2$. (See exercise 7 of Exercises 2.2.)

$$
\begin{aligned}
(A^{-1})^2A^2X &= (A^{-1})^2B \\
X &= (A^{-1})^2B, \qquad \text{if } X \text{ exists at all}
\end{aligned}
$$

The answers to (a) and (b) are different and the answers to (c) and (d) are different. In fact, if

$$A = \frac{1}{2}\begin{bmatrix} 1 & -1 \\ 1 & 1 \end{bmatrix} \quad \text{and} \quad B = \begin{bmatrix} 1 & 2 \\ 3 & 4 \end{bmatrix}$$

then

$$A^{-1} = \begin{bmatrix} 1 & 1 \\ -1 & 1 \end{bmatrix}$$

and the answers are

(a) $X = \begin{bmatrix} 4 & 6 \\ 2 & 2 \end{bmatrix}$ (b) $X = \begin{bmatrix} -1 & 3 \\ -1 & 7 \end{bmatrix}$

(c) $X = \begin{bmatrix} -2 & 10 \\ 0 & 4 \end{bmatrix}$ (d) $X = \begin{bmatrix} 6 & 8 \\ -2 & -4 \end{bmatrix}$ ■

The statements "if X exists at all" are in fact unnecessary in the example above. Multiplying an equation (on the left or on the right) by an invertible matrix does not change the solutions. The only problem in verifying this statement is defining exactly what we mean by "equation" and "solution."

By transposing everything to one side, we can write an equation in the unknown matrix X in the form $f(X) = 0$, where f is some function of the matrix X. Here "0" denotes a zero matrix of the appropriate shape.

For example, the four equations of Example 2.24 can be written

(a) $f(X) = AX - B = 0$ (b) $f(X) = XA - B = 0$

(c) $f(X) = AXA - B = 0$ (d) $f(X) = A^2X - B = 0$

By a *solution* of an equation $f(X) = 0$ we mean any matrix S such that $f(S) = 0$. Equations may have no solutions, one solution, several solutions, or an infinity of solutions. The familiar rules from elementary algebra about the number of solutions of an equation do not usually hold for matrix equations. (See exercise 4.)

PROPOSITION 2.13

1. If A has a left inverse, then multiplying an equation *on the left* by A does not change the set of solutions.

2. If A has a right inverse, then multiplying an equation *on the right* by A does not change the set of solutions.

Proof Let the equation be $f(X) = 0$. If S is a solution, then $f(S) = 0$, hence $Af(S) = A0 = 0$, and so S is a solution of $Af(X) = 0$. (This shows that multiplying by *any* A never *decreases* the number of solutions.)

Conversely, if S is a solution of $Af(X) = 0$, we have $Af(S) = 0$. If A has a left inverse L, then $L(Af(S)) = 0$, hence $(LA)f(S) = 0$ or $f(S) = 0$. Thus S is a solution of $f(X) = 0$. ∎

Since the matrix A^{-1} of Example 2.24 has a left inverse (namely A), the equation $AX = B$ has the same solutions as the equation $X = A^{-1}B$.

If A does not have a left inverse, however, left multiplication by A may introduce extra solutions.

EXAMPLE 2.25

$$A = \begin{bmatrix} -1 & 1 \\ 3 & -4 \\ 3 & -2 \end{bmatrix}, \quad B = \begin{bmatrix} 1 \\ 2 \\ 3 \end{bmatrix}, \quad L = \begin{bmatrix} 8 & 1 & 2 \\ 3 & 0 & 1 \end{bmatrix}$$

L is a left inverse for A. Solve the equation $AX = B$.

Solution Multiplying on the left by L, we have

$$AX = B$$

$$LAX = LB$$

$$X = LB = \begin{bmatrix} 8 & 1 & 2 \\ 3 & 0 & 1 \end{bmatrix}\begin{bmatrix} 1 \\ 2 \\ 3 \end{bmatrix} = \begin{bmatrix} 16 \\ 6 \end{bmatrix}$$

Checking, we have

$$A\begin{bmatrix} 16 \\ 6 \end{bmatrix} = \begin{bmatrix} -1 & 1 \\ 3 & -4 \\ 3 & -2 \end{bmatrix}\begin{bmatrix} 16 \\ 6 \end{bmatrix} = \begin{bmatrix} -10 \\ 24 \\ 36 \end{bmatrix} \neq B$$

Thus an extra solution has been introduced. The original equation has no solution. ∎

In the example above L itself does not have a left inverse, although A is a right inverse for L. (In fact if L had a left inverse also, then L would be invertible and hence, by Proposition 2.5, square.)

Additive terms in the equations pose no problem.

EXAMPLE 2.26

$$L = \begin{bmatrix} 3 & 1 & 1 & 0 \\ -3 & -3 & -2 & -1 \\ 2 & 2 & 1 & 0 \end{bmatrix}, \quad A = \begin{bmatrix} 1 & 1 & -1 \\ 0 & 1 & 0 \\ -2 & -4 & 3 \\ 1 & 1 & -3 \end{bmatrix}, \quad B = \begin{bmatrix} 1 & 1 \\ -1 & 0 \\ 0 & -2 \\ -2 & -3 \end{bmatrix},$$

$$C = \begin{bmatrix} -1 & 1 \\ 1 & 0 \\ 1 & -4 \\ 2 & -2 \end{bmatrix}$$

Using the fact that $LA = I$, solve $AX + B = C$.

Solution If X exists, then

$$\begin{aligned} AX &= C - B \\ LAX &= L(C - B) \\ X &= L(C - B) \end{aligned}$$

and so the only possible solution is

$$\begin{aligned} X &= \begin{bmatrix} 3 & 1 & 1 & 0 \\ -3 & -3 & -2 & -1 \\ 2 & 2 & 1 & 0 \end{bmatrix} \left(\begin{bmatrix} -1 & 1 \\ 1 & 0 \\ 1 & -4 \\ 2 & -2 \end{bmatrix} - \begin{bmatrix} 1 & 1 \\ -1 & 0 \\ 0 & -2 \\ -2 & -3 \end{bmatrix} \right) \\ &= \begin{bmatrix} 3 & 1 & 1 & 0 \\ -3 & -3 & -2 & -1 \\ 2 & 2 & 1 & 0 \end{bmatrix} \begin{bmatrix} -2 & 0 \\ 2 & 0 \\ 1 & -2 \\ 4 & 1 \end{bmatrix} = \begin{bmatrix} -3 & -2 \\ -6 & 3 \\ 1 & -2 \end{bmatrix} \end{aligned}$$

If X is a solution, then, substituting into the original equation, we have

$$\begin{bmatrix} 1 & 1 & -1 \\ 0 & 1 & 0 \\ -2 & -4 & 3 \\ 1 & 1 & -3 \end{bmatrix} \begin{bmatrix} -3 & -2 \\ -6 & 3 \\ 1 & -2 \end{bmatrix} + \begin{bmatrix} 1 & 1 \\ -1 & 0 \\ 0 & -2 \\ -2 & 3 \end{bmatrix} = \begin{bmatrix} -1 & 1 \\ 1 & 0 \\ 1 & -4 \\ 2 & 2 \end{bmatrix}$$

$$\begin{bmatrix} -10 & 3 \\ -6 & 3 \\ 33 & -14 \\ 12 & 7 \end{bmatrix} + \begin{bmatrix} 1 & 1 \\ -1 & 0 \\ 0 & -2 \\ -2 & 3 \end{bmatrix} = \begin{bmatrix} -1 & 1 \\ 1 & 0 \\ 1 & -4 \\ 2 & 2 \end{bmatrix}$$

$$\begin{bmatrix} -9 & 4 \\ -7 & 3 \\ 33 & -16 \\ 14 & 4 \end{bmatrix} = \begin{bmatrix} -1 & 1 \\ 1 & 0 \\ 1 & -4 \\ 2 & -2 \end{bmatrix}$$

which is false. Thus the equation has no solution. ■

EXAMPLE 2.27 Matrices L, A are as in Example 2.26.

$$B = \begin{bmatrix} 8 & 3 & 1 & 2 \\ -2 & 0 & 0 & -1 \end{bmatrix}, \qquad C = \begin{bmatrix} 22 & 1 & 2 & -2 \\ 4 & 0 & 1 & -2 \end{bmatrix}$$

Solve the equation $XL + B = C$.

Solution A is a right inverse for L.

$$\begin{aligned} XL &= C - B \\ XLA &= (C - B)A \\ X &= (C - B)A \end{aligned}$$

The only possible solution is

$$X = \left(\begin{bmatrix} 22 & 1 & 2 & -2 \\ 4 & 0 & 1 & -2 \end{bmatrix} - \begin{bmatrix} 8 & 3 & 1 & 2 \\ -2 & 0 & 0 & -1 \end{bmatrix} \right) \begin{bmatrix} 1 & 1 & -1 \\ 0 & 1 & 0 \\ -2 & -4 & 3 \\ 1 & 1 & -3 \end{bmatrix}$$

$$= \begin{bmatrix} 8 & 4 & 1 \\ 3 & 1 & 0 \end{bmatrix}$$

If X is a solution, then substituting into the original equation we have

$$\begin{bmatrix} 8 & 4 & 1 \\ 3 & 1 & 0 \end{bmatrix} \begin{bmatrix} 3 & 1 & 1 & 0 \\ -3 & -3 & -2 & -1 \\ 2 & 2 & 1 & 0 \end{bmatrix} + \begin{bmatrix} 8 & 3 & 1 & 2 \\ -2 & 0 & 0 & -1 \end{bmatrix} = \begin{bmatrix} 22 & 1 & 2 & -2 \\ 4 & 0 & 1 & -2 \end{bmatrix}$$

$$\begin{bmatrix} 14 & -2 & 1 & -4 \\ 6 & 0 & 1 & -1 \end{bmatrix} + \begin{bmatrix} 8 & 3 & 1 & 2 \\ -2 & 0 & 0 & -1 \end{bmatrix} = \begin{bmatrix} 22 & 1 & 2 & -2 \\ 4 & 0 & 1 & -2 \end{bmatrix}$$

$$\begin{bmatrix} 22 & 1 & 2 & -2 \\ 4 & 0 & 1 & -2 \end{bmatrix} = \begin{bmatrix} 22 & 1 & 2 & -2 \\ 4 & 0 & 1 & -2 \end{bmatrix}$$

Thus X is a solution in this case. ∎

In special cases one may left-multiply by a matrix that does not have a left inverse without introducing extra solutions. The next proposition provides an example. Notice, however, that the proposition is no longer true if the equation $AX = 0$ is replaced by $AX = B$. The proposition provides an explicit formula for a left inverse.

PROPOSITION 2.14 Let A be a matrix, X an unknown vector. The equation

$$AX = 0$$

has precisely the same solutions as the equation

$$A^T AX = 0.$$

In particular, A has a left inverse if and only if $A^T A$ is invertible. In this case $(A^T A)^{-1} A^T$ is a left inverse for A.

Proof We have to show that multiplying by A^T has not introduced any extra solutions. Let S be a solution of $A^T AX = 0$. Then $A^T AS = 0$ and hence $S^T A^T AS = 0$.

But for any matrix V, $V^T V = 0$ if and only if $V = 0$ (exercise 11 of Exercises 2.1), because

$$\begin{aligned} &(V^T V)[i; i] \\ &= (V^T[i;] + . \times V[;i]) \\ &= V[;i] + . \times V[;i] \\ &= \sum_h V[h; i]^2 \\ &= 0 \end{aligned}$$

which means that $V[h; i] = 0$ for all h and all i.

Thus the two equations have the same solutions. By Propositions 2.9 and 2.11 A has a left inverse if and only if $A^T A$ has a left inverse. Since $A^T A$ is always square, it has a left inverse if and only if it is invertible (Proposition 2.5).

The equation $(A^T A)^{-1} A^T A = I$ shows that $(A^T A)^{-1} A^T$ is a left inverse for A. ∎

DEFINITION 2.9 If A has a left inverse, then the particular left inverse $(A^TA)^{-1}A^T$ is called the *pseudo-inverse* and will be written ⌹A. The product $(A^TA)^{-1}A^TB$ will be written B⌹A.

This is the left inverse computed by domino and used in Section 2.2. The "division" "B⌹A" has A on the wrong side — but it is the form the machine recognizes. The left inverse ⌹A has several nice properties that single it out from the other left inverses of a matrix A. The most important property, which makes ⌹A quite useful in statistical and curve-fitting applications, is given in the next proposition. A calculus-based proof will be found in Section 2.5* and a linear algebra proof is given in Chapter 5.

Assume that A has a left inverse. By Proposition 2.6, if $AX = B$ has solutions, then it has only the solution $X = LB$, where L is *any* left inverse of A. This means that as L varies through left inverses, the product LB remains constant. If $AX = B$ has no solutions, however, then LB is free to vary with L and will do so.

Recall that, for a vector V, V +.*2 is the sum of the squares of the components of V.

It follows that $0 = V$ +.*2 only when $V = 0$.

PROPOSITION 2.15 Let A be a matrix with linearly independent columns. Let B be a vector with (ρA)[1] = ρB. Let $f(X) = (AX - B)$ +.*2. The unique value of X that minimizes $f(X)$ is $X = B$⌹A. ■

Notice that $f(X) \geq 0$ and $f(X) = 0$ only if X is a solution of $AX = B$.

DEFINITION 2.10 By a *least-squares solution* of $AX = B$ is meant any S for which $f(S)$ is a minimum. If A has linearly independent columns, then there is only one least-squares solution and it is B⌹A or $(A^TA)^{-1}A^TB$.

Least-Squares Lines

Although a proof of Proposition 2.15 must wait until later, it is not hard to see that it gives the usual formulas for the least-squares straight line [Equations (1.4)].

Given the data points (x_1, y_1), (x_2, y_2), . . . , (x_n, y_n), we wish to find the straight line $y = a + bx$ that minimizes the quantity

$$\sum_{i=1}^{n} d_i^2 = \sum_{i=1}^{n} (y_i - a - bx_i)^2$$

Now if the points all lie on a line, then the coefficients a, b are the solution of the system of linear equations.

$$\begin{aligned} a + bx_1 &= y_1 \\ a + bx_2 &= y_2 \\ &\vdots \\ a + bx_n &= y_n \end{aligned} \quad \text{or} \quad \begin{bmatrix} 1 & x_1 \\ 1 & x_2 \\ \vdots & \vdots \\ 1 & x_n \end{bmatrix} \begin{bmatrix} a \\ b \end{bmatrix} = \begin{bmatrix} y_1 \\ \vdots \\ \vdots \\ y_n \end{bmatrix} \quad \text{or} \quad AX = B$$

where $B = (y_1, \ldots, y_n)$ and $A = (x_1, \ldots, x_n) \circ . * 0\ 1$. The function being minimized is

$$f(X) = \sum_{i=1}^{n} (y_i - a - bx_i)^2$$

the sum of the deviations squared.

Multiplying through by A^T, we obtain

$$A^TAX = \begin{bmatrix} 1 & \ldots & 1 \\ x_1 & \ldots & x_n \end{bmatrix} \begin{bmatrix} 1 & x_1 \\ \vdots & \vdots \\ 1 & x_n \end{bmatrix} \begin{bmatrix} a \\ b \end{bmatrix} = \begin{bmatrix} 1 & \ldots & 1 \\ x_1 & \ldots & x_n \end{bmatrix} \begin{bmatrix} y_1 \\ \vdots \\ y_n \end{bmatrix} = A^TB$$

$$A^TAX = \begin{bmatrix} n & \Sigma\, x_i \\ \Sigma\, x_i & \Sigma\, x_i^2 \end{bmatrix} \begin{bmatrix} a \\ b \end{bmatrix} = \begin{bmatrix} \Sigma\, y_i \\ \Sigma\, x_i y_i \end{bmatrix} = A^TB$$

Solving the latter equation,

$$\begin{aligned} X &= (A^TA)^{-1}A^TB \\ &= \begin{bmatrix} n & \Sigma\, x_i \\ \Sigma\, x_i & \Sigma\, x_i^2 \end{bmatrix}^{-1} \begin{bmatrix} \Sigma\, y_i \\ \Sigma\, x_i y_i \end{bmatrix} \\ &= \frac{1}{n \Sigma\, x_i^2 - (\Sigma\, x_i)^2} \begin{bmatrix} \Sigma\, x_i^2 & -\Sigma\, x_i \\ -\Sigma\, x_i & n \end{bmatrix} \begin{bmatrix} \Sigma\, y_i \\ \Sigma\, x_i y_i \end{bmatrix} \end{aligned}$$

or

$$\begin{bmatrix} a \\ b \end{bmatrix} = \frac{1}{n \Sigma\, x_i^2 - (\Sigma\, x_i)^2} \begin{bmatrix} (\Sigma\, y_i)(\Sigma\, x_i^2) - (\Sigma\, x_i)(\Sigma\, x_i y_i) \\ n \Sigma\, x_i y_i - (\Sigma\, x_i)(\Sigma\, y_i) \end{bmatrix}$$

These are the usual formulas.

In calculating least-squares straight lines by hand, one simply sets up the equation $AX = B$, multiplies through on the left by A^T, and applies the formula for the inverse of a 2-by-2 matrix (Proposition 2.7).

EXAMPLE 2.28 Find the least-squares straight line fitting the data points (2, 1), (3, 3), (7, 3), (9, 7), and (8, 5). Sketch the resulting straight line with the data points.

Solution The matrix equation is

$$\begin{bmatrix} 1 & 2 \\ 1 & 3 \\ 1 & 7 \\ 1 & 9 \\ 1 & 8 \end{bmatrix} \begin{bmatrix} a \\ b \end{bmatrix} = \begin{bmatrix} 1 \\ 3 \\ 3 \\ 7 \\ 5 \end{bmatrix}$$

Hence

$$\begin{bmatrix}1&1&1&1&1\\2&3&7&9&8\end{bmatrix}\begin{bmatrix}1&2\\1&3\\1&7\\1&9\\1&8\end{bmatrix}\begin{bmatrix}a\\b\end{bmatrix}=\begin{bmatrix}1&1&1&1&1\\2&3&7&9&8\end{bmatrix}\begin{bmatrix}1\\3\\3\\7\\5\end{bmatrix}$$

or

$$\begin{bmatrix}5&29\\29&207\end{bmatrix}\begin{bmatrix}a\\b\end{bmatrix}=\begin{bmatrix}19\\135\end{bmatrix}$$

or

$$\begin{bmatrix}a\\b\end{bmatrix}=\frac{1}{5\cdot 207-29^2}\begin{bmatrix}207&-29\\-29&5\end{bmatrix}\begin{bmatrix}19\\135\end{bmatrix}$$

or

$$\begin{bmatrix}a\\b\end{bmatrix}=\tfrac{1}{194}\begin{bmatrix}18\\124\end{bmatrix}=\begin{bmatrix}.093\\.64\end{bmatrix}$$

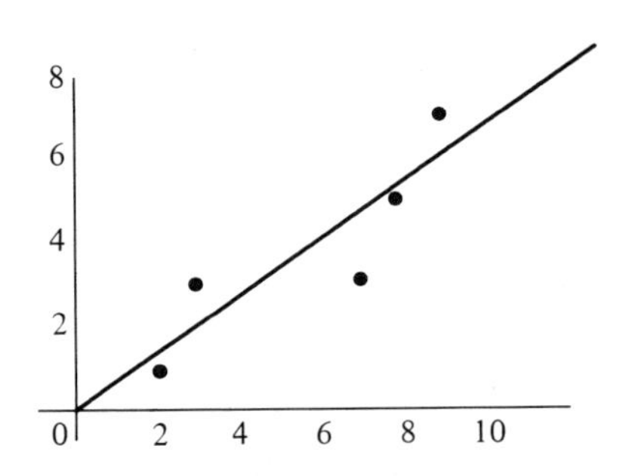

When using a machine, of course, one uses domino. ■

EXAMPLE 2.29 Compute the coefficients of the line of Example 2.28 in APL.

```
      X←2 3 7 9 8
      Y←1 3 3 7 5
      +A← X∘.*0 1
 1.000   2.000
 1.000   3.000
 1.000   7.000
 1.000   9.000
 1.000   8.000
      Y⌹A
0.0928   0.639
```

■

More generally, the least-squares polynomial of degree k is the function of the form $p(x) = a_0 + a_1x + a_2x^2 + \cdots + a_kx^k$ that minimizes

$$f(x) = \sum_{i=1}^{n} (y_i - a_0 - a_1x_i - \cdots - a_kx_ik)^2$$

If $k < n$, then $p(x)$ is uniquely determined. The least-squares straight line is the case $k = 1$.

PROPOSITION 2.16 Let $X = (x_1, \ldots, x_n)$, $y = (y_1, \ldots, y_n)$ be vectors. The coefficients of the least-squares polynomial of degree k for the data points $(x_1, y_1), \ldots, (x_n, y_n)$ are the components of the vector

```
Y⌹X∘.*0,⍳k
```
■

EXAMPLE 2.30 Fit a least-squares cubic to the data of Example 2.28. Sketch the result.

```
      X
2 3 7 9 8
      Y
1 3 3 7 5
      +COEFF←Y⌹X∘.*0 1 2 3
¯8.39  7.22  ¯1.43  0.0907
```

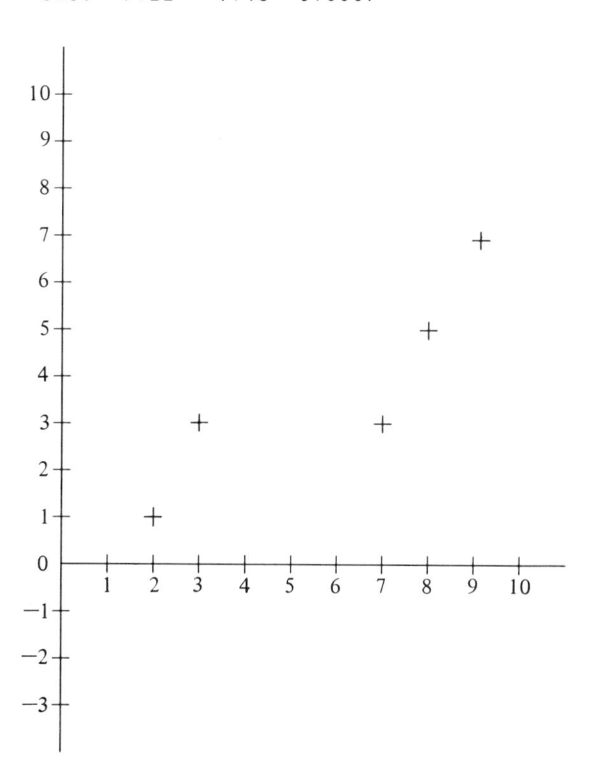

To sketch the result we find the value of the polynomial at the points $x = 0, 1, 2, 3, 4, 5, 6, 7, 8, 9$. Notice that this calculation is easily accomplished via matrix multiplication

```
      (0 1 2 3 4 5 6 7 8 9∘.*0 1 2 3)+.×COEFF
¯8.39  ¯2.51  1.07  2.88  3.47  3.38  3.17  3.36  4.52  7.17
```

(To facilitate such computations the indeterminate form 0^0 is taken to be 1 in APL.)

The graph is

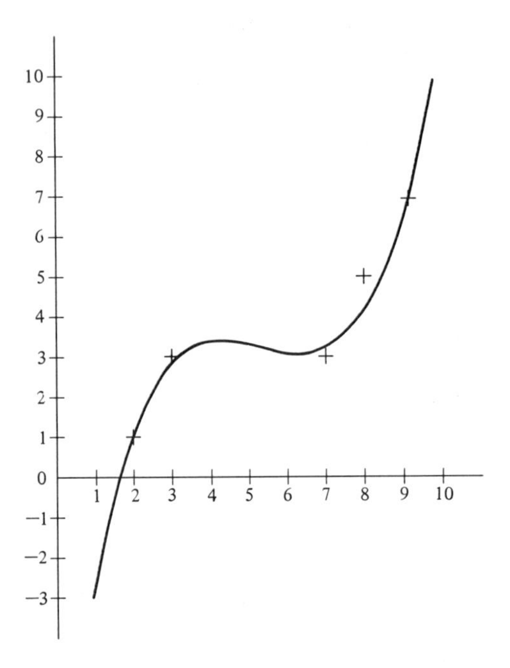

■

Multiple Regression

The technique above applies to much more than polynomial fits. In applications it often happens that more than one variable is being measured. For example, a variable w may be thought to depend on variables x, y, z via a relation of the form $w = c_1 + c_2x + c_3y + c_4z$. The data points are then vectors $(x_1, y_1, z_1, w_1), \ldots, (x_n, y_n, z_n, w_n)$. If one sets up the system

$$AX = \begin{bmatrix} 1 & x_1 & y_1 & z_1 \\ 1 & x_2 & y_2 & z_2 \\ 1 & x_3 & y_3 & z_3 \\ \vdots & \vdots & \vdots & \vdots \\ 1 & x_n & y_n & z_n \end{bmatrix} \begin{bmatrix} c_1 \\ c_2 \\ c_3 \\ c_4 \end{bmatrix} = \begin{bmatrix} w_1 \\ w_2 \\ w_3 \\ \vdots \\ w_n \end{bmatrix} = B$$

which has a solution when the equations $w_i = c_1 + c_2x_i + c_3y_i + cy_i$ are satisfied exactly, then the expression B⌹A gives the vector of coefficients c_1, c_2, c_3, c_4 that minimize the expression

$$f(x) = \sum_{1}^{n} (w_i - c_1 - c_2 x_i - c_3 v_2 - c_4 z_i)^2$$

These c_i's are called the *multiple-regression* coefficients.

EXAMPLE 2.31 The following table is derived from the data following exercise 61 of Exercises 1.2.

TABLE 2.1

Miles	*Days*	*Gallons*
203	6	14.6
202	2	13.0
199	2	13.4
236	4	14.9
175	8	13.8
160	12	14.5
146	7	13.6

The columns give the number of miles, days, and gallons used between fill-ups. Find the multiple-regression equation for distance traveled as a function of time and fuel.

Solution

```
      DAYS
6 2 2 4 8 12 7
      GALLONS
14.6  13  13.4  14.9  13.8  14.5  13.6
      MILES
203 202 199 236 175 160 146

      +A←1,∘2 7⍴DAYS,GALLONS
  1.0000   6.0000  14.6000
  1.0000   2.0000  13.0000
  1.0000   2.0000  13.4000
  1.0000   4.0000  14.9000
  1.0000   8.0000  13.8000
  1.0000  12.0000  14.5000
  1.0000   7.0000  13.6000

      +C←MILES⌹A
¯203.6  ¯8.396  31.6
```

Thus the regression equation is

$$\text{miles} = -203.6 - 8.396\ (\text{days}) + 31.6\ (\text{gallons})$$

The sum of the square deviations and the deviations themselves are

```
      (MILES−A+.×C)+.*2
764.6

      MILES−A+.×C
¯4.379  11.6  ¯4.04  2.348  9.694  6.156  ¯21.38
```

■

EXERCISES 2.3

Use the following matrices for exercises 1 and 2.

$$A = \begin{bmatrix} 1 & 0 & -1 \\ -1 & 1 & -1 \\ -2 & 0 & 3 \end{bmatrix}, \quad B = \begin{bmatrix} 1 & 0 & 1 \\ 0 & 1 & 0 \\ 2 & 0 & -2 \end{bmatrix}, \quad C = \begin{bmatrix} 0 & 1 & 0 \\ 1 & 0 & 2 \\ 0 & 1 & 0 \end{bmatrix}$$

$$D = \begin{bmatrix} 1 & 0 \\ 0 & 1 \\ -2 & 0 \\ 0 & 0 \\ 1 & 0 \end{bmatrix}, \quad E = \begin{bmatrix} 0 & 0 \\ -2 & 0 \\ 0 & 0 \\ 1 & -2 \\ 0 & 1 \end{bmatrix}, \quad F = \begin{bmatrix} 2 \\ -1 \\ -4 \\ -5 \\ 5 \end{bmatrix}, \quad G = \begin{bmatrix} 2 & -3 \\ 1 & 0 \end{bmatrix}$$

$$H = [3 \quad 4], \quad J = \begin{bmatrix} 1 & 2 \\ 0 & 1 \end{bmatrix}, \quad K = \begin{bmatrix} 1 & -1 \\ 1 & -1 \end{bmatrix}, \quad L = \begin{bmatrix} -7 & -2 & -1 & 0 & 2 \\ 9 & 2 & 2 & 0 & -1 \end{bmatrix}$$

$$M = \begin{bmatrix} 1 & 2 & 0 & 2 & 2 \\ 17 & 5 & 3 & 1 & -2 \end{bmatrix}, \quad N = \begin{bmatrix} 1 & -2 \\ 0 & 1 \end{bmatrix}, \quad P = \begin{bmatrix} 0 & 0 & -1 \\ -1 & 0 & -1 \\ -2 & 0 & 2 \end{bmatrix}$$

$$Q = [1 \quad -1 \quad -1 \quad -1 \quad -3 \quad 0], \quad R = \begin{bmatrix} 1 & 0 \\ -2 & 1 \\ -2 & 0 \\ 1 & -2 \\ 1 & 1 \end{bmatrix},$$

$$S = \begin{bmatrix} -2 & 2 & 2 & 2 & 6 & 0 \\ -2 & 2 & 2 & 2 & 6 & 0 \\ 7 & -7 & -7 & -7 & -21 & 0 \end{bmatrix} \quad T = [-4 \quad -2 \quad -10 \quad -2 \quad 3 \quad -5]^T,$$

$$U = \begin{bmatrix} 3 & 0 & 1 \\ 5 & 1 & 2 \\ 2 & 0 & 1 \end{bmatrix}, \quad V = \begin{bmatrix} 0 & 2 \\ -1 & -1 \\ 1 & -6 \\ 3 & -3 \\ 0 & 3 \end{bmatrix} \quad W = \begin{bmatrix} 8 & 4 & 1 & 2 & 0 \\ 8 & 3 & 1 & 1 & -1 \end{bmatrix}$$

1. Verify the following statements.
   (a) R has left inverses L and M.

(b) L has right inverses R and V.

(c) U and A are inverses of each other.

(d) N and J are inverses of each other.

(e) Q is a left inverse for T.

2. Using the results of exercise 1, solve for X and check the solution.

(a) $AX + B = C$ (b) $XA + B = C$ (c) $XL = 0$

(d) $XL = W$ (e) $RXL = I$ (f) $RXL = 0$

(g) $DX + EX = F$ (h) $XG + H = XK$ (i) $PXQ + XQ = S$

Hint: Use Proposition 2.8.

3. (a) Show that if L_1, L_2 are two left inverses of a matrix A and $Z = B(L_1 - L_2)$, where B is any matrix for which the product is defined, then $ZA = 0$.

(b) Find a matrix $Z \neq 0$ such that $ZR = 0$.

(c) Find a *square* matrix Z such that $ZR = 0$.

(d) Find a matrix Z $(\neq I)$ such that $ZR = R$.

4. Show that the "quadratic equation" $X^2 = I$ has an infinite number of solutions when X and I are 2 by 2.

5. Find the least-squares solution of $AX = B$. Is the least-squares solution a solution?

(a) $A = \begin{bmatrix} 1 & 0 \\ 0 & 0 \\ 0 & 1 \end{bmatrix}, \quad B = \begin{bmatrix} 2 \\ 0 \\ 3 \end{bmatrix}$ (b) $A = \begin{bmatrix} 1 & 2 \\ -2 & 1 \\ 2 & 2 \end{bmatrix}, \quad B = \begin{bmatrix} 3 \\ -1 \\ 4 \end{bmatrix}$

(c) $A = \begin{bmatrix} 1 & 1 \\ 1 & -1 \\ 1 & 1 \\ 1 & -1 \end{bmatrix}, \quad B = \begin{bmatrix} 1 \\ 0 \\ 0 \\ 0 \end{bmatrix}$ (d) $A = \begin{bmatrix} 1 & 1 \\ 0 & 1 \\ 1 & 0 \\ 0 & 0 \end{bmatrix}, \quad B = \begin{bmatrix} 0 \\ 0 \\ 0 \\ 1 \end{bmatrix}$

(e) $A = \begin{bmatrix} 1 & 1 & -1 \\ 1 & 0 & 0 \\ 1 & -1 & 0 \\ 1 & 1 & 1 \\ 1 & -1 & 0 \end{bmatrix}, \quad B = \begin{bmatrix} -2 \\ 1 \\ 3 \\ 0 \\ 3 \end{bmatrix}$

(See exercise 11 of Exercises 2.2.)

6. (Computer assignment) Use ▤ to find a least-squares solution of $AX = B$. Is the least-squares solution a solution?

(a) $A = \begin{bmatrix} -53 & -68 & 1 \\ -46 & -79 & -7 \\ 81 & -93 & 8 \\ 242 & -814 & -12 \end{bmatrix}, \quad B = \begin{bmatrix} 84 \\ 105 \\ 275 \\ 1858 \end{bmatrix}$

(b) $A = \begin{bmatrix} -82 & -56 & 40 \\ -60 & 97 & -42 \\ 26 & 74 & -92 \\ -41 & 99 & 82 \end{bmatrix}, \quad B = \begin{bmatrix} -66 \\ -96 \\ 22 \\ 52 \end{bmatrix}$

(c) $A = \begin{bmatrix} 40 & 18 \\ 74 & -2 \\ -408 & -61 \\ 32 & 33 \\ 226 & -205 \end{bmatrix}, \quad B = \begin{bmatrix} -5 \\ -69 \\ 383 \\ -97 \\ 324 \end{bmatrix}$

(d) $A = \begin{bmatrix} -268 & 386 & 213 & -1074 \\ 92 & -27 & -58 & -27 \\ 53 & -88 & -38 & -91 \\ -69 & 60 & 49 & -89 \\ -5 & -82 & -1 & -27 \end{bmatrix}, \quad B = \begin{bmatrix} -633 \\ -88 \\ -194 \\ -9 \\ -103 \end{bmatrix}$

(e) $A = \begin{bmatrix} -97 & -26 & 83 & 96 \\ -27 & 86 & 90 & 50 \\ -251 & -1842 & 225 & -387 \\ -32 & 16 & 18 & 22 \\ -326 & -118 & 559 & 157 \\ -24 & 61 & -45 & 52 \\ -12 & 80 & 2 & 63 \end{bmatrix}, \quad B = \begin{bmatrix} -60 \\ 19 \\ -1099 \\ 90 \\ 46 \\ 19 \\ -39 \end{bmatrix}$

7. (Computer assignment) Do exercise 60 in Exercises 1.2, using ⊡.

8. (Computer assignment) Do exercise 68 in Exercises 1.2, using ⊡.

9. (Computer assignment) In Example 2.31, the sum of deviations squared for miles as a linear function of days and gallons was about 765. What is the sum of the deviation if one writes miles as a function of gallons per day used between fill-ups?

10. (Computer assignment) Consider the data of Table 1.3 in Exercises 1.3. Find the regression equation of weight gained as a linear function of height at the beginning of the experiment, weight at the beginning of the experiment, and fitness per kilogram body weight [see exercise 31(b) in Exercises 1.3]. Do this for the control group and the experimental group separately. Find the sum of the square deviations in each case.

11. (Computer assignment) The table is, approximately, the sine curve on the interval zero to 2π. Sketch the curve along with the least-squares line, parabola, and cubic approximations.

x	0.00	.31	.63	.94	1.26	1.57	1.88	2.20	2.51	2.83	3.14
y	0.00	.31	.59	.81	.95	1.00	.95	.81	.59	.31	0.00

x	3.46	3.77	4.08	4.40	4.71	5.03	5.34	5.65	5.97	6.28
y	−.31	−.59	−.81	−.95	−1.00	−.95	−.81	−.59	−.31	0.00

12. (Computer assignment) From the table estimate the production of steel in 1960 by fitting a least-squares straight line to the data and using the value given by the line for 1960.

Year	1946	1947	1948	1949	1950	1951	1952	1953	1954	1955	1956
Tons of steel	66.6	84.9	88.6	78.0	96.8	105.2	93.2	111.6	88.3	117.0	115.2

13. (Computer assignment) The following table gives age, height, and weight for thirteen children.

Age	8	10	6	11	8	7	10	9	10	6	12	9	11
Height	56	52	49	61	51	49	56	48	59	40	63	55	63
Weight	64	56	55	66	54	59	76	57	71	50	75	68	75

(a) Find the regression equation for age as a function of height and weight.

(b) Find the regression equation for height as a function of age and weight.

(c) Find the regression equation for weight as a function of age and height.

14. (Computer assignment) *Least-squares fit of* $y = ae^{bx}$: If we find that the points $(x_1, y_1), (x_2, y_2), \ldots$ lie on the curve $y = ae^{bx}$, then $y_i = ae^{bx_i}$, $i = 1, 2, \ldots$. We can convert this to a linear system by taking the natural logarithm of both sides of the equations: $\ln y_i = \ln a + bx_i$. Thus we may estimate $\ln a$ and b by fitting a straight line to data points $(x_i, \ln y_i)$. Fit an exponential curve $y = ae^{bx}$ to the data.

x	.1	.2	.3	.4	.5	1	2	3	4
y	14.3	6	3	2.5	2	2	.47	.3	.2

Plot the data points and the curve on the same graph.

15. (Computer assignment) *Least-squares fit of* $y = ax^b$: If the point (x_i, y_i) lies on the curve $y = ax^b$, then $y_i = ax_i^b$. Taking logarithms as in exercise 14, we have $\ln y_i = a + b \ln x_i$. Thus we may estimate b and $\ln a$ by fitting a straight line to the data points $(\ln x_i, \ln y_i)$. Fit a curve of the form $y = ax^b$ to the data of exercise 14. Plot the curve you obtain on the same graph as the plot of exercise 14.

2.4 Affine Functions, Quadratic Forms

The material of this section is needed for the optional sections of this chapter. The material is not needed for Chapter 3. If the optional sections are not to be covered, then the material may be postponed until Chapter 4.

We will use the symbol $\mathbf{R}^n$ to denote the set of vectors v with $n = {}_\rho v$. We will denote the set of scalars (real numbers) by **R.**

DEFINITION 2.11 A *function* $f: \mathbf{R}^n \to \mathbf{R}^m$ is a rule that assigns a vector in $\mathbf{R}^m$ to each of a set of vectors in $\mathbf{R}^n$. The set of vectors in $\mathbf{R}^n$ to which the rule applies is called the *domain* of f. The collection of vectors in $\mathbf{R}^m$ that results from applying f to the domain is called the *image*.

EXAMPLE 2.32 Define $f: \mathbf{R}^5 \to \mathbf{R}^3$ by $f(v) = v[1\ 3\ 5]$. For example, $f((1, 7, 9, 6, -3)) = (1, 9, -3)$. The domain of f is all of $\mathbf{R}^5$, since the rule applies to all v of length 5. The range is all of $\mathbf{R}^3$, since any vector in $\mathbf{R}^3$ can result. This result can happen in many ways — for example, $(x, y, z) = f((x, 1, y, 2, z)) = f((x, 0, y, 0, z))$, and so on. ■

EXAMPLE 2.33 $f: \mathbf{R}^4 \to \mathbf{R}^4$ by $f(v) = v{*}.5$. For example, $f((1, 4, 9, 16)) = (1, 2, 3, 4)$. But a vector such as $(1, 2, -3, 7)$ is not in the domain of f, since $\sqrt{-3}$ is not a real number.

```
      1 2 ¯3 7*.5
DOMAIN ERROR
      1 2 ¯3 7*.5
```

The image of f is any vector in $\mathbf{R}^4$ with nonnegative components, since for such vectors $(a, b, c, d) = f((a, b, c, d){*}2) = (\sqrt{a^2}, \sqrt{b^2}, \sqrt{c^2}, \sqrt{d^2})$. ■

Notation It is convenient to eliminate the inner set of parentheses when writing functions $f: \mathbf{R}^n \to \mathbf{R}^m$. We shall write, for example, $f(1, 4, 9, 16)$ for $f((1, 4, 9, 16))$.

EXAMPLE 2.34 Given an equation $AX = B$ with A a matrix and X, B vectors, a least-squares "solution" was defined in Section 2.3 as a vector X that minimized the scalar

$$f(X) = (B - A + . \times X) +.{*}2$$

If A is m by n, then X is in $\mathbf{R}^n$ and $f: \mathbf{R}^n \to \mathbf{R}$. ■

EXAMPLE 2.35 Suppose that one has the problem of solving a set of simultaneous nonlinear equations — for example,

$$\begin{aligned} x^2 + y^2 + z^2 &= 1 \\ xyz &= 1 - x \\ x \sin y - y \sin z &= \tan x \end{aligned}$$

In Chapter 3 a theory of simultaneous *linear* equations is developed, but no such theory exists for nonlinear equations. In general, the best one can do is try numerical methods. Most canned programs for solving simultaneous nonlinear equations require that the problem first be transformed into a standard form. This form is

$$f(X) = 0; \qquad f: \mathbf{R}^n \to \mathbf{R}^n$$

In the case above, for example, one possible f is given by

$$f(x, y, z) = (1 - x^2 - y^2 - z^2, xyz + x - 1, \tan x - x \sin y + y \sin z). \quad ■$$

EXAMPLE 2.36 A standard system of differential equations from mathematical ecology is

$$\frac{dQ}{dt} = rQ - aPQ$$

$$\frac{dP}{dt} = -sP + bPQ$$

Here P is a population of predators ("wolves") and Q is a population of prey ("rabbits"). The term PQ is assumed to be proportional to the rate at which the predators encounter the prey.

Thus the first equation says that the rate of increase in the rabbit population depends on two factors. It increases with the number of rabbits but decreases with the number of wolf-rabbit encounters.

To analyze such systems of differential equations either theoretically or numerically, we put them into a standard form:

$$\frac{d}{dt} Y = f(Y), \qquad f\colon \mathbf{R}^n \to \mathbf{R}^n$$

where the vector Y is differentiated componentwise. In the equation above we have

$$Y = (P, Q), \qquad \frac{d}{dt} Y = \left(\frac{dP}{dt}, \frac{dQ}{dt}\right)$$

and $f\colon \mathbf{R}^2 \to \mathbf{R}^2$ is given by $f(P, Q) = (rQ - aPQ, -sP + bPQ)$. ∎

EXAMPLE 2.37 Curves in the plane are often written parametrically. For example,

$$\left.\begin{aligned} y &= 1 + 3t \\ x &= 2 - \ \ t \end{aligned}\right\}, \qquad -\infty < t < \infty$$

represents a line in the plane, for if we eliminate the parameter t we find

$$y = 1 + 3(2 - x)$$

or

$$y = -3x + 7.$$

The parametric form of the line can be considered a function $f\colon \mathbf{R} \to \mathbf{R}^2$ (t is a scalar) given by the formula

$$f(t) = (2 - t, 1 + 3t)$$

Here the domain of f is all scalars and the image of f is the line $y = -3x + 7$. The point $f(t)$ sweeps out this line as t sweeps from $-\infty$ to ∞ (Figure 2.1). ∎

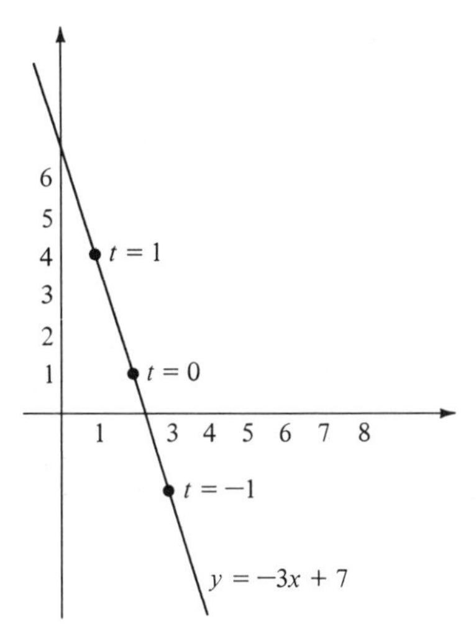

FIGURE 2.1

Linear algebra is concerned with those functions $f\colon \mathbf{R}^n \to \mathbf{R}^m$ which can be handled with matrix algebra.

DEFINITION 2.12 An *affine* (or *nonhomogeneous*) *linear function* $f\colon \mathbf{R}^n \to \mathbf{R}^m$ is a function of the form

$$Y = f(X) = B + AX$$

where A is an m-by-n matrix and B is a vector in $\mathbf{R}^m$.

DEFINITION 2.13 An affine linear function $Y = B + AX$ is called simply *linear* if $B = 0$. Thus a *linear function* $f\colon \mathbf{R}^n \to \mathbf{R}^m$ is a function of the form

$$Y = AX$$

If $f(X) = B + AX$, we will call B the *constant term* and AX the *linear term*.

Note: We often abbreviate the term "affine linear function" to "affine function."

EXAMPLE 2.38 The function of Example 2.37 can be taken to be affine if we take the parameter t to be a vector with one component instead of a scalar. Then, writing vectors as column vectors,

$$Y = \begin{bmatrix} x \\ y \end{bmatrix} = f((t)) = \begin{bmatrix} 2 \\ 1 \end{bmatrix} + \begin{bmatrix} -1 \\ 3 \end{bmatrix}[t]$$

that is,

$$X = (t), \qquad A = \begin{bmatrix} -1 \\ 3 \end{bmatrix}, \qquad B = (2, 1), \quad \text{and} \quad Y = (x, y). \qquad \blacksquare$$

For the next example we find it useful to confuse **R** (scalars) with $\mathbf{R}^1$ (single-component vectors).

EXAMPLE 2.39 A real-valued affine function of a real variable $f: \mathbf{R} \to \mathbf{R}$ is a function of the form $y = f(x) = ax + b$. Thus the affine functions $\mathbf{R} \to \mathbf{R}$ are those functions whose graph is a straight line. This graph intersects the y axis at the point $(0, b)$. Thus a function $f: \mathbf{R} \to \mathbf{R}$ is linear if and only if the graph of f is a straight line through the origin. $\blacksquare$

Examples 2.38 and 2.39 give two ways of using affine functions to represent a line in the plane. In Example 2.38 the line is the image of an affine function $f: \mathbf{R} \to \mathbf{R}^2$. In Example 2.39 the line is the graph of a function. In Chapter 4 the method of Example 2.38 will be extended to higher-dimensional spaces.

If a function $f: \mathbf{R}^n \to \mathbf{R}^m$ is expressed in formulas involving the components of $X = (x_1, x_2, \ldots, x_n)$, then f is affine if and only if these formulas involve only constant terms and terms of the form ax_i, where a is a constant. Terms of the form ax_i are called *first-degree* terms, constants are *zero-degree* terms. In the formula $f(X) = B + AX$ the constant terms make up the vector B and the terms of the forms ax_i come from the linear part AX. If there are no constant terms ($B = 0$), then the function is linear.

EXAMPLE 2.40 The function $f: \mathbf{R}^5 \to \mathbf{R}^3$ of Example 2.32 is affine. In fact it is linear. The definition is $f(X) = X[1 \;\; 3 \;\; 5]$. We can write this as

$$f\left(\begin{bmatrix} x_1 \\ x_2 \\ x_3 \\ x_4 \\ x_5 \end{bmatrix}\right) = \begin{bmatrix} x_1 \\ x_3 \\ x_5 \end{bmatrix}$$

The only terms that appear are the first-degree terms x_1, x_3, and x_5, and so the function is linear. $\blacksquare$

In Example 2.40 the function is linear but the matrix A is not obvious. The next proposition gives a method of computing A as well as a simple characterization of linear functions by two identities.

PROPOSITION 2.17 Let $f\colon \mathbf{R}^n \to \mathbf{R}^m$.

1. If f is linear, then the identities

$$f(X_1 + X_2) = f(X_1) + f(X_2)$$
$$f(\alpha X) = \alpha f(X)$$

hold for all vectors X, X_1, X_2 in $\mathbf{R}^n$ and all scalars α.

2. Conversely, if the two identities of statement 1 hold for all X, X_1, X_2 in $\mathbf{R}^n$ and all scalars α, then f is linear and

$$Y = f(x) = AX$$

where the matrix A is defined by $A[;i] = f(I[;i])$. Here I is the n-by-n identity matrix.

Proof Statement 1 is just the definition of linear function plus Proposition 2.8.
To check statement 2, notice that if the identities hold, then

$$f(\alpha_1 X_1 + \alpha_2 X_2 + \cdots + \alpha_n X_n) = \alpha_1 f(X_1) + \alpha_2 f(X_2) + \cdots + \alpha_n f(X_n)$$

for any collection of vectors X_i and scalars α_i (see Exercise 34). Thus, if $X = (x_1, x_2, \ldots, x_n)$, we apply the definition of matrix-vector multiplication to obtain

$$\begin{aligned} f(X) &= f(IX) = f(x_1 I[;1] + x_2 I[;2] + \cdots + x_n I[;n]) \\ &= x_1 f(I[;1]) + x_2 f(I[;2] + \cdots + x_n f(I[;n]) \\ &= x_1 A[;1] + x_2 A[;2] + \cdots + x_n A[;n] \\ &= AX \end{aligned}$$

■

EXAMPLE 2.41 Returning to Example 2.40, we see that

$$f(I[;1]) = \begin{bmatrix} 1 \\ 0 \\ 0 \end{bmatrix}, \quad f(I[;2]) = \begin{bmatrix} 0 \\ 0 \\ 0 \end{bmatrix}, \quad f(I[;3]) = \begin{bmatrix} 0 \\ 1 \\ 0 \end{bmatrix},$$

$$f(I[;4]) = \begin{bmatrix} 0 \\ 0 \\ 0 \end{bmatrix}, \quad f(I[;5]) = \begin{bmatrix} 0 \\ 0 \\ 1 \end{bmatrix}$$

Thus,

$$f(X) = \begin{bmatrix} 1 & 0 & 0 & 0 & 0 \\ 0 & 0 & 1 & 0 & 0 \\ 0 & 0 & 0 & 0 & 1 \end{bmatrix} \begin{bmatrix} x_1 \\ x_2 \\ x_3 \\ x_4 \\ x_5 \end{bmatrix} = \begin{bmatrix} x_1 \\ x_3 \\ x_5 \end{bmatrix}$$

■

The point of Proposition 2.17 that we would like to emphasize is that a linear function $f: \mathbf{R}^n \to \mathbf{R}^m$ is completely known when its action on the n vectors $I[;1], \ldots, I[;n]$ is known. In particular, the matrix A is unique. A general affine function is determined by its action on $n + 1$ vectors — the columns of I plus the zero vector. To see this, notice that the constant term of $f(X) = B + AX$ is simply $f(0)$, and so computing $f(0)$ gives us B. To get A we apply Proposition 2.17 to the linear function

$$g(X) = f(X) - f(0) = AX$$

EXAMPLE 2.42 Is the function $f: \mathbf{R}^3 \to \mathbf{R}^2$ given by

$$f\left(\begin{bmatrix} x \\ y \\ z \end{bmatrix}\right) = \begin{bmatrix} 3 + 12x - 7y \\ 2z + 7x - 4 \end{bmatrix}$$

affine? If so, find A and B.

Solution Only constant and first-degree terms appear, and so it is affine.

$$B = f(0) = f\left(\begin{bmatrix} 0 \\ 0 \\ 0 \end{bmatrix}\right) = \begin{bmatrix} 3 \\ -4 \end{bmatrix} \quad \text{and} \quad AX = f(x) - f(0) = g(x) = \begin{bmatrix} 12x - 7y \\ 7x - 4 \end{bmatrix}$$

is the linear part.

$$g\left(\begin{bmatrix} 1 \\ 0 \\ 0 \end{bmatrix}\right) = \begin{bmatrix} 12 \\ 7 \end{bmatrix}, \quad g\left(\begin{bmatrix} 0 \\ 1 \\ 0 \end{bmatrix}\right) = \begin{bmatrix} -7 \\ 0 \end{bmatrix}, \quad g\left(\begin{bmatrix} 0 \\ 0 \\ 1 \end{bmatrix}\right) = \begin{bmatrix} 0 \\ 2 \end{bmatrix}$$

and so

$$A = \begin{bmatrix} 12 & -7 & 0 \\ 7 & 0 & 2 \end{bmatrix}$$

and

$$f(X) = \begin{bmatrix} 3 \\ -4 \end{bmatrix} + \begin{bmatrix} 12 & -7 & 0 \\ 7 & 0 & 2 \end{bmatrix} X \quad \blacksquare$$

Statements about affine functions can often be verified with a little matrix algebra.

EXAMPLE 2.43 Show that if $f: \mathbf{R}^n \to \mathbf{R}^m$ and $g: \mathbf{R}^m \to \mathbf{R}^p$ are affine, then the composite function $g \circ f: \mathbf{R}^n \to \mathbf{R}^p$ is also affine.

Solution Recall that the *composite* function $g \circ f$ is defined by $g \circ f(X) = g(f(X))$. Suppose that $f(x) = B + AX$, $g(Y) = D + CY$. Then $g \circ f(x) = g(f(x)) =$

$D + Cf(X) = D + C(B + AX) = (D + CB) + (CA)X$. And thus $g \circ f(X) = F + EX$, where $F = D + CB$ is a vector in $\mathbf{R}^p$ and $E = CA$ is an n-by-p matrix. ■

QUADRATIC FUNCTIONS

The affine functions $f\colon \mathbf{R}^n \to \mathbf{R}^m$ are those functions containing only terms of degree zero (constants) and one (ax_i). If we replace $\mathbf{R}^m$ by $\mathbf{R}$, then we may include *second-degree* terms, by which we mean terms of the form $ax_i x_j$, as well.

In order to write second-degree equations in matrix terms we use the identity given in the next proposition. The proof of Proposition 2.18 is a variant of the calculation used in the proof of Proposition 2.3 and is left to the reader. Notice that if A is an n-by-n matrix and $n = \rho X$, then $n = \rho A{+}.{\times}X$ and $X{+}.{\times}A{+}.{\times}X$ is a scalar.

PROPOSITION 2.18 Let A be an n-by-n matrix and let $X = (x_1, \ldots, x_n)$. Then

$$(X{+}.{\times}A{+}.{\times}X) = \sum_{i=1}^{n} \sum_{j=1}^{n} A[i;j]x_i x_j \qquad ■$$

To write the expression $X{+}.{\times}A{+}.{\times}X$ in matrix multiplication (rather than matrix-vector multiplication) terms, we take X to be a column vector. Then AX is also n by 1 and X^TAX is the 1-by-1 matrix with entry equal to the scalar $X{+}.{\times}\mathrm{A}{+}.{\times}X$.

To see just what the double sum of Proposition 2.18 looks like, consider the 2-by-2 case.

$$\begin{bmatrix} x_1 \\ x_2 \end{bmatrix}^T \begin{bmatrix} a & b \\ c & d \end{bmatrix} \begin{bmatrix} x_1 \\ x_2 \end{bmatrix} = [x_1 \quad x_2] \begin{bmatrix} x_1 a + x_2 b \\ x_1 c + x_2 d \end{bmatrix}$$

$$= ax_1^2 + bx_1x_2 + cx_2x_1 + dx_2^2$$

Since $x_1x_2 = x_2x_1$, this can be shortened to

$$\begin{bmatrix} x_1 \\ x_1 \end{bmatrix}^T \begin{bmatrix} a & b \\ c & d \end{bmatrix} \begin{bmatrix} x_1 \\ x_2 \end{bmatrix} = ax_1^2 + (b + c)x_1x_2 + dx_2^2$$

In this form it is clear that different matrices A give the same result X^TAX as long as the sum $A[1;2] + A[2;1]$ remains the same. For example,

$$\begin{bmatrix} -1 & 2 \\ 2 & 3 \end{bmatrix}, \quad \begin{bmatrix} 1 & 4 \\ 0 & 3 \end{bmatrix}, \quad \begin{bmatrix} 1 & 0 \\ 4 & 3 \end{bmatrix}, \quad \begin{bmatrix} 1 & 104 \\ -100 & 3 \end{bmatrix}$$

all give the same result: $X^TAX = X_1^2 + 4x_1x_2 + 3x_2^2$.

We take advantage of this wide choice of A's to choose A symmetric. We do this by choosing $A[1;2] = A[2;1]$.

$$[x_1 \quad x_2]\begin{bmatrix} a & b \\ b & c \end{bmatrix}\begin{bmatrix} x_1 \\ x_2 \end{bmatrix} = ax_1^2 + 2bx_1x_2 + cx_2^2$$

PROPOSITION 2.19 Let A be n by n, $B = \frac{1}{2}(A + A^T)$, and let X be a column vector. Then B is symmetric and

$$X^TAX = X^TBX$$

B is the unique symmetric matrix for which the equation above holds for all X.

Proof The symmetry of B is left as exercise 35, the uniqueness as exercse 36.
First notice that X^TAX is certainly symmetric because it is a 1-by-1 matrix. Thus

$$\begin{aligned} X^TAX &= (X^TAX)^T \\ &= X^TA^T(X^T)^T \qquad \text{by Proposition 2.2} \\ &= X^TA^TX \end{aligned}$$

Then

$$\begin{aligned} X^TBX &= X^T\tfrac{1}{2}(A + A^T)X \\ &= \tfrac{1}{2}X^T(AX + A^TX) \qquad \text{by Proposition 2.8} \\ &= \tfrac{1}{2}(X^TAX + X^TA^TX) \\ &= \tfrac{1}{2}(X^TAX + X^TAX) \\ &= X^TAX. \end{aligned}$$

■

It is the fact that second-degree terms can be handled by symmetric matrices that makes symmetric matrices important.

DEFINITION 2.14 The function $f\colon \mathbf{R}^n \to \mathbf{R}$ is an *affine quadratic* function if it can be written in the form

$$f(X) = c + BX + X^TAX$$

where c is a scalar, B is a matrix, and A is a symmetric matrix (the vector X is here confused with a single-column matrix). The *constant* (or *zero-degree*) term is c, the *linear* (or *first-degree*) term is BX, and the *quadratic* (or *second-degree*) term is X^TAX.

If the constant and first-degree terms are missing ($c = 0$, $B = 0$), then f is called a *quadratic form*.

Notice that matrix B has but a single row in the definition above.

EXAMPLE 2.44 Consider the function of Example 2.34:

$$f(X) = (B - AX) +.* 2$$

Here B and X are vectors; A is a matrix. The function is clearly a sum of squares of linear and constant terms, and so it should be a quadratic function. To verify this fact take the vectors involved to be single-column matrices and use the identity $(V +.* 2) = V +.\times V$. Then

$$\begin{aligned} f(X) &= (B - AX)^T(B - AX) \\ &= (B - AX)^T B - (B - AX)^T AX \\ &= (B^T - X^T A^T)B - (B^T - X^T A^T)AX \\ &= B^T B - X^T A^T B - B^T AX + X^T A^T AX \end{aligned}$$

The term $B^T B$ is the constant term; the quadratic term is $X^T(A^T A)X$. Now $X^T A^T B$ is a 1-by-1 matrix, hence it is symmetric. Thus $X^T A^T B = (X^T A^T B)^T = B^T(A^T)^T(X^T)^T = B^T AX$. So that

$$f(X) = B^T B - (2B^T A)X + X^T(A^T A)X$$

an affine quadratic function. ■

EXAMPLE 2.45 Is the function $f: \mathbf{R}^3 \to \mathbf{R}$ defined by

$$f(x, y, z) = x^2 - 2xy + 4z - 1$$

a quadratic function? If so, write the function in matrix form.

Solution Only constant linear and quadratic terms appear, and so the function is quadratic. The constant term is $c = -1$. The linear part is $g(x, y, z) = 4z$. Now $g(1, 0, 0) = 0$, $g(0, 1, 0) = 0$, $g(0, 0, 1) = 4$, so $B = [0\ \ 0\ \ 4]$ by Proposition 2.17. The quadratic part is $x^2 - 2xy$. Using Proposition 2.18 with $x = x_1$, $y = x_2$, $z = x_3$, we see that the quadratic part is $A[1; 1]x^2 + -(A[1; 2] + A[2; 1])xy$, and so $A[1; 1] = 1$, $A[1; 2] = A[2; 1] = -1$, and

$$A = \begin{bmatrix} 1 & -1 & 0 \\ -1 & 0 & 0 \\ 0 & 0 & 0 \end{bmatrix}$$

That is,

$$f(X) = -1 + [0\ \ 0\ \ 4]\begin{bmatrix} x \\ y \\ z \end{bmatrix} + [x\ \ y\ \ z]\begin{bmatrix} 1 & -1 & 0 \\ -1 & 0 & 0 \\ 0 & 0 & 0 \end{bmatrix}\begin{bmatrix} x \\ y \\ z \end{bmatrix}$$ ■

EXERCISES 2.4

In exercises 1 through 28 decide if the function is linear, affine, quadratic, a quadratic form, or otherwise. In the first four cases write the matrix form of the equation.

1. $f\left(\begin{bmatrix} x \\ y \end{bmatrix}\right) = \begin{bmatrix} 2x - 3y \\ 2x + 3y \end{bmatrix}$

2. $f\left(\begin{bmatrix} x \\ y \end{bmatrix}\right) = \begin{bmatrix} 4x \\ 2y - x \end{bmatrix}$

3. $f\left(\begin{bmatrix} x \\ y \end{bmatrix}\right) = \begin{bmatrix} 1 + 4x - 12y \\ 2 + 6x + 7y \end{bmatrix}$

4. $f(x, y, z) = \begin{bmatrix} 2x - 3y + 4z \\ x - y + 1 \end{bmatrix}$

5. $f(x, y, z) = 1 + 2x + 3y + 4z$

6. $f(x, y, z, w) = xyzw$

7. $f(x, y, z, w) = xy + zw$

8. $f(x, y, z) = (x + y)^2 - (3 + z)^2$

9. $f(x, y, z) = (1 + 2x + 3y)(4 + x - z)$

10. $f\left(\begin{bmatrix} x \\ y \\ z \end{bmatrix}\right) = \begin{bmatrix} z \\ x \\ y \end{bmatrix}$

11. $f\left(\begin{bmatrix} x \\ y \\ z \end{bmatrix}\right) = \begin{bmatrix} 1 + x \\ 1 + y \\ 1 + z \end{bmatrix}$

12. $f(X) = X$, X in $\mathbf{R}^n$

13. $f(X) = -X$, X in $\mathbf{R}^n$

14. $f(X) = 3X$, X in $\mathbf{R}^n$

15. $f(X)$ is $X{*}2$, X in $\mathbf{R}^n$

16. $f(X)$ is $(X{*}2)[1]$, X in $\mathbf{R}^n$
17. $f(X)$ is $X[1\ \ 3\ \ 5] + X[2\ \ 4\ \ 6]$ for X in $\mathbf{R}^6$
18. $f(X)$ is `,⍉2 3⍴X` for X in $\mathbf{R}^6$
19. $f(X)$ is `,(⍳4)∘.×X` for X in $\mathbf{R}^3$
20. $f(X)$ is `,(⍳4)∘.-X` for X in $\mathbf{R}^3$
21. $f(X)$ is `,(⍳4)∘.÷X` for X in $\mathbf{R}^3$
22. $f(X)$ is `X+.×X` for X in $\mathbf{R}^n$
23. $f(X)$ is `1+X[1]+X+.×X` for X in $\mathbf{R}^n$
24. $f(X) = (0, 0)$ for X in $\mathbf{R}^n$
25. $f\colon \mathbf{R}^2 \to \mathbf{R}^2$ by $f(x, y) = [x\ \ y]\begin{bmatrix} a & b \\ c & d \end{bmatrix}$
26. $f\colon \mathbf{R}^n \to \mathbf{R}^m$ is `X+.×M`, M an n-by-m matrix.
27. $f\colon \mathbf{R}^n \to \mathbf{R}^n$ is `(X+.×M)+N+.×X`, M and N are n-by-n matrices.
28. Sketch the images of the following affine functions $f\colon \mathbf{R} \to \mathbf{R}^2$.

    (a) $f(t) = \begin{bmatrix} 2 - 3t \\ 1 + 2t \end{bmatrix}$ (b) $f(t) = \begin{bmatrix} t \\ -t \end{bmatrix}$ (c) $f(t) = \begin{bmatrix} 1 + t \\ 1 + t \end{bmatrix}$

    (d) $f(t) = \begin{bmatrix} t \\ t \end{bmatrix}$ (e) $f(t) = \begin{bmatrix} -6 - 2t \\ -6 - 2t \end{bmatrix}$

29. Show that if $f\colon \mathbf{R}^n \to \mathbf{R}^m$ and $g\colon \mathbf{R}^m \to \mathbf{R}^p$ are linear, then $g \circ f\colon \mathbf{R}^n \to \mathbf{R}^p$ is linear.
30. Let $f, g\colon \mathbf{R}^n \to \mathbf{R}^m$ and show that
    (a) If f, g are affine, then $f \pm g$ is affine.
    (b) If f, g are linear, then $f \pm g$ is linear.
31. Show that if $f\colon \mathbf{R}^n \to \mathbf{R}^m$ is affine and if $g\colon \mathbf{R}^m \to \mathbf{R}$ is quadratic, then $g \circ f\colon \mathbf{R}^n \to \mathbf{R}$ is quadratic.
32. Show that if $f, g\colon \mathbf{R}^n \to \mathbf{R}$ are affine, then the function $h(X) = f(X)g(X)$ is quadratic. Write $h(X)$ in matrix form.
33. Let $f\colon \mathbf{R}^n \to \mathbf{R}^m$ by $Y = f(X) = B + AX$.
    (a) Assuming that L is a left inverse of A, solve for X in terms of Y.
    (b) Show that, if L is a left inverse of A, then the affine function $g(Y) = LY - LB$ is a left inverse of f in the sense that $g \circ f(X) = X$.

(c) Show that if there is an affine function $g(Y) = D + CY$ such that $g \circ f(X) = X$ for all X, then A has a left inverse.

(d) Show that if there is an affine function $g(Y) = D + CY$ such that $g \circ f(X) = X$ and $f \circ g(Y) = Y$ for all X and Y, then A is invertible.

(e) Show that if A is invertible, then there is an affine function g such that $g \circ f(x) = x$ and $f \circ g(Y) = Y$ for all X and Y.

34. Let f: $\mathbf{R}^n \to \mathbf{R}^m$ satisfy the identities of Proposition 2.17, statement 1. Show that

$$f(\alpha_1 X_1 + \alpha_2 X_2 + \cdots + \alpha_n X_n) = \alpha_1 f(X_1) + \alpha_2 f(X_2) + \cdots + \alpha_n f(X_n).$$

Hint: Let $A = \alpha_1 X_1$, $B = \alpha_2 X_2 + \cdots + \alpha_n X_n$ and apply statement 1 to $f(A + B)$.

35. (a) Show that if A is a matrix and α is a scalar, then $(\alpha A)^T = \alpha A^T$.

(b) Show that if A is a square matrix, then $\frac{1}{2}(A + A^T)$ is symmetric.

36. Let A, B be n-by-n symmetric matrices. Show that if $X^T AX = X^T BX$ for all X, then $A = B$.

Hint: Take X to be $k{=}\iota n$ for $k = 1, 2, \ldots, n$ first and then use the sum of two such X's.

2.5* Multivariate Calculus Derivatives, Maxima and Minima

This section assumes some familiarity with partial differentiation. The object is to rephrase some results about partial derivatives in matrix terms. This results in a considerable simplification of formulas.

Definition 2.15 A function is said to be of class C^n if all its partial derivatives of order n exist and are continuous.

First consider a scalar-valued function f: $\mathbf{R}^n \to \mathbf{R}$. Writing $y = f(X) = f(x_1, \ldots, x_n)$, f has n first partials

$$\frac{\partial f}{\partial x_1}, \frac{\partial f}{\partial x_2}, \ldots, \frac{\partial f}{\partial x_n}$$

We collect these into a single vector and call it the derivative of f.

Definition 2.16 Let f: $\mathbf{R}^n \to \mathbf{R}$ be a class C^1 at p in $\mathbf{R}^n$. The *derivative* of f at p, denoted $Df(p)$, is the vector

$$\left(\frac{\partial f}{\partial x_1}(p), \frac{\partial f}{\partial x_2}(p), \ldots, \frac{\partial f}{\partial x_n}(p)\right)$$

This derivative is also called the *gradient* of f. If p is allowed to vary, then one has a function Df: $\mathbf{R}^n \to \mathbf{R}^n$.

Next consider the case of a function $f\colon \mathbf{R}^n \to \mathbf{R}^m$. Such a function may be considered as m scalar-valued functions:

$$f(X) = (f_1(X), f_2(X), \ldots, f_m(X))$$

where $f_k(X)$ is the kth component of the vector $f(X)$ — that is, $f_k(X) = f(X)[k]$.

DEFINITION 2.17 Let $f\colon \mathbf{R}^n \to \mathbf{R}^m$ be of class C^1 at p in $\mathbf{R}^n$. The *derivative* of f at p is the m-by-n matrix $Df(p)$ defined by

$$Df(p)[k;] = Df_k(p)$$

where $f_k(X) = f(X)[k]$.

The derivative $Df(p)$ is also called the *Jacobian matrix* of f at p. Simplifying the notation by suppressing the p's, we have

$$Df[i;j] = \frac{\partial f_i}{\partial x_j}$$

If we write $Y = f(X)$, then in the two cases defined above ($f\colon \mathbf{R}^n \to \mathbf{R}$ and $f\colon \mathbf{R}^n \to \mathbf{R}^m$) we have $(\rho Df) = (\rho Y), \rho X$. In the latter case this is clear. In the former case ρY is $\iota 0$ and so ρDf is ρX. For a function $f\colon \mathbf{R} \to \mathbf{R}$ we take Df to be df/dx. In this case ρX is $\iota 0$ as well, and the formula holds in all cases.

EXAMPLE 2.46 Compute the derivative of the function $f\colon \mathbf{R}^2 \to \mathbf{R}^3$, where

$$f(x, y) = (x^2 + y^2, \sin xy, x - y)$$

Solution $f_1(x, y) = x^2 + y^2$, $f_2(x, y) = \sin xy$, $f_3(x, y) = x - y$, so

$$\begin{aligned} Df_1 &= (2x, 2y) \\ Df_2 &= (y \cos xy, x \cos xy) \\ Df_3 &= (1, -1) \end{aligned}$$

and

$$Df = \begin{bmatrix} 2x & 2y \\ y \cos xy & x \cos xy \\ 1 & -1 \end{bmatrix} \qquad \blacksquare$$

PROPOSITION 2.20 Let $f\colon \mathbf{R}^n \to \mathbf{R}^m$ be the affine function $f(X) = B + AX$; then $Df = A$.

Proof $f_i(X) = (B + AX)[i] = B[i] + \Sigma_k A[i; k]x_k$ and $\partial f_i/\partial x_j = A[i; j]$. $\blacksquare$

The rules for differentiating sums and products easily generalize.

PROPOSITION 2.21 Let $f, g: \mathbf{R}^n \to \mathbf{R}^m$ be C^1 functions, α a scalar.

1. $D(f+g) = Df + Dg$.
2. $D(\alpha f) = \alpha Df$.
3. $D(f +.\times g) = (f +.\times Dg) + g +.\times Df$.

Proof Statements 1 and 2 are left as Exercises 22 and 23. Let $h = f +.\times g$. Then $h: \mathbf{R}^n \to \mathbf{R}$ and $h(X) = \Sigma f_i(X) g_i(X)$.

$$(Dh)[j] = \frac{\partial}{\partial x_j} \sum_i f_i g_i$$

$$= \sum_i \frac{\partial}{\partial x_j} f_i g_i$$

$$= \sum_i \left(g_i \frac{\partial f_i}{\partial x_j} + f_i \frac{\partial g_i}{\partial x_j} \right)$$

$$= \sum_i g_i \frac{\partial f_i}{\partial x_j} + \sum_i f_i \frac{\partial g_i}{\partial x_j}$$

$$= (g +.\times Df)[j] + (f +.\times Dg)[j] \qquad \blacksquare$$

EXAMPLE 2.47 Compute Df for $f(X) = X +.* 2$.

First Solution $f(X) = \Sigma_i x_i^2$. Thus $\partial f / \partial x_j = 2x_j$, and so $Df: \mathbf{R}^n \to \mathbf{R}^n$ by $Df(X) = 2X$.

Second Solution $f(X) = X +.\times X$. By Proposition 2.20 the derivative of $g(X) = X$ is the n-by-n identity matrix I, so that by Proposition 2.21 $Df = (g +.\times I) + g +.\times I$ or $Df(X) = 2g(X) = 2X$. $\blacksquare$

EXAMPLE 2.48 In Section 2.4 we wrote a general quadratic function

$$f(X) = C + BX + X^T AX \qquad (A = A^T)$$

This form involves confusing the vector X with a single-column matrix and then taking the 1-by-1 matrices BX and $X^T AX$ to be scalars. To calculate the derivative of f we need to be a bit more careful. Thus take C a scalar, B a vector, X a vector, and $A = A^T$ a matrix. Then the general quadratic function $f: \mathbf{R}^n \to \mathbf{R}$ is

$$f(X) = C + (B +.\times X) + X +.\times A +.\times X.$$

The functions $C + B +.\times X$, $A +.\times X$, and $X = I +.\times X$ are all affine functions to which Proposition 2.20 applies. Thus

$$\begin{aligned} Df(X) &= D(C + B +.\times X) + D(X +.\times A +.\times X) \\ &= B + ((A +.\times X) +.\times DX) + (DA +.\times X) +.\times X \\ &= B + (A +.\times X) + X +.\times A \\ &= B + 2AX \qquad \text{(matrix-vector multiplication)} \end{aligned}$$

The last step follows, since $A = A^T$ implies $(A +.\times X) = X +.\times A$ (exercise 24). ∎

Propositions 2.20 and 2.21 are simply a matter of notation. Proofs of deeper results about derivatives are beyond the scope of this text. The next proposition, the chain rule, is proved in texts on advanced calculus.

PROPOSITION 2.22 (*Chain rule*) Let $f: \mathbf{R}^n \to \mathbf{R}^m$ be of class C^1 at p and $g: \mathbf{R}^m \to \mathbf{R}^q$ be of class C^1 at $f(p)$. Then the composite function $g \circ f: \mathbf{R}^n \to \mathbf{R}^q$ is of class C^1 at p and

$$(Dg \circ f)(p) = Dg(f(p)) +.\times Df(p) \qquad ∎$$

Note: If Dg is a scalar, replace $+.\times$ by $\times$. The APL default definition of scalar $+.\times$ vector is not wanted here.

EXAMPLE 2.49 Compute the derivative of $f: \mathbf{R}^n \to \mathbf{R}$ given by $f(X) = (B - AX) +.* 2$, B a vector and A a matrix.

Solution This is the function that is minimized by a least-squares solution of $AX = B$. If we set $g(X) = B - AX$ and $h(Y) = Y +.* 2$, then $f = h \circ g$. The derivative of g is given by Proposition 2.20, and the derivative of h was computed in Example 2.47. Thus,

$$\begin{aligned} Df(X) &= Dh(g(X)) +.\times Dg(X) \\ &= 2(B - AX) +.\times A \qquad ∎ \end{aligned}$$

The expression $(B - AX) +.\times A$ is a vector-matrix multiplication that was defined in Section 1.3 but not used. It may be rewritten as $A^T(B - AX)$ (exercise 24).

The proof of the next proposition may be found in texts on advanced calculus, as will a rigorous definition of "interior."

PROPOSITION 2.23 (*Max − min test*) Let $f: \mathbf{R}^n \to \mathbf{R}$ be of class C^1 and let p be a point in the interior of the domain of f. If f has a local maximum or a local minimum at p, then

$$Df(p) = 0 \qquad ∎$$

DEFINITION 2.18 A *critical* point of a C^1 function $f\colon \mathbf{R}^n \to \mathbf{R}^m$ is a point p for which $Df(p) = 0$.

EXAMPLE 2.50 Find the critical points of

$$f(x, y) = 4 + 6x + 2y + 8x^2 + 10xy + y^2$$

Solution f is a quadratic function,

$$\begin{aligned} f(x, y) &= 4 + [6 \quad 2]\begin{bmatrix} x \\ y \end{bmatrix} + [x \quad y]\begin{bmatrix} 8 & 5 \\ 5 & 1 \end{bmatrix}\begin{bmatrix} x \\ y \end{bmatrix} \\ &= C + BX + X^T AX \end{aligned}$$

By Example 2.48, $Df(x, y) = B + 2AX$. Thus the critical points are the solutions of $B + 2AX = 0$ or

$$AX = -\frac{1}{2}B.$$

The matrix A is invertible and

$$\begin{aligned} X &= -\frac{1}{2}A^{-1}B \\ &= -\frac{1}{2}\cdot\frac{1}{8 - 25}\begin{bmatrix} 1 & -5 \\ -5 & 8 \end{bmatrix}\begin{bmatrix} 6 \\ 2 \end{bmatrix} \\ &= \frac{1}{34}\begin{bmatrix} -4 \\ -14 \end{bmatrix} = -\begin{bmatrix} \frac{2}{17} \\ \frac{7}{17} \end{bmatrix} \end{aligned}$$

The only critical point is -2 7÷17. ■

It is not immediately obvious if the critical point of the last example gives a maximum, a minimum, or neither a maximum nor a minimum. There is a generalization of the second-derivative test for maxima and minima to the present context. First we need to define the second derivative.

Second Derivatives

In this section we will compute second derivatives only for scalar-valued functions $f\colon \mathbf{R}^n \to \mathbf{R}$. In this case the definition is clear. The first derivative $Df(X)$ is a function $Df\colon \mathbf{R}^n \to \mathbf{R}^n$ and we define the second derivative to be

$$D^2 f = D(Df)$$

This definition implies the formula given in the next proposition, which is left as exercise 27.

PROPOSITION 2.24 Let $f: \mathbf{R}^n \to \mathbf{R}$ be of class C^2. The second derivative of f is a symmetric matrix and satisfies

$$D^2f[i;j] = \frac{\partial^2 f}{\partial x_i \, \partial x_j} \qquad \blacksquare$$

This second derivative is also called the *Hessian* matrix. The symmetry comes from the fact that for C^2 functions,

$$\frac{\partial^2 f}{\partial x_i \, \partial x_j} = \frac{\partial^2 f}{\partial x_j \, \partial x_i}$$

This last fact is proved in advanced calculus texts.

EXAMPLE 2.51 Compute the second derivative of the general quadratic function

$$f(X) = C + BX + X^T AX$$

Solution We saw in Example 2.50 that $Df(X) = B + 2AX$, an affine function, hence $D^2f(X) = 2A$. $\blacksquare$

The proof of the next proposition, the second-derivative test, is part of the subject matter of advanced calculus.

PROPOSITION 2.25 (*Second-derivative test*) Let $f: \mathbf{R}^n \to \mathbf{R}$ be of class C^2 and let p be a critical point of $f(Df(p) = 0)$. Assume that the Hessian matrix $D^2f(p)$ is invertible. Let $Q(X) = X +.\times D^2f(p) +.\times X$. Then the behavior of f at p is given by the behavior of Q at $X = 0$. In particular,

1. If Q has a maximum at $X = 0$, then f has a local maximum at $X = p$.
2. If Q has a minimum at $X = 0$, then f has a local minimum at $X = p$.
3. If Q has neither a minimum nor a maximum at $X = 0$, then f has neither a maximum nor a minimum at $X = p$.

If $D^2f(p)$ is singular, the test fails. $\blacksquare$

In general the behavior of the function $Q(X)$ of Proposition 2.25 is not obvious from a casual inspection of the Hessian matrix. Methods for analyzing such functions will be developed in Section 5.2. Those methods are not always needed, however. We can, for example, use Proposition 2.25 to prove the least-squares approximation result, Proposition 2.15.

PROPOSITION 2.26 Let A be an m-by-n matrix with linearly independent columns, B a vector in $\mathbf{R}^m$. The function

$$f(X) = (B - AX) +.* 2$$

has a unique minimum at $X = (A^TA)^{-1}A^TB$, the unique solution of $A^TAX = A^TB$.

Proof By Example 2.49 and the following remark, the critical points are given by $A^T(B - AX) = 0$ or

$$A^TAX = A^TB$$

If A has linearly independent columns, then A^TA is invertible, so that the only critical point is $X = (A^TA)^{-1}A^TB$. Since $Df(X) = -2A^T(B - AX) = -2A^TB + 2A^TAX$, we have $D^2f(X) = 2A^TA$. Thus

$$\begin{aligned} Q(X) &= 2X^T(A^TA)X \\ &= 2(AX) +.\times AX \\ &= 2(AX) +.* 2 \\ &\geq 0 \end{aligned}$$

and $Q(0) = 0$. Thus f has a minimum at $B⌹A$. ∎

In Proposition 2.25 $Q(0) = 0$. It follows that $Q(X)$ has a maximum at $X = 0$ if and only if $Q(X) \leq 0$ for all X and $Q(X)$ has a minimum at $X = 0$ if and only if $Q(X) \geq 0$ for all X.

EXAMPLE 2.52 Discuss the critical points of the function

$$f(x, y) = 3 + 4x - 6y + 2xy$$

Solution The function $f(x, y)$ is a quadratic function. Applying Example 2.51, we have

$$Df(x, y) = \begin{bmatrix} 4 + 2y \\ -6 + 2x \end{bmatrix}, \qquad D^2f(x, y) = 2\begin{bmatrix} 0 & 1 \\ 1 & 0 \end{bmatrix}$$

There is a single critical point at $(3, -2)$ and

$$Q(x, y) = 4xy$$

Since $Q(1, 1) = 4 > 0$ and $Q(1, -1) = -4 < 0$, the function $Q(X)$ has neither max nor min at $X = 0$, hence the function $f(x, y)$ has neither max nor min at point $p = (3, -2)$. ∎

EXERCISES 2.5*

In exercises 1 through 10 compute the derivatives of the given functions.

1. $f(x, y) = x + y + 1$
2. $f(x, y) = x^2 + y + 1$
3. $f(x, y, z) = (x + y, x - z)$
4. $f(x, y, z) = (\cos x, \sin y, \tan z)$
5. $f(t) = (\cos t, \sin t, t)$
6. $f\colon \mathbf{R}^n \to \mathbf{R}^n$ by $f(X) = 3X$
7. $f\colon \mathbf{R}^n \to \mathbf{R}$ by $f(X) =$ X+.*3
8. $f\colon \mathbf{R}^4 \to \mathbf{R}^6$ by $f(X) =$,⍉ 2 3 ⍴ X
9. $f\colon \mathbf{R}^n \to \mathbf{R}^n$ by $f(X) =$ 1⍉X.
10. $f\colon \mathbf{R}^n \to \mathbf{R}^m$ by $f(X) =$ X +.× A, A an n-by-m matrix
11. Let α be a real number. Show that D(X+.*α) is α×X*α − 1.
12. Let α be a real number. Show that D(X*α) is the matrix α×I×(2⍴X)⍴X*α − 1, where I is an identity matrix.
13. Show that D(*X) = I×(*X)∘.×X*0, where I is an identity matrix.
14. Use the chain rule and exercise 11 to compute the derivative of $f(X) =$ (B + AX)+.*α, α a real number.
15. Use the chain rule to compute the derivative of $f(X) = \ln$ (X+.*2).

In exercises 16 through 23 discuss the critical points of the given functions.

16. $f(x, y) = 1 + x^2 + y^2$
17. $f(x, y) = 1 + x^2 - y^2$
18. $f(x, y, z) = x^2 + y^2 - z^2 + x + 4$
19. $f(x, y) = x^3 + y^3$
20. $f(x, y) = x^4 + y^4 - 2(x^2 + y^2)$
21. $f(X) =$ X+.*3
22. $f(x, y) = xy$
23. $f(x, y) = xe^y + ye^x$ at $p = (-1, -1)$
24. Prove statement 1 of Proposition 2.21.
25. Prove statement 2 of Proposition 2.21.
26. Let A be m by n and let $m =$ ⍴V. Show that

(V +.× A) = (⍉A) +.× V

In particular, if A is symmetric, (V +.× A) = A +.× V.

27. Let $f\colon \mathbf{R}^n \to \mathbf{R}$. Show that $D^2f[i; j] = \partial^2 f/\partial x_i\, \partial x_j$.

2.6* Multivariate Calculus Linearization, Newton's Method

DEFINITION 2.19 Let $f\colon \mathbf{R}^n \to \mathbf{R}^m$ be a function of class C^1. The *affine approximation* or *linearization* of f at point P in $\mathbf{R}^n$ is the affine function

$$A(X) = f(p) + Df(p)(X - p)$$

Notice that $A(p) = f(p)$ and $DA(p) = Df(p)$. The function A is the unique affine transformation with the same value at p as f and the same derivative at p as f.

In advanced calculus texts the Taylor expansion of f at p is defined. The affine approximation here defined coincides with the first two terms of the Taylor expansion. For real-valued functions we can write one more term in matrix notation. The next proposition is easily checked if one has the definition of higher-dimensional Taylor series at hand.

PROPOSITION 2.27 Let $f\colon \mathbf{R}^n \to \mathbf{R}$ be a function of class C^2 at p in $\mathbf{R}^n$. The first three terms of the Taylor expansion of f are

$$Q(X) = f(p) + Df(p)(X - p) + \frac{1}{2!}(X - p)^T D^2 f(p)(X - p) \qquad \blacksquare$$

Generally speaking, a nonlinear problem is "linearized" by replacement of some function f by its affine approximation A at some point p. We will give two examples of this: Newton's method for approximating roots and the linearization of a differential equation.

Newton's Method

Suppose that one has n nonlinear equations in n unknowns to solve. In Example 2.35 it was shown that such a problem can be put in the form

$$f(X) = 0, \qquad f\colon \mathbf{R}^n \to \mathbf{R}^n$$

Further, if $f(X) = B + AX$ and A is invertible, then we can solve $f(X) = 0$. The answer is $X = -B \boxminus A$.

Newton's method seeks to approximate a solution ever more closely, starting with an initial (we hope educated) guess X_0.

Given an approximation X_n to a solution of $f(X) = 0$, the next approximation is the root of the affine approximation to f at X_n.

$$\begin{aligned}
A(X) &= f(X_n) + Df(X_n)(X - X_n) \\
&= 0 \\
Df(X_n)(X - X_n) &= -f(X_n) \\
X - X_n &= (Df(X_n))^{-1} f(X_n) \\
X_{n+1} &= X \\
&= X_n - (Df(X_n))^{-1} f(X_n) \\
&= X_n - f(X_n) \boxminus Df(X_n)
\end{aligned}$$

It is shown in numerical analysis texts that X_n converges to a root of f if the initial approximation is close enough.

If $f\colon \mathbf{R} \to \mathbf{R}$, then the graph of $A(x) = f(x_n) + f'(x_n)(x - x_n)$ is a line tangent to the graph of $y = f(x)$ at the point $(x_n, f(x_n))$. This is illustrated in Figure 2.2a. Point x_{n+1} is where this line crosses the axis. Problems with the method are illustrated in Figures 2.2b and 2.2c.

It is impractical to carry out actual calculations until the techniques of recur-

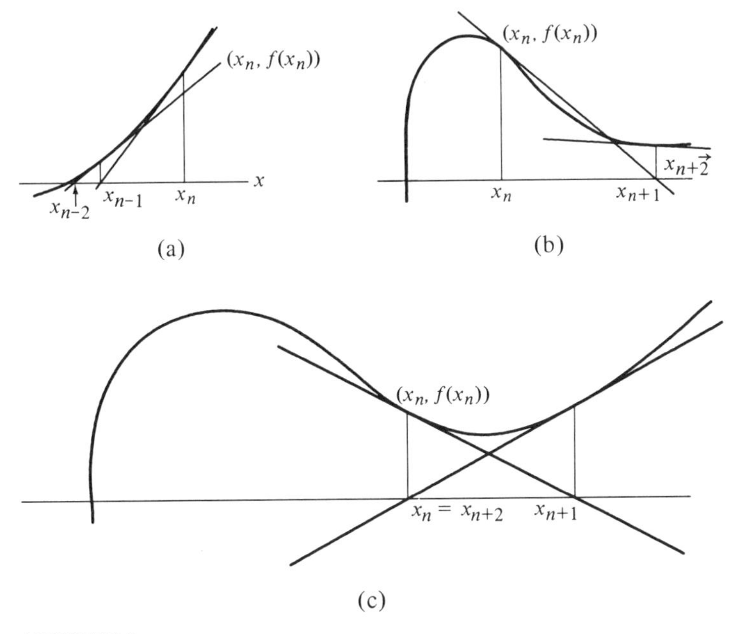

FIGURE 2.2

sive function writing in APL are covered in Chapter 3. The calculations for the next example will be found in Section 3.7*.

EXAMPLE 2.53 Consider the system of simultaneous nonlinear equations

$$\frac{x^2}{16} + \frac{y^2}{4} = 1$$
$$x^2 - y^2 = 1$$

These equations have been put in canonical form for easy recognition. The first is an ellipse with center at (0, 0), x intercepts ± 4, and y intercepts ± 2. The second is a hyperbola with asymptotes $y = \pm x$ and x intercepts ± 1 (see Figure 2.3). There are four solutions, placed symmetrically around the origin: $(\pm 2, \pm\sqrt{3})$.

Clearing the denominators in the first equation, we get the equivalent system

$$x^2 + 4y^2 - 16 = 0$$
$$x^2 - y^2 - 1 = 0$$

or $f(x, y) = 0$, where $f\colon \mathbf{R}^2 \to \mathbf{R}^2$ by

$$f(x, y) = \begin{bmatrix} x^2 + 4y^2 - 16 \\ x^2 - y^2 - 1 \end{bmatrix}$$

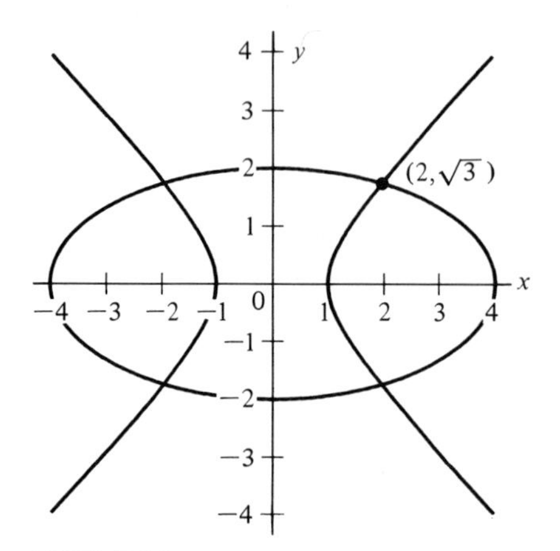

FIGURE 2.3

Thus

$$Df(x, y) = \begin{bmatrix} 2x & 8y \\ 2x & -2y \end{bmatrix}$$

Starting with $X_0 = (1, 1)$, we have

$$\begin{aligned} X_1 &= X_0 - Df(X_0)^{-1}f(X_0) \\ &= \begin{bmatrix} 1 \\ 1 \end{bmatrix} - \begin{bmatrix} 2 & 8 \\ 2 & -2 \end{bmatrix}^{-1} \begin{bmatrix} -11 \\ -1 \end{bmatrix} \\ &= \begin{bmatrix} 1 \\ 1 \end{bmatrix} + \frac{1}{20} \begin{bmatrix} -2 & -8 \\ -2 & 2 \end{bmatrix} \begin{bmatrix} -11 \\ -1 \end{bmatrix} \\ &= \begin{bmatrix} 1 \\ 1 \end{bmatrix} + \frac{1}{20} \begin{bmatrix} 30 \\ 20 \end{bmatrix} \\ &= \begin{bmatrix} 2.5 \\ 2 \end{bmatrix} \end{aligned}$$

Continuing in this manner, we obtain the following table (rounded off to three digits) (this table is computed in Chapter 3, Section 3.7*):

n	x_n	$x_n - (2, \sqrt{3})$
0	(1, 1)	(−1, −.732)
1	(2.5, 2)	(.5, .268)
2	(2.05, 1.75)	(.05, .0179)
3	(2.00, 1.73)	$(6.1E - 4, 9.2E - 5)$
4	(2.00, 1.73)	$(9.29E - 8, 2.45E - 9)$
5	(2.00, 1.73)	$(2.17E - 15, 5.2E - 18)$

If one starts with $X_0 = (.1, .1)$, one first jumps away to (20, 15.1) and then comes back within about 10^{-15} of the answer in ten steps.

Notice that

$$(Df(x, y))^{-1} = \frac{1}{20xy}\begin{bmatrix} -2x & -8y \\ 2x & 2y \end{bmatrix}$$

does not exist on the x axis ($y = 0$) or the y axis ($x = 0$) and hence the method breaks down here. ■

EXAMPLE 2.54 We will use Newton's method to derive an algorithm for estimating square roots on a calculator with only the four functions $+, -, \times, \div$.

To find the square root of a is to find a root of $f(x) = x^2 - a$. Here f: $\mathbf{R} \to \mathbf{R}$, and so $Df(x) = f'(x)$ and the Newton iteration becomes

$$x_{n+1} = x_n - \frac{f(x_n)}{f'(x_n)} = x_n - \frac{x_n^2 - a}{2x_n}$$

or

$$x_{n+1} = \frac{1}{2}\left(\frac{a}{x_n} + x_n\right)$$

For $a = 2$ we have, starting with $x_1 = 1$,

n	x_n	x_n^2
1	1	1
2	1.5	2.25
3	1.41667	2.007
4	1.4142156	2.000057
5	1.4142135	1.9999998

For a calculator with a single memory this calculation can be quite rapid. First store x_1 in the memory, then cycle through the sequence: a, $\div$, memory recall, $+$, memory recall, $\div$, 2, memory store. After each cycle, x_{n+1} is stored in the memory. ■

We will return to Newton's method in Chapter 3, where it will be automated.

DIFFERENTIAL EQUATIONS

A standard technique used for studying differential equations is *linearization*.

Suppose that a differential equation has been put in the standard form

$$\frac{d}{dt} Y = f(Y), \qquad f\colon \mathbf{R}^n \to \mathbf{R}^n$$

All differential equations can be put in this form. (This is illustrated in the examples below. If t appears explicitly, however, then there are no critical points — see exercise 19.)

Notice first that if Y_0 is a root of the equation

$$f(Y) = 0$$

then $Y(t) \equiv Y_0$ is a solution of the differential equation, because

$$\frac{d}{dt} Y_0 = 0 = f(Y_0)\dagger$$

Points Y_0 are called *critical points* of the differential equation. We study the behavior of the differential equation, *near* Y_0, by solving the differential equation

$$\frac{d}{dy} Y = a(Y)$$

where $a(Y)$ is the affine approximation of f at Y_0.

Now $a(Y) = f(Y_0) + Df(Y_0)(Y - Y_0) = Df(Y_0)(Y - Y_0)$, since $f(Y_0) = 0$. Introduce the new variable $H = Y - Y_0$ or $Y = Y_0 + H$. Then

$$\frac{d}{dt} Y = \frac{d}{dt}(Y_0 + H) = \frac{d}{dt} Y_0 + \frac{d}{dt} H = \frac{d}{dt} H$$

and so one has

$$\frac{d}{dt} H = Df(Y_0)H$$

This is a *linear* differential equation. A technique for solving linear differential equations is given in Chapter 7, Section 7.6*.

EXAMPLE 2.55 Find the critical points of the differential equation

$$\frac{dx}{dt} = ax - x^2$$

Write the linearized differential equation for each critical point.

†Recall that if

$$y = \begin{bmatrix} y_1 \\ \vdots \\ y_n \end{bmatrix} \quad \text{then} \quad \frac{d}{dt} y = \begin{bmatrix} dy_1/dt \\ \vdots \\ dy_n/dt \end{bmatrix}$$

Solution The critical points are the solutions of

$$f(x) = ax - x^2 = 0$$

or $x = 0$, $x = a$. In this case $f: \mathbf{R} \to \mathbf{R}$, and $Df(x) = f'(x) = a - 2x$.
$x = 0$: $h = x - 0 = x$, and the linearized equation is

$$\frac{dh}{dt} = Df(0)h = ah \qquad \text{(the solution is } ce^{at}\text{)}$$

$x = a$: $h = x - a$, and the linearized equation is

$$\frac{dh}{dt} = Df(a)h = -ah \qquad \text{(the solution is } ce^{-at}\text{)} \qquad \blacksquare$$

EXERCISES 2.6*

In exercises 1 through 5 find the affine approximation to the given function f at the point p indicated. For functions $\mathbf{R}^n \to \mathbf{R}$ and $\mathbf{R} \to \mathbf{R}^n$ take Df to be a single-row matrix and single-column matrix, respectively.

1. $f(x, y) = (x^2 + y^2, x^2 - y^2)$, $p = (0, 0)$, $p = (1, 1)$
2. $f(x, y, z) = (xy, xz, yz)$, $p = (0, 0, 0)$, $p = (1, 1, 1)$
3. $f(X) = B + AX$, any p
4. $f(X) = X+.*2$, any p
5. $f(t) = (\cos t, \sin t, t)$, $p = 0$, $p = 2\pi$

Let $f: \mathbf{R} \to \mathbf{R}^n$ for $n = 2, 3$. Then the image of f is a curve in $\mathbf{R}^2$ or $\mathbf{R}^3$. The derivative $Df(P)$ is a vector with n components, and in this case we take the affine approximation to be $A(t) = f(p) + tDf(p)$. The image of $A(t)$ is a straight line. It is the line tangent to the curve at point $f(p)$. In exercises 6 through 10 sketch the curves and the lines tangent to them at the point indicated.

6. $f(t) = (t, t^2)$, $p = f(1) = (1, 1)$
7. $f(t) = (\cos t, \sin t)$, $p = f(\pi/4) = (1, 1) \div \sqrt{2}$
8. $f(t) = t(\cos t, \sin t)$, $p = f(\pi/2)$
9. $f(t) = (\cos t, \sin t, t)$, $p = f(0)$
10. $f(t) = (\cosh t, \sinh t)$, $p = f(0)$

In exercises 11 through 15 a system of simultaneous nonlinear equations is given. Using Newton's method, compute X_1 from the given X_0. (A computer is necessary for exercises 14 and 15.)

11. $x^2 + 4y^2 - 16 = 0$
    $x^2 - y^2 - 1 = 0$, $x_0 = (.1, .1)$
12. $x^2 - y = 0$
    $x + y = 1$, $x_0 = (0,1)$
13. $\tan x = x$, $x_0 = 1$
14. $x^2 + y^2 + z^2 = 1$
    $x^2 + y^2 - z^2 = 0$
    $x + y = 1$, $x_0 = (1, 0, 1)$

15. $$\begin{aligned} x^2 - y^2 + z^2 - w^2 &= 4 \\ \sin(xyzw) &= 0 \\ x + y + z + w &= 0 \\ x + y \ln zw &= 4, \quad \mathbf{x}_0 = (1, 2, 3, 4) \end{aligned}$$

In exercises 16 through 18 find the critical points of the given differential equation. Write the linearized differential equation for each critical point.

16. $$\frac{dx}{dt} = x(y - 1)$$
$$\frac{dy}{dt} = y(x - 1)$$

17. $$\frac{d^2z}{dt^2} = z^2 - 2z + 1$$

18. $$\frac{dx}{dt} = x - x^2$$

19. Write the differential equation $d^2x/dt^2 = tx$ in vector form by taking $y_1 = t$, $y_2 = x$, $y_3 = dx/dt$. Show that there are no critical points.

20. Devise an algorithm to compute cube roots on a four-function calculator with memory.

CHAPTER THREE

Systems of Linear Equations

In this chapter we will develop the theory of general systems of linear equations. The tool we will use is the row-echelon form of a matrix. Until now we have been unable to solve systems of linear equations for which the columns of the coefficient matrix are linearly dependent. The solutions of such systems can be read off from the row echelon form of the augmented matrix of the system. The solution technique, known as *Gaussian reduction,* is developed in Section 3.2. The theory associated with the echelon form of a matrix is developed in Section 3.3, and the solution process is coded for machine execution in Section 3.5.

The APL necessary to automate Gaussian reduction is developed in Sections 3.1 and 3.4. The rest of the sections are optional; they will not be needed later in the text. The APL techniques developed in Section 3.4 make it easy to compute high powers of a matrix, and in Section 3.6* we sketch two applications: linear difference equations and Markov chain matrices. In Section 3.7* we give some indication of what can be done with *nonlinear* equations. This section includes coding that makes the Newton's method algorithm of Section 2.6* much easier to use. Finally, in Section 3.8* we give an example of what can be done when the automatic Gaussian reduction developed in Section 3.5 does not work well.

In Chapter 2, two important results were left unproved: (1) the result that a matrix has a left inverse if and only if it has linearly independent columns (Proposition 2.9) and (2) the fact that an invertible matrix must be square (Proposition 2.5). These statements are proved in Section 3.3.

3.1 APL Functional Notation

Recall the general definition of a function.

DEFINITION 3.1 A *function, f,* is a rule (i.e., an unambiguous procedure) that assigns to each element of a set D, called the *domain* of the function, a unique element of a set R, called the *range* of the function. If y in R is assigned to x in D, we write $y = f(x)$.

In elementary mathematics the sets D and R consist of real numbers and the functions are usually given by a formula such as $y = f(x) = \sin x$ or $y = f(x) = \sqrt{1 - x^2}$. We will have need of much more general types of functions and in fact have already encountered them. For example, the transpose function ⍉ assigns to each matrix A a unique matrix A^T. The domain of ⍉ is the set of all matrices, and the range of ⍉ is also the set of all matrices. Another example is the index generator ⍳. The domain of ⍳ is the set of nonnegative integers (0, 1, 2, 3, . . .) and the range is, say, the set of all vectors.

The range of a function is to some extent a matter of choice. The smallest possible range is called the *image*. The image consists of all y in R of the form $y = f(x)$ for some x in the domain D. Thus, since A = ⍉ ⍉A, the image of ⍉ is the set of all matrices, but the image of ⍳ is much smaller than the set of all vectors.

Procedures for defining and using a large class of functions are part of APL and we will make constant use of this fact from now on.

For example, we will often need identity matrices of various sizes. Recall that the n-by-n identity matrix is defined by $I[j; k] = 1$ if $j = k$ and is zero otherwise. That is, $I \leftarrow J$ ∘.= K where $J \leftarrow K \leftarrow$ ⍳n.

It becomes tedious to go through this procedure each time. We need a function, call it ID rather than f, for which $ID(n)$ is the n-by-n identity matrix. This function is defined in APL by

```
      ∇ Z←ID N
[1]       Z←(⍳N)∘.=⍳N
      ∇
```

Function definitions are set off by a pair of dels (∇). The first line, which is unnumbered, is called the *header*. In the example above the header is `Z←ID N`. The header indicates the way the function is to be used. The *function name* is ID, the *argument* is N, and the *result* is Z. In conventional mathematical notation this header would be $Z = ID(N)$.

The numbered lines of the function — there is only one line in this example — define the result, Z, in terms of the given argument, N.

Notice that parentheses are not needed in APL. The expressions `ID(3)` and `ID 3` produce the same result.

```
        ID 3
1    0    0
0    1    0
0    0    1
```

To enter a function such as ID one types a del followed by the header. The del switches APL from calculator mode to function definition mode.

The machine replies by giving the line number [1], and the user types the first line. The machine continues to supply line numbers until the user types a ∇ instead of a new line. The second ∇ switches the machine back to calculator mode, and the function is ready for use. (Procedures for correcting mistakes, deleting lines, and inserting lines vary from machine to machine.)

Note: The function *ID* will be used from now on. A copy should be saved in your machine's workspace file. Useful functions should be added to this file as you work through the text.

Example 3.1 Use APL notation to define the following function.

Function name: *NORM*
Function input: Any vector *V*
Function output: The square root of the sum of the squares of the components of *V*

Solution

```
      ∇Z←NORM V
[1]   Z←(V+.*2)*÷2
[2]   ∇
        NORM 1 2 3
3.742
```

■

Example 3.2 Use APL notation to define the following function:

Name: *TOTWO*
Input: Any nonnegative integer n

Output: $\sum_{k=0}^{n} \left(\frac{1}{2}\right)^k$

Solution

```
      ∇T←TOTWO N
[1]   T←.5+.*0,⍳N
[2]   ∇

        TOTWO 3
1.875
        TOTWO 5
1.96875
        TOTWO 10
1.999023438
        TOTWO 100
2
```

$\left[\text{The correct answer is } 2 - \left(\frac{1}{2}\right)^{99}.\right]$ ■

In more complex computations it is often convenient to temporarily store intermediate results. For example, suppose we wish to approximate e^x by the first

26 terms of the McLaurin expansion of e^x:

$$\sum_{k=0}^{25} \frac{x^k}{k!}$$

The most convenient way to compute this is

```
K←0 , ι25
+/(X*K)÷!K
```

An APL function that does this is

```
    ∇ Y←EXP X;K
[1]    K←ι25
[2]    Y←+/(X*K)÷!K
    ∇
```

The variable K is called a *local variable.* The local variable exists only while the function is running. When the function finishes, the local variable disappears. We define the local variables by listing them after the argument, separated by semicolons, as in the expression

```
∇ Z←FCN X;A;B;C
```

which is the header of a function with three local variables A, B, and C.

EXAMPLE 3.3 In Section 1.3 we gave the computations necessary to change a matrix of raw scores to a matrix of Z-scores. Write an APL function, *Z-SCORE,* which takes a matrix of raw scores as an argument and returns a matrix of standard scores.

Solution

```
    ∇ Z←ZSCORE A ;MEANS;SD
[1]    MEANS←(+⌿A)÷(ρA)[1]
[2]    A←A-(ρA)ρMEANS
[3]    SD←((+⌿A*2)÷(ρA)[1])*÷2
[4]    Z←A÷(ρA)ρSD
    ∇

       A
  60   120
  61   120
  62   135
  63   135
```

```
64   130
65   135
66   150
68   140
69   170
70   145
71   160
72   160
74   160
75   160
76   175

      ZSCORE A
¯1.530 ¯1.570
¯1.340 ¯1.570
¯1.140 ¯0.675
¯0.939 ¯0.675
¯0.741 ¯0.973
¯0.542 ¯0.675
¯0.344  0.219
 0.052 ¯0.377
 0.251  1.410
 0.450 ¯0.079
 0.648  0.814
 0.847  0.814
 1.240  0.814
 1.440  0.814
 1.640  1.710
```

This is the same answer we obtained in Section 1.3. ■

The argument and result of a function are also local variables and are not confused with workspace variables of the same name. In the last example the local variable *A* is changed on line [2] of *ZSCORE*, but the original matrix of raw data, also called *A*, is not changed.

The examples above are monadic functions. Dyadic function definitions are also permitted. The header of a dyadic function takes the form

```
∇ Z←X FCN Y
```

and corresponds to the standard mathematical notation $z = \text{fcn}(x, y)$. The variable *X* is the *left argument* and the variable *Y* is the *right argument*. Local variables are defined as with monadic functions.

```
∇ Z←X FCN Y;A;B;C
```

The next example gives a function that is very useful in a variety of contexts.

EXAMPLE 3.4 Given an interval $[a, b]$ and a number n, we wish to divide the interval $[a, b]$ into n subintervals of equal length with endpoints $a = x_0 < x_1 < x_2 < \cdots < x_n = b$. Notice that there are $n + 1$ endpoints.

Name:	*CHOP*
Left argument:	The number of subintervals
Right argument:	The endpoints of the interval to be subdivided — a vector with two components: (a, b)
Result:	The vector of endpoints of the subintervals

Solution If $[a, b]$ is to be divided into n subintervals, then the length of each of these subintervals is $h = (b - a) \div n$ and the endpoints are $x_k = a + kh$, $k = 0, 1, 2, \ldots, n$.

In APL notation this becomes

```
K←0,⍳N
X←A+K×(B−A)÷N
```

Hence the function:

```
    ∇ X←N CHOP AB
[1]    X←AB[1]+(0,⍳N)×−−/AB÷N
    ∇

       10 CHOP 0 1
0  0.1  0.2  0.3  0.4  0.5  0.6  0.7  0.8  0.9  1
```
■

The function *CHOP* is useful for graphing. Suppose we wish to plot $y = f(x) = x^3 - 1$ on the interval $[-1, 1]$.

```
       +X←10 CHOP ¯1 1
¯1  ¯0.8  ¯0.6  ¯0.4  ¯0.2  ¯1.73E¯18  0.2  0.4  0.6  0.8  1
       +Y←¯1+X*3
¯2  ¯1.51  ¯1.22  ¯1.06  ¯1.01  ¯1  ¯0.992  ¯0.936  ¯0.784  ¯0.488
      ¯5.2E¯18
```

(The numbers `X[6]` and `Y[11]` above are zero plus machine error.) These points can now be plotted by hand or passed on to a plotting program.

The polynomial $x^3 - 1$ has an especially simple form. For a general polynomial a slightly more elaborate expression of Y in terms of X is needed.

EXAMPLE 3.5

Name:	AT
Left argument:	Vector of coefficients of a polynomial, $p(x) = a_0 + a_1x + \cdots + a_nx^n$, in the order $a_0, a_1, \ldots, a_n$

Right argument: Vector of x values $x_1, x_2, \ldots, x_m$
Result: Vector of values $y_i = p(x_i)$

Solution $y_i = p(x_i) = a_0 + a_1x_i + a_2x_i^2 + \cdots + a_nx_i^n = A[i;] +.\times P$ where P is the vector of coefficients and A is the matrix with $i;j$ component equal to x_i^j. Notice that the degree of the polynomial p is $^{-}1 + \rho P$.

```
    ∇ Y←P AT X
[1]    Y←(X∘.*¯1+⍳⍴P)+.×P
    ∇
```

The polynomial $x^3 - 1$ of Example 3.4 is represented by the vector `¯1 0 0 1`.

```
        ¯1 0 0 1 AT 10 CHOP ¯1 1
¯2  ¯1.51  ¯1.22  ¯1.06  ¯1.01  ¯1  ¯0.992  ¯0.936  ¯0.784  ¯0.488
     ¯5.2E¯18
```

Notice that the function AT also works when X is a single scalar. In this case `X∘.*¯1+⍳⍴P` is a vector. ■

Defined functions may call other defined functions without limitation, and this is very useful. It means that a function need be defined only once — not redefined in the body of other functions that may use it.

Consider, for example, the function $ZSCORE$ defined in Example 3.3 above. On line `ZSCORE [1]` the average of the row measurements is computed. A function to compute means is useful in itself.

```
    ∇ M←AVE A
[1]    M←(+/A)÷(⍴A)[1]
    ∇
```

Similarly, a function for computing standard deviations is useful. The standard deviation squared is just the average of the squared differences from the mean.

```
    ∇ S←SD A
[1]    S←(AVE(A−(⍴A)⍴AVE A)*2)*÷2
    ∇
```

The function $ZSCORE$ may now be rewritten

```
    ∇ Z←ZSCORE A
[1]    Z←(A−(⍴A)⍴AVE A)÷(⍴A)⍴SD A
    ∇
```

Using the same example as before:

```
      A
60    120
61    120
62    135
63    135
64    130
65    135
66    150
68    140
69    170
70    145
71    160
72    160
74    160
75    160
76    175

         AVE A
67.7     146

         SD A
5.04     16.8

         ZSCORE A
¯1.530  ¯1.570
¯1.340  ¯1.570
¯1.140  ¯0.675
¯0.939  ¯0.675
¯0.741  ¯0.973
¯0.542  ¯0.675
¯0.344   0.219
 0.052  ¯0.377
 0.251   1.410
 0.450  ¯0.079
 0.648   0.814
 0.847   0.814
 1.240   0.814
 1.440   0.814
 1.640   1.710
```

These are, of course, the same results that we obtained before. The new version of *ZSCORE,* however, is clearer, and we also have two other useful functions, *AVE* and *SD.*

EXERCISES 3.1

In exercises 1 through 7 a function $y = f(x)$ is given. Define a monadic APL function with

Argument: A vector $X = (x_1, \ldots, x_n)$
Result: The vector $Y = (y_1, \ldots, y_n)$ with $y_i = f(x_i)$

A sample table of values is given for each function so that the answer may be checked at a terminal, if one is available.

1. $f(x) = 1 + x$

x	1	2	3
y	2	3	4

2. $f(x) = \dfrac{1}{1 - x + 4x^5}$

x	0	1	200
y	1	.25	$7.8E - 13$

3. $f(x) = \cos(x) + \sin(x)$

x	1	2	3
y	1.4	.49	$-.85$

4. $f(x) = \sqrt{1 + x^2 - 6x^3}$

x	-3	-2	-1	0
y	13.1	7.28	2.83	1

5. $f(x) = e^{-x^2}$

x	0	.1	.5
y	1	.99	.779

6. $f(x) = \sqrt{1 - \sinh^2 x}$

x	0	.4	.8
y	1	.91	.46

7. $f(x) = \log_3 \sqrt{(x^2 + 1)/(x^2 - 1)}$

x	2	3	4
y	.23	.10	.057

In exercises 8 through 11 a function $z = f(x, y)$ is given. Define a dyadic APL function with

Left argument: A vector $X = (x_1, x_2, \ldots, x_n)$
Right argument: A vector $Y = (y_1, y_2, \ldots, y_n)$
Result: The vector $z = (z_1, z_2, \ldots, z_n)$ with $z_i = f(x_i, y_i)$

A sample table of values is given for each function so that the answer may be checked at a terminal if one is available.

8. $f(x, y) = x^2 + y^2 - 1$

x	1	2	3
y	4	5	6
z	16	28	44

9. $f(x, y) = x^2 + \cos y + 10$

x	1	2	3
y	4	5	6
z	10.35	14.28	19.96

10. $f(x, y) = \sin y \cos x$

x	1	2	3
y	4	5	6
z	$-.41$	.40	.28

11. $f(x, y) = x^y - y^x$

x	1	2	3	4
y	4	5	6	π
z	-3	7	513	-19.5

12. A triangle with base b and height h has area $\frac{1}{2}bh$.

(a) Define a dyadic APL function:

Name: *AREA*
Left argument: A vector of bases $(b_1, b_2, \ldots, b_n)$
Right argument: A vector of heights $(h_1, h_2, \ldots, h_n)$
Result: The vector of areas $\frac{1}{2}b_i h_i$

(b) Define a dyadic APL function:

Name: *TABLE*
Left argument: A vector of bases $(b_1, b_2, \ldots, b_n)$
Right argument: A vector of heights $(h_1, h_2, \ldots, h_n)$
Result: A table (matrix) of values where the entry in the $i; j$ position is the area of a triangle with base b_i and height h_j

13. Define a monadic APL function with

Name: *TRIGTABLE*
Argument: A vector of angles $(\theta_1, \theta_2, \ldots, \theta_n)$ in radians
Result: A table (matrix) of values of the functions sin, cos, tan, sec, csc, cot, at the angles $\theta_1, \ldots, \theta_n$

14. If a triangle has sides of length a, b, c, then the area of the triangle is $\sqrt{s(s-a)(s-b)(s-c)}$, where $s = \frac{1}{2}(a + b + c)$.

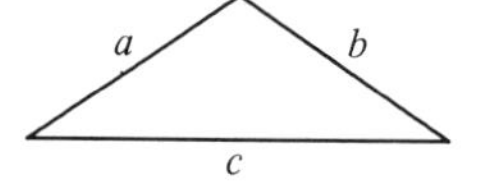

(a) Define a monadic APL function with

Name: *AREA2*
Argument: The vector of side lengths (a, b, c)
Result: The area of the triangle with sides a, b, c

(b) Define a monadic APL function with

Name: *AREA3*
Argument: A matrix T with three columns.
Result: The vector with ith component equal to the area of a triangle with side lengths equal to the components of $T[i;]$.

15. (a) Define a function for making least-squares polynomial fits to measured data. The function should have

Name: *DEGFIT*
Left argument: The degree of the polynomial to be fitted, a nonnegative integer

Right argument: A matrix consisting of the data points. The first column consists of the x coordinates of the measured points and the second column of the corresponding y coordinates

Result: The coefficients of the least-squares polynomial for the data, constant term first. (See Proposition 2.16.)

(Computer assignment) Use the function of part (a) to do the following exercises from Section 2.3.

(b) Exercise 7. (c) Exercise 8. (d) Exercise 10.
(e) Exercise 11. (f) Exercise 12.

16. Let $x_1, \ldots, x_n$ be a set of measurements with mean $\bar{x}$. The rth moment about the mean is defined to be the number

$$m_r = \frac{(x_1 - \bar{x})^r + (x_2 - \bar{x})^r + \cdots + (x_n - \bar{x})^r}{n}$$

Here, m_2 is the variance (m_3 is called *skewness* and m_4 is *kurtosis*). Using the function *AVE* defined after Example 3.5, define a dyadic APL function with

Name: *MOM*
Right argument: r
Left argument: The vector $(x_1, \ldots, x_n)$
Result: m_r
Test data: For data vector 2 3 7 8 10 one has variance = 9.2, skewness = −3.6, and kurtosis = 122.

Simpson's rule is an approximation for the integral

$$\int_a^b f(s)\, ds$$

given in most elementary calculus books.

Divide the interval $[a, b]$ into n subintervals ($X \leftarrow n$ *CHOP* a, b), where n is even, and define Y by $Y[i] = f(X[i])$. Let h be $(b - a)/n$. The approximation given by Simpson's rule is written conventionally as

$$\frac{h}{3}(y_1 + 4y_2 + 2y_3 + 4y_4 + 2y_5 + \cdots + 4y_{n-2} + 2y_{n-1} + 4y_n + y_{n+1})$$

or

```
H×((1,((¯2+⍴Y)⍴4 2),1)+.×Y)÷3
```

17. Define a monadic APL function with

Name: *SIN*
Argument: A number x
Result: The Simpson's rule approximation to $\int_0^x \cos u\, du$ with 500 subintervals

(Computer assignment) How closely does *SIN* x approximate $\sin(x)$ for $x = 0, \pi/2, \pi, 2\pi$?

3.2 Solving General Linear Equations

In Chapter 2 we discussed the equation

$$AX = B$$

under the assumption that A possessed a left inverse. In that case we saw that the only possible solution was given by $X = (A^TA)^{-1}A^TB$ or B ⌹ A. No algorithm for computing $(A^TA)^{-1}$ was given for A with more than two columns, however.

In this section we will develop a procedure — Gaussian reduction to echelon form — to solve $AX = B$ for general A and to compute a left inverse when one exists.

As an introduction to the method, consider the following system of equations

$$\begin{aligned} x \phantom{{}+ y} + 2z &= 4 \\ -x + y + z &= 1 \\ x + 2y + 8z &= 14 \end{aligned} \qquad \text{or} \qquad \begin{bmatrix} 1 & 0 & 2 \\ -1 & 1 & 1 \\ 1 & 2 & 8 \end{bmatrix} X = \begin{bmatrix} 4 \\ 1 \\ 14 \end{bmatrix}$$

If we add the first equation to the second and subtract the first equation from the third, we obtain

$$\begin{aligned} x + 2z &= 4 \\ y + 3z &= 5 \\ 2y + 6z &= 10 \end{aligned} \qquad \text{or} \qquad \begin{bmatrix} 1 & 0 & 2 \\ 0 & 1 & 3 \\ 0 & 2 & 6 \end{bmatrix} X = \begin{bmatrix} 4 \\ 5 \\ 10 \end{bmatrix}$$

If next we multiply the second equation by -2 and add the result to the third, we obtain

$$\begin{aligned} x + 2z &= 4 \\ y + 3z &= 5 \\ 0 &= 0 \end{aligned} \qquad \text{or} \qquad \begin{bmatrix} 1 & 0 & 2 \\ 0 & 1 & 3 \\ 0 & 0 & 0 \end{bmatrix} X = \begin{bmatrix} 4 \\ 5 \\ 0 \end{bmatrix}$$

The solutions of the last set of equations are easily described. There are an infinity of solutions. If z is given any arbitrary value, say t, then the values of the other variables are determined.

$$\begin{aligned} x &= 4 - 2t \\ y &= 5 - 3t \\ z &= t \end{aligned} \qquad \text{or} \qquad X = \begin{bmatrix} 4 \\ 5 \\ 0 \end{bmatrix} + t \begin{bmatrix} -2 \\ -3 \\ 1 \end{bmatrix}$$

Does this give the solutions of the original set of equations? In this case it is easy to check:

$$\begin{bmatrix} 1 & 0 & 2 \\ -1 & 1 & 1 \\ 1 & 2 & 8 \end{bmatrix} \left(\begin{bmatrix} 4 \\ 5 \\ 0 \end{bmatrix} + t \begin{bmatrix} -2 \\ -3 \\ 1 \end{bmatrix} \right) = \begin{bmatrix} 1 & 0 & 2 \\ -1 & 1 & 1 \\ 1 & 2 & 8 \end{bmatrix} \begin{bmatrix} 4 \\ 5 \\ 0 \end{bmatrix} + t \begin{bmatrix} 1 & 0 & 2 \\ -1 & 1 & 1 \\ 1 & 2 & 8 \end{bmatrix} \begin{bmatrix} -2 \\ -3 \\ 1 \end{bmatrix}$$

$$= \begin{bmatrix} 4 \\ 1 \\ 14 \end{bmatrix} + t \begin{bmatrix} 0 \\ 0 \\ 0 \end{bmatrix} = \begin{bmatrix} 4 \\ 1 \\ 14 \end{bmatrix}$$

We do not yet know, however, that these are all the solutions. Two questions need to be answered: Will the method always work? What precisely is the method?

Proposition 2.13 shows that multiplying the equation $AX = B$ on the left by the matrix P will not change the set of solutions as long as P has a left inverse. The manipulations above can be interpreted as such multiplications.

In the first step of the process above the first equation was left unchanged, the second equation was replaced by the second equation plus the first, and the third equation was replaced by the third equation minus the first.

Consider the matrix

$$P_1 = \begin{bmatrix} 1 & 0 & 0 \\ 1 & 1 & 0 \\ -1 & 0 & 1 \end{bmatrix}$$

If we multiply the matrix version of the original set of equations by P_1, we have

$$\begin{bmatrix} 1 & 0 & 0 \\ 1 & 1 & 0 \\ -1 & 0 & 1 \end{bmatrix} \begin{bmatrix} 1 & 0 & 2 \\ -1 & 1 & 1 \\ 1 & 2 & 8 \end{bmatrix} X = \begin{bmatrix} 1 & 0 & 0 \\ 1 & 1 & 0 \\ -1 & 0 & 1 \end{bmatrix} \begin{bmatrix} 4 \\ 1 \\ 14 \end{bmatrix}$$

or

$$\begin{bmatrix} 1 & 0 & 2 \\ 0 & 1 & 3 \\ 0 & 2 & 6 \end{bmatrix} X = \begin{bmatrix} 4 \\ 5 \\ 10 \end{bmatrix}$$

Further, the matrix P_1 is invertible with

$$P_1^{-1} = \begin{bmatrix} 1 & 0 & 0 \\ -1 & 1 & 0 \\ 1 & 0 & 1 \end{bmatrix}$$

And so by Proposition 2.13 we have not changed the set of solutions.

In the second step multiply the second equation by -2 and add it to the third equation. This corresponds to multiplying by

$$P_2 = \begin{bmatrix} 1 & 0 & 0 \\ 0 & 1 & 0 \\ 0 & -2 & 1 \end{bmatrix}$$

which is invertible with inverse

$$P_2^{-1} = \begin{bmatrix} 1 & 0 & 0 \\ 0 & 1 & 0 \\ 0 & 2 & 1 \end{bmatrix}$$

Multiplying the matrix form of the second set of equations by P_2 gives

$$\begin{bmatrix} 1 & 0 & 0 \\ 0 & 1 & 0 \\ 0 & -2 & 1 \end{bmatrix} \begin{bmatrix} 1 & 0 & 2 \\ 0 & 1 & 3 \\ 0 & 2 & 6 \end{bmatrix} X = \begin{bmatrix} 1 & 0 & 0 \\ 0 & 1 & 0 \\ 0 & -2 & 1 \end{bmatrix} \begin{bmatrix} 4 \\ 5 \\ 10 \end{bmatrix}$$

$$\begin{bmatrix} 1 & 0 & 2 \\ 0 & 1 & 3 \\ 0 & 0 & 0 \end{bmatrix} X = \begin{bmatrix} 4 \\ 5 \\ 0 \end{bmatrix}$$

Thus the final set of equations has the same solutions as the original set.

We could accomplish the procedure in one step by multiplying the matrix version of the original set of equations by

$$G = P_2 P_1 = \begin{bmatrix} 1 & 0 & 0 \\ 1 & 1 & 0 \\ -3 & -2 & 1 \end{bmatrix}$$

which is invertible with inverse

$$G^{-1} = (P_2 P_1)^{-1} = P_1^{-1} P_2^{-1} = \begin{bmatrix} 1 & 0 & 0 \\ -1 & 1 & 0 \\ 1 & 2 & 1 \end{bmatrix}$$

The matrices P_1, P_2 are called *pivot* matrices. Here is how they are used in general. Consider the matrix

$$M = \begin{bmatrix} 1 & -4 & 1 \\ -1 & 2 & 1 \\ 0 & 3 & 1 \end{bmatrix}$$

We wish to use the 3 in the 3;2 position to replace the 4 and 2 in the column $M[;2]$

by zeros. We do this by adding appropriate multiples of row $M[3;]$ to rows $M[2;]$ and $M[1;]$. This operation is called *pivoting on the* 3;2 *entry*. In this case we need $-\frac{2}{3}$ times $M[3;]$ added to $M[2;]$ and $\frac{4}{3}$ times $M[3;]$ added to $M[1;]$. We can accomplish this by multiplying by the matrix

$$P = \begin{bmatrix} 1 & 0 & \frac{4}{3} \\ 0 & 1 & -\frac{2}{3} \\ 0 & 0 & 1 \end{bmatrix}$$

$$\begin{bmatrix} 1 & 0 & \frac{4}{3} \\ 0 & 1 & -\frac{2}{3} \\ 0 & 0 & 1 \end{bmatrix} \begin{bmatrix} 1 & -4 & 1 \\ -1 & 2 & 1 \\ 0 & 3 & 1 \end{bmatrix} = \begin{bmatrix} 1 & 0 & \frac{7}{3} \\ -1 & 0 & \frac{1}{3} \\ 0 & 3 & 1 \end{bmatrix}$$

A pivot matrix is a modified identity matrix, I, in which the off-diagonal elements of a single column have been changed. If one wishes to pivot on the $i;j$ entry of the matrix M, then the off-diagonal entries of $I[;i]$ are replaced by the entries of $-M[;j] \div M[i;j]$ with the ith entry dropped out. We obtain the inverse of a pivot matrix by changing the signs of the off-diagonal entries.

EXAMPLE 3.6 Let

$$M = \begin{bmatrix} 1 & 2 & -4 & 1 \\ 2 & -1 & 6 & -3 \\ 4 & 1 & 2 & 7 \end{bmatrix}$$

Write the pivot matrices, and their inverses, for pivoting on the 2;1, 1;2, 3;3, and 2;4 entries.

Solution To pivot on 2;1: Starting with ID 3 (see Section 3.1), we must replace the off-diagonal entries of $(ID\ 3)[;2]$ by $-M[;1] \div M[2;1]$ or $-\ 1\ 2\ 4 \div 2$ with the second entry dropped out.

$$P = \begin{bmatrix} 1 & -\frac{1}{2} & 0 \\ 0 & 1 & 0 \\ 0 & -2 & 1 \end{bmatrix} \quad \text{and} \quad P^{-1} = \begin{bmatrix} 1 & \frac{1}{2} & 0 \\ 0 & 1 & 0 \\ 0 & 2 & 1 \end{bmatrix}$$

To pivot on 1;2: This time the off-diagonal entries of $(ID\ 3)[;1]$ are replaced by $-M[;2] \div M[1;2]$ with the first entry dropped out.

$$P = \begin{bmatrix} 1 & 0 & 0 \\ \frac{1}{2} & 1 & 0 \\ -\frac{1}{2} & 0 & 1 \end{bmatrix} \quad \text{and} \quad P^{-1} = \begin{bmatrix} 1 & 0 & 0 \\ -\frac{1}{2} & 1 & 0 \\ \frac{1}{2} & 0 & 1 \end{bmatrix}$$

To pivot on 3;3:

$$P = \begin{bmatrix} 1 & 0 & 2 \\ 0 & 1 & -3 \\ 0 & 0 & 1 \end{bmatrix}, \qquad P^{-1} = \begin{bmatrix} 1 & 0 & -2 \\ 0 & 1 & 3 \\ 0 & 0 & 1 \end{bmatrix}$$

To pivot on 2;4:

$$P = \begin{bmatrix} 1 & \frac{1}{3} & 0 \\ 0 & 1 & 0 \\ 0 & \frac{7}{3} & 1 \end{bmatrix}, \qquad P^{-1} = \begin{bmatrix} 1 & -\frac{1}{3} & 0 \\ 0 & 1 & 0 \\ 0 & -\frac{7}{3} & 1 \end{bmatrix} \qquad \blacksquare$$

By multiplying by pivot matrices, an equation of the form $AX = B$ can be transformed into an equation with self-evident solutions. This is because multiplication by a pivot matrix eliminates a variable from all but one of the equations of the corresponding system of linear equations.

EXAMPLE 3.7 Solve the system of equations

$$\begin{aligned} x_1 + x_2 \phantom{{}+ x_3} + 2x_4 &= 6 \\ 3x_1 + 2x_2 + x_3 + x_4 &= 5 \\ 9x_1 + 5x_2 + 4x_3 + x_4 &= 11 \\ 4x_1 + 2x_2 + 2x_3 &= 4 \end{aligned}$$

Solution The corresponding matrix system is

$$\begin{bmatrix} 1 & 1 & 0 & 2 \\ 3 & 2 & 1 & 1 \\ 9 & 5 & 4 & 1 \\ 4 & 2 & 2 & 0 \end{bmatrix} X = \begin{bmatrix} 6 \\ 5 \\ 11 \\ 4 \end{bmatrix}$$

Pivoting on the 1;1 entry, we have

$$\begin{bmatrix} 1 & 0 & 0 & 0 \\ -3 & 1 & 0 & 0 \\ -9 & 0 & 1 & 0 \\ -4 & 0 & 0 & 1 \end{bmatrix} \begin{bmatrix} 1 & 1 & 0 & 2 \\ 3 & 2 & 1 & 1 \\ 9 & 5 & 4 & 1 \\ 4 & 2 & 2 & 0 \end{bmatrix} X = \begin{bmatrix} 1 & 0 & 0 & 0 \\ -3 & 1 & 0 & 0 \\ -9 & 0 & 1 & 0 \\ -4 & 0 & 0 & 1 \end{bmatrix} \begin{bmatrix} 6 \\ 5 \\ 11 \\ 4 \end{bmatrix}$$

or

$$\begin{bmatrix} 1 & 1 & 0 & 2 \\ 0 & -1 & 1 & -5 \\ 0 & -4 & 4 & -17 \\ 0 & -2 & 2 & -8 \end{bmatrix} X = \begin{bmatrix} 6 \\ -13 \\ -43 \\ -20 \end{bmatrix}$$

Notice the form of the corresponding system of linear equations:

$$\begin{aligned} x_1 + \quad x_2 \qquad\quad + \ 2x_4 &= \ \ 6 \\ -x_2 + \ x_3 - \ \ 5x_4 &= -13 \\ -4x_2 + 4x_3 - 17x_4 &= -43 \\ -2x_2 + 2x_3 - \ \ 8x_4 &= -20 \end{aligned}$$

The variable x_1 has been eliminated from all but the first equation. Pivoting on any position in the second column will eliminate x_2 from all but one equation. If we pivot on 1;2, however, x_1 will be reintroduced into the other equations. Therefore we should pivot on the 2;2, 3;2, or 4;2 positions. To avoid fractions we pivot on the 2;2 position. In this case we first divide by $-1 = M[3; 3]$ and then change signs:

$$\begin{bmatrix} 1 & 1 & 0 & 0 \\ 0 & 1 & 0 & 0 \\ 0 & -4 & 1 & 0 \\ 0 & -2 & 0 & 1 \end{bmatrix} \begin{bmatrix} 1 & 1 & 0 & 2 \\ 0 & -1 & 1 & -5 \\ 0 & -4 & 4 & -17 \\ 0 & -2 & 2 & -8 \end{bmatrix} X = \begin{bmatrix} 1 & 1 & 0 & 0 \\ 0 & 1 & 0 & 0 \\ 0 & -4 & 1 & 0 \\ 0 & -2 & 0 & 1 \end{bmatrix} \begin{bmatrix} 6 \\ -13 \\ -43 \\ -20 \end{bmatrix}$$

or

$$\begin{bmatrix} 1 & 0 & 1 & -3 \\ 0 & -1 & 1 & -5 \\ 0 & 0 & 0 & 3 \\ 0 & 0 & 0 & 2 \end{bmatrix} X = \begin{bmatrix} -7 \\ -13 \\ 9 \\ 6 \end{bmatrix}$$

Now the variables x_1 and x_2 each appear in only one equation. The variable x_3 appears in two, but trying to eliminate it from one of the equations at this point would put x_1 or x_2 back into another equation. We may eliminate the variable x_4 from all but one equation by pivoting on the 3;4 or 4;4 position. Pivoting on the 3;4 position, we have

$$\begin{bmatrix} 1 & 0 & 1 & 0 \\ 0 & 1 & \frac{5}{3} & 0 \\ 0 & 0 & 1 & 0 \\ 0 & 0 & -\frac{2}{3} & 1 \end{bmatrix} \begin{bmatrix} 1 & 0 & 1 & -3 \\ 0 & -1 & 1 & -5 \\ 0 & 0 & 0 & 3 \\ 0 & 0 & 0 & 2 \end{bmatrix} X = \begin{bmatrix} 1 & 0 & 1 & 0 \\ 0 & 1 & \frac{5}{3} & 0 \\ 0 & 0 & 1 & 0 \\ 0 & 0 & -\frac{2}{3} & 1 \end{bmatrix} \begin{bmatrix} -7 \\ -13 \\ 9 \\ 6 \end{bmatrix}$$

or

$$\begin{bmatrix} 1 & 0 & 1 & 0 \\ 0 & -1 & 1 & 0 \\ 0 & 0 & 0 & 3 \\ 0 & 0 & 0 & 0 \end{bmatrix} X = \begin{bmatrix} 2 \\ 2 \\ 9 \\ 0 \end{bmatrix}$$

Things cannot be improved further by pivoting. The corresponding system of equations is

$$\begin{aligned} x_1 \quad + x_3 \quad &= 2 \\ -x_2 + x_3 \quad &= 2 \\ 3x_4 &= 9 \\ 0 &= 0 \end{aligned} \qquad \text{or} \qquad \begin{aligned} x_1 \quad + x_3 \quad &= \ \ 2 \\ x_2 + x_3 \quad &= -2 \\ x_4 &= \ \ 3 \end{aligned}$$

There are infinitely many solutions. Once a value for the variable x_3 is chosen, the other variables are determined. Setting $x_3 = t$, say, we have

$$\begin{aligned} x_1 &= \ \ 2 - t \\ x_2 &= -2 - t \\ x_3 &= t \\ x_4 &= 3 \end{aligned} \qquad \text{or} \qquad X = \begin{bmatrix} 2 \\ -2 \\ 0 \\ 3 \end{bmatrix} + t \begin{bmatrix} -1 \\ -1 \\ 1 \\ 0 \end{bmatrix} \qquad \blacksquare$$

Besides pivot matrices, we will use two other types of matrices.

A *switch* matrix is obtained by interchanging two rows of an identity matrix. If S is the switch matrix obtained by interchanging rows i and j of $ID\ n$, then SM is just M with rows i and j interchanged. Notice that $S = S^{-1}$.

For example, if we interchange rows 2 and 4 in ID 4 to get S,

$$SM = \begin{bmatrix} 1 & 0 & 0 & 0 \\ 0 & 0 & 0 & 1 \\ 0 & 0 & 1 & 0 \\ 0 & 1 & 0 & 0 \end{bmatrix} \begin{bmatrix} m_{11} & m_{12} & m_{13} & \cdots \\ m_{21} & m_{22} & m_{23} & \cdots \\ m_{31} & m_{32} & m_{33} & \cdots \\ m_{41} & m_{42} & m_{43} & \cdots \end{bmatrix} = \begin{bmatrix} m_{11} & m_{12} & m_{13} & \cdots \\ m_{41} & m_{42} & m_{43} & \cdots \\ m_{31} & m_{32} & m_{33} & \cdots \\ m_{21} & m_{22} & m_{23} & \cdots \end{bmatrix}$$

A *multiplier* matrix is obtained by replacing a diagonal entry of $ID\ n$ by a nonzero scalar λ. If L is a multiplier matrix with $\lambda = L[i;\,i]$, then LM is M with $M[i;]$ replaced by $\lambda M[i;]$:

$$\begin{bmatrix} 1 & 0 & 0 \\ 0 & 1 & 0 \\ 0 & 0 & \lambda \end{bmatrix} \begin{bmatrix} m_{11} & m_{2}i_{2} & m_{13} & \cdots \\ m_{21} & m_{22} & m_{23} & \cdots \\ m_{31} & m_{32} & m_{33} & \cdots \end{bmatrix} = \begin{bmatrix} m_{11} & m_{12} & m_{13} & \cdots \\ m_{21} & m_{22} & m_{23} & \cdots \\ \lambda m_{31} & \lambda m_{32} & \lambda m_{33} & \cdots \end{bmatrix}$$

A multiplier matrix is invertible. If L is defined by $L[i;\,i] = \lambda$, then L^{-1} is the multiplier matrix with λ^{-1} in the $i;i$ position.

We will refer to pivot, switch, and multiplier matrices collectively as *elementary matrices.* The inverse of an elementary matrix is an elementary matrix of the same type.

Gaussian reduction uses elementary matrices to reduce a system of linear equations to a standard form. The standard form we will use is called the *echelon* form.

DEFINITION 3.2 A matrix E is said to be in *row-echelon form* if

1. The zero rows of E, if any, are at the bottom; that is, no row of zeros has a nonzero row below it.
2. The first entry in each nonzero row is a 1. These 1's are called *leading ones* and the columns containing the leading ones are called *pivot columns.*
3. The leading 1 is the only nonzero entry in a pivot column.
4. The leading 1 in a given row is to the right of the leading 1's in the rows above; that is, if $E[i;j]$ and $E[k;l]$ are leading 1's and $i < k$, then $j < l$.

Identity matrices and zero matrices are in row echelon form. The matrix

$$\begin{bmatrix} 1 & 0 & 1 & 0 \\ 0 & -1 & 1 & 0 \\ 0 & 0 & 0 & 3 \\ 0 & 0 & 0 & 0 \end{bmatrix}$$

from Example 3.7 is not in row echelon form because the first nonzero entries in rows 2 and 3 are not 1's. If we multiply by the appropriate multiplier matrices, we can get an echelon form

$$\begin{bmatrix} 1 & 0 & 0 & 0 & 1 \\ 0 & 1 & 0 & 0 & 0 \\ 0 & 0 & \frac{1}{3} & 0 & 0 \\ 0 & 0 & 0 & 1 & 0 \end{bmatrix} \begin{bmatrix} 0 & 0 & 0 & 1 \\ -1 & 0 & 0 & 0 \\ 0 & 1 & 0 & 0 \\ 0 & 0 & 1 & 0 \end{bmatrix} \begin{bmatrix} 0 & 1 & 0 \\ -1 & 1 & 0 \\ 0 & 0 & 3 \\ 0 & 0 & 0 \end{bmatrix} = \begin{bmatrix} 1 & 0 & 1 & 0 \\ 0 & 1 & -1 & 0 \\ 0 & 0 & 0 & 1 \\ 0 & 0 & 0 & 0 \end{bmatrix}$$

The pivot columns are 1, 2, and 4.

In row-reducing a system of equations $AX = B$, we will not carry the whole equation along. We will just work with the *partitioned matrix* A , B, which is usually written $[A \mid B]$. By the definition of matrix multiplication we have $(M +.\times N)[;k] = M +.\times N[;k]$. This means in particular that

$$P[A \mid B] = [PA \mid PB]$$

Instead of multiplying the equation $AX = B$ by the elementary matrix P, we can multiply the partitioned matrix $[A \mid B]$ by P.

The product of the elementary matrices used to reduce the system can be useful. We will keep track of this product by using the partitioned matrix $[A \mid B \mid I]$, where $I \leftarrow ID(\rho A)[1]$. Then, as we multiply by elementary matrices, we have

$$P_1[A \mid B \mid I] = [P_1A \mid P_1B \mid P_1]$$
$$P_2[P_1A \mid P_1B \mid P_1] = [P_2P_1A \mid P_2P_1B \mid P_2P_1]$$

and so on, and the last block accumulates the products of the elementary matrices.

The complete description of the algorithm as we will use it is:

Gaussian reduction Given the matrix equation $AX = B$, form the partitioned matrix $[A|B|I]$, where I is an identity matrix with the same number of rows as A.

By multiplying by elementary matrices, obtain the partitioned matrix $[E|F|G]$, where E is in echelon form. The matrix G is invertible, and multiplying $AX = B$ on the left by G gives $EX = F$. The solutions of $EX = F$ are determined by inspection.

EXAMPLE 3.8 Solve the system of equations

$$\begin{aligned} x - y - 2z &= -5 \\ -2x + 3y + 5z &= 9 \\ x - 2y - 3z &= 0 \\ -2y - 2z &= 2 \end{aligned}$$

by Gaussian reduction. Use G to check the arithmetic.

Solution Begin with the partitioned matrix

$$\left[\begin{array}{rrr|r|rrrr} 1 & -1 & -2 & -5 & 1 & 0 & 0 & 0 \\ -2 & 3 & 5 & 9 & 0 & 1 & 0 & 0 \\ 1 & -2 & -3 & 0 & 0 & 0 & 1 & 0 \\ 0 & -2 & -2 & 2 & 0 & 0 & 0 & 1 \end{array}\right]$$

Pivoting on 1;1,

$$\left[\begin{array}{rrrr} 1 & 0 & 0 & 0 \\ 2 & 1 & 0 & 0 \\ -1 & 0 & 1 & 0 \\ 0 & 0 & 0 & 1 \end{array}\right]\left[\begin{array}{rrr|r|rrrr} 1 & -1 & -2 & -5 & 1 & 0 & 0 & 0 \\ -2 & 3 & 5 & 9 & 0 & 1 & 0 & 0 \\ 1 & -2 & -3 & 0 & 0 & 0 & 1 & 0 \\ 0 & -2 & -2 & 2 & 0 & 0 & 0 & 1 \end{array}\right]$$

$$= \left[\begin{array}{rrr|r|rrrr} 1 & -1 & -2 & -5 & 1 & 0 & 0 & 0 \\ 0 & 1 & 1 & -1 & 2 & 1 & 0 & 0 \\ 0 & -1 & -1 & 5 & -1 & 0 & 1 & 0 \\ 0 & -2 & -2 & 2 & 0 & 0 & 0 & 1 \end{array}\right]$$

Pivoting on 2;2,

$$\left[\begin{array}{rrrrr} 1 & 1 & 0 & 0 & 1 \\ 0 & 1 & 0 & 0 & 0 \\ 0 & 1 & 1 & 0 & 0 \\ 0 & 2 & 0 & 1 & 0 \end{array}\right]\left[\begin{array}{rr|r|rrrr} -1 & -2 & -5 & 1 & 0 & 0 & 0 \\ 1 & 1 & -1 & 2 & 1 & 0 & 0 \\ -1 & -1 & 5 & -1 & 0 & 1 & 0 \\ -2 & -2 & 2 & 0 & 0 & 0 & 1 \end{array}\right]$$

$$= \left[\begin{array}{ccc|c|cccc} 1 & 0 & -1 & -6 & 3 & 1 & 0 & 0 \\ 0 & 1 & 1 & -1 & 2 & 1 & 0 & 0 \\ 0 & 0 & 0 & 4 & 1 & 1 & 1 & 0 \\ 0 & 0 & 0 & 0 & 4 & 2 & 0 & 1 \end{array}\right]$$

$$= [E \mid F \mid G]$$

The equation $EX = F$ has no solution. Indeed the corresponding system of equations is

$$\begin{aligned} x - z &= -6 \\ y + z &= -1 \\ 0 &= 4 \qquad (\text{i.e., } 0 \cdot x + 0 \cdot y + 0 \cdot z = 4) \\ 0 &= 0 \end{aligned}$$

and the third equation is not satisfied for any values of x, y, z.

To check the arithmetic using G, one multiplies $AX = B$ on the left by G. If the result is not $EX = F$, then an arithmetic mistake has been made.

$$G[A \mid B] = \begin{bmatrix} 3 & 1 & 0 & 0 \\ 2 & 1 & 0 & 0 \\ 1 & 1 & 1 & 0 \\ 4 & 2 & 0 & 1 \end{bmatrix} \left[\begin{array}{ccc|c} 1 & -1 & -2 & -5 \\ -2 & 3 & 5 & 9 \\ 1 & -2 & -3 & 0 \\ 0 & -2 & -2 & 2 \end{array}\right]$$

$$= \left[\begin{array}{ccc|c} 1 & 0 & -1 & -6 \\ 0 & 1 & 1 & -1 \\ 0 & 0 & 0 & 4 \\ 0 & 0 & 0 & 0 \end{array}\right] = [E \mid F] \qquad \blacksquare$$

Once an invertible matrix G has been found for which $E = GA$ is in echelon form, then this G may be used to solve equations of the form $AX = B$ for different matrices B.

EXAMPLE 3.9 Solve

$$\begin{bmatrix} 1 & -1 & -2 \\ -2 & 3 & 5 \\ 1 & -2 & -3 \\ 0 & -2 & -2 \end{bmatrix} X = \begin{bmatrix} -1 \\ 4 \\ -3 \\ -4 \end{bmatrix}$$

Solution The matrix on the left is the same as the matrix of Example 3.8. Thus we multiply the equation $AX = B$ above by the matrix G computed in Example 3.8. We do not need to perform the computation GA, because we know that

$GA = E$. Multiplication on the left by G gives

$$\begin{bmatrix} 1 & 0 & -1 \\ 0 & 1 & 1 \\ 0 & 0 & 0 \\ 0 & 0 & 0 \end{bmatrix} X = \begin{bmatrix} 3 & 1 & 0 & 0 \\ 2 & 1 & 0 & 0 \\ 1 & 1 & 1 & 0 \\ 4 & 2 & 0 & 1 \end{bmatrix} \begin{bmatrix} -1 \\ 4 \\ -3 \\ -4 \end{bmatrix} = \begin{bmatrix} 1 \\ 2 \\ 0 \\ 0 \end{bmatrix}$$

The corresponding system of equations is

$$\begin{aligned} x - z &= 1 \\ y + z &= 2 \\ 0 &= 0 \\ 0 &= 0 \end{aligned}$$

The variable z can be chosen arbitrarily and then the other variables are determined. Say $z = t$; then

$$\begin{aligned} x &= 1 + t \\ y &= 2 - t \\ z &= t \end{aligned} \qquad \text{or} \qquad X = \begin{bmatrix} 1 \\ 2 \\ 0 \end{bmatrix} + t \begin{bmatrix} 1 \\ -1 \\ 1 \end{bmatrix} \qquad \blacksquare$$

The above pattern is a general one. The variables corresponding to nonpivot columns of E can be chosen arbitrarily. Once these variables are assigned values, the variables corresponding to the pivot columns are determined.

Suppose, for example, that in solving a system $AX = B$, the partitioned matrix $[A \mid B]$ has been reduced to

$$[E \mid F] = \left[\begin{array}{cccccc|c} 1 & 0 & 0 & 0 & 2 & 0 & 5 \\ 0 & 1 & 1 & 0 & 3 & 0 & 6 \\ 0 & 0 & 0 & 1 & 4 & 0 & 7 \\ 0 & 0 & 0 & 0 & 0 & 1 & 8 \end{array}\right]$$

pivot columns

There are four leading 1's in this echelon form in positions 1;1, 2;2, 3;4, and 4;6. The pivot columns are thus $E[;\ 1\ 2\ 4\ 6]$. Notice that $E[;3]$ is not a pivot column, since it does not contain a leading 1. Let $X = (x_1, x_2, x_3, x_4, x_5, x_6)$. The nonpivot columns of E are $E[;\ 3\ 5]$, and x_3, x_5 may be chosen arbitrarily, say $x_3 = t_1$ and $x_5 = t_2$. The system of linear equations corresponding to $[E \mid F]$ then becomes

$$\begin{aligned} x_1 + 2t_2 &= 5 \\ x_2 + t_1 + 3t_2 &= 6 \\ x_4 + 4t_2 &= 7 \\ x_6 &= 8 \end{aligned}$$

which gives

$$\begin{aligned} x_1 &= 5 - 2t_2 \\ x_2 &= 6 - t_1 - 3t_2 \\ x_3 &= t_1 \\ x_4 &= 7 - 4t_2 \\ x_5 &= t_2 \\ x_6 &= 8 \end{aligned} \qquad \text{or} \qquad X = \begin{bmatrix} 5 \\ 6 \\ 0 \\ 7 \\ 0 \\ 8 \end{bmatrix} + t_1 \begin{bmatrix} 0 \\ -1 \\ 1 \\ 0 \\ 0 \\ 0 \end{bmatrix} + t_2 \begin{bmatrix} -2 \\ -3 \\ 0 \\ -4 \\ 1 \\ 0 \end{bmatrix}$$

In the last form, notice that if B is changed in the original equation, only the first term of X is affected. The terms involving the arbitrary parameters are unchanged. If $B = 0$, then $F = GB = 0$, and the first term of X disappears. This means that the terms involving the arbitrary parameters are the solutions of $AX = 0$. The vectors multiplied by the arbitrary parameters are linearly independent (exercise 67).

COMPRESSION, TAKE, DROP

We need some notation for deleting entries from vectors and also for deleting rows and columns from matrices.

One way to delete an entry from a vector is to use *compression.* The compression function is denoted by /. The form is L/V, where V is the vector and L is a vector of zeros and ones (such vectors are sometimes called *logical vectors* or *masks*). The entries of V corresponding to the zeros of L are deleted.

```
      1 0 1 / 1 2 3       delete second component
1 3
      0 1 0 / 1 2 3       delete first and third components
2
      1 1 1 / 1 2 3       delete no components
1 2 3
      0 0 0 / 1 2 3       delete all components (leaving ⍳0)

      ⍴ 0 0 0 / 1 2 3
0
```

The logical vectors L are usually created by using the logical or relational functions

```
      2 ≠ ⍳3
1 0 1
```

Another selection function is *take,* denoted by ↑. The expression 3↑V means "Take the first three components of V" and ¯3↑V means "Take the last three components." The function *drop,* denoted by ↓, works similarly.

```
      3↑⍳5
1 2 3
```

```
      ¯3↑⍳5
4 5 6
      3↓⍳5
4 5
      ¯3↓⍳5
1 2
```

If A is a matrix, then L/A drops columns and $L⌿A$ drops rows. The expression 1 2↑A means "Take 1 row and 2 columns" and 0 ¯4↓A means "drop 4 columns from the right end of the matrix."

Now consider the definition of the pivot matrix P needed to pivot on the 1;3 entry, say, of a matrix A.

Start with an identity matrix with the same number of rows, N, as A:

```
P←ID N←1↑⍴A
```

Next we need to change the column P[;1]. Each entry but the first should be replaced by the corresponding entry of $-A[;3] \div A[1;3]$.

```
P[(1≠⍳N)/⍳N;1]←(1≠⍳N)/−A[;3]÷A[1;3]
```

We can now write a function that performs the pivot operation on a given matrix. The right argument will be the matrix being row-reduced and the left argument will be a vector giving the pivot position.

```
    ∇ Z←X PIVOT A ;P;N
[1]    P←ID N←1↑⍴A
[2]    P[(X[1]≠⍳N)/⍳N;X[1]]←(X[1]≠⍳N)/−A[;X[2]]÷A[X[1];X[2]]
[3]    Z←P+.×A
    ∇

      +A←3 3⍴⍳9
 1  2  3
 4  5  6
 7  8  9
      2 2 PIVOT A
¯6.00E¯1    3.47E¯18    6.00E¯1
 4.00E0     5.00E0      6.00E0
 6.00E¯1    1.39E¯17   ¯6.00E¯1
```

Notice the round-off error in column 2 of the result. The precise form of this column will vary with the computer used. Since the 1;2 and 3;2 entries are less than ⎕CT times the entries of A, we may take them to be zero. (Recall from Section 2.2 that ⎕CT is the fuzz factor, here taken to be about 10^{-15}.)

A function that interchanges rows $X[1]$ and $X[2]$ of the matrix A is

```
    ∇ Z←X SWITCH A
[1]    A[X;]←A[X[2 1];]
[2]    Z←A
    ∇

       A
 1  2  3
 4  5  6
 7  8  9
       1 3 SWITCH A
 7  8  9
 4  5  6
 1  2  3
```

Notice that a switch matrix need not be created. Any function that performs the same task as multiplying by a switch matrix will do. The third kind of elementary matrix is the multiplier matrix. We need multiplier matrices only to create leading 1's when the first nonzero entry of a row is not equal to 1. For our purposes, then, it is sufficient to have a function that will put a 1 in a given position, the $i;j$ position, say, by dividing the ith row by the $i;j$ entry. Call the function *LDR* for leader.

```
    ∇ Z←X LDR A
[1]    A[X[1];]←A[X[1];]÷A[X[1];X[2]]
[2]    Z←A
    ∇

       A
 1  2  3
 0  5  6
 0  8  9
       2 2 LDR A
 1.000  2.000  3.000
 0.000  1.000  1.200
 0.000  8.000  9.000
```

Note: Copies of the functions *PIVOT*, *SWITCH*, and *LDR* should be saved in a workspace file for future use.

EXAMPLE 3.10 Solve the system

$$\begin{aligned} x + 2y + 3z &= 6 \\ 4x + 5y + 6z &= 7 \\ 7x + 8y + 9z &= 8 \end{aligned}$$

by row reduction. Check your answer.

Solution This is of the form $AX = B$ with

```
        A
  1  2  3
  4  5  6
  7  8  9
        B
6  7  8
```

Form the partitioned matrix $[A|B|I] = M$.

```
      +M ←(A,B),ID 3
1  2  3  6  1  0  0
4  5  6  7  0  1  0
7  8  9  8  0  0  1
```

If we use the seven to clean up the first column, we get

```
      +M1←3 1 PIVOT M
1.73E¯18   8.57E¯1   1.71E0   4.86E0   1.00E0   0.00E0   ¯1.43E1
6.94E¯18   4.29E¯1   8.57E¯1  2.43E0   0.00E0   1.00E0   ¯5.71E¯1
7.00E0     8.00E0    9.00E0   8.00E0   0.00E0   0.00E0    1.00E0
```

The entries $M1[1\ \ 2;1]$ are small enough to be assumed zero. We wish to reduce A to echelon form, so that

```
      +M2←1 1 LDR 1 3 SWITCH M1
1.00E0     1.14E0    1.29E0    1.14E0   0.00E0   0.00E0    1.43E¯1
6.94E¯18   4.29E¯1   8.57E¯1   2.43E0   0.00E0   1.00E0   ¯5.71E¯1
1.73E¯18   8.57E¯1   1.71E0    4.86E0   1.00E0   0.00E0   ¯1.43E¯1
```

Next pivot on the 2;2 entry and set the leading 1.

```
      +M3←2 2 LDR 2 2 PIVOT M2
 1.00E0     1.73E¯18   ¯1.00E0    ¯5.33E0    0.00E0   ¯2.67E0    1.67E0
 1.62E¯17   1.00E0      2.00E0     5.67E0    0.00E0    2.33E0   ¯1.33E0
¯1.21E¯17   8.67E¯19    1.73E¯18   4.16E¯17  1.00E0   ¯2.00E0    1.00E0
```

Assuming the small entries to be zero, A has been reduced to the echelon form

$$\begin{bmatrix} 1 & 0 & -1 \\ 0 & 1 & 2 \\ 0 & 0 & 0 \end{bmatrix}$$

We check this using the matrix G.

```
      +G←3 ¯3↑M3
0.000 ¯2.670  1.670
0.000  2.330 ¯1.330
1.000 ¯2.000  1.000

      G+.×A,B
 1.00E0      0.00E0    ¯1.00E0    ¯5.33E0
 2.78E¯17    1.00E0     2.00E0     5.67E0
¯6.94E¯18    1.39E¯17   1.39E¯17   4.16E¯17
```

Taking the small entries to be zero, A is in echelon form, and so we assume that the calculation is correct.

There is one nonpivot column, $E[;3]$. Setting $z = t$, we have

$$X = \begin{bmatrix} x \\ y \\ z \end{bmatrix} = \begin{bmatrix} -5.33 \\ 5.67 \\ 0 \end{bmatrix} + t \begin{bmatrix} 1 \\ -2 \\ 1 \end{bmatrix} \quad \blacksquare$$

It is instructive to see just what happens in the last example if the entry `M3[3;3] = 1.73E¯18` is assumed nonzero. Then, pivoting on the 3;3 position and setting the 3;3 entry equal to 1, we get

```
      +M4←3 3 LDR 3 3 PIVOT M3
¯6.00E0   5.00E¯1   0.00E0   1.87E1   5.76E17  ¯1.15E18   5.76E17
 1.40E1   0.00E0    0.00E0  ¯4.23E1  ¯1.15E18   2.31E18  ¯1.15E18
¯7.00E0   5.00E¯1   1.00E0   2.40E1   5.76E17  ¯1.15E18   5.76E17
```

Clearly something is wrong, because the first two columns have lost their nice form. The check also shows that something is wrong.

```
      +G←3 ¯3↑M4
 5.76E17   ¯1.15E18    5.76E17
¯1.15E18    2.31E18   ¯1.15E18
 5.76E17   ¯1.15E18    5.76E17

      G+.×A,B
¯2.000   4.000   2.000   20.000
 8.000   0.000   0.000  ¯32.000
¯4.000   8.000   8.000   24.000
```

This matrix is not in row echelon form. Therefore an error has been made.

We will use G as a check even though the problem is apparent in `M4`. We need G for other reasons, and other codings of *PIVOT* can produce reasonable-looking matrices `M4` (see exercise 66).

Machine Computation Check If GA is close to an echelon form, accept the result.

The term "close" is not precisely defined and can depend on several factors, such as experimental error in measured data. For the most part, in this text E is close to an echelon form if one gets an echelon form by setting equal to zero those components of E which have an absolute value smaller than the entries of A×⎕CT.

Inverses

Gaussian reduction can be used to compute left inverses. If a matrix A is row-reduced to $E = GA$ and E has no nonpivot columns, then a left inverse for A may be extracted from G.

First an observation about matrix multiplications. It follows immediately from Proposition 2.1 that $(M +.\times N)[i;] = M[i;] +.\times N$. Thus if we take the partitioned matrix $M \leftarrow P,[1]Q$, which is often written

$$M = \left[\frac{P}{Q}\right]$$

we have

$$MN = \left[\frac{P}{Q}\right]N = \left[\frac{PN}{QN}\right]$$

Now if the echelon form E has only pivot columns, then

$$GA = E = \left[\frac{I}{0}\right]$$

where I is an identity and 0 is a zero matrix. If A and hence E are square, the matrix 0 does not appear and $E = I$. Suppose that I is k by k and let $G_1 = G[\iota k;]$. Let G_2 be the remaining rows of G. Then

$$\left[\frac{I}{0}\right] = E = GA = \left[\frac{G_1}{G_2}\right]A = \left[\frac{G_1A}{G_2A}\right]$$

hence $G_1A = I$, $G_2A = 0$. In particular, G_1 is a left inverse for A.

It follows from Propositions 3.3 and 3.4 of the next section that this procedure will always work; the echelon form of a matrix is unique and if A has a left inverse, then E has only pivot columns.

For this application of Gaussian reduction we do not have an equation $AX = B$ but simply a matrix A. Form the partitioned matrix $[A \mid I]$ and row-reduce to $[E \mid G]$. If E has only pivot columns, then a left inverse may be extracted from G. If E has some nonpivot columns, then A does not have a left inverse.

EXAMPLE 3.11 Does the matrix

$$A = \begin{bmatrix} 1 & -1 & -1 \\ -2 & 3 & 1 \\ 1 & -2 & 1 \\ 0 & -2 & 2 \end{bmatrix}$$

possess left inverses?

Solution Begin by pivoting on the 1;1 entry:

$$\begin{bmatrix} 1 & 0 & 0 & 0 \\ 2 & 1 & 0 & 0 \\ -1 & 0 & 1 & 0 \\ 0 & 0 & 0 & 1 \end{bmatrix}\left[\begin{array}{rrr|rrrr} 1 & -1 & -1 & 1 & 0 & 0 & 0 \\ -2 & 3 & 1 & 0 & 1 & 0 & 0 \\ 1 & -2 & 1 & 0 & 0 & 1 & 0 \\ 0 & -2 & 2 & 0 & 0 & 0 & 1 \end{array}\right]$$

$$= \left[\begin{array}{rrr|rrrr} 1 & -1 & -1 & 1 & 0 & 0 & 0 \\ 0 & 1 & -1 & 2 & 1 & 0 & 0 \\ 0 & -1 & 2 & -1 & 0 & 1 & 0 \\ 0 & -2 & 2 & 0 & 0 & 0 & 1 \end{array}\right]$$

Next, pivot on the 2;2 entry:

$$\begin{bmatrix} 1 & 1 & 0 & 0 \\ 0 & 1 & 0 & 0 \\ 0 & 1 & 1 & 0 \\ 0 & 2 & 0 & 1 \end{bmatrix}\left[\begin{array}{rrr|rrrr} 1 & -1 & -1 & 1 & 0 & 0 & 0 \\ 0 & 1 & -1 & 2 & 1 & 0 & 0 \\ 0 & -1 & 2 & -1 & 0 & 1 & 0 \\ 0 & -2 & 2 & 0 & 0 & 0 & 1 \end{array}\right]$$

$$= \left[\begin{array}{rrr|rrrr} 1 & 0 & -2 & 3 & 1 & 0 & 0 \\ 0 & 1 & -1 & 2 & 1 & 0 & 0 \\ 0 & 0 & 1 & 1 & 1 & 1 & 0 \\ 0 & 0 & 0 & 4 & 2 & 0 & 1 \end{array}\right]$$

Finally, pivot on the 3;3 entry:

$$\begin{bmatrix} 1 & 0 & 2 & 0 \\ 0 & 1 & 1 & 0 \\ 0 & 0 & 1 & 0 \\ 0 & 0 & 0 & 1 \end{bmatrix}\left[\begin{array}{rrr|rrrr} 1 & 0 & -2 & 3 & 1 & 0 & 0 \\ 0 & 1 & -1 & 2 & 1 & 0 & 0 \\ 0 & 0 & 1 & 1 & 1 & 1 & 0 \\ 0 & 0 & 0 & 4 & 2 & 0 & 1 \end{array}\right]$$

$$= \left[\begin{array}{rrr|rrrr} 1 & 0 & 0 & 5 & 3 & 2 & 0 \\ 0 & 1 & 0 & 3 & 2 & 1 & 0 \\ 0 & 0 & 1 & 1 & 1 & 1 & 0 \\ 0 & 0 & 0 & 4 & 2 & 0 & 1 \end{array}\right] = [E|G]$$

$$E = \left[\begin{array}{ccc} 1 & 0 & 0 \\ 0 & 1 & 0 \\ 0 & 0 & 1 \\ \hline 0 & 0 & 0 \end{array}\right] \quad \text{so} \quad G = \left[\begin{array}{c} G_1 \\ \hline G_2 \end{array}\right] = \left[\begin{array}{cccc} 5 & 3 & 2 & 0 \\ 3 & 2 & 1 & 0 \\ 1 & 1 & 1 & 0 \\ \hline 4 & 2 & 0 & 1 \end{array}\right]$$

and

$$G_1 = \begin{bmatrix} 5 & 3 & 2 & 0 \\ 3 & 2 & 1 & 0 \\ 1 & 1 & 1 & 0 \end{bmatrix}$$

is a left inverse for A.

Check

$$G_1A = \begin{bmatrix} 5 & 3 & 2 & 0 \\ 3 & 2 & 1 & 0 \\ 1 & 1 & 1 & 0 \end{bmatrix} \begin{bmatrix} 1 & -1 & -1 \\ -2 & 3 & 1 \\ 1 & -2 & 1 \\ 0 & -2 & 2 \end{bmatrix} = \begin{bmatrix} 1 & 0 & 0 \\ 0 & 1 & 0 \\ 0 & 0 & 1 \end{bmatrix} \quad ■$$

The process above does not produce the left inverse $(A^TA)^{-1}A^T$ computed by ⌹. Using the matrix of the last example, in fact,

```
        A
 1  ¯1  ¯1
¯2   3   1
 1  ¯2   1
 0  ¯2   2
      ⌹A
 0.047   0.524   2.000  ¯1.240
¯0.047   0.476   1.000  ¯0.762
¯0.143   0.429   1.000  ¯0.286
```

The left inverse G_1 computed above is not unique. A different choice of pivots can produce a different G_1. The algorithm used by ⌹, however, does not involve pivoting. It reduces linear systems by using the elementary reflections defined in Chapter 5. The resulting algorithm is less sensitive to roundoff error than Gaussian reduction.

EXERCISES 3.2

Let
$$A = \begin{bmatrix} 1 & 3 & -1 & 2 \\ 4 & 0 & 1 & -1 \\ 3 & 1 & 2 & \frac{1}{2} \\ 0 & 1 & -\frac{1}{3} & 2 \end{bmatrix}$$

Write the pivot matrix to pivot on the given position.

1. 1;1 entry
2. 1;2 entry
3. 2;1 entry
4. 2;4 entry
5. 3;4 entry
6. 4;3 entry.

Is the given matrix in row echelon form? If so, identify the pivot columns.

7. $\begin{bmatrix} 1 & 0 & 0 \\ 0 & 0 & 1 \\ 0 & 0 & 0 \end{bmatrix}$

8. $\begin{bmatrix} 1 & 0 & 0 \\ 0 & 0 & 0 \\ 0 & 0 & 0 \end{bmatrix}$

9. $\begin{bmatrix} 1 & 2 & 3 \\ 0 & 1 & 0 \\ 0 & 0 & 0 \end{bmatrix}$

10. $\begin{bmatrix} 1 \\ 0 \\ 0 \\ 0 \end{bmatrix}$

11. $\begin{bmatrix} 0 & 0 & 1 & 2 & 3 \end{bmatrix}$

12. $\begin{bmatrix} 1 & 2 & 0 & 3 & 0 \\ 0 & 0 & 1 & 4 & 0 \\ 0 & 0 & 0 & 0 & 0 \\ 0 & 0 & 0 & 0 & 1 \end{bmatrix}$

13. $\begin{bmatrix} 0 & 0 \\ 0 & 0 \end{bmatrix}$

14. $\begin{bmatrix} 1 & 0 & 1 & 0 \\ 0 & 0 & 0 & 1 \\ 0 & 0 & 0 & 1 \end{bmatrix}$

15. $\begin{bmatrix} 1 & 2 & 0 & 3 & 0 & 0 \\ 0 & 0 & -1 & 4 & 0 & 0 \\ 0 & 0 & 0 & 0 & 1 & 0 \\ 0 & 0 & 0 & 0 & 0 & 1 \\ 0 & 0 & 0 & 0 & 0 & 0 \end{bmatrix}$

16. Show that if E is in row echelon form, then 1 1 ↓ E is in row-echelon form.

Without computation, write the inverses of the elementary matrices given in exercises 17 through 25.

17. $\begin{bmatrix} 1 & 0 & 0 \\ 0 & 1 & 0 \\ 0 & 0 & 2 \end{bmatrix}$

18. $\begin{bmatrix} 1 & 0 & 0 \\ 0 & \frac{1}{2} & 0 \\ 0 & 0 & 1 \end{bmatrix}$

19. $\begin{bmatrix} 1 & 0 & 0 \\ 0 & -1 & 0 \\ 0 & 0 & 1 \end{bmatrix}$

20. $\begin{bmatrix} 0 & 0 & 1 \\ 0 & 1 & 0 \\ 1 & 0 & 0 \end{bmatrix}$

21. $\begin{bmatrix} 1 & 0 & 0 & 0 \\ 0 & 0 & 0 & 1 \\ 0 & 0 & 1 & 0 \\ 0 & 1 & 0 & 0 \end{bmatrix}$

22. $\begin{bmatrix} 1 & 0 & 0 \\ 0 & 1 & 0 \\ 0 & 0 & 1 \end{bmatrix}$

23. $\begin{bmatrix} 1 & 0 & 2 \\ 0 & 1 & 3 \\ 0 & 0 & 1 \end{bmatrix}$

24. $\begin{bmatrix} 1 & -\frac{1}{2} & 0 & 0 \\ 0 & 1 & 0 & 0 \\ 0 & 3 & 1 & 0 \\ 0 & 1 & 0 & 1 \end{bmatrix}$

25. $\begin{bmatrix} 1 & 1 & 0 & 0 \\ 0 & 1 & 0 & 0 \\ 0 & 1 & 1 & 0 \\ 0 & 1 & 0 & 1 \end{bmatrix}$

Solve the linear systems in exercises 26 through 35 either by hand or using the functions *PIVOT, SWITCH,* and *LDR*.

26. $\begin{aligned} x + y + z &= 9 \\ 2x + y + z &= 11 \\ 2y + z &= 10 \end{aligned}$

27. $\begin{aligned} x - 2y - 2z &= -4 \\ x - y - z &= 1 \\ -x + 3y + 3z &= 11 \end{aligned}$

28. $\begin{aligned} 9x + 5y + 5z &= 33 \\ x + y + z &= 5 \\ 3x + 2y + 2z &= 12 \end{aligned}$

29. $\begin{aligned} -3x_1 + x_2 + 2x_3 + 5x_4 &= 5 \\ -x_1 + 3x_2 + x_3 + 11x_4 &= 20 \\ -3x_1 + x_3 - 2x_4 &= 0 \\ 2x_1 + 2x_2 + 10x_4 &= 22 \end{aligned}$

30. $$\begin{aligned} x_1 + x_2 + x_3 + 9x_4 &= 18 \\ x_1 + 2x_2 + x_3 + 12x_4 &= 24 \\ -x_1 - 2x_2 \qquad - 8x_4 &= -17 \\ x_1 + x_2 + 2x_3 + 13x_4 &= 25 \end{aligned}$$

31. $$\begin{aligned} 2x_1 + 4x_2 + x_3 + 8x_4 &= 16 \\ 2x_1 + 4x_2 + 2x_3 + 12x_4 &= 22 \\ x_1 + 2x_2 + x_3 + 6x_4 &= 18 \\ 2x_1 + 4x_2 \qquad + 4x_4 &= 10 \end{aligned}$$

32. $$\begin{aligned} 3x_1 + 6x_2 - 2x_3 - 2x_4 &= 3 \\ x_1 + 2x_2 + 4x_3 + 18x_4 &= 29 \\ 3x_1 + 6x_2 - x_3 + 2x_4 &= 9 \\ x_1 + 2x_2 - x_3 - 2x_4 &= -1 \end{aligned}$$

33. $$\begin{aligned} 2x + 2y + 3z &= 22 \\ 3x + 3y + 3z &= 32 \\ 2x + 3y + 3z &= 25 \\ 4x + 3y + z &= 21 \\ -2x - y + z &= 3 \end{aligned}$$

34. $$\begin{aligned} 3x + 9y + 3z &= 45 \\ 3y + z &= 13 \\ 7y + 5z &= 41 \\ x + 5y + 2z &= 25 \\ 3y + 2z &= 17 \end{aligned}$$

35. $$\begin{aligned} x_1 - x_2 - x_3 - x_4 - x_5 \qquad &= -1 \\ x_2 + 3x_3 + 5x_4 + 7x_5 \qquad &= 9 \\ -2x_1 + 3x_2 + 5x_3 + 7x_4 + 9x_5 + x_6 &= 21 \\ -x_1 \qquad - 2x_3 - 4x_4 - 6x_5 + x_6 &= 2 \\ -2x_1 + x_2 - x_3 - 3x_4 - 5x_5 + x_6 &= 3 \\ x_1 - 2x_2 - 4x_3 - 6x_4 - 8x_5 - 2x_6 &= -30 \end{aligned}$$

In exercises 36 through 44 use Gaussian reduction to compute a left inverse, if one exists, of the matrix A, either by hand or using the functions *PIVOT*, *SWITCH*, and *LDR*.

36. $A = \begin{bmatrix} 1 & 1 & -1 \\ 1 & 2 & -2 \\ 1 & -1 & 2 \end{bmatrix}$ 37. $A = \begin{bmatrix} 1 & -1 & 0 \\ -1 & 2 & -2 \\ -2 & 2 & 0 \end{bmatrix}$ 38. $A = \begin{bmatrix} 1 & 1 & -4 \\ -2 & -1 & 6 \\ 1 & 2 & -5 \end{bmatrix}$

39. $A = \begin{bmatrix} 1 & -1 & 3 \\ 0 & 1 & -2 \\ -2 & 1 & -3 \\ -1 & -1 & 0 \\ 1 & -3 & 8 \end{bmatrix}$ 40. $A = \begin{bmatrix} 1 & -1 & 0 \\ -2 & 3 & 1 \\ -1 & 2 & 2 \\ 1 & -1 & -1 \\ -2 & 0 & -4 \end{bmatrix}$ 41. $A = \begin{bmatrix} 1 & 1 \\ 0 & 1 \\ 1 & 2 \\ -1 & -2 \\ 0 & 0 \\ -1 & -2 \end{bmatrix}$

42. $\begin{bmatrix} 1 & 1 & 5 \\ 0 & 1 & 3 \\ -2 & -4 & -16 \\ 0 & -2 & -6 \\ -2 & -4 & -16 \end{bmatrix}$ 43. $\begin{bmatrix} 1 & 0 & -2 & -1 \\ -2 & 1 & 2 & -2 \\ 1 & 1 & -3 & -4 \\ -1 & 1 & 1 & -1 \end{bmatrix}$ 44. $\begin{bmatrix} 1 & -2 & 5 & -4 \\ -2 & 5 & -12 & 9 \\ 0 & -2 & 5 & -2 \\ 1 & -2 & 4 & -3 \end{bmatrix}$

In exercises 45 through 49 use *SWITCH*, *PIVOT*, and *LDR* to compute the echelon form of the given matrix. Check your answers.

45. $\begin{bmatrix} 47 & 50 & 244 & 30 \\ 55 & 84 & 362 & 64 \\ 75 & 9 & 177 & 18 \end{bmatrix}$ 46. $\begin{bmatrix} 8 & -766 & 342 & 74 \\ -32 & 75 & -17 & -35 \\ -46 & -79 & 60 & -34 \\ 446 & 636 & -523 & 341 \end{bmatrix}$

47. $\begin{bmatrix} 39 & 3 & 14 & -17 & -15 & 30 \\ 355 & -97 & 494 & 94 & -3 & -30 \\ -59 & 11 & -67 & -69 & 111 & -150 \end{bmatrix}$

48. $\begin{bmatrix} 57 & 99 & 51 & 98 \\ 294 & 338 & 146 & -245 \\ 6 & 32 & 20 & -49 \\ 150 & 390 & 222 & -294 \end{bmatrix}$

49. $\begin{bmatrix} -71 & 353 & 435 & 47 & -48 & -97 \\ -83 & -61 & -81 & 67 & 84 & -1 \\ -79 & 77 & 91 & -932 & -420 & 688 \\ 67 & -491 & -607 & 409 & 192 & -295 \end{bmatrix}$

Compute by hand the results of the expressions in exercises 50 through 65.

50.
```
V←3 1 6 2 7
(3=ιρV)/V
```

51.
```
V←3 1 6 2 7
(3=ιρV)/ιρV
```

52.
```
V←3 1 6 2 7
V[(3=ιρV)/ιρV]
```

53.
```
V←1 ¯2 6 ¯4 ¯5
(V<0)/V
```

54.
```
V←1 ¯2 6 ¯4 ¯5
(V≥0)/V
```

55.
```
V←ι20
((V<15)∧V>5)/V
```

56.
```
A←3 3ρι9
1 1↑A
```

57.
```
A←3 3ρι9
1 1↓A
```

58.
```
A←3 3ρι9
2 1↑A
```

59.
```
A←3 3ρι9
0 2↓A
```

60.
```
A←3 3ρι9
2 ¯2↑A
```

61.
```
A←3 3ρι9
2 ¯2↓A
```

62.
```
A←3 3ρι9
0 3↑A
```

63.
```
A←3 3ρι9
0 0↑A
```

64.
```
A←3 3ρι9
3 2↓A
```

65.
```
A←3 3ρι9
0 0↓A
```

66. (Computer assignment) a version of *PIVOT* that does not create a pivot matrix is

```
    ∇ Z←X PIVOT2 A ;E
[1]   E←X[1]≠ι(ρA)[1]
[2]   Z←A-(E×A[;X[2]])∘.×A[X[1];]÷A[X[1];X[2]]
    ∇
```

To pivot on $i;j$ this version creates a matrix whose rows are the proper multiples of row $A[i;]$ and then subtracts this matrix from A.

(a) Redo Example 3.10 using *PIVOT*2 instead of *PIVOT*. How do the new matrices $M1$, $M2$, $M3$ compare with the $M1$, $M2$, $M3$ in Example 3.10?

(b) Create $M4$ by pivoting $M3$ on the 3;3 entry. Is $M4$ an echelon form?

(c) If $M4$ were an echelon form, then $M4[;4]$ would be a solution of the original system. Is $M4[;4]$ a solution of the original system?

(d) Use the new G to check the accuracy of $M4$.

67. In solving $AX = B$, one row reduces $[A \mid B]$ to an echelon form E. The solutions are then written in the form

$$X = v_0 + t_1 v_1 + t_2 v_2 + \cdots + t_k v_k$$

or $X = v_0 + PT$, where $P[;i] = v_i$ and $T[i] = t_i$. There is one parameter t_i for each nonpivot column of E. Suppose that the nonpivot columns of E are $E[;V]$. Show that

```
P[V;]=IDρV
```

and then apply Proposition 2.11 to show that the columns of P are linearly independent.

3.3 The Echelon Form of a Matrix

The importance of the echelon form of a matrix is this: The echelon form of A is an explicit list of the linear relationships among the columns of A.

Consider a typical echelon form:

$$E = \begin{bmatrix} 1 & 0 & 2 & 0 & 4 & 0 \\ 0 & 1 & 3 & 0 & 5 & 0 \\ 0 & 0 & 0 & 1 & 6 & 0 \\ 0 & 0 & 0 & 0 & 0 & 1 \\ 0 & 0 & 0 & 0 & 0 & 0 \end{bmatrix}$$

The pivot columns are $E[;\ 1\ 2\ 4\ 6]$. The nonpivot columns of E are linear combinations of the pivot columns — the preceding pivot columns, in fact. Let $V = (1, 2, 4, 6)$. Then

$$E[;3] = 2E[;1] + 3E[;2] = E[;1\quad 2]\begin{bmatrix} 2 \\ 3 \end{bmatrix} = E[;V]\begin{bmatrix} 2 \\ 3 \\ 0 \\ 0 \end{bmatrix}$$

$$E[;5] = 4E[;1] + 5E[;2] + 6E[;4] = E[;1\quad 2\quad 4]\begin{bmatrix} 4 \\ 5 \\ 6 \end{bmatrix} = E[;V]\begin{bmatrix} 4 \\ 5 \\ 6 \\ 0 \end{bmatrix}$$

Further, the pivot columns are linearly independent. This follows easily from Proposition 2.11, which states that the columns of A are linearly independent if and only if the equation $AX = 0$ has only the trivial solution $X = 0$. For, letting $I = ID\ 4$, we have, if $E[;V]X = 0$,

$$0 = E[;V]X = \begin{bmatrix} I \\ 0 \end{bmatrix} X = \begin{bmatrix} IX \\ 0X \end{bmatrix} = \begin{bmatrix} X \\ 0 \end{bmatrix}$$

hence $X = 0$.

Now, suppose that E was derived from A by row-reduction — that is, $E = GA$, where G is a product of elementary matrices. Since G is a product of elementary matrices, G is square and invertible. If $G = P_n P_{n-1} \ldots P_1$, where the P_i are elementary matrices, then $G^{-1} = P_1^{-1} P_2^{-1} \ldots P_n^{-1}$, also a product of elementary matrices, and $A = G^{-1}E$. It follows that the columns of $A[;V]$ are also linearly independent.

If

$$A[;V]X = 0$$

then

$$GA[;V]X = 0$$

or

$$E[;V]X = 0$$

and so $X = 0$. Further,

$$A[;3] = (G^{-1}E)[;3] = G^{-1}E[;3] = G^{-1}E[;V]\begin{bmatrix}2\\3\\0\\0\end{bmatrix}$$

$$= A[;V]\begin{bmatrix}2\\3\\0\\0\end{bmatrix} = A[;1 \quad 2]\begin{bmatrix}2\\3\end{bmatrix}$$

or $A[;3] = 2A[;1] + 3A[;2]$.

Similarly,

$$A[;5] = 4A[;1] + 5A[;2] + 6A[;4]$$

We record these facts for future reference.

PROPOSITION 3.1 Let A be row-reduced to E, a matrix in echelon form. Let V be the vector of indices of the pivot columns of E (in ascending order). Then

1. The columns of $A[;V]$ are linearly independent.
2. Given a column index j, let W be the components of V less than or equal to j (i.e., $W \leftarrow (V \leq J)/V$) and let C be the first ρW components of $E[;j]$ (i.e., $C \leftarrow (\rho W) \uparrow E[;J]$). Then

$$A[;j] = A[;W]C \qquad \blacksquare$$

This proposition will apply to all cases, such as E a matrix of zeros, if some conventions are observed.

In the case that E is an m-by-n zero matrix, we take V to be $\iota 0$, the vector without components. Then $E[;V]$ and $A[;V]$ are m by 0 and represent an empty set of vectors in $\mathbf{R}^m$. We consider an empty set of vectors to be linearly independent. We also consider matrices without rows or without columns (i.e., $0\vee.=\rho A$) to be in echelon form.

These conventions are not arbitrary; rather they are, for the most part, variants of the convention that a sum over an empty index set is zero ($0=+/\iota 0$, see exercise 50 in Exercises 1.2 and exercise 8 in Exercises 2.1). Such conventions avoid the separate consideration of special cases and are quite helpful in coding functions for machine execution.

EXAMPLE 3.12 Extract a linearly independent set of vectors from the set

$$\{(0,0,0),\ (2,1,1),\ (0,1,0),\ (4,5,2)\}$$

and express the other vectors as linear combinations of the linearly independent set.

Solution Set up the vectors as columns of a matrix

```
      A
0  2  0  4
0  1  1  5
0  1  0  2
```

and row-reduce to echelon form:

```
      E
0  1  0  2
0  0  1  3
0  0  0  0
```

The pivot columns are $E[;1\ \ 2]$, and so a linearly independent set is $A[;1\ \ 2]$ or (2, 1, 1) and (0, 1, 0). From $E[;4]$ we see that $(4, 5, 2) = 2(2, 1, 1) + 3(0, 1, 0)$ or

```
      V←2 3

      A[;V]+.×2 3
4  5  2
```

or we can apply the formulas of Proposition 3.1:

```
      W←(V≤4)/V
      C←(ρW)↑E[;4]
      A[;W]+.×C
4  5  2
```

The other vector, (0, 0, 0), can of course be expressed as $(0,0,0) = 0 \cdot (2, 1, 1) + 0 \cdot (0, 1, 0)$. Proposition 3.1, on the other hand, says that it should be a linear combination of the elements of $A[;V]$ that precede it. The set of such vectors is empty, but the formulas still apply.

```
      A[;1]
0  0  0
      W←(V≤1)/V
      ρW
0
```

```
      C←(ρW)↑E[;1]
      ρC
0
      A[;W]+.×C
0 0 0
```
■

From Proposition 3.1 we immediately have

PROPOSITION 3.2
1. Any set of $n + 1$ vectors in $\mathbf{R}^n$ is linearly dependent.
2. If $v_1, \ldots, v_n$ are linearly independent vectors in $\mathbf{R}^n$, then every vector in $\mathbf{R}^n$ is a unique linear combination of $v_1, \ldots, v_n$.

Proof Let the $n + 1$ vectors form the columns of A and row-reduce to an echelon form E. The matrix E is n by $n + 1$, and, since two leading 1's cannot occur in the same row, there are at most n pivot columns. Hence there is at least one nonpivot column. This proves statement 1. For statement 2 let $v_i = A[;i]$ and $v = A[;\ n + 1]$. Then the first n columns of E are pivot columns and the last is a nonpivot column and hence a linear combination of the pivot columns. If we interpret E as the echelon form of the system $A[;\iota n]X = v$, we see this solution is unique. ■

EXAMPLE 3.13 Show that every vector in $\mathbf{R}^3$ is a linear combination of the vectors $v_1 = (1, 1, 1)$, $v_2 = (1, 2, -1)$, and $v_3 = (2, 3, 1)$. Write the vectors $u = (19, 30, 4)$, $v = (23, 36, 5)$, and $w = (27, 42, 6)$ as linear combinations of v_1, v_2, v_3.

Solution Imitating the proof of Proposition 3.2, we make v_1, v_2, v_3, u, v, w the columns of a matrix A and row-reduce the matrix to echelon form. By the methods of Section 3.2 we get

$$GA = \begin{bmatrix} 5 & -3 & -1 & 1 \\ 2 & -1 & -1 & 1 \\ -3 & 2 & 1 & 1 \end{bmatrix} \begin{bmatrix} 1 & 2 & 19 & 23 & 27 \\ 2 & 3 & 30 & 36 & 42 \\ -1 & 1 & 4 & 5 & 6 \end{bmatrix}$$

$$= \begin{bmatrix} 1 & 0 & 0 & 1 & 2 & 3 \\ 0 & 1 & 0 & 4 & 5 & 6 \\ 0 & 0 & 1 & 7 & 8 & 9 \end{bmatrix} = E$$

Thus $A[;1\ 2\ 3]$ are three linearly independent vectors in $\mathbf{R}^3$, hence every vector in $\mathbf{R}^3$ is a unique linear combination of v_1, v_2, v_3. In particular,

$$\begin{aligned} u &= v_1 + 4v_2 + 7v_3 \\ v &= 2v_1 + 5v_2 + 8v_3 \\ w &= 3v_1 + 6v_2 + 9v_3 \end{aligned}$$ ■

Now we wish to prove a basic result: a matrix has one and only one row echelon form. To prove that every matrix has an echelon form we must show that the row-reduction process works. Since anyone who has row-reduced a few matrices knows that the process will work, a formal proof of the fact may seem a waste of time. There is, however, a practical reason for writing down a proof. The tasks of proving that row-reduction works, describing the row-reduction process in detail, and coding row-reduction for machine execution are closely related.

PROPOSITION 3.3 Every matrix may be row-reduced to a unique row echelon form.

Proof By the term "row-reduce A" we mean "multiply A by a sequence of elementary matrices, $G = P_n P_{n-1} \dots P_1$, so that $E = GA$ is in row echelon form."

If A is a zero matrix, then it is already in row echelon form, and G is an identity matrix that we consider to be an elementary matrix (it multiplies rows by 1). If $A \neq 0$, then there is a first column, $A[;k]$ say, which is nonzero. Suppose that $A[r;k] \neq 0$. Let P_1 be the switch matrix that interchanges rows 1 and r, let P_2 be the pivot matrix for the $1;k$ entry of $P_1 A$, and let P_3 be the multiplier matrix that sets the $1;k$ entry of $P_2 P_1 A$ to 1. Let $A_1 = P_3 P_2 P_1 A$.

We have now completed one cycle of the row-reduction process and have the following setup with $l = 1$ and $p = k$.

(i) The first p columns of A_l are an echelon form with l nonzero rows.

(ii) If the matrix B_l, obtained by dropping the first l rows and p columns from A_l, is a zero matrix, then A_l is in row echelon form.

To check statement (i) notice that the only nonzero entry of $A_1[;\iota k]$ is 1 in the $1;k$ position. So the conditions of Definition 3.1 certainly hold for $A[1;\iota k]$. Further, if $B_1 = 0$, then $A_1[1;]$ is the only nonzero row of A_1, and so the conditions hold for A_1 as well.

Now suppose that conditions (i) and (ii) hold for some $l > 1$. If $B_l \neq 0$, we carry out another cycle of the row-reduction process. Suppose that $B_l[;k]$ is the first nonzero column of B_l. Suppose that $B_l[r;k] \neq 0$. Use P_1 to interchange rows $l + 1$ and $l + r$ in A_l, use P_2 to pivot on the $l + 1; p + k$ entry of $P_1 A_l$, and use P_3 to set the $l + 1; p + k$ entry of $P_2 P_1 A_l$ equal to 1, giving $A_{l+1} = P_3 P_2 P_1 A_l$.

Condition (i) holds for A_{l+1} with l replaced by $l + 1$ and p replaced by $p + k$. First, since $B_l[;k]$ is the first nonzero column of B_l, $A_{l+1}[;\ \iota(p + k)]$ has only $l + 1$ nonzero rows and $A_{l+1}[l + 1;\ \iota(p + k)]$ is zero except for a 1 in the last position. This 1 is the only nonzero entry in its column. Since the entries of $A_l[\iota l;\ \iota(p + k - 1)]$ have not changed, (i) is true.

If $B_{l+1} = 0$, $A_{l+1}[l + 1;]$ is the last nonzero row and A_{l+1} is in row echelon form.

Recall that a matrix without rows or without columns is a zero matrix. Thus we will get $B_l = 0$ when we run out of rows or columns, if not before, and so the process must end. The matrix A has then been row-reduced.

Next we must show that there is only one echelon form. That is, if $E_1 = G_1A$ and $E_2 = G_2A$ are in row-reduced echelon form, where G_1 and G_2 are products of elementary matrices, then $E_1 = E_2$. Notice that we do not assert that G_1 and G_2 are equal. During each cycle of the row-reduction process one has a choice of pivots, and different choices produce different G's.

Now $A = G_1^{-1}E_1 = G_2^{-1}E_2$, and so $E_2 = G_2G_1^{-1}E_1$. Since G_1 and G_2 are products of elementary matrices, so too is $G_2G_1^{-1}$, and so E_1 may be row-reduced to E_2, and similarly E_2 may be row-reduced to E_1. This means that we may apply Proposition 3.1 with either one of the matrices E_i as A and the other as E.

If $E_1 \neq E_2$, then there is a first column, the kth say, at which they differ. Now $E_1[;k]$ and $E_2[;k]$ cannot both be nonpivot columns, because then, according to Proposition 3.1, they would be linear combinations, with identical coefficients, of the previous columns in which E_1 and E_2 do not differ. If one, say $E_1[;k]$, is a nonpivot column and $E_2[;k]$ is a pivot column, then $E_1[;k]$ is independent of the previous pivot columns, and so, by Proposition 3.1, $E_2[;k]$ must also be independent of the previous columns. But if $E_2[;k]$ is a nonpivot column, it is not independent of the previous pivot columns. The matrices E_1 and E_2 may be exchanged in this argument. Thus, we cannot have one of $E_i[;k]$ a pivot column and the other a nonpivot column. The only remaining choice is that both $E_1[;k]$ and $E_2[;k]$ are pivot columns and hence can differ only in the position of the leading 1. But if $E_1[;\ \iota(k-1)] = E_2[;\ \iota(k-1)]$ has r nonzero rows, then this leading 1 must appear in the $(r+1)$st row in both matrices. This contradiction shows that $E_1 = E_2$. ■

Now we are in a position to prove Propositions 2.5 and 2.9. The next proposition includes Proposition 2.9.

PROPOSITION 3.4 Let A be a matrix. The following statements are all true or all false.

1. The matrix A has a left inverse.
2. The row echelon form of A is $\left[\frac{I}{0}\right]$, where I is an identity matrix.
3. The columns of A are linearly independent.

Proof It was shown in Section 2.3 that statement 1 implies statement 3. Suppose that statement 3 is true. Let E be the row-echelon form of A, $E[;V]$ the pivot columns. Now if there were a nonpivot column $E[;j]$, then $A[;j]$ would be a linear combination of the columns of $A[;V]$. This cannot be if statement 3 is true. Thus E has no nonpivot columns. That is, $E = \left[\frac{I}{0}\right]$.

We saw in Section 3.2 that if statement 2 is true, then statement 1 is true. In fact, if $GA = E = \left[\frac{I}{0}\right]$, then $G[\iota 1 \downarrow \rho A;]$ is a left inverse for A. ■

The next proposition includes Proposition 2.5.

PROPOSITION 3.5 Let A be a matrix

1. An invertible matrix is a product of elementary matrices. In particular, an invertible matrix is square.
2. If A is square and has a left inverse, then A is invertible.
3. If A is square and has a right inverse, then A is invertible.

Proof

1. Suppose that A is invertible. By Proposition 3.4 there is a product of elementary matrices G such that

$$GA = \left[\frac{I}{0}\right]$$

Since G is square, A and

$$\left[\frac{I}{0}\right]$$

have the same shape. Thus we wish to show that the 0 does not actually appear. Multiplying on the right by A^{-1} gives

$$G = GAA^{-1} = \left[\frac{I}{0}\right]A^{-1} = \left[\frac{A^{-1}}{0}\right]$$

Now G is invertible, and so

$$I = GG^{-1} = \left[\frac{A^{-1}}{0}\right]G^{-1} = \left[\frac{A^{-1}G^{-1}}{0}\right]$$

So the rows of zeros do not appear (they do not appear in I) and $A^{-1}G^{-1} = I$—that is, $A = G^{-1}$. But G is a product of elementary matrices and hence A is also. If $G = P_nP_{n-1} \ldots P_1$, then $G^{-1} = P_1^{-1}P_2^{-1} \ldots P_n^{-1}$.

2. Since A has a left inverse, there is a product of elementary matrices G such that

$$GA = \left[\frac{I}{0}\right]$$

by Proposition 3.4, and since A is square, the rows of zeros do not appear. Thus $GA = I$, hence $A = G^{-1}GA = G^{-1}I = G^{-1}$ and hence A is invertible with $A^{-1} = G$.

3. Statement 3 follows from statement 2 by taking transposes. ∎

The echelon form of a matrix is unique. Thus no matter how we row-reduce A to E, we always obtain the same number of pivot columns.

DEFINITION 3.3 The *rank* of a matrix is the number of pivot columns in the row echelon form.

The rank of A is the minimum number of columns needed to generate all the columns of A as linear combinations.

In the examples that follow and in the exercises at the end of this section a number of row-reductions must be performed. These computations are tedious if done by hand or even using the functions *SWITCH, PIVOT,* and *LDR* defined in Section 3.2. In Section 3.5 the advanced function-writing techniques discussed in Section 3.4 are used to define two functions, *GAUSS* and *ECHELON*. The expression $G \leftarrow GAUSS\ A$ defines G to be an invertible matrix such that $E = GA$ is in row echelon form. The expression $E \leftarrow ECHELON\ A$ defines E to be the echelon form of A. We will use these expressions in the examples below to briefly indicate that a row reduction has been performed.

EXAMPLE 3.14 Find a linear independent subset of the set of vectors

$$\{(3, 5, -2, 4), (9, 15, -6, 12), (2, 1, 0, 0), \\ (-1, -4, 2, -4), (2, 8, -4, 8), (0, 2, -1, 2)\}$$

Express the other vectors as linear combinations of the independent vectors.

Solution Store the vectors as columns of a matrix

```
      A
 3    9   2  ¯1   2   0
 5   15   1  ¯4   8   2
¯2   ¯6   0   2  ¯4  ¯1
 4   12   0  ¯4   8   2
```

and reduce to echelon form:

```
        ECHELON A
 1.000E0      3.000E0      0.000E0   ¯1.000E0     2.000E0     6.939E¯18
¯2.776E¯17   ¯6.939E¯17    1.000E0    1.000E0    ¯2.000E0    ¯6.939E¯18
¯3.469E¯17   ¯5.551E¯17    0.000E0    2.082E¯17  ¯4.163E¯17   1.000E0
 0.000E0      0.000E0      0.000E0    0.000E0     0.000E0     0.000E0
```

Assuming the small numbers to be zero, the pivot columns are $E[;1\ \ 3\ \ 6]$. Thus an independent set is

$$A[;1\quad 3\quad 6] = \begin{bmatrix} 3 & 2 & 0 \\ 5 & 1 & 2 \\ -2 & 0 & -1 \\ 4 & 0 & 2 \end{bmatrix}$$

The display merely indicates that *to three significant digits*

$$\begin{bmatrix} 9 \\ 15 \\ -6 \\ 12 \end{bmatrix} = A[;2] = 3A[;1] + 0 \cdot A[;3] + 0 \cdot A[;6] = \begin{bmatrix} 3 & 2 & 0 \\ 5 & 1 & 2 \\ -2 & 0 & -1 \\ 4 & 0 & 2 \end{bmatrix} \begin{bmatrix} 3 \\ 0 \\ 0 \end{bmatrix}$$

$$\begin{bmatrix} -1 \\ -4 \\ 2 \\ -4 \end{bmatrix} = A[;4] = -A[;1] + A[;3] + 0 \cdot A[;6] = \begin{bmatrix} 3 & 2 & 0 \\ 5 & 1 & 2 \\ -2 & 0 & -1 \\ 4 & 0 & 2 \end{bmatrix} \begin{bmatrix} -1 \\ 1 \\ 0 \end{bmatrix}$$

$$\begin{bmatrix} 2 \\ 8 \\ -4 \\ 8 \end{bmatrix} = A[;5] = 2A[;1] - 2A[;3] + 0 \cdot A[;6] = \begin{bmatrix} 3 & 2 & 0 \\ 5 & 1 & 2 \\ -2 & 0 & -1 \\ 4 & 0 & 2 \end{bmatrix} \begin{bmatrix} 2 \\ -2 \\ 0 \end{bmatrix}$$

where vectors have been written as column vectors. Carrying out the computations shows the equations to be exactly true. ■

Many superficially different problems can be reduced to the kind of computation done in the last example. As an illustration of this, we will show how to give a rigorous meaning, for linear equations at least, to such often-heard statements as "Two equations in three unknowns leave one degree of freedom" and "We have four equations and three unknowns, hence the system is overdetermined and has no solution."

To avoid confusion, notice that although we begin with the familiar system

$$AX = B$$

we do *not* now reduce the augmented matrix $[A \mid B]$. Rather we now reduce the matrix $[A \mid B]^T$. This is no way to solve the system of linear equations, but it does yield the theoretical result we are after.

First, notice that we have been treating individual linear equations of the form

$$a_1x_1 + a_2x_2 + \cdots + a_nx_n = b$$

as though they were vectors. In the row-reduction process this equation becomes a *row* vector in the augmented matrix $[A \mid B]$. It is multiplied and added as though it were the vector

$$(a_1, a_2, \ldots, a_n, b)$$

in $\mathbf{R}^{n+1}$.

We say that a linear equation is a *linear combination* of other linear equations if the statement is true for the corresponding vectors. A system of linear equations in which some equations are linear combinations of the others is called *redundant*. If the equations are independent, the system is *irredundant*.

EXAMPLE 3.15 From the set of linear equations

$$\begin{aligned}
-460x_1 + 76x_2 + 67x_3 - 541x_4 + 219x_5 &= 313\\
-285x_1 + 35x_2 + 40x_3 - 240x_4 + 110x_5 &= 135\\
155x_1 + 83x_2 - 9x_3 - 673x_4 + 157x_5 &= 424\\
-102x_1 + 68x_2 - 34x_3 + 272x_4 + 170x_5 &= 306
\end{aligned}$$

extract an irredundant set. Express the remaining equations, if any, as a linear combination of the irredundant set.

Solution We set up the corresponding vectors as columns of a matrix A. Notice that A is the *transpose* of the partitioned matrix used to solve the system.

```
      A
¯460 ¯285  155 ¯102
  76   35   83   68
  67   40   ¯9  ¯34
¯541 ¯240 ¯673  272
 219  110  157  170
 313  135  424  306
```

Reduce this matrix to echelon form:

```
     ECHELON A
 1.000E0      0.000E0      5.230E0      2.602E¯18
¯6.939E¯18    1.000E0     ¯8.986E0      1.041E¯17
 5.421E¯19   ¯4.066E¯19    5.963E¯18    1.000E0
¯1.943E¯16   ¯3.469E¯17   ¯4.025E¯16    1.388E¯16
¯1.110E¯16    5.551E¯17   ¯1.971E¯15    1.388E¯16
 5.551E¯17    0.000E0      2.220E¯16   ¯5.551E¯17
```

The pivot columns are $E[;1\ 2\ 4]$ and, to three significant figures, $E[;3] = 5.230\ E[;1] - 8.986\ E[;2]$.

Thus equations 1, 2, and 4 form an irredundant set and equation 3 is, approximately, 5.230 times equation 1 minus 8.986 times equation 2. ∎

The next proposition is a fundamental result about the rank of a matrix. The proof must wait until the concept of a vector space has been developed.

PROPOSITION 3.6 The matrices A and A^T have the same rank. ∎

Assuming 3.6 is true, we have

PROPOSITION 3.7

1. Any system of $n + 2$ linear equations in n unknowns is redundant.

2. An irredundant system of $n + 1$ linear equations in n unknowns has no solutions.

3. An irredundant system of k linear equations in n unknowns with $n \leq k$ has solutions. If $n = k$, there is a unique solution; otherwise the solutions involve $n - k$ arbitrary parameters.

Proof Given k equations in n unknowns, we have k vectors in $\mathbf{R}^{n+1}$. If $k > n + 1$, these vectors are linearly dependent by Proposition 3.2, and this proves statement 1.

Assume now that the k equations are irredundant and are written in the matrix form $AX = B$. Then the partitioned matrix $[A \mid B]$ is k by $n + 1$ and the matrix $[A \mid B]^T$ has rank k. Thus by Proposition 3.6 the matrix $[A \mid B]$ has rank k. This means there are no rows of zeros at the bottom. If $k = n + 1$, this means the echelon form of $[A \mid B]$ is

$$\left[\begin{array}{c|c} * & * \\ \hline 0 & 1 \end{array}\right]$$

and hence the original system can be reduced to a system containing the equation $0 = 1$. This proves statement 2. If $k \leq n$, then we have solutions, and there is an arbitrary parameter for each nonpivot column. ∎

EXERCISES 3.3

In exercises 1 through 5, (a) show that every vector in $\mathbf{R}^n$ is a linear combination of the vectors $v_1, v_2, \ldots, v_n$; (b) write the vectors $w_1, w_2, \ldots, w_m$ as linear combinations of $v_1, v_2, \ldots, v_n$.

1. $v_1 = (1, 1)$, $v_2 = (1, 2)$
   $w_1 = (6, 11)$, $w_2 = (8, 14)$, $w_3 = (10, 17)$, $w_4 = (12, 20)$
2. $v_1 = (2, -1)$, $v_2 = (-1, 1)$
   $w_1 = (-3, 4)$, $w_2 = (-2, 4)$, $w_3 = (-1, 4)$, $w_4 = (0, 4)$
3. $v_1 = (1, -2, 1)$, $v_2 = (1, -1, 0)$, $v_3 = (0, -1, 2)$
   $w_1 = (0, -2, 3)$, $w_2 = (0, 2, -3)$, $w_3 = (-2, 2, 1)$
4. $v_1 = (-2, 3, 1)$, $v_2 = (-2, 2, 1)$, $v_3 = (-1, 1, 1)$
   $w_1 = (-1, 2, 1)$, $w_2 = (1, -2, -1)$, $w_3 = (3, -4, -1)$
5. $v_1 = (-2, 1, 1, 0)$, $v_2 = (1, 2, -1, 3)$, $v_3 = (1, 0, 0, 1)$, $v_4 = (0, 1, -1, 1)$
   $w_1 = (3, -3, 2, -2)$, $w_2 = (1, -2, 1, -1)$, $w_3 = (3, -1, 1, 0)$

In exercises 6 through 10 compute the rank of the given matrix.

6. $\begin{bmatrix} 1 & 0 & 0 \\ 1 & 1 & 1 \\ 1 & 1 & 1 \end{bmatrix}$

7. $\begin{bmatrix} 1 & -1 & 2 \\ 0 & 2 & -4 \\ 0 & 1 & -2 \end{bmatrix}$

8. $\begin{bmatrix} 1 & 0 & 2 & 0 \\ 1 & 1 & 5 & -2 \\ -2 & 1 & -1 & -1 \\ 1 & -2 & -4 & 5 \end{bmatrix}$

9. $\begin{bmatrix} 1 & 0 & 2 & 0 \\ -7 & 5 & 1 & 0 \\ 3 & -1 & 3 & 1 \\ -6 & 3 & -3 & -1 \end{bmatrix}$

10. 1 0 1 0 ρ 0

In exercises 11 through 15 find a linearly independent subset of the given set of vectors and express the other vectors as linear combinations of this subset.

11. $v_1 = (1, -2)$, $v_2 = (2, -4)$, $v_3 = (0, 1)$, $v_4 = (3, -2)$

12. $v_1 = (2, -1)$, $v_2 = (4, -2)$, $v_3 = (-1, 1)$, $v_4 = (2, 1)$

13. $v_1 = (1, 1, 0)$, $v_2 = (-1, 0, -2)$, $v_3 = (-2, -1, -2)$
$v_4 = (-3, -2, -2)$, $v_5 = (-1, -1, 1)$, $v_6 = (-5, -2, -2)$

14. $v_1 = (-2, -1, -2)$, $v_2 = (3, 1, 2)$, $v_3 = (5, 2, 4)$
$v_4 = (7, 3, 6)$, $v_5 = (1, 0, 1)$, $v_6 = (9, 1, 6)$

15. $v_1 = (1, -2, -1, -2)$, $v_2 = (0, 1, 1, -1)$, $v_3 = (2, -6, -4, -2)$
$v_4 = (-1, 2, 1, 2)$

For exercises 16 through 20: (a) Extract an irredundant set of equations from the given set. (b) Without solving the irredundant system, state whether it has no solution, a unique solution, or an infinity of solutions. If the latter, how many arbitrary parameters are there in the solutions?

16.
$$\begin{aligned} x + y &= -2 \\ -7x - 7y &= 14 \\ x + 2y &= -2 \\ 5x + 8y &= -10 \\ -2x - 2y &= 5 \end{aligned}$$

17.
$$\begin{aligned} 6x - y &= 2 \\ -42x + 7y &= -14 \\ -x + y &= 0 \\ 9x + y &= 4 \\ 2x &= 1 \end{aligned}$$

18.
$$\begin{aligned} x - 2z &= -1 \\ -2x + y + 2z &= 2 \\ -3x + y + 4z &= 3 \\ -4x + y + 6z &= 4 \\ x + y - 4z &= -1 \end{aligned}$$

19.
$$\begin{aligned} -3x + y - 4z &= -2 \\ x + 2y - z &= 0 \\ 4x + y + 3z &= 2 \\ 7x + 7z &= 4 \\ -8x + 5y - 13z &= -6 \end{aligned}$$

20.
$$\begin{aligned} 7x_1 - x_2 + x_3 - x_4 &= 1 \\ -x_1 + x_2 - 2x_3 - x_4 &= 0 \\ 11x_1 + x_2 - 4x_3 - 5x_4 &= 2 \\ -23x_1 + 5x_2 - 7x_3 + x_4 &= -3 \\ 2x_1 - x_2 + x_3 &= 0 \\ -x_1 + x_2 + x_3 + x_4 &= 0 \end{aligned}$$

21. Show that a rank 1 matrix A can be written as $V \circ . \times W$ for suitable vectors V and W. Hint: Write $A = G^{-1}E$. How many columns of G^{-1} are relevant?

22. Show that A and B have the same echelon form if and only if there is an invertible matrix F such that $B = FA$.

23. (Computer assignment) If Proposition 2.11 is to hold when $0 = 1 \downarrow \rho A$, then the solution of $AX = 0$ must be the "zero vector" $\iota 0$. Set $A \leftarrow 10\ 0\rho 0$ and compute ⌹A and $(10\rho 0)$⌹A at a terminal.

3.4 Branching and Recursion

In this section we discuss some advanced APL function-writing techniques. The object is to develop enough APL to enable us to write a function *ECHELON* that will take a matrix for an argument and return the echelon form of the matrix. The material on branching and comparison tolerance below will allow us to do this in Section 3.5. The material on recursion is not needed for 3.5 but will be needed in later chapters.

BRANCHING

In a complex process, such as row-reduction, not all operations are known in advance. For example, one does not know where the third pivot column in a row-reduction will occur until the first and second pivot operations have been performed. A function to row-reduce a matrix must be able to take different actions for different matrices. This is the purpose of branching or GO TO instructions in computer languages.

Let us look at a simple example. In applications one often encounters "piecewise" functions — functions that are given by different formulas over different intervals, such as

$$f(x) = \begin{cases} 0 & \text{if } x < 0 \\ x^2 & \text{if } x \geq 0 \end{cases} \tag{3.1}$$

For example, a circuit element might give no output for negative voltage input and a nonlinear output for positive voltage input.

An APL function capable of applying one formula when $x \geq 0$ and another formula when $x < 0$ could be used to compute a function such as $f(x)$. In APL such functions are written using the symbol →, called *branch* or *GO TO*.

The expression

```
→3
```

in an APL function means "GO TO line 3 instead of the next line." Now → is not quite a normal APL function, but it obeys the same rule as a monadic APL function: it operates on the result of the expression to its right. For example,

```
→0 1 0/1 2 3
```

means "*GO TO* line 2." This means that one can jump to different lines, depending on circumstances. Before looking at examples of this, we need to be more precise about the operation of the monadic function →.

1. The right argument of → is a scalar or vector.
2. →0 or "*GO TO* line zero" means "The computation has been completed, return the result." Nonexistent line numbers have the same effect. If n is not a line number for the function, then →n is the same as →0.
3. →⍳0 has no effect. The function just passes on to the next line as if the → were not there.
4. If V is a vector but not ⍳0, then →V is the same as →V[1]. The other components of V, if any, are ignored.

For example,

```
→(X<0)/0
```

will be →0 if X is less than zero and →⍳0 if X is not less than zero. Thus the expression reads "Stop if X is less than zero, otherwise, go on," and the function $f(x)$ of Equation (3.1) can be written using branching as

```
    ∇ Z←F X
[1]   Z←0
[2]   →(X<0)/0
[3]   Z←X*2
    ∇
```

Some further examples are given below.

EXAMPLES 3.16 Use branching to define an APL function to compute

$$f(x) = \begin{cases} x^3 & \text{if } x \leq 1 \\ x^2 & \text{if } x > 1 \end{cases}$$

Solution

```
    ∇ Z←F X
[1]   Z←X*3
[2]   →(X≤1)/0
[3]   Z←X*2
    ∇
```

■

EXAMPLE 3.17 Use branching to define an APL function to compute

$$f(x) = \begin{cases} 0 & \text{if } x \leq 0 \\ x & \text{if } 0 < x \leq 1 \\ 1 & \text{if } x > 1 \end{cases}$$

Solution

```
    ∇ Z←F X
[1]    Z←0
[2]    →(X≤0)/0
[3]    Z←X
[4]    →(X≤1)/0
[5]    Z←1
    ∇
```

■

EXAMPLE 3.18 Use branching to define an APL function to compute

$$f(x, y) = \begin{cases} 0 & \text{if } 0 \leq x \leq 1 \text{ and } 0 \leq y \leq 1 \\ xy(1 - x)(1 - y) & \text{otherwise} \end{cases}$$

Solution

```
    ∇ Z←X F Y
[1]    Z←0
[2]    →((0≤X)∧(X≤1)∧(0≤Y)∧Y≤1)/0
[3]    Z←X×Y×(1-X)×1-Y
    ∇
```

■

In none of the examples above was a jump to an actual line number used. The "stop or continue" form used above suffices for most simple situations. A version of Example 3.16, the version most beginners would write, can be used to illustrate branching to a line number:

```
    ∇ Z←F X
[1]    →(X≤1)/4
[2]    Z←X*3
[3]    →0
[4]    Z←X*2
    ∇
```

(This version first checks to see if $X \geq 1$ or $X < 1$ and then applies the appropriate formula. As the given solution to Example 3.16 shows, it is usually simpler to assume one case — for example, $X < 1$ — apply the appropriate formula, and then change to another formula if necessary.)

Expressions of the form →(X≥0)/4 are often inconvenient, because editing the function may cause the line numbers to change. Then each line containing a → must be checked to see if it is still valid. We can avoid this problem by labeling statements. A preferred version of *F* is

```
    ∇ Z←F X
[1]     →(X≤1)/POS
[2]     Z←X*3
[3]     →0
[4]     POS:Z←X*2
    ∇
```

The line [4]POS:Z←X*2 contains the label POS. A *label* is a variable name beginning a line and separated from the rest of the line by a colon (:). It is simply a special type of local variable that does not need to be declared in the header and whose value is always the line number to its left. If this last version of *F* is edited and line numbers are changed, the branch statement on line [1] need not be changed.

Comparison Tolerance (Fuzz)

Machine computations are not exact. Numbers that should be zero are merely small.

```
      1-3×÷3
1.734723476E¯18
```

The display above will vary from machine to machine, but it will rarely yield precisely zero. On the other hand, one would like an expression such as $1 = 3(\frac{1}{3})$ to test out to be true anyway.

```
      1=3×÷3
1
```

This is accomplished by allowing the function = (as well as >, ≤, and so on) to ignore small enough differences. The size of the differences ignored is set by the "system variable" ⎕CT, whose size in turn varies with the particular machine and can be changed by the user. To find out its value for a particular machine one simply types ⎕CT. The number ⎕CT is called the *comparison tolerance* or *fuzz*. It can be thought of as the largest number for which the expression

```
1=1+⎕CT
```

is "true."

```
      1=1+⎕CT
1
```

For example, if ⎕CT is 10^{-4}, then 1 will be considered equal to 1.0001.

```
      ⎕CT←1E¯4
      1=1.0001
1
```

⎕CT may take any value between 0 and 1, 0 included and 1 excluded.

More generally, `A=B` will return a 1 as long as $|A - B|$ is less than `M×⎕CT`,† where M is the absolute value of A or the absolute value of B, whichever is greater (`M←⌈/|A,B`).

If `A=B` returns a `1`, we say that A is *fuzzily equal* to B.

EXAMPLE 3.19 Set `⎕CT←1E¯4`. Which of the following pairs of numbers are fuzzily equal?

(a) 1 and 1.0001.
(b) 100 and 100.01.
(c) 10^{-5} and 10^{-6}.
(d) 0 and 10^{-100}.

Solutions

(a) If `A←1` and `B←1.0001`, then $B - A$ is 10^{-4}, M is 1.0001, and `M×⎕CT` is thus a bit more than 10^{-4}. A and B are fuzzily equal.

(b) `A←100`, `B←100.01`, $B - A$ is 10^{-2}, M is 100.01, and so `M×⎕CT` is a bit larger than 10^{-2}. A and B are fuzzily equal.

(c) `A←.00001`, `B←.000001`, and $A - B$ is .000009. M is 10^{-5}, and so `M×⎕CT` is 10^{-9}, which is less than `9E¯6`. A and B are *not* fuzzily equal.

(d) `A←0`, `B←1E¯100`, and $B - A$ is 10^{-100}. M is also 10^{-100}, and so `M×⎕CT` is 10^{-104}, which is less than 10^{-100}. A is not fuzzily equal to B. ■

Example 3.19(d) illustrates a general fact. The expression `0=A` returns a 1 only if A is precisely zero. Thus `1=1+⎕CT` returns a `1`, but `0=⎕CT` returns a `0`.

We will confine our use of fuzzy equality for the most part to a single form. When we wish to test if the quantity B is negligible compared to the quantity A, we will use the expression

```
A=A+B
```

DEFINITION 3.4 We say that B is *negligible compared to A* if A is fuzzily equal to $A + B$.

We have used the setting `⎕CT←1E¯4` for illustration only. The fuzz should almost never be changed from its default value. This value varies from machine to machine and is set to the value thought to be the best choice for the majority of computations.

†This is not quite true. We are ignoring complications that arise from the machine representation of numbers.

RECURSION

Mathematical induction is a concept that is particularly important for linear algebra. Proofs and definitions often proceed by induction. For example, the proof of the existence of the echelon form of a matrix (Proposition 3.3) was an inductive proof. Inductive definitions are often called *recursive,* especially in a computer science context. An APL function that is defined by mathematical induction is called a *recursive function.*

A simple example is the factorial function, written $!n$ in APL notation and $n!$ otherwise. The definition for n an integer is

$$n! = n(n-1)(n-2) \cdot \cdots \cdot 2 \cdot 1, \qquad n > 0 \quad \text{and} \quad 0! = 1$$

or $\times/\iota n$.

This function can be defined inductively as

$$f(n) = \begin{cases} 1 & \text{if } n = 0 \\ n \cdot f(n-1) & \text{if } n > 0 \end{cases}$$

Such definitions have a strict form consisting of two parts.

1. The specification of the starting value.
2. The expression of the nth value in terms of previous values.

The APL version of the function $f(n)$ is

```
    ∇ Z←FACT N
[1]   Z←1
[2]   →(N=0)/0
[3]   Z←N×FACT N−1
    ∇
```

The specification of the starting value is on line [1]. The function continues only if $n > 0$. On line [3] the function is called again with a smaller value of n. Without the starting value the function would call itself forever — or until it filled the workspace.

The above function is not necessary in APL, since the factorial is monadic !. However, no primitive function will give us the nth power of a square matrix. Notice that we could define A^n as

$$A^n = \begin{cases} I & \text{if } n = 0 \\ A \cdot A^{n-1} & \text{if } n > 0 \end{cases}$$

where $I = ID\ 1 \uparrow \rho A$.

```
    ∇ Z←A TOTHE N
[1]     Z←ID 1↑ρA
[2]     →(0=N)/0
[3]     Z←A+.×A TOTHE N−1
    ∇
```

We will show in Section 3.6* how powers of a matrix arise in applications. First, however, we give some more examples of recursive functions.

EXAMPLE 3.20 Compute x_{50} if

$$x_n = \begin{cases} 1 & \text{if } n = 1 \\ x_{n-1}\left(1 - \dfrac{x_{n-1}}{2}\right) & \text{if } n > 1 \end{cases}$$

Solution

```
    ∇ Z←F N
[1]     Z←1
[2]     →(N=1)/0
[3]     Z←F N−1
[4]     Z←Z×1−Z÷2
    ∇

        F 50
0.03652
```

■

EXAMPLE 3.21 Let x_n be defined as in Example 3.20. Compute the vector $(x_1, x_2, \ldots, x_{10})$ and the vector $(x_{40}, \ldots, x_{50})$.

Solution Write a function *VF N* that computes the vector $x_1, \ldots, x_n$:

```
    ∇ Z←VF N
[1]     Z←,1
[2]     →(N=1)/0
[3]     Z←VF N−1
[4]     Z←Z,Z[N−1]×1−Z[N−1]÷2
    ∇

        VF 10
1  0.5  0.375  0.3047  0.2583  0.2249  0.1996  0.1797  0.1636
    0.1502  39
        ↓VF 50
0.04489  0.04388  0.04292  0.042  0.04112  0.04027  0.03946
    0.03868  0.03793  0.03721  0.03652
```

On line [3] of the solution `VF N−1` or $(x_1, x_2, \ldots, x_{n-1})$ is computed. Then `VF N` is just $(x_1, x_2, \ldots, x_{n-1}, x_{n-1}(1 - x_{n-1}/2))$. x_n appears to be a decreasing sequence. ■

A more natural coding for the function F in Example 3.20 would be

```
    ∇ Z←F1 N
[1]   Z←1
[2]   →(N=1)/0
[3]   Z←(F1 N-1)×1-(F1 N-1)÷2
    ∇
```

This function will work, but $F1$ costs more to use than F, because it involves more function calls and hence more computation.

How many more function calls? Let us let $c(n)$ be the number of times F is called to compute $F\ n$ and $c1(n)$ be the number of times $F1$ is called to compute $F1\ n$.

First we get a formula for $c(n)$. This is not particularly hard. $c(1) = 1$, because F stops on line [2]. If $n > 1$, then $F\ n$ involves the original function call, and then on line [3] we get $c(n - 1)$ more function calls. Thus

$$c(n) = \begin{cases} 1 & \text{if } n = 1 \\ 1 + c(n-1) & \text{if } n > 1 \end{cases}$$

Thus $c(1) = 1$, $c(2) = 1 + 1 = 2$, $c(3) = 1 + 2 = 3$, and $c(n) = n$.

Next we calculate $c1(n)$. Again $c1(1) = 1$. For $c1(n)$ we have the original function call plus $2c1(n - 1)$ function calls on line [3]. Thus

$$c1(n) = \begin{cases} 1 & \text{if } n = 1 \\ 1 + 2c1(n-1) & \text{if } n > 1 \end{cases}$$

A formula for $c1(n)$ is not so obvious, but we can get an idea of how it grows easily enough.

```
    ∇ Z←VC1 N
[1]   Z←,1
[2]   →(N=1)/0
[3]   Z←VC1 N-1
[4]   Z←Z,1+2×Z[N-1]
    ∇

      VC1 20
1  3  7  15  31  63  127  255  511  1023  2047  4095  8191  16380
      32770  65540  131100  262100  524300  1049000
```

Thus to compute x_{20} the function F uses twenty function calls and the function $F1$ uses more than a million.

A formula for $c1(n)$ may be found using the methods of Section 3.6*.

EXERCISES 3.4

Use branching to define APL functions to compute the functions given in exercises 1 through 8.

1. $f(x) = \begin{cases} x & \text{if } x \le 1 \\ x^2 & \text{if } x > 1 \end{cases}$

2. $f(x) = \begin{cases} \sqrt{1 - x^2} & \text{if } |x| \le 1 \\ \sqrt{x^2 - 1} & \text{if } |x| > 1 \end{cases}$

3. $f(x) = \begin{cases} -x - 1 & \text{if } x \le -1 \\ -\sqrt{1 - x^2} & \text{if } -1 \le x \le 1 \\ x - 1 & \text{if } x \ge 1 \end{cases}$

4. $f(x) = \begin{cases} -x - \frac{\pi}{2} & \text{if } x \le -\frac{\pi}{2} \\ \cos x & \text{if } -\frac{\pi}{2} < x < \frac{\pi}{2} \\ x - \frac{\pi}{2} & \text{if } x \ge \frac{\pi}{2} \end{cases}$

5. $f(x) = \begin{cases} x \text{ if } n - 1 < x \le n, & n \text{ even} \\ -x \text{ if } n - 1 < x \le n, & n \text{ odd} \end{cases}$

6. $f(x, y) = \begin{cases} x + y & \text{if } |x| + |y| < 1 \\ x^2 + y^2 & \text{otherwise} \end{cases}$

7. $f(x, y) = \begin{cases} xy & \text{if } 3x^2 + 4y^2 - 6xy \le 3 \\ 0 & \text{otherwise} \end{cases}$

8. $f(x, y) = \begin{cases} 0 & \text{if } x^2 + y^2 \ge 1 \\ \sqrt{1 - x^2} & \text{otherwise} \end{cases}$

For exercises 9 through 17 assume that `⎕CT←1E¯4`.

9. Are 10,000 and 10,001 fuzzily equal?
10. Are $-10{,}000$ and $-10{,}001$ fuzzily equal?
11. Are 2 and 2.0001 fuzzily equal?
12. Are 1 and 1.0002 fuzzily equal?
13. Is $-.1$ negligible compared to 10,000?
14. Is $-.1$ negligible compared to 1000?
15. Is .1 negligible compared to 1000?
16. Is 10^{-15} negligible compared to 10^{-10}?
17. The set of points fuzzily equal to 10,000 form an interval $[a, b]$. Find a and b. Is 10,000 in the center of this interval?

    Hint: Write $b = 10{,}000 + \varepsilon$ and find an upper bound for ε.

(Computer assignment) In exercises 18 through 25 compute $x_1, x_2, \ldots, x_N$ for the given N.

18. $x_1 = 2$, $x_n = x_{n-1}^2 - x_{n-1}^3$, $N = 7$

19. $x_1 = 2$, $x_n = x_{n-1} + 2x_{n-1}^2$, $N = 9$

20. $x_1 = 1$, $x_n = 2x_{n-1}(1 - x_{n-1}/10)$, $N = 10$

21. $x_1 = 1$, $x_n = x_{n-1} + \cos x_{n-1}$, $N = 5$

22. $x_1 = 0,\ x_2 = 1$, $x_n = 3x_{n-1} - 4x_{n-2}$, $N = 20$

23. $x_1 = 1,\ x_2 = 1$, $x_n = x_{n-1} + 4x_{n-2}$, $N = 20$

24. $x_1 = 1,\ x_2 = 2,\ x_3 = 3$, $x_n = x_{n-1} - 2x_{n-2} + 3x_{n-3}$, $N = 20$

25. $x_1 = 1,\ x_2 = 2,\ x_3 = 3$, $x_n = x_{n-1}^2 - x_{n-3}^3$, $N = 12$

26. Compute $\begin{bmatrix} .9996 & .0002 \\ -.0006 & 1.0003 \end{bmatrix}^n$ for $n = 2000;\ 10{,}000;\ 100{,}000$.

3.5 Automating Gaussian Reduction

If A is a matrix, then there is an invertible matrix G such that $E = GA$ is in echelon form. In this section we will define a function *GAUSS* such that $G = GAUSS\ A$ and a function *ECHELON* such that $E = ECHELON\ A$.

Machine arithmetic is not exact, and during row-reductions the errors may sometimes accumulate in such a way as to produce a totally erroneous answer. For this reason our automatic version of Gaussian reduction will have an error check built in, the same error check we used in Section 3.2. To compute the echelon form of a matrix A we form the partitioned matrix $[A \mid I]$ and row-reduce to $[E \mid G]$, where E is an echelon form. The function *ECHELON* will not return E, however. The function *ECHELON* will return GA. If GA is "sufficiently close" to an echelon form, we accept the answer. If it is not close enough for our liking, other methods must be tried. These methods involve numerical analysis beyond the scope of this text. If A is a small matrix, however, then step-by-step use of the functions *SWITCH, PIVOT,* and *LDR* will often reveal the problem.

Since G is an invertible matrix, GA has the same echelon form as A (exercise 22 of Exercises 3.3), and a second application of the functions will occasionally produce a better-looking answer.

The main function to be defined is *GAUSS*. In fact, echelon is just

```
    ∇ Z←ECHELON A
[1]     Z←(GAUSS A)+.×A
    ∇
```

PARTIAL PIVOTING

In row-reducing small matrices by hand, one usually pivots on whichever entry involves the least work — the entry that involves the simplest fractions, for example. In automatic processes, however, one attempts to choose pivots in such a way as to minimize the arithmetic errors committed by the machine. The strategy we will employ is called *partial pivoting*. In partial pivoting one uses row interchanges (i.e., the function *SWITCH*) to bring the entry of largest magnitude (absolute value) to the pivot position. (*Full pivoting* involves column interchanges as well.)

For example, suppose that after the first column of A has been cleaned up, one has

$$\left[\begin{array}{ccc|c} 1 & 100 & 30 & \\ 0 & 1 & 6 & \\ 0 & 14 & -20 & * \\ 0 & -50 & 7 & \end{array}\right]$$

Row-reducing by hand, one would now pivot on the 2;2 position. Employing the partial-pivoting strategy, however, one first switches rows 2 and 4 to put the largest (in absolute value) available number into the pivot position. The entry in the 1;2 position is larger, but of course it is not available — using it would disturb the form of the first column.

Partial pivoting is not difficult to code into a row-reduction function. One must search for a nonzero pivot anyway, and so take the largest available magnitude to pivot on.

DYADIC ⍳

To implement the partial-pivoting strategy we must be able to locate that component of a vector with the largest absolute value. Finding the largest absolute value in a vector is not difficult. If *V* is a vector, then |*V* gives the absolute value of each component, and a reduction using the maximum function will pick out the largest entry: ⌈/|*V*.

This is not quite what we want, however. It gives us the size of the desired pivot quantity but does not tell us what row the quantity appears in. We can find the row index by using the equality function and compression,† but it is simpler to use the dyadic function ⍳, called *index-of*. The left argument of ⍳ must be vector *V*. If the right argument of ⍳ is a scalar *A*, then *V*⍳*A* gives the index of the *first* occurrence of *A* in *V*. If *A* does not occur as a component of *V*, then *V*⍳*A* is 1 + ⍴*V*.

```
      1 ¯2 3⍳3
3
      1 ¯2 3⍳¯2
2
      1 ¯2 3⍳2
4
```

If *A* is an array other than a scalar, then the action is componentwise.

```
      A
1  2
3  4
      1 ¯2 3⍳A
1  4
3  4
```

Thus, to find the index of the (first-occurring) component of largest absolute value in the vector V we may use the expression

```
V⍳⌈/|V
```

Next we ask just what is meant by "zero" in the machine version of row-reduction. Consider the row-reduction:

†If L←⌈/|1↓V, then 1↑(L=1↓V)/⍳¯1+⍴V is the row index.

```
      A
1  2  3
4  5  6
7  8  9

      +A←1 1 LDR 1 1 PIVOT 1 3 SWITCH A
1.0E0        1.1E0        1.3E0
6.9E¯18      4.3E¯1       8.6E¯1
1.7E¯18      8.6E¯1       1.7E0

      +A←2 2 LDR 2 2 PIVOT 2 3 SWITCH A
1.0E0        1.7E¯18     ¯1.0E0
2.0E¯18      1.0E0        2.0E0
6.1E¯18      8.7E¯19      1.7E¯18
```

(These numbers will vary with the brand of computer used.)

We want our Gaussian reduction function to stop at this point, because the last row is "small enough" to be assumed zero. The definition of "small enough" we will use is *negligible compared to A*. In Section 3.4 we defined a number C to be negligible compared to B if the expression B=B+C is true in the sense that the result is a 1. (This depends upon the value of ⎕CT.)

Of course A is a matrix, not a single number, and so we will use the largest magnitude in A to set the scale. In the case above the largest magnitude in the *original A* is 9. The expression 9=9+A shows which components of the current A are negligible compared to 9.

```
      9=9+A
0  1  0
1  0  0
1  1  1
```

GAUSS

We are now in a position to define the function *GAUSS*. It is

```
    ∇ G←GAUSS A ;S;P;L;T;B;V;K;R
[1]    S←⌈/,|A
[2]    P←L←0
[3]    T←ρA
[4]    A←A,ID 1↑T
[5]   CYCLE:V←+⌿B←|(L,P)↓T↑A
[6]    →((ρV)<K←(S≠S+V)⍳1)/END
[7]    R←B[;K]⍳⌈/B[;K]
[8]    A←((L+1),P+K)LDR((L+1),P+K)PIVOT(L+1,R)SWITCH A
[9]    →CYCLE,(L←L+1),P←P+K
[10]  END:G←(-T[1 1])↑A
    ∇
```

The function closely follows the proof of Proposition 3.3 — the proof that the row-reduction process works. In that proof at the end of the lth cycle we had a matrix A_l such that

(i) The first p columns of A_l are an echelon form with nonzero rows.

(ii) If the matrix B_l obtained by dropping the first l rows and p columns from A_l is a zero matrix, then A_l is in row echelon form.

In the function *GAUSS* the matrix A is the augmented matrix $[A_l \mid G_l]$, where G_l is the product of the elementary matrices used for the first l cycles. We obtain the matrix B from B_l by replacing each component by its absolute value.

On line [1] the scale is set for fuzzy comparisons. The scalar S is the largest absolute value in A. The numbers l and p are initialized to zero on line [2], and A is augmented by an identity matrix on line [4].

According to the proof of Proposition 3.3 we look at B_l. If $B_l = 0$, we are done. If $B_l \neq 0$, we must find the first nonzero column of B_l. Lines [5] and [6] accomplish this. The vector V is obtained by summing the columns of B, the absolute value of B_l. Zero columns of B_l give rise to zero components of V. On line [6] K is defined to be the index of the first component of V that is *not* negligible compared to S.

If $B_l = 0$, then K is $1 + \rho V$ and we are done. In this case the function jumps to line [10] and returns G_l. If $B_l \neq 0$, then $B_l[;K]$ is the first nonzero column of B_l. On line [7] the largest component of $B[;K]$ is located and the cycle is completed on line [8]. On line [9] the numbers l and p are updated and the function begins the next cycle. (Recall that $\rightarrow W$ looks only at the first component of the vector W.)

The functions are simple to use.

```
      A
1  2  3
4  5  6
7  8  9

      ⍞G←GAUSS A
¯1.30  0.00  0.33
 1.20  0.00 ¯0.17
¯0.50  1.00 ¯0.50

      G+.×A
 1.0E0        0.0E0       ¯1.0E0
 1.7E¯18      1.0E0        2.0E0
 3.5E¯18      0.0E0        3.5E¯18

      ECHELON A
 1.0E0        0.0E0       ¯1.0E0
 1.7E¯18      1.0E0        2.0E0
 3.5E¯18      0.0E0        3.5E¯18
```

Writing down the solution of a system of linear equations, given the echelon form, is a special case of the process known as back-substitution. A back-substitution function is defined in Chapter 5.

If we wish to use the function *ECHELON* within another function, we need an automatic way of checking the acceptability of the result of *ECHELON*. A way of doing this is provided in exercise 13.

EXERCISES 3.5

(Computer assignment) In exercises 1 through 5 use *ECHELON* to solve the system of equations.

1. $$\begin{aligned} x + 2y + 3z &= 4 \\ 5x + 6y + 7z &= 8 \\ 9x + 10y + 11z &= 12 \end{aligned}$$

2. $$\begin{aligned} 2x_1 - x_2 + x_3 + 3x_4 &= 5 \\ -x_1 + x_2 + x_3 + x_4 &= 1 \\ x_1 - x_2 - x_3 - x_4 + x_5 &= 7 \end{aligned}$$

3. $$\begin{aligned} 5x + y + 13z + 25t &= 2 \\ y + 3z + 5t &= 0 \\ 2x \quad + 4z + 8t &= 1 \end{aligned}$$

4. $$\begin{aligned} 219x + 330y + 609z &= 861 \\ -583x - 206y - 489z &= -1004 \\ -23x + 86y + 75z &= 65 \\ -64x - 92y + 6z &= -35 \\ 13x - 78y + 43z &= 66 \end{aligned}$$

5. $$\begin{aligned} -1281x_1 + 1123x_2 - 559x_3 - 1496x_4 &= -237 \\ -71x_1 + 50x_2 + 61x_3 - 15x_4 &= 52 \\ 43x_1 + 37x_2 - 24x_3 + 92x_4 &= 99 \\ 22x_1 + 74x_2 - 21x_3 - 15x_4 &= 47 \\ 89x_1 - 60x_2 + 96x_3 + 73x_4 &= 20 \end{aligned}$$

(Computer assignment) In exercises 6 through 10 use Gauss to compute the inverse, if one of the specified type exists.

6. $$A = \begin{bmatrix} 1 & 2 & 3 \\ 4 & 5 & 6 \\ 7 & 8 & -9 \end{bmatrix}$$
Inverse

7. $$\begin{bmatrix} 1 & 2 \\ 4 & 5 \\ 7 & 8 \end{bmatrix}$$
Left inverse

8. $$\begin{bmatrix} 1 & 2 & 3 \\ 4 & 5 & 6 \end{bmatrix}$$
Right inverse

9. $$\begin{bmatrix} 434 & -220 & 250 \\ -66 & 39 & -40 \\ -28 & 53 & -30 \\ 124 & 109 & 10 \end{bmatrix}$$
Left inverse

10. $$\begin{bmatrix} -104 & -375 & 395 & 52 \\ -32 & 7 & 93 & -84 \\ -2 & -99 & 29 & 76 \end{bmatrix}$$
Right inverse

11. (a) Let S be a scalar and V a vector. Show that the expression

```
S∧.=S+V
```

is 1 if and only if the components of V are negligible compared to S.

(b) Let S be a scalar and A a matrix. Show that the expression

```
+/S∧.=S+A
```

counts the number of columns of A that are negligible compared to S and that

```
+/(S+A)∧.=S
```

does the same for the rows of A.

(c) Show that the expression

```
+/(S+A)∨.≠S
```

counts the number of rows of A that are *not* negligible compared to S.

12. (Computer assignment) From exercise 11(c), the function

```
    ∇ Z←RANK A;S
[1]   S←⌈/,|A
[2]   Z←+/(S+ECHELON A)∨.≠S
    ∇
```

estimates the rank of a matrix A. Compute `ECHELON A` and `RANK A` for A equal to

(a) `3 3ρι9` (b) `1E9×3 3ριι9` (c) `1E20×3 3ρι9`

13. (a) Show that if E is an echelon form and any columns of zeros are deleted from E, the resulting matrix is still in echelon form.

(b) Show that if E is an echelon form, so is 1 1 ↓ E.

(c) Show that the function

```
    ∇ Z←S ECHCHK E
[1]   E←(S∨.≠S+E)/E
[2]   Z←1
[3]   →(0∨.=ρE)/0
[4]   Z←(1=E[1;1])∧S∧.=S+1↓E[;1]
[5]   Z←Z∧S ECHCHK 1 1↓E
    ∇
```

returns a 1 if E is an echelon form after components negligible with respect to S have been set to zero and returns a 0 otherwise.

14. (Computer assignment) The n-by-n Hilbert Matrix H is defined by $H[i;j] = 1/(i+j-1)$.

(a) Write a function *HILB* such that *HILB* n is the n-by-n Hilbert matrix.

(b) Does the function *ECHELON* row-reduce *HILB* 20 to an acceptable echelon form? (Don't print it out; compute `1 ECHCHK E←ECHELON HILB 20`, where *ECHCHK* is defined in exercise 13.)

(c) Let `EE←ECHELON ECHELON HILB 20`. Is `EE` an echelon form? [See part (b).]

(d) The Hilbert matrix is nonsingular, and the truncated version stored in the machine is probably nonsingular also. What is `RANK EE`, where `EE` is from part (c) and *RANK* is from exercise 12?

3.6* Powers of Matrices

In Section 3.5 we indicated that the problem of analyzing the powers of a matrix was an important one. In this section we will give some reasons for its importance.

We begin with a simple formula.

PROPOSITION 3.8 Let a, v_1 be vectors in $\mathbf{R}^n$ and let A be an n-by-n matrix. Define a sequence of vectors v_n by

$$v_n = \begin{cases} v_1 & \text{if } n = 1 \\ a + Av_{n-1} & \text{if } n > 1 \end{cases}$$

Then

$$v_n = (I + A + \cdots + A^{n-2})a + A^{n-1}v_1, \qquad n > 1$$

If $I - A$ is invertible, then

$$v_n = (I - A)^{-1}(I - A^{n-1})a + A^{n-1}v_1, \qquad n > 1$$

Proof First we show that

$$v_n = (I + A + \cdots + A^{n-2})a + A^{n-1}v_1$$

The proof, of course, is by induction. If $n = 2$, this is

$$v_2 = Ia + Av_1 = a + Av_1$$

which is correct. So assume the formula is correct for v_{n-1}. Then

$$\begin{aligned} v_n &= a + Av_{n-1} \\ &= a + A((I + A + \cdots + A^{n-3})a + A^{n-2}v_1) \\ &= (I + A + \cdots + A^{n-2})a + A^{n-1}v_1 \end{aligned}$$

and so the formula is also correct for n.

Now let $S = I + A + \cdots + A^{n-2}$. Then

$$AS = A + A^2 + \cdots + A^{n-2} + A^{n-1}$$

and

$$S - AS = (I - A)S = I - A^{n-1}$$

Thus, if $I - A$ is invertible,

$$S = (I - A)^{-1}(I - A^{n-1}) \qquad \blacksquare$$

EXAMPLE 3.22 In Section 3.4 we computed the first twenty values of

$$c1(n) = \begin{cases} 1 & \text{if } n = 1 \\ 1 + 2c1(n-1) & \text{if } n > 1 \end{cases}$$

If we take $A = 2$, $a = 1$, and $v_1 = 1$ then

$$\begin{aligned} c1(n) &= (1-2)^{-1}(1-2^{n-1}) + 2^{n-1} \\ &= 2^{n-1} - 1 + 2^{n-1} \\ &= 2^n - 1 \end{aligned}$$

■

Proposition 3.8 applies to two wide classes of problems.

LINEAR DIFFERENCE EQUATIONS

A kth-order linear difference equation is a recursively defined sequence

$$x_n = a_1 x_{n-k} + a_2 x_{n-k+1} + \cdots + a_k x_{n-1}$$

The starting values $x_1, x_2, \ldots, x_k$ must be given.

For example, the *Fibionacci numbers* are defined by

$$x_1 = 1, \qquad x_2 = 1, \qquad x_n = x_{n-2} + x_{n-1}$$

Thus $k = 2$ and $a_1 = a_2 = 1$.

To apply Proposition 3.8 we set $v_1 = (x_1, \ldots, x_k)$, the starting values, and let $v_n = (x_n, x_{n-1}, \ldots, x_{n-k})$. Then

$$v_n = \begin{bmatrix} x_n \\ \vdots \\ \\ x_{n-k+1} \end{bmatrix} = \left[\begin{array}{ccc|c} a_k & a_{k-1} & \cdots & a_1 \\ \hline & & & 0 \\ & I & & \vdots \\ & & & 0 \end{array}\right] \begin{bmatrix} x_{n-1} \\ x_{n-2} \\ \vdots \\ x_{n-k} \end{bmatrix} = Av_{n-1}$$

Thus $v_n = A^n v_1$, and the study of such difference equations is reduced to the study of A^n. To study A^n one uses the eigenvalues of A, defined in Chapter 7.

To compute the sequence $x_1, x_2, \ldots, x_n, \ldots$, however, it is easiest to use the original formulation, which states that x_n is

$$(a_1, a_2, \ldots, a_k) +.\times (x_{n-k}, x_{n-k+1}, \ldots, x_{n-1})$$

EXAMPLE 3.23 Compute the first twenty Fibionacci numbers.

Solution

```
    ∇ Z←FIB N
[1]    Z←N↑1 1
[2]    →(N≤2)/0
[3]    Z←FIB N-1
[4]    Z←Z,1 1+.×¯2↑Z
    ∇

       FIB 20
1  1  2  3  5  8  13  21  34  55  89  144  233  377  610  987
      1597  2584  4181  6765
```

■

STOCHASTIC MATRICES

A matrix is called *stochastic* if its entries are nonnegative and the sum of entries in each row is 1. Such matrices arise often in applications. The entry $A[i;j]$ usually gives the probability that if the system being studied is in state i, it will move to state j. The number $A[i;j]$ is called a *transition probability*.

For example, suppose that a large number of fleas are hopping about on a designer sheet done in, say, three colors. Suppose that $A[i;j]$ is the probability that a flea sitting on color i will have moved to color j at the end of 1 second. Now assume that there are n_1 fleas on color 1, n_2 fleas on color 2, and n_3 fleas on color 3. One second later the distribution of fleas will be

$$\begin{array}{ll} n_1A[1;1] + n_2A[2;1] + n_3A[3;1] & \text{fleas on color 1} \\ n_1A[1;2] + n_2A[2;2] + n_3A[3;2] & \text{fleas on color 2} \\ n_1A[1;3] + n_2A[2;3] + n_3A[3;3] & \text{fleas on color 3} \end{array}$$

If we let the vector $v = (n_1, n_2, n_3)$ denote the situation in which there are n_i fleas on color i, then after 1 second the distribution of fleas is $v +.\times A$.

We can put this in the form of Proposition 3.8 by taking transposes, but it is hardly necessary. The distribution of fleas after n seconds is $v +.\times A^n$, and we analyze the system by analyzing the powers of the matrix A.

The matrix A is called a *Markov chain matrix*. It can be shown that the rows of A^n often all approach the same vector as $n \to \infty$. This vector usually represents the steady-state behavior of the system being analyzed.

EXAMPLE 3.24 Let

$$A = \begin{bmatrix} \frac{1}{2} & 0 & \frac{1}{2} \\ \frac{1}{3} & \frac{1}{3} & \frac{1}{3} \\ \frac{1}{5} & \frac{1}{5} & \frac{3}{5} \end{bmatrix}$$

Does A^n seem to approach a limit as $n \to \infty$?

Solution Using the function *TOTHE* defined in Section 3.4, we have

```
      A
0.5000 0.0000 0.5000
0.3333 0.3333 0.3333
0.2000 0.2000 0.6000

      +A20←A TOTHE 20
0.3158 0.1579 0.5263
0.3158 0.1579 0.5263
0.3158 0.1579 0.5263
```

So the rows of A^{20} are all the same to four digits. Let us see if A^{40} is fuzzily equal to A^{20}.

```
      A20=A20+.×A20
0  0  0
0  0  0
0  0  0
```

Thus there are differences between A^{40} and A^{20}. We can try a somewhat higher power:

```
      +A400←A20 TOTHE 20
0.3158 0.1579 0.5263
0.3158 0.1579 0.5263
0.3158 0.1579 0.5263

      A400=A400+.×A400
 1  1  1
 1  1  1
 1  1  1
```

So A^{400} is fuzzily equal to A^{800} (⎕*CT* is here about $3E^{-}15$). ∎

EXERCISES 3.6*

For exercises 1 through 4 write the difference equation in matrix form by finding the v_1, a, and A of Proposition 3.8.

1. $x_1 = 1,\ x_2 = 1;\ x_n = x_{n-1} - x_{n-2},\ n > 2$
2. $x_1 = 3,\ x_2 = 2,\ x_3 = 1;\ x_n = 2 - x_{n-2} + x_{n-3},\ n > 3$
3. $x_1 = 1,\ x_2 = 2;\ x_n = -x_{n-1} + x_{n-2} - 5,\ n > 2$
4. $x_1 = x_2 = 0,\ x_3 = 7;\ x_n = x_{n-2}$
5. $x_1 = a,\ x_2 = b,\ x_n = \sqrt{2}x_{n-1} - x_{n-2}$. Show that $x_n = x_{n+8}$ for all n. Hint: What is A^8?

6. Use Proposition 3.8 to compute x_9 for exercise 3 above.

7. Use Proposition 3.8 to compute x_9 for exercise 2 above.

8. (Computer assignment) For the matrices A below does A^n seem to converge to a matrix with constant rows as $n \to \infty$? (Notice that the rows sum to 1.)

$$\text{(a)}\quad A = \begin{bmatrix} .01 & .05 & .94 \\ .3 & .7 & 0 \\ .45 & .35 & .20 \end{bmatrix} \qquad \text{(b)}\quad A = \begin{bmatrix} 1 & 2 & 3 & -5 \\ 4 & 5 & 6 & -14 \\ 7 & 8 & 9 & -23 \\ 10 & 11 & 12 & -32 \end{bmatrix}$$

9. (Computer assignment) In the discussion following Example 3.23 assume that, of the fleas on color 1, 5 percent move to color 2 and 5 percent move to color 3. Of the fleas on color 2, 5 percent move to color 1 and 20 percent move to color 3. Of the fleas on color 3, 10 percent move to color 1 and 20 percent move to color 2. What is the long-term distribution of fleas?

3.7* Nonlinear Equations

Nonlinear equations are much more difficult than linear equations. This section presents two procedures for estimating roots of nonlinear equations. The first method, sectioning, is quite satisfactory for finding roots of $f(x) = 0$ when $f\colon \mathbf{R} \to \mathbf{R}$ and is easily computed. The second, Newton's method, applies to functions $f\colon \mathbf{R}^n \to \mathbf{R}^n$. Newton's method was discussed in Section 2.6*, and the discussion there assumes familiarity with partial differentiation. In this section we define a function that makes Newton's method easier to use. A fully automatic Newton's method is not attempted.

SECTIONING

Let $f\colon \mathbf{R} \to \mathbf{R}$ and suppose we wish a root of $f(x)$. Assume that we can find numbers a and b, with $a < b$, such that $f(a)$ and $f(b)$ have different signs. If the function f is continuous, then by the intermediate-value theorem of elementary calculus there is a number ξ, $a \leq \xi \leq b$ such that $f(\xi) = 0$.

To estimate ξ we chop the interval $[a, b]$ into subintervals — 100 or 1000 subintervals, say† — and find the first subinterval on which f changes sign, call it $[a_1, b_1]$ (see Figure 3.1). The process can be repeated with $[a_1, b_1]$ replacing $[a, b]$ until ξ is located as accurately as desired. To subdivide $[a, b]$ into subintervals we can use the function *CHOP* from Example 3.4. Suppose that $f\colon \mathbf{R} \to \mathbf{R}$ has been defined as an APL function called *FCN* that works componentwise on vectors. Then

```
Y←FCN X←100 CHOP a , b
```

†Depending upon the storage available in your workspace.

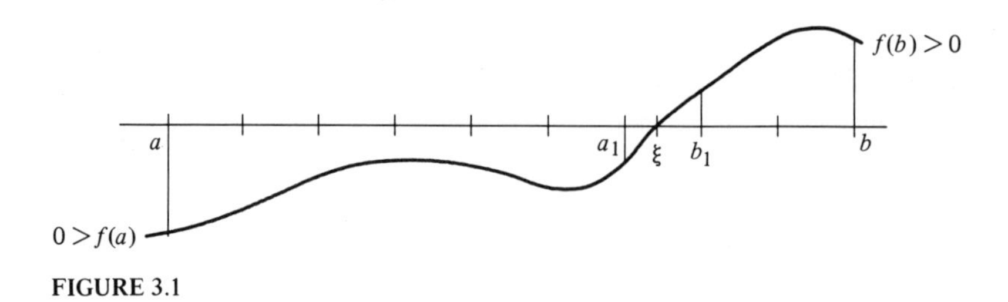

FIGURE 3.1

defines X to be the vector of endpoints of the subintervals and Y to be the vector of values of f at these points. To locate the subinterval $[a_1, b_1]$ we must find the first sign change in the vector Y.

Now the monadic function $\times$ gives the sign of a number $\times A$ is -1 if A is negative, 1 if A is positive, and 0 if A is zero. If k is such that $(\times Y[k-1]) \neq \times Y[k]$ is the first sign change, then k is $((\times Y) \neq \times Y[1])\iota 1$.† Thus a_1, b_1 is just

```
X[¯1 0+((×Y)≠×Y[1])⍳1]
```

Suppose we decide to accept as a root any x such that $|f(x)| < E$ for a given small number E. Then the roots s, if any, among the components of X are given by the compression

```
Z←(E>|Y)/X
```

If there are no "roots," then Z is $\iota 0$.

The function $CENTSECT$ takes E and $I = (a, b)$ as arguments and returns at least one "root," provided $f(a)$ and $f(b)$ have different signs.

```
    ∇ Z←E CENTSECT  I ;X;Y
[1]   L:Y←FCN X←100 CHOP I
[2]     I←X[¯1 0+((×Y)≠×Y[1])⍳1]
[3]     →(0=⍴Z←(E>|Y)/X)/L
    ∇
```

EXAMPLE 3.25 Find the roots of $f(x) = x^2 - \sin x$.

Solution $f(x) = 0$ if and only if $x^2 = \sin x$. Sketching the curves $y = x^2$ and $y = \sin x$ (Figure 3.2) indicates that in addition to the root $x = 0$, there is a root between 0 and π.

†The dyadic function ι is discussed in Section 3.5.

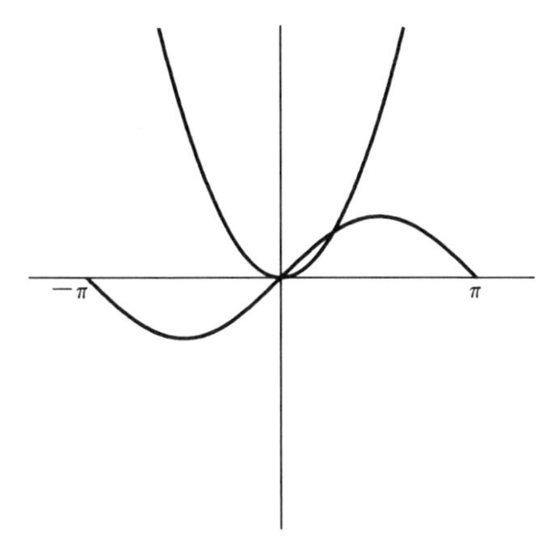

FIGURE 3.2

First define the function f:

```
    ∇ Z←FCN X
[1]     Z←(X*2)-1○X
    ∇
```

Next we need a starting interval $[a, b]$. For b we can take π. We need $a > 0$, otherwise we may just get $x = 0$, which we already know. Try the interval $[.1, \pi]$.

```
      FCN .1,○1
¯0.08983  9.87
```

The signs of $f(a)$ and $f(b)$ are different, and so $[.1, \pi]$ will do. Taking $E = 10^{-10}$, we have

```
      +ANS←1E¯10 CENTSECT .1,○1
0.8767

      FCN ANS
¯3.492E¯11
```

Notice that if we try $E = 10^{-15}$ we get

```
+ANS←1E¯15 CENTSECT  .1,○1
0.8767  0.8767  0.8767  0.8767  0.8767  0.8767
```

This indicates six valid cases. The range of x values is

```
      -/ANS[1 6]
¯1.521E¯15
```

so they differ only in the fifteenth decimal place. All meet our criterion for a root: $|f(x)| < 10^{-15}$.

```
      FCN ANS
¯7.936E¯16  ¯4.562E¯16  ¯1.162E¯16  2.212E¯16  5.612E¯16
      9.003E¯16
```

■

Many prefer a function that stops when the *interval* is smaller than a specified E. In this case one tests the length of `I` in line [3] of `CENTSECT` above and returns to `L` if `E≤|-/I` (see exercise 1).

EXAMPLE 3.26 Find an approximate root of $1 + x + x^2 - 5x^4$ with $E = 10^{-10}$.

Solution A definition of *FCN* in this case is

```
   ∇ Z←FCN X
[1]   Z←(X∘.*0,⍳4)+.×1 1 1 0 ¯5
   ∇
```

To find an initial interval we first look at $f(k)$, $k = -10, -9, \ldots, 10$:

```
      FCN 20 CHOP ¯10 10
¯49910  ¯32730  ¯20420  ¯11960  ¯6449  ¯3104  ¯1267  ¯398  ¯77
      ¯4  1  ¯2  ¯73  ¯392  ¯1259  ¯3094  ¯6437  ¯11950  ¯20410
      ¯32710  ¯49890
```

From the sign changes we see that there is a root in $[-1, 0]$ and another in $[0, 1]$. In $[-1, 0]$ we have

```
      +ANS←1E¯10 CENTSECT ¯1 0
¯0.6256

      FCN ANS
¯1.758E¯11
```

and in $[0, 1]$ we have

```
      +ANS←ANS,1E¯10  CENTSECT  0 1
¯0.6256  0.846  0.846  0.846  0.846  0.846  0.846  0.846  0.846
      0.846  0.846  0.846  0.846  0.846  0.846  0.846  0.846  0.846
      0.846  0.846  0.846  0.846
       FCN ANS
¯1.758E¯11 9.528E¯11  8.586E¯11  7.644E¯11  6.702E¯11  5.76E¯11
      4.818E¯11  3.876E¯11  2.934E¯11  1.992E¯11  1.05E¯11
      1.076E¯12  ¯8.344E¯12  ¯1.776E¯11  ¯2.718E¯11  ¯3.66E¯11
       4.602E¯11  ¯5.544E¯11  ¯6.486E¯11  ¯7.428E¯11  ¯8.37E 11
      ¯9.312E¯11
```

■

NEWTON'S METHOD

Let f: $\mathbf{R}^n \to \mathbf{R}^n$ be of class C'. In Section 2.6* the Newton's method iteration for a root of f was defined as

$$x_n = x_{n-1} - f(x_{n-1}) ⌹ Df(x_{n-1})$$

where x_{n-1} is an approximation to a root and x_n is the next approximation. Here Df is the derivative (= Jacobian matrix) of f.

A convenient way to use this formula is to write a function that will display the result of N iterations as the rows of a matrix. Let the vector x_0 be the initial guess at a root. An APL version of the formula above is

```
    ∇ Z←N NEWTON X0
[1]   Z←(1,ρ,X0)ρX0
[2]   →(N=0)/0
[3]   Z←(N−1)NEWTON X0
[4]   Z←Z,[1]Z[N;]−(FCN Z[N;])⌹DFCN Z[N;]
    ∇
```

The functions f and Df must be defined as the APL functions *FCN* and *DFCN*.

The next example was presented earlier as Example 2.53.

EXAMPLE 3.27 Estimate the first-quadrant solution of the simultaneous equations

$$\frac{x^2}{16} + \frac{y^2}{4} = 1 \qquad \text{(an ellipse)}$$

$$x^2 - y^2 = 1 \qquad \text{(a hyperbola)}$$

Solution The solution is, from Example 2.53, $(2, \sqrt{3})$.

To define the function f: $\mathbf{R}^2 \to \mathbf{R}^2$ we first multiply the first equation by 16. Then f: $\mathbf{R}^2 \to \mathbf{R}^2$ is given by

$$f(x, y) = \begin{bmatrix} x^2 + 4y^2 - 16 \\ x^2 - y^2 - 1 \end{bmatrix}$$

or

```
    ∇ Z←FCN X
[1]   Z←(2 3ρ1 4 ¯16 1 ¯1 ¯1)+.×(X,1)*2
    ∇
```

and

$$Df = \begin{bmatrix} 2x & 8y \\ 2x & -2y \end{bmatrix}$$

or

```
    ∇ Z←DFCN X
[1]   Z←2 2ρ2 8 2 ¯2×X,X
    ∇
```

We start with an initial estimate of $x_0 = (1, 1)$ and compute six iterations.

```
      +ANS←6 NEWTON 1 1
 1.0000   1.0000
 2.5000   2.0000
 2.0500   1.7500
 2.0010   1.7320
 2.0000   1.7320
 2.0000   1.7320
 2.0000   1.7320

      FCN ANS[6;]
8.674E¯15   8.639E¯15
```

Thus after six iterations we are quite close. Since in this case the answer is known to be $(2, \sqrt{3})$, we can find the actual error.

```
      ANS-(ρANS)ρ2,3*÷2
¯1.000E0       ¯7.321E¯1
 5.000E¯1       2.679E¯1
 5.000E¯2       1.795E¯2
 6.098E¯4       9.205E¯5
 9.292E¯8       2.446E¯18
 2.161E¯15      5.204E¯18
 3.469E¯18      3.469E¯18
```

Since the APL processor used for this example carries only eighteen digits, we cannot expect to do better. ■

Newton's method usually converges quadratically, once a sufficiently close estimate is obtained. Quadratic convergence means that if the order of magnitude of the error at step n is E, then the order of magnitude of the error at step $n + 1$ is CE^2 for some constant C.

EXERCISES 3.7*

1. Write a version of *CENTSECT* that stops when a root of $f(x)$ is trapped in an interval of length less than E.

In exercises 2 through 5 find a root of the given function.

2. $f(x) = 1 + x + x^3$
3. $f(x) = \frac{1}{2}x - \text{Tan}^{-1} x$
4. $f(x) = 1 + x^{1/2} - x^{3/5}$ $(x \geq 0)$
5. $f(x) = x + \cos x - \sinh x$

In exercises 6 through 9 you may estimate starting values by looking up the general shape of the named curves in a mathematical handbook.

6. Use Newton's method to find where the bifolium $(x^2 + y^2)^2 = 3x^2y$ cuts the astroid $x^{2/3} + y^{2/3} = 1$.

7. Use Newton's method to find where the Cassinian oval $(x^2 + y^2 + 1)^2 - 4x^2 = \frac{1}{16}$ $(x < 0)$ cuts the strophoid $y^2(1 - x) = x^2(1 + x)$.

8. Find where the folium of Descartes $x^3 + y^3 - 3xy = 0$ cuts the witch of Agnesi $x^2y + 4y = 8$:

   (a) By solving the equation of the witch for y, substituting into the equation of the folium and using *CENTSECT*.

   (b) By Newton's method.

9. Find the four points common to the ellipsoid

$$x^2 + \frac{y^2}{4} + \frac{z^2}{9} = 1$$

the hyperboloid of two sheets described by $z^2 - x^2 - y^2 = 1$, and the elliptic cylinder $9(x - 1)^2 + y^2 = 9$.

10. Find the maxima and minima, if any, of the function

$$f(x, y) = x^4 + 2x^2y^2 + y^4 - 4x^2 + 6$$

Hint: Set the derivative equal to zero.

3.8* Natural Cubic Splines (A Symmetric Tridiagonal System)

It would seem that Section 3.5 finishes off the problem of solving linear equations. All one has to do is drop the augmented matrix into *ECHELON* and read off the answer. This is far from true. Often in practice one encounters situations in which *GAUSS* simply will not do. These situations call for special methods, adapted to the individual problem. This section is devoted to one example that often arises; the natural cubic spline.

Cubic splines are used for curve fitting when there is little or no error in the data† and no reason to choose a curve of any particular form to fit the data. A cubic spline goes through the given points with as little "wiggling" as possible, subject to the constraint of having a continuous second derivative. (The term "spline" originates in a drafting instrument used to draw a smooth curve through a set of points plotted on a drafting board.)

A typical modern application is in the field of computer graphics. Some points are specified on a CRT screen, perhaps by touching a light pen to the screen, and the computer then draws a smooth curve through the points.

Splines are also used in cinema photographic analysis of the motions of complex systems. A complex system such as a tennis player's upper body or a karate expert's hand and forearm is carefully photographed. The changing positions of reference points are then recorded frame by frame, and cubic splines are fitted to

†There is, however, a "least-squares cubic spline" used in the presence of measurement error. This more sophisticated type of spline will not be discussed here.

the data. Since splines have continuous second derivatives, the resulting expressions can be differentiated to find the forces and accelerations on the reference points.

Splines are piecewise functions. The x axis is divided into many subintervals, and a different formula is used over each one. In the most common situation a different cubic polynomial is used between each successive pair of data points.

Suppose, for example, that the data points are $(x_1, y_1), (x_2, y_2), \ldots, (x_n, y_n)$ with $a = x_1 < x_2 < \cdots < x_n = b$. On each interval $[x_i, x_{i+1}]$ we want a cubic polynomial

$$S_i(x) = C_{1i} + C_{2i}x + C_{3i}x^2 + C_{4i}x^3, \qquad i = 1, 2, \ldots, n-1$$

The polynomials must satisfy the conditions (Figure 3.3):

(1) $\quad S_i(x_i) = y_i, \qquad i = 1, 2, \ldots, n-1$

(2) $\quad S_i(x_{i+1}) = y_{i+1}, \qquad i = 1, 2, \ldots, n-1$

(3) $\quad S_i'(x_{i+1}) = S_{i+1}'(x_{i+1}), \qquad i = 1, 2, \ldots, n-2$

(4) $\quad S_i''(x_{i+1}) = S_{i+1}''(x_{i+1}), \qquad i = 1, 2, \ldots, n-2$

From (1) and (2) we have $S_i(x_{i+1}) = y_{i+1} = S_{i+1}(x_{i+1})$, and so the resulting piecewise function has continuous first and second derivatives.

There are $n - 1$ functions $S_i(x)$ and hence $4(n - 1) = 4n - 4$ unknown constants C_{ij}. The conditions (1), (2), (3), and (4) provide $2(n - 1) + 2(n - 2) = 4n - 6$ linear equations among the C_{ij}. Thus at least two more equations are needed for a unique solution. One has a *natural* cubic spline if the two extra equations are taken to be

(5) $\quad S_1''(x_1) = 0$

(6) $\quad S_{n-1}''(x_n) = 0$

It can be shown that the $4n - 4$ equations provided by conditions (1) through

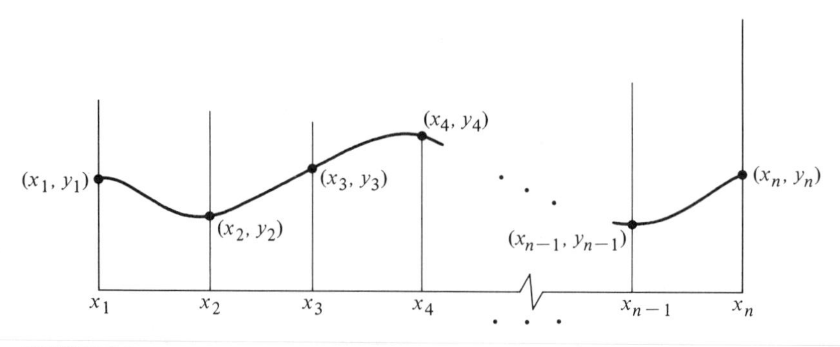

FIGURE 3.3

(6) uniquely determine the unknown coefficients C_{ij}. In a typical application, however, n may be more than 100, and so the matrix is larger than 400 by 400. Many APL systems simply cannot work with such large matrices, and *ECHELON* is useless for the problem.

The matrix of the system is large but *sparse*. It is mostly zeros. Even though it is larger than 400 by 400, each row contains at most six nonzero entries. To solve such systems one devises schemes, adapted to the particular problem, which solve the systems without storing the zeros.

Most schemes for cubic splines first reduce the problem from that of a sparse $4n - 4$ by $4n - 4$ system to a sparse $n - 1$ by $n - 1$ system in which the unknowns are the numbers $S_i'(x_i)$ or $S_i''(x_i)$. We shall use an $n - 1$ by $n - 1$ system with unknowns

$$s_i = S_i''(x_i), \qquad i = 1, 2, \ldots, n - 1$$

Since we want a natural cubic spline, we have $s_1 = 0$ and $s_n = 0$.

Since S_i is a cubic, $S_i''(x)$ is the straight line from (x_i, s_i) to (x_{i+1}, s_{i+1}), and hence we can write it in the form

$$S_i''(x) = \frac{1}{\Delta x_i}[s_{i+1}(x - x_i) - s_i(x - x_{i+1})], \qquad i = 1, 2, \ldots, n - 1$$

Here $\Delta x_i = x_{i+1} - x_i$. Notice that $S_i''(x)$ is linear in x, $S_i''(x_i) = s_i$, and $S_i''(x_{i+1}) = s_{i+1}$, and so the formula is a correct one.

If we know the s_i's, it is not difficult to get an expression for $S_i(x)$. Integrating twice and choosing the integration constants cleverly gives

$$S_i(x) = \frac{1}{6\,\Delta x_i}[s_{i+1}(x - x_i)^3 - s_i(x - x_{i+1})^3] + c_i(x - x_i) + d_i(x - x_{i+1}) \quad (3.2)$$

$$i = 1, 2, \ldots, n - 1$$

Using the conditions (1) $[S_i(x_i) = y_i]$ and (2) $[S_i(x_{i+1}) = y_{i+1}]$ in this expression gives

$$c_i = \left(y_{i+1} - \frac{(\Delta x_i)^2}{6} s_{i+1}\right) \div \Delta x_i, \qquad d_i = -\left(y_i - \frac{(\Delta x_i)^2}{6} s_i\right) \div \Delta x_i$$

Suppose we let X, Y, S be vectors with components (x_i), (y_i), and (s_i), respectively. Further, define an APL function Δ by

```
    ∇ Z←Δ X ;K
[1]   K←ι ¯1+ρX
[2]   Z←X[K+1]-X[K]
    ∇
```

Then the constants c_i and d_i are given by

```
C←((1↓Y)÷Δ X)−(1↓S)×Δ X÷6
D←−((¯1↓Y)÷Δ X)−(¯1↓S)×Δ X÷6
```

Now let us return to the problem of computing the $n - 2$ numbers s_i, $i = 2, 3, \ldots, n - 1$. We have used conditions (1) and (2) to compute the c_i and d_i. Condition (4) was used to write $S_i''(x)$ in terms of s_i and s_{i+1}. We must use condition (3) to determine the s_i.

Differentiating $S_i(x)$ and using the expressions for c_i and d_i,

$$S_i'(x) = \frac{1}{2\,\Delta x_i}\left[s_{i+1}(x - x_i)^2 - s_i(x - x_{i+1})^2\right] + \frac{y_{i+1} - y_i}{\Delta x_i} - \frac{\Delta x_i}{6}(s_{i+1} - s_i)$$

Condition (3) states that $S_i'(x_{i+1}) = S_{i+1}'(x_{i+1}), i = 1, 2, \ldots, n - 2$. Hence we have the following equations for the unknown quantities s_i:

$$\Delta x_i s_i + 2(\Delta x_i + \Delta x_{i+1})s_{i+1} + \Delta x_{i+1}s_{i+2} = 6\left[\frac{\Delta y_{i+1}}{\Delta x_{i+1}} - \frac{\Delta y_i}{\Delta x_i}\right] = b_i$$

where $\Delta y_i = y_{i+1} - y_i$. Since $s_1 = 0$, the augmented matrix of this system is

$$\left[\begin{array}{cccc|c} 2(\Delta x_1 + \Delta x_2) & \Delta x_2 & 0 & \ldots & b_1 \\ \Delta x_2 & 2(\Delta x_2 + \Delta x_3) & \Delta x_3 & \ldots & b_2 \\ 0 & \Delta x_3 & 2(\Delta x_3 + \Delta x_4) & \ldots & b_3 \\ 0 & 0 & \Delta x_4 & & \\ \vdots & \vdots & \vdots & & \vdots \end{array}\right]$$

The coefficient matrix is symmetric and *tridiagonal;* that is, only the main diagonal and the minor diagonals directly above and below are nonzero.

This coefficient matrix, you will notice, is sparse. Of the $(n - 2)^2$ entries only $(n - 1) + 2(n - 2) = 3n - 5$ are nonzero and, since it is symmetric, only $(n - 1) + (n - 2) = 2n - 3$ numbers need be stored to keep track of the coefficient matrix.

The next task is to write a special APL function to solve this special system. The system is of the form $AX = B$, where

$$A = \begin{bmatrix} a_{11} & a_{12} & 0 & 0 & & 0 \\ a_{12} & a_{22} & a_{23} & 0 & & 0 \\ 0 & a_{23} & a_{33} & a_{34} & & \vdots \\ 0 & 0 & a_{34} & a_{44} & \ddots & \\ & \vdots & & & a_{n-1,n-1} & a_{n-1,n} \\ 0 \ldots & 0 \ldots & & & a_{n,n-1} & a_{nn} \end{bmatrix}, \qquad B = \begin{bmatrix} b_1 \\ b_2 \\ \vdots \\ \\ b_n \end{bmatrix}$$

Suppose that we clean up the first column of the augmented matrix by pivoting on

a_{11} [in our case $a_{11} = 2(\Delta x_1 + \Delta x_2) \neq 0$]:

$$\left[\begin{array}{c|cccccc|c} 1 & a_{12}/a_{11} & 0 & . & . & . & . \; 0 & b_1/a_{11} \\ \hline 0 & a_{22} - a_{12}^2/a_{11} & a_{23} & 0 & & & & b_2 - a_{12}b_1/a_{11} \\ 0 & a_{23} & a_{33} & \ddots & & & & b_3 \\ 0 & 0 & & & \ddots & & & \vdots \\ 0 & 0 & & & & \ddots & & \\ \vdots & \vdots & & & & & & \\ 0 & . & & & & & . & \end{array}\right]$$

The submatrix in the box is again symmetric and tridiagonal. Further, in our case,

$$a_{22} - \frac{a_{12}^2}{a_{11}} = \frac{4\,\Delta x_1(\Delta x_2 + \Delta x_3) + 3\,\Delta x_2^2}{2(\Delta x_1 + \Delta x_2)} > 0$$

so we can pivot next on the 2;2 entry. It can be shown for the cubic spline matrix† that row interchanges will never be necessary. This means that the matrix is nicely set up for an inductive solution. If we assume that we already know the solution of such a system when it is $n - 1$ by $n - 1$, then we already have the values of the variables $x_2, x_3, \ldots, x_n$, and the value of x_1 is just

$$x_1 = (b_1 - a_{12}x_2) \div a_{11}$$

To get the induction started, we notice that the solution of a 1-by-1 system is just $x_1 = b_1 \div a_{11}$.

In the APL function *TRIDI* the left argument is a matrix with three columns. The first column is the diagonal $a_{11}, a_{22}, a_{33}, \ldots, a_{nn}$. The second column is $a_{12}, a_{23}, \ldots, a_{n-1,n}, 0$ (the zero is just padding), and the third column is $b_1, b_2, \ldots, b_n$.

```
    ∇ Z←TRIDI A
[1]    Z←÷/A[1;3 1]
[2]    →(1=1↑ρA)/0
[3]    A[2;1 3]←A[2;1 3]-A[1;2 3]×A[1;2]÷A[1;1]
[4]    Z←TRIDI 1 0↓A
[5]    Z←((A[1;3 2]÷A[1;1])+.×1,-1↑Z),Z
    ∇
```

On line [1] Z is set equal to $b_1 \div a_{11}$. On line [2] we stop if A has only one row. Otherwise line [3] replaces a_{22} by $a_{22} - a_{12}^2/a_{11}$ and replaces b_2 by $b_2 - a_{12}b_1/a_{11}$. On line [4] we pick up the solution of the $(n - 1)$-by-$(n - 1)$ system and on line [5] set $x_1 = (b_1 - a_{12}x_2) \div a_{11}$.

† See, for example, L. F. Shampine and R. C. Allen, *Numerical Computing: An Introduction* (Philadelphia: W. B. Saunders, 1973), p. 57.

If the available storage is small, we can always rewrite a recursive function using loops (exercise 5).

The data we are given to begin with consist of a vector X of the numbers $x_1, x_2, \ldots, x_n$ (in ascending order) and the corresponding vector of Y coordinates $y_1, y_2, \ldots, y_n$. From these two vectors we must first build the matrix A to be used as the right argument of *TRIDI*. A has $n - 2$ rows and 3 columns. Define H by

```
H←Δ X
```

Then H is the vector with components $H[i] = \Delta x_i$ and length $n - 1$. We have the identities

```
A[;1]=2×(¯1↓H)+1↓H
A[;2]=1↓H
```

This last expression makes $A[n - 2;2] = \Delta x_{n-1}$ rather than 0, but this component is just padding — it is not used in any of the computations. Finally, observe that the expression

$$b_i = 6\left[\frac{\Delta y_{i+1}}{\Delta x_{i+1}} - \frac{\Delta y_i}{\Delta x_i}\right]$$

translates into the identity

```
A[;3]=6×Δ(Δ Y)÷Δ X
```

The function

```
    ∇ A←X SETUP Y ;H
[1]   A←(2×(¯1↓H)+1↓H),[1.5]1↓H←Δ X
[2]   A←A,6×Δ(Δ Y)÷H
    ∇
```

will be useful for setting up the matrix A. (The expression "`,[1.5]`" is the *lamination* function. It catenates two vectors into the columns of a matrix.)

Finally, we have the question of what form the storage of the cubic polynomials should take.

In Equation (3.2) above we wrote the ith polynomial as

$$S_i(x) = a_i(x - x_i)^3 + b_i(x - x_{1+1})^3 + c_i(x - x_i) + d_i(x - x_{i+1})$$

where

$$a_i = \frac{s_{i+1}}{6\,\Delta x_i}, \qquad b_i = \frac{-s_i}{6\,\Delta x_i}, \qquad c_i = \frac{y_{i+1}}{\Delta x_i} - \frac{s_{i+1}\,\Delta x_i}{6}, \qquad d_i = \frac{-y_i}{\Delta x_i} + \frac{s_i\,\Delta x_i}{6}$$

If we store a_i, b_i, c_i, d_i as the ith row of a matrix CF and if T is any vector of points in the interval $[x_i, x_{i+1}]$, then S_i can be evaluated on all the components of T simultaneously by the expression

```
((T*3),T←T∘.-X[i+0 1])+.×CF[i;]
```

For this reason we store the coefficients in the $(n - 1)$-by-4 matrix CF with $CF[i;] = (a_i, b_i, c_i, d_i)$. The function *SPLINE* below takes the data vectors X and Y as input and returns CF.

```
    ∇ CF←X SPLINE Y ;S;H
[1]    S←0,(TRIDI X SETUP Y),0
[2]    CF←((1↓S),[1.5]-¯1↓S)÷6×H←H,[1.5]H←Δ X
[3]    CF←CF,(((1↓Y),[1.5]-¯1↓Y)÷H)-(H*2)×CF
    ∇
```

The function we use to compute points in the interval $[x_i, x_{i+1}]$ is

```
    ∇ Z←C EVALAT T
[1]    Z←((T*3),T←T∘.-(1↑T),¯1↑T)+.×C
    ∇
```

The left argument C is $CF[i;]$, the first component of the vector T is x_i, and the last component is x_{i+1}.

EXAMPLE 3.28 The three points (0, 0), (1, 1), (2, 8) lie on the cubic $f(x) = x^3$, but $f(x)$ is *not* the natural cubic spline through the three points. Although f satisfies conditions (1) through (5) [with $s_1(x) = s_2(x) = f(x)$], we have $f''(2) = 12 \neq 0$, so condition (6) is violated.

The natural cubic spline is given by

```
      +CF←0 1 2 SPLINE 0 1 8
1.500  0.000 ¯0.500  0.000
0.000 ¯1.500  8.000  0.500
```

That is,

$$S(x) = \begin{cases} \frac{3}{2}x^3 - \frac{1}{2}x & 0 \leq x \leq 1 \\ -\frac{3}{2}(x - 2)^3 + 8(x - 1) + \frac{1}{2}(x - 2), & 1 \leq x \leq 2 \end{cases}$$

and it is not difficult to see that $S(x)$ satisfies conditions (1) through (6).

To sketch $S(x)$ we compute some intermediate values at intervals of $\Delta X = .1$.

```
      CF[1;] EVALAT 10 CHOP 0 1
0  ¯0.0485  ¯0.088  ¯0.11  ¯0.104  ¯0.0625  0.024  0.164  0.368
      0.643  1
```

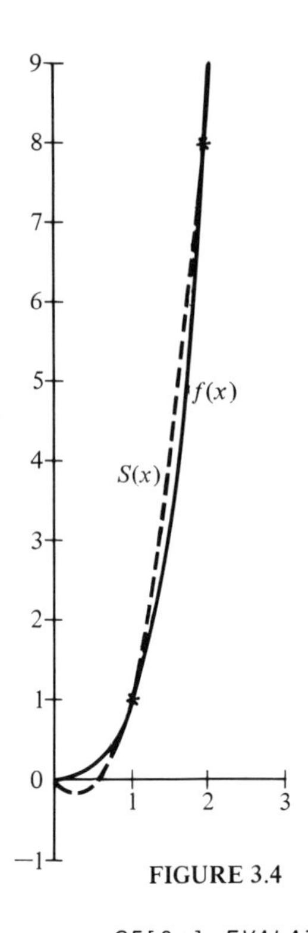

FIGURE 3.4

```
        CF[2;] EVALAT 10 CHOP 1 2
1  1.44  1.97  2.56  3.22  3.94  4.7  5.49  6.31  7.15  8
```

The graphs of $f(x)$ and $S(x)$ on $[0, 2]$ are sketched in Figure 3.4. ■

In many applications, such as computer graphics, one wants more general curves than the graphs of functions. In particular, one often wants closed curves (such as circles and ellipses). In such cases one uses a parametric spline:

$$S_i(t) = \begin{bmatrix} x_i(t) \\ y_i(t) \end{bmatrix} = \begin{bmatrix} a_i + b_i t + c_i t^3 + d_i t^4 \\ e_i + f_i t + g_i t^3 + h_i t^4 \end{bmatrix}$$

For a discussion of such splines, complete with APL functions for computing them, see R. G. Selfridge, "Splines and Graphs," *APL Quote-Quad,* **8**:4 (June 1978), 29–33.

EXERCISES 3.8*

1. Verify that the function $S(x)$ of Example 3.28 satisfies conditions (1) through (4).

(Computer assignment) In exercises 2, 3, and 4 fit a natural cubic spline to the data and sketch the resulting graph.

2. The data of Examples 2.28 and 2.30.

3. The data of Exercise 12 of Section 2.3.

4. The data of Exercise 14 of Section 2.3.

5. Write a version of *TRIDI* that is not recursive. [*Hint:* You will need two loops. The first should in effect reduce a symmetric tridiagonal matrix plus an augmentation column to the form

$$\left[\begin{array}{cccccc|c} 1 & \rho_1 & 0 & & \dots & 0 & \tau_1 \\ 0 & 1 & \rho_2 & 0 & \dots & 0 & \tau_2 \\ & & & \ddots & & & \\ & & & & \ddots & 1 & \tau_n \end{array}\right]$$

The second loop should then solve the latter system by starting with $x_n = \tau_n$ and working back up.]

6. Show that if the numbers Δx_i are all equal, then the tridiagonal coefficient matrix may be represented by the matrix

```
(N,N)⍴4 1,((N−2)⍴0),1
```

CHAPTER FOUR

Geometry and Coordinate Systems

For vectors with two and three components the algebraic manipulations of the previous chapter may be given geometric interpretations. This is done in the first three sections of this chapter. The connection between the geometry, which is restricted to two and three dimensions, and the algebra, which has no such restriction, is then used to extend the geometric concepts to higher dimensions. The geometry involved is technically known as *affine geometry*. *Euclidean geometry*, which involves the notions of distance, angle, and congruence, will be taken up in Chapter 5.

In Section 4.1 vector addition and scalar-vector multiplication are interpreted geometrically for vectors in $\mathbf{R}^2$ and $\mathbf{R}^3$. Geometric questions of the type, "What is the intersection of the two planes?" are answered by using the machinery of Chapter 3 to row-reduce a system of linear equations.

Affine coordinate systems are introduced in Section 4.2, and coordinate-change formulas are derived. These coordinate-change formulas are used in Section 4.3 to give a detailed analysis of quadratic functions of two variables. This material will be used in Chapter 5 to analyze quadratic functions of n variables. The analysis of the general quadratic function (a process generally known as "diagonalizing a symmetric matrix") is the basis for the second-derivative test for maxima and minima and for the statistical techniques of principal-component analysis and factor analysis.

In Section 4.4 the geometric notions of the first two sections are formally extended to higher dimensions, and the set of solutions of a system of linear equations in n variables is shown to form a "flat" in $\mathbf{R}^n$. The coordinate-change formulas developed in Section 4.2 are also extended to $\mathbf{R}^n$.

In Section 4.5 the notions of translations and parallelism in $\mathbf{R}^n$ give rise to the definition of a "subspace" of $\mathbf{R}^n$. Special subspaces associated to a matrix are defined and used to prove that a matrix and its transpose have the same rank. Many of the results in succeeding chapters will be phrased in terms of subspaces.

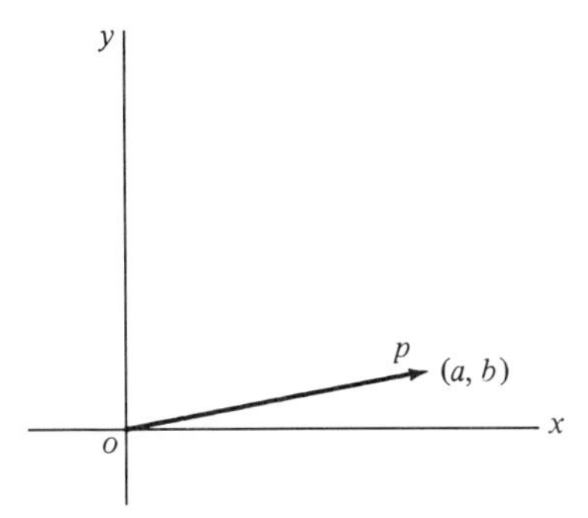

FIGURE 4.1

4.1 Geometric Vectors, Lines, and Planes

Points in the plane of Euclidean geometry may be identified with vectors in $\mathbf{R}^2$. To do this, one chooses an orthogonal pair of axes and assigns to each point p a pair of coordinates (a, b). The point p may then be identified with the vector (a, b) in $\mathbf{R}^2$.

This identification is far from an empty formalism. Historically, vector algebra has its origins in the Euclidean geometry of the plane and space.† To form the connection with geometry, we identify the vector (a, b) with the directed line segment or "arrow" from the origin of the coordinate system to the point p (Figure 4.1). Once this identification has been made, two of our basic operations with vectors can be given purely geometric interpretations. They are addition of vectors and the multiplication of a vector by a scalar.

VECTOR ADDITION

Suppose that p has coordinates (a, b) and q has coordinates (c, d). The sum of the vectors is $(a, b) + (c, d) = (a + c, b + d)$. Form the parallelogram determined by the points p, q, and the origin. The shaded triangles in Figure 4.2 are then congruent. From this it can be deduced that the fourth vertex of the parallelogram has coordinates $(a + c, b + d)$. Thus the geometric "sum" of the two vectors corresponds to the fourth vertex of the parallelogram.

There are two useful ways of looking at this geometric version of vector addition. These are pictured in Figure 4.3, where we use the symbol p to denote any of the three entities: the point p, the coordinate pair $(a, b) = p$, and the directed line segment from the origin to the point p. In Figure 4.3(a) the sum of the line segments p and q is the line segment that makes up the diagonal of the parallelogram.

†First, Gauss interpreted complex numbers as "vectors" in the plane. Next, Hamilton extended the ideas to three-dimensional space with his system of quaternions. The algebra of quaternions then evolved into the vector analysis of physics and engineering.

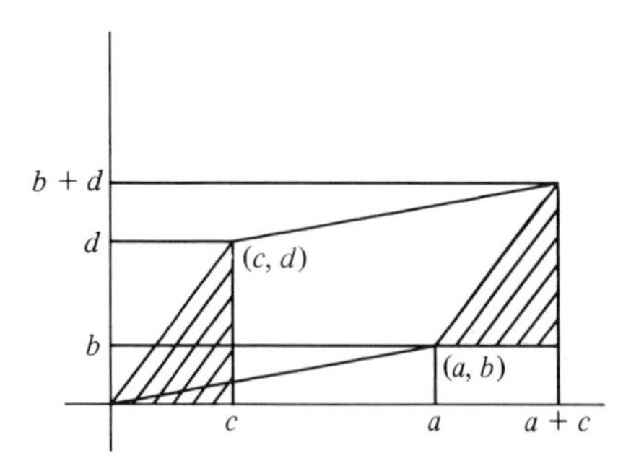

FIGURE 4.2

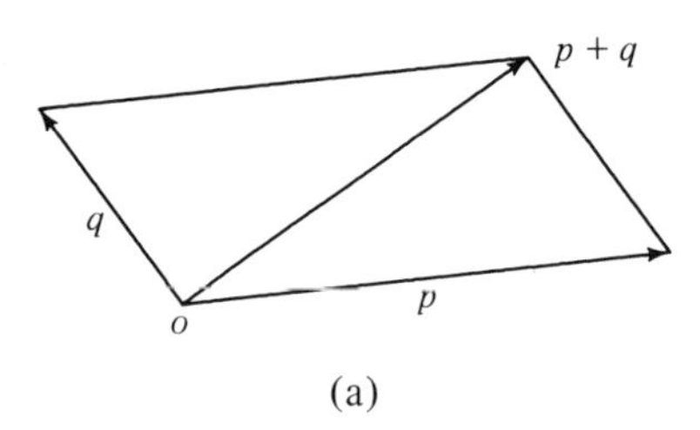

(a)

(b)

FIGURE 4.3

In Figure 4.3(b) the parallelogram has been abbreviated to a triangle. To form $p + q$, the vector q has been picked up and, without changing its length or direction, moved so that its tail coincides with the head of p. The vector $p + q$ is then the line segment from the tail of p to the head of q.

Figure 4.3(b) comes from a different definition of geometric vector than the one we are using. In this alternate development a vector is something with "length" and "direction." Two vectors in different regions of space are then considered "equal" if they have the same length and direction. In this development a vector has no fixed position but may be drawn anywhere.

Formally we will consider our vectors to have their tails firmly nailed to the origin.† We feel free to move them about for heuristic purposes, however.

SCALAR MULTIPLICATION

Let the point p have coordinates (a, b) and let λ be a scalar. The product of the scalar λ and the vector (a, b) is the vector $(\lambda a, \lambda b)$. The corresponding points are plotted in Figure 4.4. The length of $(\lambda a, \lambda b)$ is

$$\sqrt{(\lambda a)^2 + (\lambda b)^2} = |\lambda| \sqrt{a^2 + b^2}$$

†We are, of course, free to choose the position of the origin.

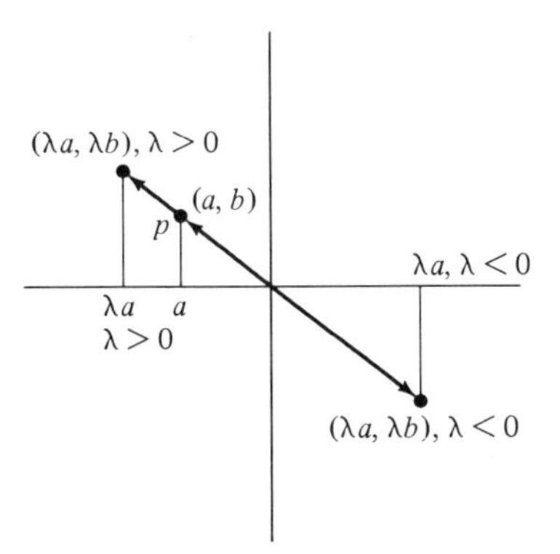

FIGURE 4.4

and the slopes of the line segments are

$$\frac{b}{a} = \frac{\lambda b}{\lambda a} \qquad (\lambda \neq 0)$$

Thus scalar multiplication for $\lambda > 0$ simply stretches or shrinks the length of the segment by the factor λ without changing the slope. If $\lambda < 0$, there is reversal of direction as well; that is, the line segment is first stretched or shrunk by the factor $|\lambda|$ and then, if $\lambda < 0$, it is reflected through the origin.

The situation is precisely the same in three-dimensional Euclidean space (Figure 4.5). Choose three mutually perpendicular axes meeting in a single point as a coordinate system. Then each point p may be identified with a vector of coordinates (a, b, c) in $\mathbf{R}^3$ and with the directed line segment from the origin of the coordinate system to p. With these identifications, the vector operations in $\mathbf{R}^3$

$$(a, b, c) + (a', b', c') = (a + a', b + b', c + c')$$
$$\lambda(a, b, c) = (\lambda a, \lambda b, \lambda c)$$

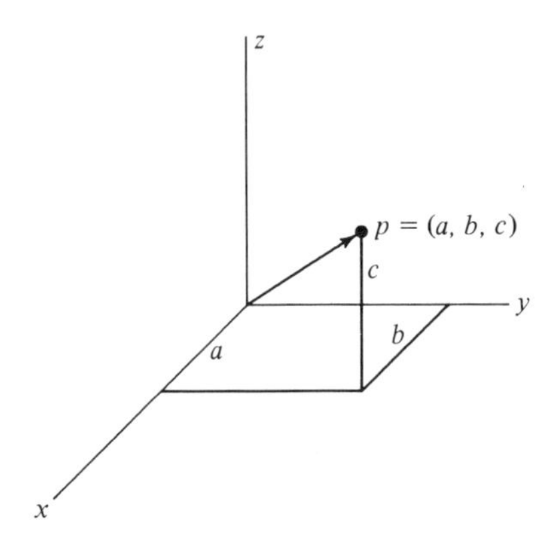

FIGURE 4.5

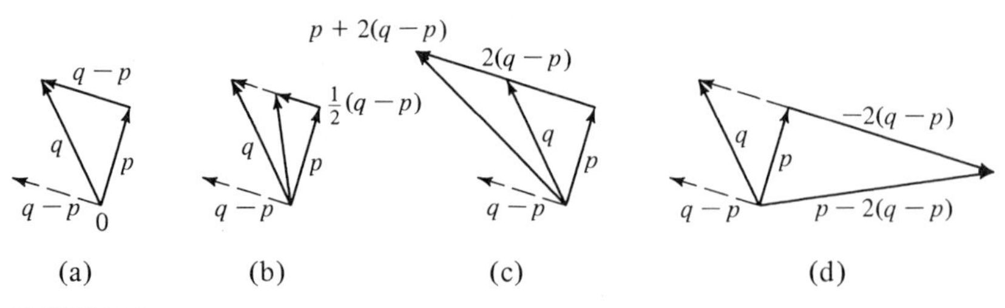

FIGURE 4.6

have exactly the same geometric meaning as in the plane. In what follows, we will assume for the most part that the vectors lie in three-dimensional space. This makes very little difference to the pictures we draw, since they are really a kind of schematic drawing. Since addition and scalar multiplication have a geometric meaning, we may dispense with the coordinate axes and still use algebraic manipulations. This is the appeal of the vector approach to elementary geometry.

We will constantly form the vector $q - p$ from the vectors p and q, and so let us look at what it is geometrically. Now, $p + (q - p) = q$ (to see this, think of them as pairs on $\mathbf{R}^2$ or triples on $\mathbf{R}^3$). So, if we take the vector $q - p$ and move it so that its tail coincides with the head of p, then the head of $q - p$ falls on the head of q [Figure 4.6(a)], and so we may think of $q - p$ as the *vector from p to q*. Since $\frac{1}{2}(q - p)$ is half as long as $q - p$, the vector $p + \frac{1}{2}(q - p)$ is halfway along the segment from p to q [Figure 4.6(b)]. The vector $p + 2(q - p)$, on the other hand, lies on the extension of this segment beyond q, and $p + (-2)(q - p)$ lies on the extension of this segment beyond p [Figures 4.6(c), (d)].

PROPOSITION 4.1 Let p and q be distinct points. As the parameter t varies, the tip of the vector

$$l(t) = p + t(q - p) = (1 - t)p + tq$$

sweeps along the line through points p and q. In particular,

for	$0 < t < 1$	$l(t)$ lies between p and q
for	$t > 1$	q lies between p and $l(t)$
for	$t < 0$	p lies between q and $l(t)$ ■

The proposition is illustrated in Figure 4.7. The vector $l(t)$ is a *parametric* representation of the line through points p and q. Notice that if $p = q$, then $p - q$ is the zero vector and $l(t) \equiv p$.

EXAMPLE 4.1

(a) Find a parametric representation of the line in the plane through points $(1, 1)$ and $(-3, 0)$.

(b) Find a parametric representation of the line in space through the points $(1, 1, 1)$ and $(-2, 1, -2)$.

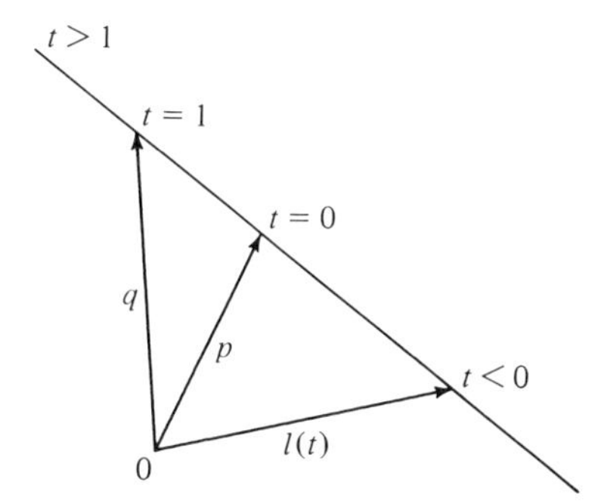

FIGURE 4.7

Solution

(a) Writing vectors as column vectors, take $p = (1, 1)$, $q = (-3, 0)$:

$$l(t) = p + t(q - p) = \begin{bmatrix} 1 \\ 1 \end{bmatrix} + t\left(\begin{bmatrix} -3 \\ 0 \end{bmatrix} - \begin{bmatrix} 1 \\ 1 \end{bmatrix}\right)$$

$$l(t) = \begin{bmatrix} 1 \\ 1 \end{bmatrix} + t\begin{bmatrix} -4 \\ -1 \end{bmatrix}$$

Notice that since

$$l(-1) = \begin{bmatrix} 1 \\ 1 \end{bmatrix} - \begin{bmatrix} -4 \\ -1 \end{bmatrix} = \begin{bmatrix} 5 \\ 2 \end{bmatrix} \quad \text{and} \quad l(2) = \begin{bmatrix} 1 \\ 1 \end{bmatrix} + 2\begin{bmatrix} -4 \\ -1 \end{bmatrix} = -\begin{bmatrix} 7 \\ 1 \end{bmatrix}$$

are points on this line, another parametric representation is

$$m(t) = \begin{bmatrix} 5 \\ 2 \end{bmatrix} + t\left(\begin{bmatrix} -7 \\ -1 \end{bmatrix} - \begin{bmatrix} 5 \\ 2 \end{bmatrix}\right) = \begin{bmatrix} 5 \\ 2 \end{bmatrix} + t\begin{bmatrix} -12 \\ -3 \end{bmatrix}$$

(b)

$$l(t) = \begin{bmatrix} 1 \\ 1 \\ 1 \end{bmatrix} + t\left(\begin{bmatrix} -2 \\ 1 \\ -2 \end{bmatrix} - \begin{bmatrix} 1 \\ 1 \\ 1 \end{bmatrix}\right) = \begin{bmatrix} 1 \\ 1 \\ 1 \end{bmatrix} + t\begin{bmatrix} -3 \\ 0 \\ -3 \end{bmatrix}$$

As in part (a), many other parametric representations are possible. ■

In the plane we commonly write a line as a Cartesian equation $y = ax + b$ (or $x = c$ for vertical lines). We easily obtain this Cartesian equation from a parametric representation by "eliminating the parameter."

For the line of Example 4.1 we have

$$\begin{bmatrix} x \\ y \end{bmatrix} = l(t) = \begin{bmatrix} 1 \\ 1 \end{bmatrix} + t\begin{bmatrix} -4 \\ -1 \end{bmatrix} = \begin{bmatrix} 1 - 4t \\ 1 - t \end{bmatrix}$$

or $x = 1 - 4t$, $y = 1 - t$. Solving for t in terms of x gives

$$y = 1 - t = 1 - \left(\frac{x-1}{-4}\right) = \frac{1}{4}x + \frac{3}{4}$$

No single equation in x, y, z will give a line in space, however, and the parametric expression is usually to be preferred.

Note: Any expression of the form

$$l(t) = p + tv, \qquad -\infty < t < \infty$$

where p and $v \neq 0$ are vectors, sweeps out a line as t varies. It is the line through p and $q = p + v$.

INTERSECTION OF LINES

Let $l(t) = p + tv$ and $m(t) = q + tw$ be parametric representations of two lines. If the lines intersect in a point r, then there are parameter values t_1, t_2 such that $l(t_1) = r = m(t_2)$.

Hence

$$p + t_1 v = q + t_2 w$$

or

$$t_1 v - t_2 w = q - p$$

or

$$[v \mid -w]\begin{bmatrix} t_1 \\ t_2 \end{bmatrix} = q - p$$

EXAMPLE 4.2 Do the following pairs of lines intersect?

(a) $$l(t) = \begin{bmatrix} 1 \\ 1 \\ 1 \end{bmatrix} + t\begin{bmatrix} 1 \\ -2 \\ 1 \end{bmatrix}, \quad m(t) = \begin{bmatrix} 6 \\ -6 \\ 3 \end{bmatrix} + t\begin{bmatrix} -1 \\ 1 \\ 0 \end{bmatrix}$$

(b) The line through $(1, 1, 1)$ and $(2, 2, 0)$ and the line through $(3, 0, 2)$ and $(6, -2, 3)$

(c) $$l(t) = \begin{bmatrix} -1 \\ 1 \end{bmatrix} + t\begin{bmatrix} 2 \\ -3 \end{bmatrix}, \quad m(t) = \begin{bmatrix} 5 \\ -8 \end{bmatrix} + t\begin{bmatrix} -4 \\ 6 \end{bmatrix}$$

Solutions

(a) If $l(t_1) = m(t_2)$, then

$$\begin{bmatrix} 1 \\ 1 \\ 1 \end{bmatrix} + t_1\begin{bmatrix} 1 \\ -2 \\ 1 \end{bmatrix} = \begin{bmatrix} 6 \\ -6 \\ 3 \end{bmatrix} + t_2\begin{bmatrix} -1 \\ 1 \\ 0 \end{bmatrix}$$

$$t_1 \begin{bmatrix} 1 \\ -2 \\ 1 \end{bmatrix} - t_2 \begin{bmatrix} -1 \\ 1 \\ 0 \end{bmatrix} = \begin{bmatrix} 6 \\ -6 \\ 3 \end{bmatrix} - \begin{bmatrix} 1 \\ 1 \\ 1 \end{bmatrix}$$

$$\begin{bmatrix} 1 & 1 \\ -2 & -1 \\ 1 & 0 \end{bmatrix} \begin{bmatrix} t_1 \\ t_2 \end{bmatrix} = \begin{bmatrix} 5 \\ -7 \\ 2 \end{bmatrix}$$

The echelon form of the augmented matrix is

```
        ECHELON  3 3ρ  1  1  5  ¯2  ¯1  ¯7  1  0  2
1.000    0.000    2.000
0.000    1.000    3.000
0.000    0.000    0.000
```

or $t_1 = 2$, $t_2 = 3$. The point of intersection is

$$\begin{bmatrix} 1 \\ 1 \\ 1 \end{bmatrix} + 2 \begin{bmatrix} 1 \\ -2 \\ 1 \end{bmatrix} = \begin{bmatrix} 3 \\ -3 \\ 3 \end{bmatrix} = \begin{bmatrix} 6 \\ -6 \\ 3 \end{bmatrix} + 3 \begin{bmatrix} -1 \\ 1 \\ 0 \end{bmatrix}$$

(b) Parametric representations of the lines are

$$l(t) = \begin{bmatrix} 1 \\ 1 \\ 1 \end{bmatrix} + t\left(\begin{bmatrix} 2 \\ 2 \\ 0 \end{bmatrix} - \begin{bmatrix} 1 \\ 1 \\ 1 \end{bmatrix} \right) = \begin{bmatrix} 1 \\ 1 \\ 1 \end{bmatrix} + t \begin{bmatrix} 1 \\ 1 \\ -1 \end{bmatrix}$$

$$m(t) = \begin{bmatrix} 3 \\ 0 \\ 2 \end{bmatrix} + t\left(\begin{bmatrix} 6 \\ -2 \\ 3 \end{bmatrix} - \begin{bmatrix} 3 \\ 0 \\ 2 \end{bmatrix} \right) = \begin{bmatrix} 3 \\ 0 \\ 2 \end{bmatrix} + t \begin{bmatrix} 3 \\ -2 \\ 1 \end{bmatrix}$$

If $m(t_2) = l(t_1)$, then

$$\begin{bmatrix} 1 \\ 1 \\ 1 \end{bmatrix} + t_1 \begin{bmatrix} 1 \\ 1 \\ -1 \end{bmatrix} = \begin{bmatrix} 3 \\ 0 \\ 2 \end{bmatrix} + t_2 \begin{bmatrix} 3 \\ -2 \\ 1 \end{bmatrix}$$

$$\begin{bmatrix} 1 & -3 \\ 1 & 2 \\ -1 & -1 \end{bmatrix} \begin{bmatrix} t_1 \\ t_2 \end{bmatrix} = \begin{bmatrix} 3 \\ 0 \\ 2 \end{bmatrix} - \begin{bmatrix} 1 \\ 1 \\ 1 \end{bmatrix} = \begin{bmatrix} 2 \\ -1 \\ 1 \end{bmatrix}$$

The augmented matrix of this system is

$$\left[\begin{array}{rr|r} 1 & -3 & 2 \\ 1 & 2 & -1 \\ -1 & -1 & 1 \end{array}\right]$$

which has echelon form *ID* 3. Thus there is no solution to $l(t_1) = m(t_2)$, and the lines do not intersect.

(c) If $l(t_1) = m(t_2)$, then

$$\begin{bmatrix} -1 \\ 1 \end{bmatrix} + t_1\begin{bmatrix} 2 \\ -3 \end{bmatrix} = \begin{bmatrix} 5 \\ -8 \end{bmatrix} + t_2\begin{bmatrix} -4 \\ 6 \end{bmatrix}$$

or

$$\begin{bmatrix} 2 & 4 \\ -3 & -6 \end{bmatrix}\begin{bmatrix} t_1 \\ t_2 \end{bmatrix} = \begin{bmatrix} 5 \\ -8 \end{bmatrix} - \begin{bmatrix} -1 \\ 1 \end{bmatrix} = \begin{bmatrix} 6 \\ -9 \end{bmatrix}$$

The augmented matrix of the system is

$$\left[\begin{array}{cc|c} 2 & 4 & 6 \\ -3 & -6 & -9 \end{array}\right]$$

which has echelon form

$$\left[\begin{array}{cc|c} 1 & 2 & 3 \\ 0 & 0 & 0 \end{array}\right]$$

This means that t_2 can be chosen arbitrarily and then $t_1 = 3 - 2t_2$. Thus $m(t_2) = l(3 - 2t_2)$, and m and l are two different parametric representations of the same straight line. ■

PARALLEL LINES

Two lines, l and m, are parallel if we can slide one, without rotating or twisting, onto the other (Figure 4.8). The idea of sliding without rotating or twisting is formalized in the following definition.

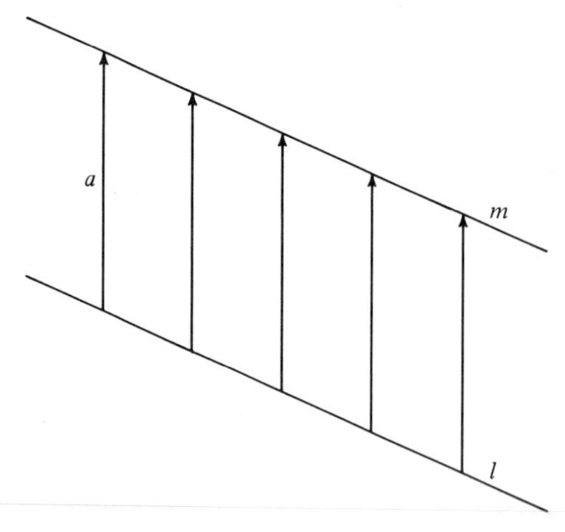

FIGURE 4.8

DEFINITION 4.1 Let a be a vector in $\mathbf{R}^n$. The affine function $T: \mathbf{R}^n \to \mathbf{R}^n$ given by $T(X) = a + X$ is called *translation by a.*

Suppose that $l(t) = p + tv$, $m(t) = q + tw$ are parametrizations of l and m and that l and m are proper lines and not points; that is, $v \neq 0$ and $w \neq 0$. If l and m are parallel, then (Figure 4.8) there will be a vector a such that

$$T(l(t)) = a + p + tv$$

is another parametrization of m. This means that a proper choice of a will cause the system

$$a + p + t_1 v = q + t_2 w$$

to have an infinite number of solutions (t_1, t_2), which in turn means that for properly chosen a, the echelon form of the matrix

$$[v \mid -w \mid q - p - a]$$

is $\begin{bmatrix} 1 & * & * \\ 0 & 0 & 0 \end{bmatrix}$ in the plane

or $\begin{bmatrix} 1 & * & * \\ 0 & 0 & 0 \\ 0 & 0 & 0 \end{bmatrix}$ in space

Thus the lines are parallel if and only if the second column of this echelon form is *not* a pivot column. By Proposition 3.1 this means that $-w$, and hence w also, is a multiple of v. We will state this result in the form that generalizes to higher dimensions.

PROPOSITION 4.2 Let $l(t) = p + tv$, $m(t) = q + tw$, represent lines in the plane or space. The lines are parallel if and only if the vectors v and w are linearly dependent. ∎

Notice that the proposition does not contain the hypothesis $v \neq 0$ and $w \neq 0$. If either vector is zero, then the two vectors are linearly dependent and one or both of the lines degenerates to a point. We will consider a point to be "parallel" to anything (point, line, plane).

Again assume that $v \neq 0$, $w \neq 0$, but now suppose that l and m are *not* parallel. Then v, w are linearly independent and the first two columns of $[v \mid -w \mid q - p]$ are pivot columns. In the plane this means that the only possible echelon form is

$$\begin{bmatrix} 1 & 0 & * \\ 0 & 1 & * \end{bmatrix}$$

and so l and m have a unique point of intersection. For distinct lines in the plane,

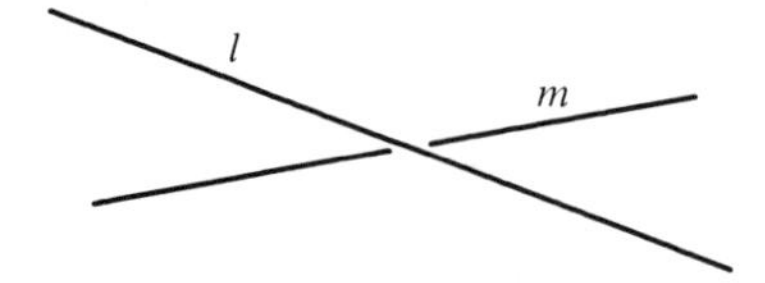

FIGURE 4.9

parallelism is equivalent to nonintersection. In space, however, there are two possible echelon forms:

$$\begin{bmatrix} 1 & 0 & * \\ 0 & 1 & * \\ 0 & 0 & 0 \end{bmatrix} \quad \text{and} \quad \begin{bmatrix} 1 & 0 & 0 \\ 0 & 1 & 0 \\ 0 & 0 & 1 \end{bmatrix}$$

In the first case there is a point of intersection, but in the second case there is no solution to the system $l(t_1) = m(t_2)$.

Two lines that are neither parallel nor intersecting are called *skew* (Figure 4.9).

EXAMPLE 4.3

(a) In Example 4.2(b) the line through (1, 1, 1) and (2, 2, 0) and the line through (3, 0, 2) and (6, −2, 3) were shown to have $[v\,|\,-w\,|\,q-p]$ matrix equal to

$$\begin{bmatrix} 1 & -3 & 2 \\ 1 & 2 & -1 \\ -1 & -1 & 1 \end{bmatrix}$$

which row-reduces to *ID* 3. Thus these lines are skew.

(b) Let

$$l(t) = \begin{bmatrix} 1 \\ 2 \end{bmatrix} + t\begin{bmatrix} 1 \\ 1 \end{bmatrix}, \qquad m(t) = \begin{bmatrix} 4 \\ 3 \end{bmatrix} + t\begin{bmatrix} -2 \\ -2 \end{bmatrix}$$

Then

$$[v\,|\,-w\,|\,q-p] = \begin{bmatrix} 1 & 2 & 3 \\ 1 & 2 & 1 \end{bmatrix}$$

and one row-reduction step gives us

$$\begin{bmatrix} 1 & 2 & 3 \\ 0 & 0 & -2 \end{bmatrix}$$

The lines are parallel and distinct (i.e., not the same line). For a nontrivial inter-

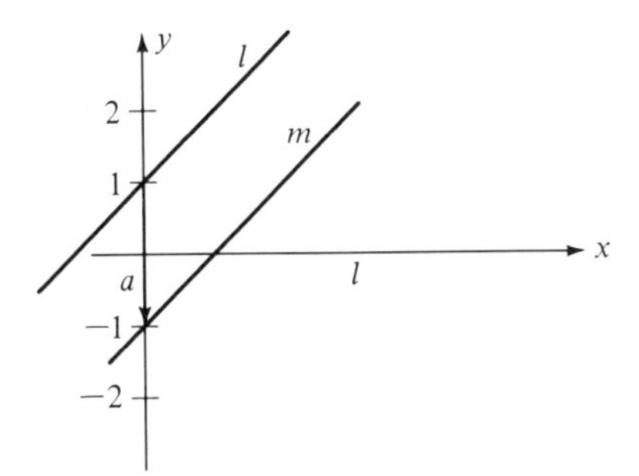

FIGURE 4.10

section the last row should be zero. This suggests adding

$$a = \begin{bmatrix} 0 \\ 2 \end{bmatrix} \text{ to } q \quad \text{or} \quad a = \begin{bmatrix} 0 \\ -2 \end{bmatrix} \text{ to } p$$

If we do the latter and look for solutions of $a + l(t_1) = m(t_2)$, we find that

$$[v \mid -w \mid q - p - a] = \begin{bmatrix} 1 & 2 & 3 \\ 1 & 2 & 3 \end{bmatrix}$$

which has echelon form

$$\begin{bmatrix} 1 & 2 & 3 \\ 0 & 0 & 0 \end{bmatrix}$$

and the two lines coincide.

The lines l and m and the vector a are sketched in Figure 4.10. ■

PLANES

Two distinct points determine a line, and three points, if they do not lie on a line, determine a plane.

Let p_0, p_1, p_2 be three noncollinear points determining a plane π. Let $v_1 = p_1 - p_0$ and $v_2 = p_2 - p_0$. Then any point p in the plane π can be written as $p = p_0 + t_1 v_1 + t_2 v_2$ for some choice of scalars t_1, t_2. To see this, consider the parallelogram (Figure 4.11) with p_0 and p as opposite vertices and sides along the lines $l_i(t) = p_0 + tv_i$ $(i = 1, 2)$. If the other two vertices are $l_1(t_1)$ and $l_2(t_2)$, then $p = p_0 + t_1 v_1 + t_2 v_2$.

If the points p_0, p_1, p_2 are all on the same line, then $l_1(t)$ and $l_2(t)$ are two parametric representations of this line, and in particular v_1 and v_2 are linearly dependent. Conversely, if v_1, v_2 is a linearly dependent pair of vectors, then l_1 and l_2 define parallel lines, both through p_0, and so p_0, $p_1 = l_1(1)$, and $p_2 = l_2(1)$ are collinear by Proposition 4.2. This gives us an easy test for collinearity.

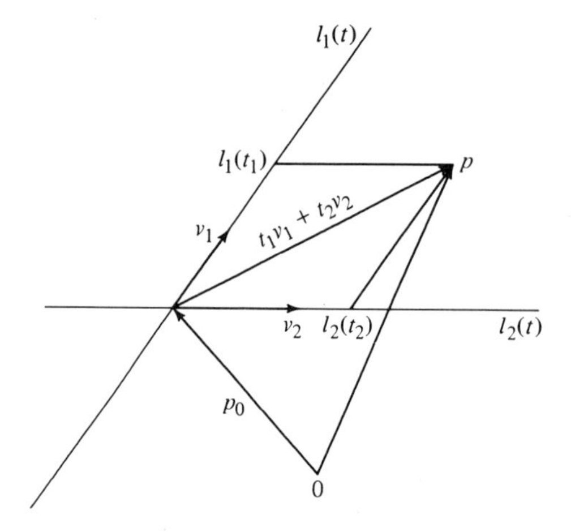

FIGURE 4.11

PROPOSITION 4.3 The points p_0, p_1, p_2 are collinear if and only if the vectors $p_1 - p_0$ and $p_2 - p_0$ are linearly dependent. ∎

EXAMPLE 4.4 Are the points (1, 1, 1), (2, 2, 0), and (−1, −1, 3) collinear?

Solution Let $p_0 = (1, 1, 1)$, $p_1 = (2, 2, 0)$, and $p_2 = (-1, -1, 3)$. Then the matrix

$$[p_1 - p_0 \mid p_2 - p_0] = \begin{bmatrix} 1 & -2 \\ 1 & -2 \\ -1 & 2 \end{bmatrix}$$

row-reduces to

$$\begin{bmatrix} 1 & -2 \\ 0 & 0 \\ 0 & 0 \end{bmatrix}$$

and hence has linearly dependent columns. Thus the points are collinear. ∎

The expression $p(t_1, t_2) = p_0 + t_1(p_1 - p_0) + t_2(p_2 - p_0)$ is called a *parametric representation* of the plane. The next proposition summarizes the discussion on planes so far.

PROPOSITION 4.4. Let p_0, p_1, p_2 be three noncollinear points. A parametric representation of the plane determined by p_0, p_1, p_2 is

$$p(t_1, t_2) = p_0 + t_1(p_1 - p_0) + t_2(p_2 - p_0)$$

Conversely, any expression of the form

$$p(t_1, t_2) = p_0 + t_1 v_1 + t_2 v_2$$

is a parametric representation of a plane, provided the vectors v_1 and v_2 are linearly independent. The plane is determined by the three noncollinear points p_0, $p_1 = p_0 + v_1$, and $p_2 = p_0 + v_2$. ∎

The computational procedure we used for the intersection of two lines extends immediately to the intersection of two planes, the intersection of a line and a plane, or even to the question of a point lying in a plane.

EXAMPLE 4.5

(a) Does the point $(6, -7, 1)$ lie in the plane with parametric representation

$$p(t_1, t_2) = \begin{bmatrix} 1 \\ 0 \\ -1 \end{bmatrix} + t_1 \begin{bmatrix} 1 \\ -2 \\ 1 \end{bmatrix} + t_2 \begin{bmatrix} 1 \\ -1 \\ 0 \end{bmatrix} ?$$

(b) Does the line

$$l(t) = \begin{bmatrix} 2 \\ -1 \\ 0 \end{bmatrix} + t \begin{bmatrix} 1 \\ -5 \\ -2 \end{bmatrix}$$

intersect the plane

$$p(t_1, t_2) = \begin{bmatrix} 1 \\ 1 \\ 1 \end{bmatrix} + t_1 \begin{bmatrix} 1 \\ -2 \\ -2 \end{bmatrix} + t_2 \begin{bmatrix} -1 \\ 3 \\ 2 \end{bmatrix} ?$$

(c) What is the intersection of the planes

$$p(t_1, t_2) = \begin{bmatrix} 1 \\ 2 \\ 3 \end{bmatrix} + t_1 \begin{bmatrix} -1 \\ 1 \\ -1 \end{bmatrix} + t_2 \begin{bmatrix} -3 \\ 2 \\ 1 \end{bmatrix}$$

$$q(s_1, s_2) = \begin{bmatrix} -6 \\ 10 \\ 10 \end{bmatrix} + s_1 \begin{bmatrix} 11 \\ -8 \\ -1 \end{bmatrix} + s_2 \begin{bmatrix} -2 \\ 1 \\ -1 \end{bmatrix} ?$$

Solutions

(a) We need to find t_1, t_2 such that

$$\begin{bmatrix} 1 \\ 0 \\ -1 \end{bmatrix} + t_1 \begin{bmatrix} 1 \\ -2 \\ 1 \end{bmatrix} + t_2 \begin{bmatrix} 1 \\ -1 \\ 0 \end{bmatrix} = \begin{bmatrix} 6 \\ -7 \\ 1 \end{bmatrix}$$

or

$$\begin{bmatrix} 1 & 1 \\ -2 & -1 \\ 1 & 0 \end{bmatrix} \begin{bmatrix} t_1 \\ t_2 \end{bmatrix} = \begin{bmatrix} 6 \\ -7 \\ 1 \end{bmatrix} - \begin{bmatrix} 1 \\ 0 \\ -1 \end{bmatrix} = \begin{bmatrix} 5 \\ -7 \\ 2 \end{bmatrix}$$

The echelon form of the augmented matrix

$$\left[\begin{array}{cc|c} 1 & 1 & 5 \\ -2 & -1 & -7 \\ 1 & 0 & 2 \end{array}\right] \quad \text{is} \quad \left[\begin{array}{cc|c} 1 & 0 & 2 \\ 0 & 1 & 3 \\ 0 & 0 & 0 \end{array}\right]$$

Thus $t_1 = 2$, $t_2 = 3$, and $(6, -7, 1) = p(2, 3)$ lies in the plane.

(b) This time we need t_1, t_2, and t such that

$$\begin{bmatrix} 1 \\ 1 \\ 1 \end{bmatrix} + t_1 \begin{bmatrix} 1 \\ -2 \\ -2 \end{bmatrix} + t_2 \begin{bmatrix} -1 \\ 3 \\ 2 \end{bmatrix} = \begin{bmatrix} 2 \\ -1 \\ 0 \end{bmatrix} + t \begin{bmatrix} 1 \\ -5 \\ -2 \end{bmatrix}$$

or

$$\begin{bmatrix} 1 & -1 & -1 \\ -2 & 3 & 5 \\ -2 & 2 & 2 \end{bmatrix} \begin{bmatrix} t_1 \\ t_2 \\ t \end{bmatrix} = \begin{bmatrix} 2 \\ -1 \\ 0 \end{bmatrix} - \begin{bmatrix} 1 \\ 1 \\ 1 \end{bmatrix} = \begin{bmatrix} 1 \\ -2 \\ -1 \end{bmatrix}$$

The echelon form of the augmented matrix

$$\left[\begin{array}{ccc|c} 1 & -1 & -1 & 1 \\ -2 & 3 & 5 & -2 \\ -2 & 2 & 2 & -1 \end{array}\right] \quad \text{is} \quad \left[\begin{array}{ccc|c} 1 & 0 & 2 & 0 \\ 0 & 1 & 3 & 0 \\ 0 & 0 & 0 & 1 \end{array}\right]$$

hence there is no solution. (The line is parallel to the plane.)

(c) We need t_1, t_2, s_1, s_2 such that

$$\begin{bmatrix} 1 \\ 2 \\ 3 \end{bmatrix} + t_1 \begin{bmatrix} -1 \\ 1 \\ -1 \end{bmatrix} + t_2 \begin{bmatrix} -3 \\ 2 \\ 1 \end{bmatrix} = \begin{bmatrix} -6 \\ 10 \\ 10 \end{bmatrix} + s_1 \begin{bmatrix} 11 \\ -8 \\ -1 \end{bmatrix} + s_2 \begin{bmatrix} -2 \\ 1 \\ -1 \end{bmatrix}$$

or

$$\begin{bmatrix} -1 & -3 & -11 & 2 \\ 1 & 2 & 8 & -1 \\ -1 & 1 & 1 & 1 \end{bmatrix} \begin{bmatrix} t_1 \\ t_2 \\ s_1 \\ s_2 \end{bmatrix} = \begin{bmatrix} -6 \\ 10 \\ 10 \end{bmatrix} - \begin{bmatrix} 1 \\ 2 \\ 3 \end{bmatrix} = \begin{bmatrix} -7 \\ 8 \\ 7 \end{bmatrix}$$

The echelon form of the augmented matrix

$$\begin{bmatrix} -1 & -3 & -11 & 2 & -7 \\ 1 & 2 & 8 & -1 & 8 \\ -1 & 1 & 1 & 1 & 7 \end{bmatrix} \quad \text{is} \quad \left[\begin{array}{cccc|c} 1 & 0 & 2 & 0 & 4 \\ 0 & 1 & 3 & 0 & 5 \\ 0 & 0 & 0 & 1 & 6 \end{array}\right]$$

Thus s_1 is arbitrary, say $s_1 = t$, then

$$\begin{bmatrix} t_1 \\ t_2 \\ s_1 \\ s_2 \end{bmatrix} = \begin{bmatrix} 4 - 2t \\ 5 - 3t \\ t \\ 6 \end{bmatrix}$$

Substituting $(s_1, s_2) = (t, 6)$ into $q(s_1, s_2)$ or $(t_1, t_2) = (4 - 2t, 5 - 3t)$ into $p(t_1, t_2)$ gives the line

$$l(t) = q(t, 6) = \begin{bmatrix} -6 \\ 10 \\ 10 \end{bmatrix} + t\begin{bmatrix} 11 \\ -8 \\ -1 \end{bmatrix} + 6\begin{bmatrix} -2 \\ 1 \\ -1 \end{bmatrix}$$
$$= \begin{bmatrix} -18 \\ 16 \\ 4 \end{bmatrix} + t\begin{bmatrix} 11 \\ -8 \\ -1 \end{bmatrix}$$

as the intersection of the two planes. ■

As with lines, two planes are parallel if there is a translation that carries one to the other. A line is parallel to a plane if there is a translation that puts the line inside the plane.

PROPOSITION 4.5

1. Let two proper planes have parametric representations

$$p(t_1, t_2) = p_0 + t_1 v_1 + t_2 v_2$$
$$q(s_1, s_2) = q_0 + s_1 w_1 + s_2 w_2$$

The planes are parallel if and only if w_1 and w_2 are linear combinations of v_1 and v_2 — that is, RANK $[v_1 | v_2 | w_1 | w_2] \leq 2$.

2. Let a plane and a line have parametric representations

$$p(t_1, t_2) = p_0 + t_1 v_1 + t_2 v_2; \qquad l(t) = q + tw$$

The line and plane are parallel if and only if the set of vectors $\{v_1, v_2, w\}$ is linearly dependent — that is, RANK $[v_1 | v_2 | w] \leq 2$.

Proof We prove statement 1 and leave statement 2 as exercise 26. The two planes are parallel if and only if we can translate one to the other. This means that a proper choice of the vector a will cause the system

$$a + p(t_1, t_2) = q(s_1, s_2) \tag{4.1}$$

to have an entire plane of solutions. This means at least two nonpivot columns in the first four columns of $[v_1|v_2|-w_1|-w_2|q_0 - p_0 - a]$. Since the planes are assumed proper (i.e., not lines or points), v_1, v_2 are linearly independent and the first two columns are pivot columns (Proposition 3.1), and so the third and fourth columns are nonpivot columns. This means (Proposition 3.1) that $-w_1$ and $-w_2$ and hence w_1 and w_2 are linear combinations of v_1 and v_2.

Conversely, if w_1 and w_2 are linear combinations of v_1 and v_2, then the third and fourth columns are nonpivot columns and the solutions of Equation (4.1) are of the form

$$\begin{bmatrix} t_1 \\ t_2 \\ s_1 \\ s_2 \end{bmatrix} = \begin{bmatrix} * \\ * \\ * \\ * \end{bmatrix} + \alpha_1 \begin{bmatrix} * \\ * \\ 1 \\ 0 \end{bmatrix} + \alpha_2 \begin{bmatrix} * \\ * \\ 0 \\ 1 \end{bmatrix} = v_0 + \alpha_1 u_1 + \alpha_2 u_2$$

with u_1, u_2 linearly independent. ∎

CARTESIAN EQUATIONS

The solutions of an equation of the form

$$ax + by + cz = d$$

form a plane in space. This is so because the echelon form of the augmented matrix $[a \ b \ c|d]$ will have one pivot column and two nonpivot columns among the first three columns (Proposition 3.7). The solutions will be of the form

$$X = \begin{bmatrix} x \\ y \\ z \end{bmatrix} = p_0 + t_1 v_1 + t_2 v_2$$

For example, if the equation is

$$y - 3z = 12$$

then the augmented matrix is $[0 \ 1 \ -3|12]$, which is already in echelon form. The nonpivot columns of $[0 \ 1 \ -3]$ are 1 and 3, and so x and z may be chosen arbitrarily, say $x = t_1$, $z = t_2$. Then $y = 12 + 3z = 12 + 3t_2$. Thus

$$\begin{bmatrix} x \\ y \\ z \end{bmatrix} = \begin{bmatrix} t_1 \\ 12 + 3t_2 \\ t_2 \end{bmatrix} = \begin{bmatrix} 0 \\ 12 \\ 0 \end{bmatrix} + t_1 \begin{bmatrix} 1 \\ 0 \\ 0 \end{bmatrix} + t_2 \begin{bmatrix} 0 \\ 3 \\ 1 \end{bmatrix}$$

The plane through the points (0, 12, 0), (1, 12, 0), and (0, 15, 1).

Again by Proposition 3.7 two irredundant equations

$$ax + by + cz = d$$
$$ex + fy + gz = h$$

will define a line in space, since the solutions, containing one arbitrary parameter, will be of the form

$$l(t) = p + tw$$

Thus row-reduction can be used to derive a parametric representation of a line or plane from Cartesian equations. Row-reduction may also be used to go from a parametric representation to Cartesian equations.

Example 4.6 Find a pair of Cartesian equations for the line

$$l(t) = \begin{bmatrix} 2 \\ 3 \\ 5 \end{bmatrix} + t \begin{bmatrix} 1 \\ 4 \\ 6 \end{bmatrix}$$

Solution Two points that determine the line are

$$l(0) = \begin{bmatrix} 2 \\ 3 \\ 5 \end{bmatrix}, \qquad l(1) = \begin{bmatrix} 3 \\ 7 \\ 11 \end{bmatrix}$$

We need numbers a, b, c, d such that $l(0)$ and $l(1)$ are solutions of $ax + by + cz = d$.

Write this as

$$-d + ax + by + cz = 0$$

and seek solutions of

$$\begin{bmatrix} -1 & 2 & 3 & 5 \\ -1 & 3 & 7 & 11 \end{bmatrix} \begin{bmatrix} d \\ a \\ b \\ c \end{bmatrix} = 0$$

The matrix

$$\begin{bmatrix} -1 & 2 & 3 & 5 \\ -1 & 3 & 7 & 11 \end{bmatrix} \text{ row-reduces to } \begin{bmatrix} 1 & 0 & 5 & 7 \\ 0 & 1 & 4 & 6 \end{bmatrix}$$

So letting $c = t_1$, $d = t_2$, we have

$$\begin{bmatrix} d \\ a \\ b \\ c \end{bmatrix} = t_1 \begin{bmatrix} -5 \\ -4 \\ 1 \\ 0 \end{bmatrix} + t_2 \begin{bmatrix} -7 \\ -6 \\ 0 \\ 1 \end{bmatrix}$$

Thus we have an infinite number of coefficient sets (d, a, b, c) to choose from. We need two irredundant equations, however; that is, the coefficient vectors must be linearly independent. Taking $(t_1, t_2) = (-1, 0)$ and $(0, -1)$ gives

$$\begin{aligned} 4x - y \quad &= 5 \\ 6x \quad - z &= 7 \end{aligned}$$

as a pair of equations with solutions

$$l(t) = \begin{bmatrix} 2 \\ 3 \\ 5 \end{bmatrix} + t \begin{bmatrix} 1 \\ 4 \\ 6 \end{bmatrix} \qquad \blacksquare$$

EXERCISES 4.1

In exercises 1 through 22 give the "incidence relations" for the points, lines, or planes. That is, does the point lie on the line or plane; do the lines intersect, are they parallel, coincident; does the line intersect the plane, lie on the plane, and so forth?

1. line: $l(t) = \begin{bmatrix} 1 \\ 1 \\ 1 \end{bmatrix} + t \begin{bmatrix} 1 \\ 2 \\ 3 \end{bmatrix}$
   point: $(1, 0, 1)$

2. line: through $(1, 1, 1)$ and $(2, 3, 4)$
   point: $(\frac{1}{2}, 0, -\frac{1}{2})$

3. line: through $(1, 1)$ and $(3, 7)$
   point: $(2, 4)$

4. line: $\begin{bmatrix} -6 \\ -8 \end{bmatrix} + t \begin{bmatrix} 1 \\ -2 \end{bmatrix}$
   line: $\begin{bmatrix} -10 \\ 3 \end{bmatrix} + s \begin{bmatrix} 2 \\ -5 \end{bmatrix}$

5. line: through $(-8, 0)$ and $(-7, 1)$
   line: through $(-9, 2)$ and $(-8, 2)$

6. line: $\begin{bmatrix} -4 \\ -7 \end{bmatrix} + t \begin{bmatrix} 1 \\ -2 \end{bmatrix}$
   line: $\begin{bmatrix} -3 \\ 8 \end{bmatrix} + s \begin{bmatrix} -2 \\ 4 \end{bmatrix}$

7. line: through $(-1, -4, -8)$ and $(0, -5, -7)$

line: $\begin{bmatrix} 1 \\ -3 \\ -3 \end{bmatrix} + t \begin{bmatrix} 0 \\ -1 \\ -1 \end{bmatrix}$

8. line: $\begin{bmatrix} -3 \\ 0 \\ 1 \end{bmatrix} + t \begin{bmatrix} 4 \\ -2 \\ -1 \end{bmatrix}$

line: $\begin{bmatrix} -5 \\ 4 \\ 4 \end{bmatrix} + s \begin{bmatrix} 2 \\ -5 \\ -2 \end{bmatrix}$

9. line: $\begin{bmatrix} -2 \\ -1 \\ -3 \end{bmatrix} + t \begin{bmatrix} 1 \\ 1 \\ -2 \end{bmatrix}$

line: through $(-2, 0, 2)$ and $(-4, -2, 6)$

10. line: $\begin{bmatrix} -5 \\ -4 \\ -2 \end{bmatrix} + t \begin{bmatrix} 1 \\ -2 \\ 1 \end{bmatrix}$

line: $\begin{bmatrix} -8 \\ 2 \\ -5 \end{bmatrix} + s \begin{bmatrix} -2 \\ 4 \\ -2 \end{bmatrix}$

11. plane: $\begin{bmatrix} 1 \\ 1 \\ 1 \end{bmatrix} + t_1 \begin{bmatrix} 2 \\ 3 \\ 4 \end{bmatrix} + t_2 \begin{bmatrix} -5 \\ -6 \\ -7 \end{bmatrix}$

point: $(8, 9, 11)$

12. plane: $\begin{bmatrix} 1 \\ 0 \\ -1 \end{bmatrix} + t_1 \begin{bmatrix} 2 \\ -3 \\ 1 \end{bmatrix} + t_2 \begin{bmatrix} -1 \\ 0 \\ 1 \end{bmatrix}$

point: $(-2, 3, -1)$

13. plane: $\begin{bmatrix} -5 \\ -6 \\ 1 \end{bmatrix} + t_1 \begin{bmatrix} 1 \\ -2 \\ 0 \end{bmatrix} + t_2 \begin{bmatrix} 0 \\ 1 \\ -2 \end{bmatrix}$

line: $\begin{bmatrix} -11 \\ 13 \\ 9 \end{bmatrix} + t \begin{bmatrix} 2 \\ -5 \\ 1 \end{bmatrix}$

14. plane: $\begin{bmatrix} -6 \\ 1 \\ -3 \end{bmatrix} + t_1 \begin{bmatrix} 1 \\ -2 \\ 1 \end{bmatrix} + t_2 \begin{bmatrix} -1 \\ 3 \\ -1 \end{bmatrix}$

line: $\begin{bmatrix} -11 \\ 10 \\ -4 \end{bmatrix} + t \begin{bmatrix} 1 \\ -1 \\ 0 \end{bmatrix}$

15. plane: $\begin{bmatrix} 1 \\ 1 \\ -7 \end{bmatrix} + t_1 \begin{bmatrix} 1 \\ -1 \\ 0 \end{bmatrix} + t_2 \begin{bmatrix} -2 \\ 3 \\ 1 \end{bmatrix}$

line: $\begin{bmatrix} -1 \\ 3 \\ -6 \end{bmatrix} + t \begin{bmatrix} 4 \\ -7 \\ -3 \end{bmatrix}$

16. plane: through $(-8, -6, -1)$, $(-7, -6, -2)$, and $(-10, -5, -2)$
line: through $(-9, -5, -3)$ and $(-10, -4, -5)$

17. plane: $\begin{bmatrix} -2 \\ -7 \\ -5 \end{bmatrix} + t_1 \begin{bmatrix} 1 \\ -1 \\ -2 \end{bmatrix} + t_2 \begin{bmatrix} -1 \\ 2 \\ 3 \end{bmatrix}$

plane: $\begin{bmatrix} -18 \\ 14 \\ 39 \end{bmatrix} + s_1 \begin{bmatrix} 2 \\ -2 \\ -5 \end{bmatrix} + s_2 \begin{bmatrix} 14 \\ -18 \\ -38 \end{bmatrix}$

18. plane: through $(-3, -4, -4)$, $(-2, -5, -4)$, and $(-3, -3, -3)$
plane: through $(0, -16, -6)$, $(-2, -18, -10)$, and $(0, -4, 0)$

19. plane: $\begin{bmatrix} 1 \\ -2 \\ -7 \end{bmatrix} + t_1 \begin{bmatrix} 1 \\ -1 \\ 1 \end{bmatrix} + t_2 \begin{bmatrix} -1 \\ 2 \\ -2 \end{bmatrix}$

plane: $\begin{bmatrix} 13 \\ -16 \\ 14 \end{bmatrix} + s_1 \begin{bmatrix} 2 \\ -6 \\ 6 \end{bmatrix} + s \begin{bmatrix} -12 \\ 18 \\ -24 \end{bmatrix}$

20. plane: $\begin{bmatrix} 1 \\ -5 \\ -3 \end{bmatrix} + t_1 \begin{bmatrix} 1 \\ 1 \\ -1 \end{bmatrix} + t_2 \begin{bmatrix} -1 \\ 0 \\ 0 \end{bmatrix}$

plane: $\begin{bmatrix} 0 \\ -5 \\ -2 \end{bmatrix} + s_1 \begin{bmatrix} 1 \\ 2 \\ -2 \end{bmatrix} + s_2 \begin{bmatrix} 1 \\ -4 \\ 4 \end{bmatrix}$

21. plane: through $(-2, -4, -7)$, $(-1, -4, -8)$, and $(-1, -3, -10)$
plane: through $(-2, -3, -8)$, $(-7, -6, 3)$, and $(-11, -8, 11)$

22. plane: $\begin{bmatrix} -5 \\ -5 \\ -1 \end{bmatrix} + t_1 \begin{bmatrix} 1 \\ -2 \\ -1 \end{bmatrix} + t_2 \begin{bmatrix} 0 \\ 1 \\ -2 \end{bmatrix}$

plane: $\begin{bmatrix} 1 \\ -10 \\ -21 \end{bmatrix} + s_1 \begin{bmatrix} -2 \\ 1 \\ 8 \end{bmatrix} + s_2 \begin{bmatrix} -4 \\ 3 \\ 14 \end{bmatrix}$

23. Let p and q be points in the plane or space. Let s, t be numbers between 0 and 1 such that $s + t = 1$. Show that $r = sp + tq$ lies on the line segment between p and q.

Hint: Proposition 4.1.

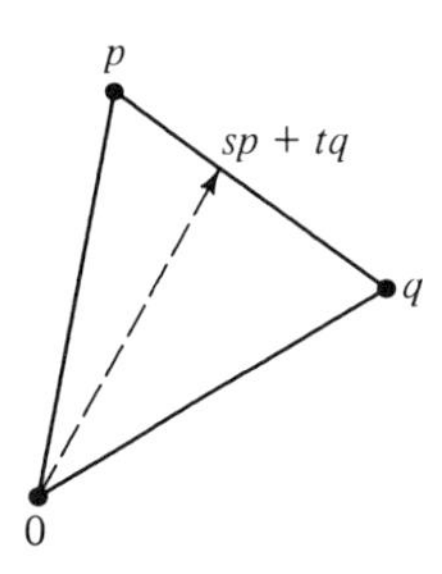

24. If the vertices of a triangle are on the tips of vectors a, b, c, then $\frac{1}{3}(a + b + c)$ is the *center of gravity* of the triangle. A *median* of a triangle is a line segment from a vertex to the midpoint of the opposite side, say, from a to $\frac{1}{2}(b + c)$ (see exercise 23). Show that the center of gravity lies on each median.

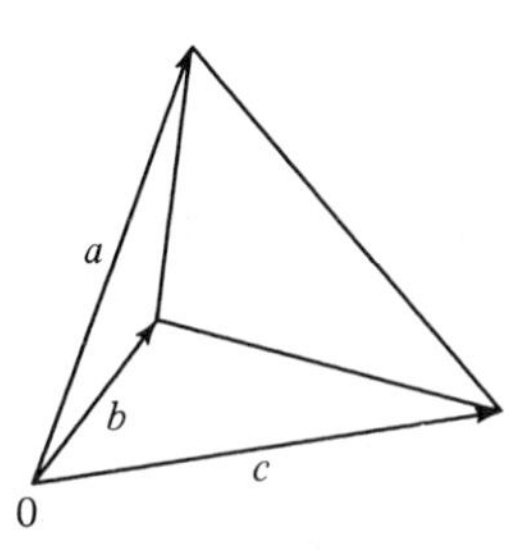

25. Suppose that the vertices of a parallelogram are at the tips of the vectors a, b, c, d. Show that $\frac{1}{2}(a + c) = \frac{1}{2}(b + d)$ and hence the diagonals of a parallelogram bisect each other (see exercise 23).

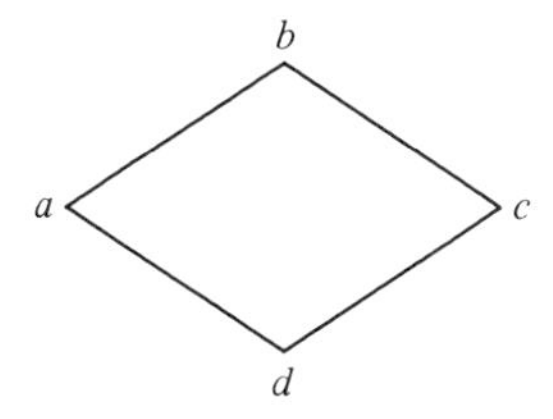

26. Prove statement 2 of Proposition 4.5.

27. Show that a line and a plane in space must either intersect or be parallel.

28. Show that the vectors u_1, u_2 defined in the proof of Proposition 4.5 are linearly independent.

4.2 Coordinate Systems in the Plane and Space

We will use coordinate systems of a type called *affine* coordinate systems. Affine coordinate systems are more general than the familiar rectangular coordinate systems but less general than curvilinear coordinate systems such as polar coordinates.

To lay out an affine grid on a plane, choose an origin and two lines intersecting at the origin for axes. The two axes need not be perpendicular (Figure 4.12). Next choose a unit point for each axis and mark off a uniform scale along each axis. The scales on the two axes need not be the same.

Last, lay out a grid of lines parallel to the chosen axes. To find the coordinates of a point, first find where the grid lines through the point cut the axis and then read the coordinates off the scale marked on the axis (see Figure 4.12).

In the familiar rectangular coordinate system, the axes are perpendicular and

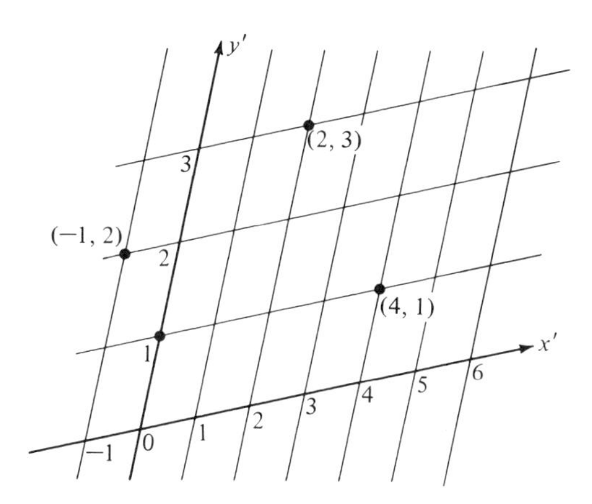

FIGURE 4.12

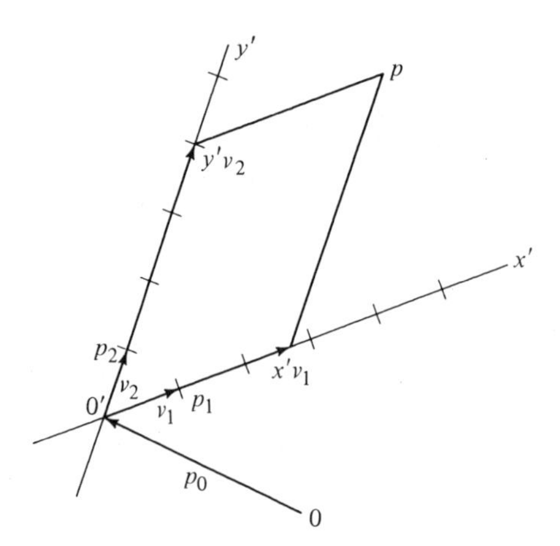

FIGURE 4.13

the scales are the same on both axes. Since an affine grid consists of parallelograms, affine coordinate systems are easily handled by vector methods.

Let a point p have coordinates $X = (x, y)^T$ in the given coordinate system and coordinates $X' = (x', y')^T$ in a new affine coordinate system. Let the point $0'$, the origin of the new coordinate system, be at p_0 and let the unit points on the new axes be at p_1 and p_2. Let $v_i = p_i - p_0$ for $i = 1, 2$. Then (see Figure 4.13) we have

$$\begin{bmatrix} x \\ y \end{bmatrix} = X = p_0 + x'v_1 + y'v_2 = p_0 + [v_1 | v_2]\begin{bmatrix} x' \\ y' \end{bmatrix}$$

or

$$X = p_0 + PX' \qquad \text{where } P = [p_1 - p_0 | p_2 - p_0]$$

Since the columns of P are linearly independent (Proposition 4.3), the matrix P is invertible and

$$X' = P^{-1}(X - p_0)$$

EXAMPLE 4.7 Define an affine coordinate system in the plane with origin at $(1, -2)$, the unit point on the x' axis at $(4, -1)$, and the unit point on the y' axis at $(8, 0)$.

(a) If p has $x'y'$ coordinates $(1, 3)$, what are the xy coordinates of p?

(b) If p has xy coordinates $(1, -1)$, what are the $x'y'$ coordinates of p?

Solution

$$P = [p_1 - p_0 | p_2 - p_0] = \left[\begin{bmatrix} 4 \\ -1 \end{bmatrix} - \begin{bmatrix} 1 \\ -2 \end{bmatrix} \middle| \begin{bmatrix} 8 \\ 0 \end{bmatrix} - \begin{bmatrix} 1 \\ -2 \end{bmatrix}\right]$$
$$= \begin{bmatrix} 3 & 7 \\ 1 & 2 \end{bmatrix}$$

(a) $\begin{bmatrix} x \\ y \end{bmatrix} = X = p_0 + PX' = \begin{bmatrix} 1 \\ -2 \end{bmatrix} + \begin{bmatrix} 3 & 7 \\ 1 & 2 \end{bmatrix}\begin{bmatrix} 1 \\ 3 \end{bmatrix}$

$$= \begin{bmatrix} 1 \\ -2 \end{bmatrix} + \begin{bmatrix} 24 \\ 7 \end{bmatrix} = \begin{bmatrix} 25 \\ 5 \end{bmatrix}$$

(b) $\begin{bmatrix} x' \\ y' \end{bmatrix} = X' = P^{-1}(X - p_0) = \begin{bmatrix} 3 & 7 \\ 1 & 2 \end{bmatrix}^{-1}\left(\begin{bmatrix} 1 \\ -1 \end{bmatrix} - \begin{bmatrix} 1 \\ -2 \end{bmatrix}\right)$

$$= \frac{1}{6 - 7}\begin{bmatrix} 2 & -7 \\ -1 & 3 \end{bmatrix}\begin{bmatrix} 0 \\ 1 \end{bmatrix}$$

$$\begin{bmatrix} x' \\ y' \end{bmatrix} = \begin{bmatrix} 7 \\ -3 \end{bmatrix}$$ ■

Notice that an affine coordinate system in the plane is completely fixed once the three points p_0, p_1, and p_2 are chosen.

The situation in space is a straightforward extension of the case of the plane. To define an affine coordinate system in space, one chooses an origin and three axes through the origin. The axes cannot, of course, all lie in the same plane (Figure 4.14). Choose unit points p_1, p_2, p_3 on the axes and mark off uniform scales.

In this case the point p with $x'y'z'$ coordinates X' has xyz coordinates

$$X = p_0 + x'v_1 + y'v_2 + z'v_3$$

$$= p_0 + [v_1 | v_2 | v_3]\begin{bmatrix} x' \\ y' \\ z' \end{bmatrix}$$

$$X = p_0 + PX$$

where $P = [v_1 | v_2 | v_3]$ and $v_i = p_i - p_0$.

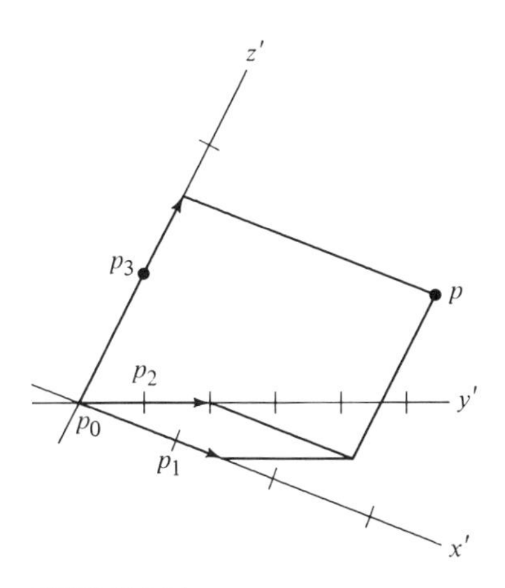

FIGURE 4.14

We have a valid coordinate system when the points p_0, p_1, p_2, and p_3 do not all lie in the same plane. It is useful to have a test for this.

PROPOSITION 4.6 The four points p_0, p_1, p_2, and p_3 are coplanar if and only if the three vectors $p_1 - p_0$, $p_2 - p_0$, and $p_3 - p_0$ are linearly dependent.

Proof This is geometrically evident. We provide an algebraic proof as well.

First assume that the points all lie in the same plane. Let $p(t_1, t_2) = q_0 + t_1 w_1 + t_2 w_2$ be a parametric representation of the plane. Let $Q = [w_1 | w_2]$. Then the plane can be written

$$p(T) = q_0 + QT$$

where $T = [t_1 \; t_2]^T$

Because the points are all in the plane, there are vectors T_0, T_1, T_2, T_3 such that $p_i = q_0 + QT_i$ and so, since

$$p_i - p_0 = q_0 + QT_i - (q_0 + QT_0) = Q(T_i - T_0)$$

then

$$P = [p_1 - p_0 | p_2 - p_0 | p_3 - p_0] = Q[T_1 - T_0 | T_2 - T_0 | T_3 - T_0] = QS$$

where P is 3-by-3, Q is 3-by-2, and S is 2-by-3. By looking at the echelon form of S we see that $SX = 0$ has an infinity of solutions, hence $PX = QSX = 0$ also has an infinity of solutions. Thus by Proposition 2.11 the columns of P are linearly dependent.

Conversely, suppose that the vectors $p_1 - p_0$, $p_2 - p_0$, $p_3 - p_0$ are linearly dependent. By renumbering the points if necessary, we may assume that $p_3 - p_0 = \alpha(p_1 - p_0) + \beta(p_2 - p_0)$. Then the expression

$$p(t_1, t_2) = p_0 + t_1(p_1 - p_0) + t_2(p_2 - p_0)$$

gives $p_0 = p(0, 0)$, $p_1 = p(1, 0)$, $p_2 = p(0, 1)$, $p_3 = p(\alpha, \beta)$. If the vectors $p_1 - p_0$ and $p_2 - p_0$ are linearly independent, then $p(t_1, t_2)$ is a parametric representation of a plane. Otherwise, it is a line or a point. In any case, $p(t_1, t_2)$ is contained in a plane for all t_1, t_2. ■

The proposition shows that the columns of P in the coordinate-change formula given above are independent, hence that P is invertible. Thus solving for X' in terms of X gives

$$X' = P^{-1}(X - p_0)$$

EXAMPLE 4.8 Define an affine coordinate system in space with origin at $(-1, -1, 2)$, the unit point on the x' axis at $(-2, 0, 2)$, the unit point on the y' axis at $(-2, -2, 3)$, and the unit point on the z' axis at $(-2, -2, 4)$.

(a) If p has $x'y'z'$ coordinates $(1, 0, 1)$, what are the xyz coordinates of P?

(b) If p has xyz coordinates $(1, 3, 2)$, what are the $x'y'z'$ coordinates of P?

Solution

$$P = [p_1 - p_0 \,|\, p_2 - p_0 \,|\, p_3 - p_0]$$
$$= \left[\begin{bmatrix} -2 \\ 0 \\ 2 \end{bmatrix} - \begin{bmatrix} -1 \\ -1 \\ 2 \end{bmatrix} \middle| \begin{bmatrix} -2 \\ -2 \\ 3 \end{bmatrix} - \begin{bmatrix} -1 \\ -1 \\ 2 \end{bmatrix} \middle| \begin{bmatrix} -2 \\ -2 \\ 4 \end{bmatrix} - \begin{bmatrix} -1 \\ -1 \\ 2 \end{bmatrix} \right]$$
$$= \begin{bmatrix} -1 & -1 & -1 \\ 1 & -1 & -1 \\ 0 & 1 & 2 \end{bmatrix}$$

(a) $$\begin{bmatrix} x \\ y \\ z \end{bmatrix} = X = p_0 + PX' = \begin{bmatrix} -1 \\ -1 \\ 2 \end{bmatrix} + \begin{bmatrix} -1 & -1 & -1 \\ 1 & -1 & -1 \\ 0 & 1 & 2 \end{bmatrix} \begin{bmatrix} 1 \\ 0 \\ 1 \end{bmatrix}$$
$$= \begin{bmatrix} -1 \\ -1 \\ 2 \end{bmatrix} + \begin{bmatrix} -2 \\ 0 \\ 2 \end{bmatrix} = \begin{bmatrix} -3 \\ -1 \\ 4 \end{bmatrix}$$

(b) Using the formula $X' = P^{-1}(X - p_0)$,

(1 3 2 - ⁻1 ⁻1 2) ⊟P

1 ⁻6 3 ■

The simplest coordinate change is a translation, moving the axes parallel to themselves to a new location (Figure 4.15). If a is the vector to the new origin, then $p_0 = a$, and if v_i is the vector from the origin to the unit point on the ith axis, then $p_i = v_i + a$ and $p_i - p_0 = v_i$. Since v_i has a 1 in the ith component and zeros elsewhere, the matrix P is an identity matrix and we have the next proposition, which applies to both the plane and space.

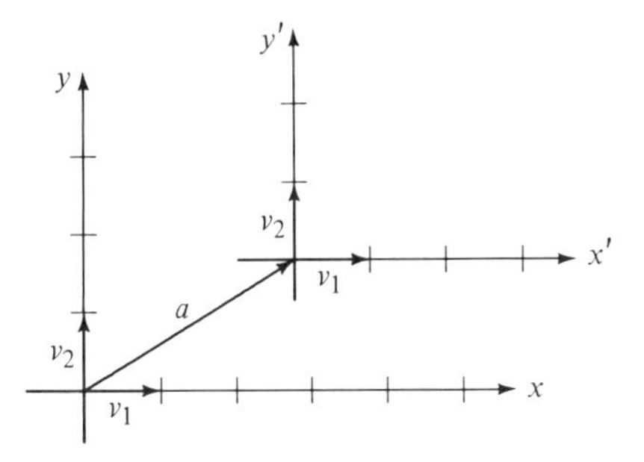

FIGURE 4.15

PROPOSITION 4.7. Let new coordinates be defined by translating the origin of coordinates to a. The coordinate-change formulas are

$$X = a + X', \qquad X' = X - a \qquad \blacksquare$$

An important type of coordinate change in the plane is a rotation of axes. Before deriving the coordinate-change formula for an axis rotation, we need some facts about rotations.

Consider the mapping $T: \mathbf{R}^2 \to \mathbf{R}^2$ that rotates points through an angle of θ radians [Figure 4.16(a)] about the origin. A parallelogram will rotate to a parallelogram, and rotations preserve lengths. Thus by Proposition 2.17 a rotation about the origin is a linear transformation, and to write its matrix we need know only the images of the points (1, 0) and (0, 1) under T. By the definition of sine and cosine we have [Figure 4.16(b)].

$$T(1, 0) = (\cos\theta, \sin\theta)$$

and, since the rotation will preserve the angle between (1, 0) and (0, 1),

$$\begin{aligned} T(0, 1) &= \left(\cos\left(\theta + \frac{\pi}{2}\right), \sin\left(\theta + \frac{\pi}{2}\right)\right) \\ &= (-\sin\theta, \cos\theta) \end{aligned}$$

Thus the rotation is given by

$$\begin{bmatrix} y_1 \\ y_2 \end{bmatrix} = Y = T(X) = \begin{bmatrix} \cos\theta & -\sin\theta \\ \sin\theta & \cos\theta \end{bmatrix} \begin{bmatrix} x_1 \\ x_2 \end{bmatrix} = R_\theta X$$

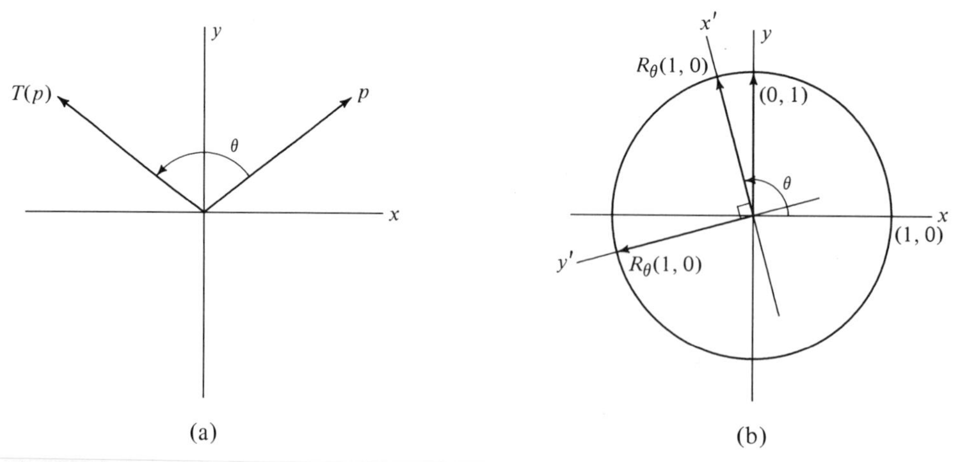

FIGURE 4.16

where the point p has coordinates (x_1, x_2). Here θ is measured in radians, $\theta \geq 0$ indicates a counterclockwise rotation, and $\theta \leq 0$ a clockwise rotation.

The inverse of a rotation matrix is easy to calculate. The inverse operation of a rotation through the angle θ is rotation through the angle $-\theta$; that is,

$$R_\theta^{-1} = R_{-\theta} = \begin{bmatrix} \cos(-\theta) & -\sin(-\theta) \\ \sin(-\theta) & \cos(-\theta) \end{bmatrix} = \begin{bmatrix} \cos\theta & \sin\theta \\ -\sin\theta & \cos\theta \end{bmatrix} = R_\theta^T$$

An APL function to produce a rotation matrix R_θ is

```
    ∇ Z←ROT X
[1]  Z←2 1∘.○X+0,○÷2
    ∇
```

where X is the angle θ in radians. We will need this function from time to time.

PROPOSITION 4.8 Define $x'y'$ coordinates in the plane by rotating the x and y axes through θ radians. Then the coordinate change formulas are

$$\begin{bmatrix} x \\ y \end{bmatrix} = \begin{bmatrix} \cos\theta & -\sin\theta \\ \sin\theta & \cos\theta \end{bmatrix} \begin{bmatrix} x' \\ y' \end{bmatrix}, \quad \begin{bmatrix} x' \\ y' \end{bmatrix} = \begin{bmatrix} \cos\theta & \sin\theta \\ -\sin\theta & \cos\theta \end{bmatrix} \begin{bmatrix} x \\ y \end{bmatrix}$$

Proof We have $p_0 = 0$,

$$p_1 = R_\theta \begin{bmatrix} 1 \\ 0 \end{bmatrix} = \begin{bmatrix} \cos\theta \\ \sin\theta \end{bmatrix}, \quad p_2 = R_\theta \begin{bmatrix} 0 \\ 1 \end{bmatrix} = \begin{bmatrix} -\sin\theta \\ \cos\theta \end{bmatrix}$$

and so

$$X = p_0 + PX' = \begin{bmatrix} \cos\theta & -\sin\theta \\ \sin\theta & \cos\theta \end{bmatrix} = R_\theta X' \qquad \blacksquare$$

EXAMPLE 4.9 Introduce $x'y'$ coordinates in the plane by translating the origin to the point (3, 3) and then rotating the axes 30° clockwise (Figure 4.17).

(a) What are the xy coordinates of the point with $x'y'$ coordinates $(\sqrt{3}, 1)$?

(b) What are the $x'y'$ coordinates of the point with xy coordinates (3, 1)?

Solution First we translate the origin of coordinates to the point (3, 3) obtaining an intermediate coordinate system, call it the $x''y''$ coordinate system. Then by Proposition 4.7 we have

$$X = \begin{bmatrix} 3 \\ 3 \end{bmatrix} + X''$$

In the $x''y''$ system we now define the $x'y'$ coordinate system by rotating the axes

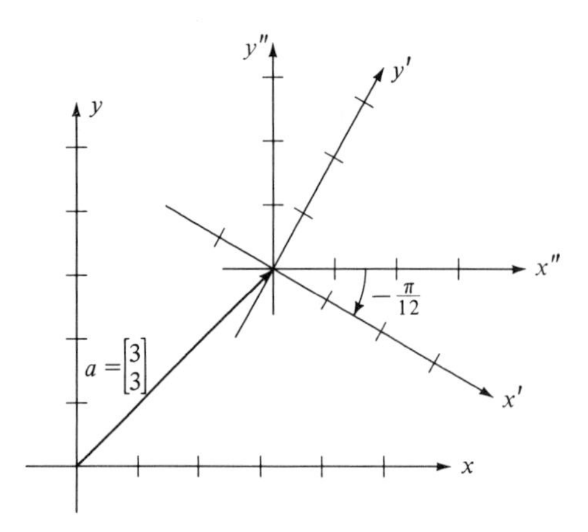

FIGURE 4.17

through 30° in the negative direction. This gives the coordinate-change formula (Proposition 4.8)

$$X'' = R_{-\pi/12}X' = \frac{1}{2}\begin{bmatrix} \sqrt{3} & 1 \\ -1 & \sqrt{3} \end{bmatrix} X'$$

Substituting the second formula into the first gives

$$\begin{bmatrix} x \\ y \end{bmatrix} = \begin{bmatrix} 3 \\ 3 \end{bmatrix} + \frac{1}{2}\begin{bmatrix} \sqrt{3} & 1 \\ -1 & \sqrt{3} \end{bmatrix}\begin{bmatrix} x' \\ y' \end{bmatrix}$$

(a)

$$\begin{bmatrix} x \\ y \end{bmatrix} = \begin{bmatrix} 3 \\ 3 \end{bmatrix} + \frac{1}{2}\begin{bmatrix} \sqrt{3} & 1 \\ -1 & \sqrt{3} \end{bmatrix}\begin{bmatrix} \sqrt{3} \\ 1 \end{bmatrix} = \begin{bmatrix} 5 \\ 3 \end{bmatrix}$$

(b) Solving the equation

$$\begin{bmatrix} 3 \\ 1 \end{bmatrix} = \begin{bmatrix} 3 \\ 3 \end{bmatrix} + \frac{1}{2}\begin{bmatrix} \sqrt{3} & 1 \\ -1 & \sqrt{3} \end{bmatrix} X'$$

gives

$$X' = \frac{1}{2}\begin{bmatrix} \sqrt{3} & -1 \\ 1 & \sqrt{3} \end{bmatrix}\begin{bmatrix} 0 \\ -2 \end{bmatrix} = \begin{bmatrix} 1 \\ -\sqrt{3} \end{bmatrix}$$

where the identity $R_\theta^{-1} = R_\theta^T$ was used to invert $R_{-\pi/12}$. ∎

AFFINE FUNCTIONS

Suppose we have an affine transformation $T: \mathbf{R}^n \to \mathbf{R}^n$, where $n = 2$ or 3 — that is, in the plane or space. Suppose that $Y = T(X) = b + AX$. If we change to X'

coordinates, where $X = p_0 + PX'$, then $Y = p_0 + PY'$ as well and

$$\begin{aligned} Y = p_0 + PY' &= b + A(p_0 + PX') \\ PY' &= (b + Ap_0 - p_0) + APX' \\ Y' &= P^{-1}(b + Ap_0 - p_0) + P^{-1}APX' \\ Y' &= q_0 + BX' \end{aligned}$$

where $B = P^{-1}AP$ and $q_0 = P^{-1}(b + (A - I)p_0)$. We record this formula for future reference.

PROPOSITION 4.9 Let $T: \mathbf{R}^n \to \mathbf{R}^n$ be an affine transformation. Let T be given by $Y = T(X) = b + AX$. Introduce X' coordinates via the formula $X = p_0 + PX'$. Then, in the new coordinate system T is given by

$$Y' = c + BX'$$

where $B = P^{-1}AP$ and $c = P^{-1}(b + (A - I)p_0)$.

In particular, if T is linear ($b = 0$) and the coordinate change leaves the origin fixed ($p_0 = 0$), we have

$$Y' = (P^{-1}AP)X' \quad \blacksquare$$

The formulas of Proposition 4.9 should not be memorized. In any particular application it is usually easier to start with the formula $X = p_0 + PX'$ and make the appropriate substitution. This is especially true if one wishes to go from X' to X coordinates. Moving from X' to X coordinates is a very common case. Typically we define a special coordinate system (the X' system) in which our problem has an especially simple form. We solve the problem in the X' coordinate system and then use the formula $X = p_0 + PX'$ to transfer the solution to the original coordinate system. The rest of the examples in this section illustrate this technique.

We may put the formulas of Proposition 4.10 in more compact form by using the *projective representation* of an affine transformation, which is developed in exercises 39, 40, and 41.

EXAMPLE 4.10 The plane is mapped onto itself by a rotation of 45° clockwise about the point (2, 2). Write a formula for this transformation.

Solution A rotation of 45° clockwise (or $-\pi/4$ radians) *about the origin* is a linear transformation with matrix

$$p = \frac{1}{\sqrt{2}}\begin{bmatrix} 1 & 1 \\ -1 & 1 \end{bmatrix}$$

Define an $x'y'$ coordinate system by translating the origin to (2, 2). Then

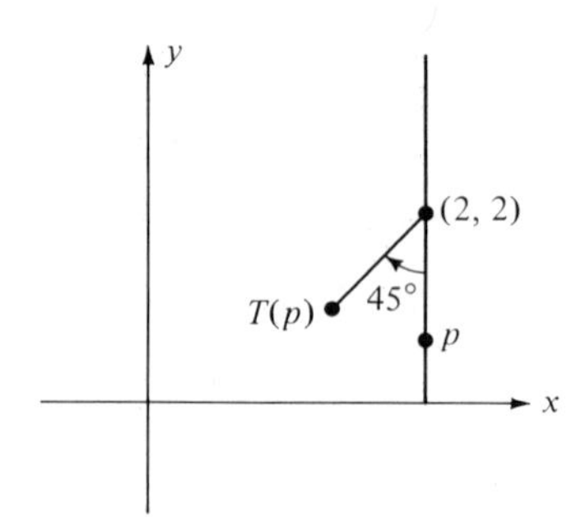

$X = (2, 2) + X'$, and the transformation is given by

$$Y' = \frac{1}{\sqrt{2}}\begin{bmatrix} 1 & 1 \\ -1 & 1 \end{bmatrix} X'$$

So, substituting $X' = X - (2, 2)$ and $Y' = Y - (2, 2)$, we have

$$Y - \begin{bmatrix} 2 \\ 2 \end{bmatrix} = \frac{1}{\sqrt{2}}\begin{bmatrix} 1 & 1 \\ -1 & 1 \end{bmatrix}\left(X - \begin{bmatrix} 2 \\ 2 \end{bmatrix}\right)$$

$$Y = \begin{bmatrix} 2 \\ 2 \end{bmatrix} - \frac{1}{\sqrt{2}}\begin{bmatrix} 1 & 1 \\ -1 & 1 \end{bmatrix}\begin{bmatrix} 2 \\ 2 \end{bmatrix} + \frac{1}{\sqrt{2}}\begin{bmatrix} 1 & 1 \\ -1 & 1 \end{bmatrix} X$$

or

$$Y = \begin{bmatrix} y_1 \\ y_2 \end{bmatrix} = \begin{bmatrix} 2 - 2\sqrt{2} \\ 2 \end{bmatrix} + \frac{1}{\sqrt{2}}\begin{bmatrix} 1 & 1 \\ -1 & 1 \end{bmatrix}\begin{bmatrix} x_1 \\ x_2 \end{bmatrix}$$
$$= T(x_1, x_2)$$

(This is a rotation of $-\pi/4$ about 0, followed by a translation.) ■

EXAMPLE 4.11 The plane is mapped into itself by reflection in the line through the origin at an angle of θ radians to the x axis. Show that this transformation, T, is linear and find a formula for it.

Solution The easiest way to show T linear is to find a formula for T and thus show that T is multiplication by a matrix.

Reflection in the *x-axis* is given by $(a, b) \mapsto (a, -b)$. So define $x'y'$ coordinates by rotating the xy axes through the angle θ. Then, in $x'y'$ coordinates, T has the formula

$$\begin{bmatrix} y'_1 \\ y'_2 \end{bmatrix} = Y' = T(X') = \begin{bmatrix} 1 & 0 \\ 0 & -1 \end{bmatrix}\begin{bmatrix} x'_1 \\ x'_2 \end{bmatrix}$$

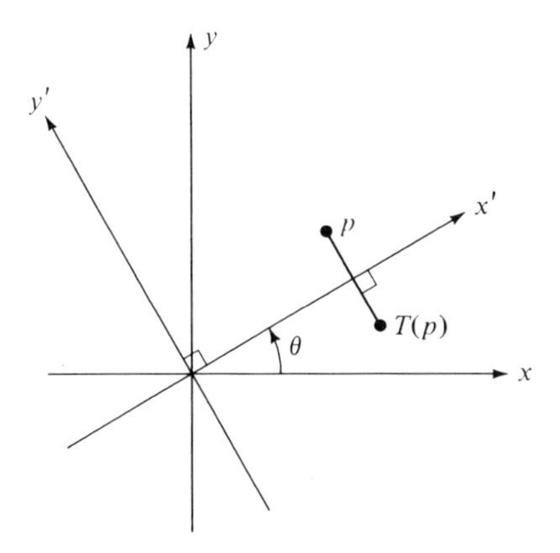

The coordinate-change formula is

$$X = \begin{bmatrix} \cos\theta & -\sin\theta \\ \sin\theta & \cos\theta \end{bmatrix} X' = R_\theta X' \quad \text{or} \quad X' = R_{-\theta} X$$

(Recall that $R_\theta^{-1} = R_{-\theta} = R_\theta^T$.) Thus

$$Y' = \begin{bmatrix} 1 & 0 \\ 0 & -1 \end{bmatrix} X', \qquad R_{-\theta} Y = \begin{bmatrix} 1 & 0 \\ 0 & -1 \end{bmatrix} R_{-\theta} X$$

or

$$Y = R_\theta \begin{bmatrix} 1 & 0 \\ 0 & -1 \end{bmatrix} R_{-\theta} X = \begin{bmatrix} \cos\theta & -\sin\theta \\ \sin\theta & \cos\theta \end{bmatrix} \begin{bmatrix} 1 & 0 \\ 0 & -1 \end{bmatrix} \begin{bmatrix} \cos\theta & \sin\theta \\ -\sin\theta & \cos\theta \end{bmatrix} X$$

$$= \begin{bmatrix} \cos^2\theta - \sin^2\theta & 2\sin\theta\cos\theta \\ 2\sin\theta\cos\theta & \sin^2\theta - \cos^2\theta \end{bmatrix} X$$

or

$$\begin{bmatrix} y_1 \\ y_2 \end{bmatrix} = \begin{bmatrix} \cos 2\theta & \sin 2\theta \\ \sin 2\theta & -\cos 2\theta \end{bmatrix} \begin{bmatrix} x_1 \\ x_2 \end{bmatrix}$$ ■

EXAMPLE 4.12 We map space to itself by projecting points parallel to the line through points (1, 1, 1) and (2, 2, 4) onto the plane through the three points (0, 0, −1), (1, 0, 0), and (0, 1, 0). Find the formula for this mapping.

Solution In any coordinate system with x' and y' axes lying in the plane and a z' axis parallel to the direction of projection, the mapping is simply $(x_1', x_2', x_3') \to (x_1', x_2', 0)$.

Such coordinate system is given by taking $p_0 = (0, 0, -1)$, $p_1 = (1, 0, 0)$, $p_2 = (0, 1, 0)$, and $p_3 = p_0 + (2, 2, 4) - (1, 1, 1) = p_0 + (1, 1, 3)$. Then

$$Y' = \begin{bmatrix} y_1' \\ y_2' \\ y_3' \end{bmatrix} = \begin{bmatrix} 1 & 0 & 0 \\ 0 & 1 & 0 \\ 0 & 0 & 0 \end{bmatrix} \begin{bmatrix} x_1' \\ x_2' \\ x_3' \end{bmatrix} = BX'$$

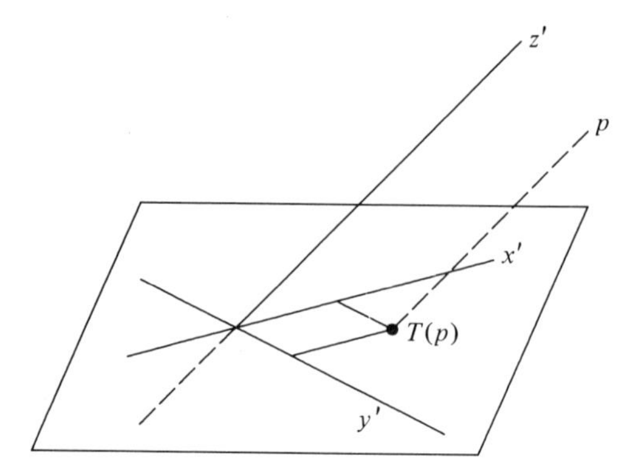

and

$$X = p_0 + [p_1 - p_0 \,|\, p_2 - p_0 \,|\, p_3 - p_0]X' = \begin{bmatrix} 0 \\ 0 \\ -1 \end{bmatrix} + \begin{bmatrix} 1 & 0 & 1 \\ 0 & 1 & 1 \\ 1 & 1 & 3 \end{bmatrix} X'$$

Then

$$X' = P^{-1}\left(X - \begin{bmatrix} 0 \\ 0 \\ -1 \end{bmatrix}\right)$$

where $P = [p_1 - p_0 \,|\, p_2 - p_0 \,|\, p_3 - p_0]$, hence

$$Y' = P^{-1}\left(Y - \begin{bmatrix} 0 \\ 0 \\ -1 \end{bmatrix}\right) = BX' = BP^{-1}\left(X - \begin{bmatrix} 0 \\ 0 \\ -1 \end{bmatrix}\right)$$

or

$$Y = \begin{bmatrix} 0 \\ 0 \\ -1 \end{bmatrix} + PBP^{-1}X - PBP^{-1}\begin{bmatrix} 0 \\ 0 \\ -1 \end{bmatrix}$$

or

$$Y = c + AX$$

where $A = PBP^{-1}$; that is,

```
      +A←P+.×B+.×⌹P
2.000   1.000  ¯1.000
1.000   2.000  ¯1.000
3.000   3.000  ¯2.000
```

and $c = (I - PBP^{-1})p_0$; that is,

```
      +C←((ID 3)−P+.×B+.×⌹P)+.×0 0 ¯1
¯1  ¯1  ¯3
```

so

$$Y = \begin{bmatrix} y_1 \\ y_2 \\ y_3 \end{bmatrix} = \begin{bmatrix} -1 \\ -1 \\ -3 \end{bmatrix} + \begin{bmatrix} 2 & 1 & -1 \\ 1 & 2 & -1 \\ 3 & 3 & -2 \end{bmatrix} \begin{bmatrix} x_1 \\ x_2 \\ x_3 \end{bmatrix} \quad \blacksquare$$

EXERCISES 4.2

In exercises 1 through 5 new affine coordinate systems are specified by giving the origin and unit points. Convert the given xyz coordinates to $x'y'z'$ coordinates and convert the given $x'y'z'$ coordinates to xyz coordinates.

1. $p_0 = (1, 1)$, $p_1 = (2, -1)$, $p_2 = (-1, 6)$

x'	0	1	-1	-2
y'	0	1	1	4

x	0	1	-1	2
y	0	1	1	4

2. $p_0 = (-1, 2)$, $p_1 = (0, 2)$, $p_2 = (-4, 3)$

x'	1	-1	1	-1
y'	1	1	-1	-1

x	1	-1	1	-1
y	1	1	-1	-1

3. $p_0 = (1, -2)$, $p_1 = (2, -6)$, $p_2 = (2, -5)$

x'	0	0	1	-1	-10
y'	0	1	0	-1	-10

x	0	0	1	1	-1
y	0	1	0	1	-1

4. $p_0 = (1, 1, 1)$, $p_1 = (2, 0, 2)$, $p_2 = (2, 1, 0)$, $p_3 = (-3, 3, 2)$

x'	0	1	-1
y'	1	1	1
z'	0	1	-1

x	0	0	1	1
y	0	0	0	1
z	0	1	0	1

5. $p_0 = (-1, -1, 2)$, $p_1 = (0, -4, 3)$, $p_2 = (-2, 3, 1)$, $p_3 = (-2, 3, 2)$

x'	0	1	-1
y'	1	1	1
z'	0	1	-1

x	0	1	1	2
y	0	0	1	0
z	0	0	1	2

In exercises 6 through 11 a coordinate-change formula $X = p_0 + PX'$ is given and a curve is defined in $x'y'$ coordinates. On an xy coordinate system plot the x' and y' axes, mark off the scales on the axes, and sketch the curve by first computing $x'y'$ coordinates of points on the curve and then plotting the corresponding xy coordinates.

6. $X = \begin{bmatrix} 2 \\ 1 \end{bmatrix} + X'; \ y' = x'$

7. $X = \begin{bmatrix} 2 \\ 1 \end{bmatrix} + \begin{bmatrix} 4 & 0 \\ 0 & 3 \end{bmatrix} X'; \ y' = x'$

8. $X = \begin{bmatrix} 0 & -1 \\ -1 & 0 \end{bmatrix} X'; \ y = x'^2$

9. $X = \begin{bmatrix} 0 & -5 \\ 1 & 0 \end{bmatrix} X'; \ y' = x'$

10. $X = \begin{bmatrix} 2 \\ 2 \end{bmatrix} + \begin{bmatrix} 2 & -1 \\ 1 & 0 \end{bmatrix} X'; \ y' = x'^2$

11. (Computer assignment) $X = \begin{bmatrix} 3 & 1 \\ -3 & 1 \end{bmatrix} X'; \ x'^2 + y'^2 = 1$

Hint: `C←2 1∘.○10 CHOP 0,○2` gives $x'y'$ coordinates for ten points on this curve, and `(2 2⍴3 1 ¯3 1)+.×C` gives the corresponding xy coordinates.

In exercises 12 through 18 a new coordinate system is described. Find the coordinate-change formula $X = p_0 + PX'$.

12. Positive x' axis is positive y axis, positive y' axis is positive x axis, no scale changes.

13. The new axes lie on the old axes but the distance from the origin to the unit point on the x' axis is double the distance on the x axis and the distance from the origin to the unit point on the y' axis is half the distance to the unit point on the y axis.

14. The new axes are obtained by rotating the old axes 30° clockwise.

15. The positive x' axis is the negative y axis, the positive y' axis is the negative x axis, and the scales are the same.

16. The new axes are obtained by moving the old axes ten units in the direction of the point (3, 4).

17. The x' axis is the line $3y + 2x = 8$ and the y' axis is the line $y = 2x$.

Note: Since the scales on the new axes are unspecified, there will be arbitrary parameters in P.

18. The x' and y' axes are obtained by rotating the xy axes 45° counterclockwise about the z axis. The z' axis is obtained by rotating the z' axis 180° about the origin (in *any* direction!).

19. Do the given sets of points all lie in the same plane?

Hint: Use Proposition 4.6.

(a) $(1, 1, 1), (2, 2, 0), (-1, 0, 2), (-4, -1, 3), (-5, -1, 3)$

(b) $(0, 1, 1), (1, 1, 2), (2, 1, 3), (1, 1, 3)$

(c) $(-1, 1, 0), (-2, 0, 0), (3, 1, 1), (10, 0, 3), (13, -1, 4)$

(d) $(-1, -1, -1), (1, -1, -2), (-1, 0, -2), (-2, -1, 0)$

In exercises 20 through 30 an affine transformation $Y = b + AX$ is given and a new coordinate system is defined by specifying the origin and unit points. Find the form $Y' = q + BX'$ of the affine function in the new coordinate system.

20. $Y = \begin{bmatrix} -1 \\ -1 \end{bmatrix} + \begin{bmatrix} 3 & -1 \\ 0 & 2 \end{bmatrix} X, \ p_0 = (1, 1), \ p_1 = (2, 2), \ p_2 = (0, 1)$

21. $Y = \begin{bmatrix} 4 \\ -5 \end{bmatrix} + \begin{bmatrix} -6 & -4 \\ 8 & 6 \end{bmatrix} X, \ p_0 = (0, 1), \ p_1 = (1, 0), \ p_2 = (-1, 3)$

22. $Y = \begin{bmatrix} 0 & -1 \\ 0 & 1 \end{bmatrix} X, \ p_0 = (2, -2), \ p_1 = (3, -3), \ p_2 = (3, -2)$

23. $Y = \begin{bmatrix} 1 \\ -3 \end{bmatrix} + \begin{bmatrix} -1 & 1 \\ 0 & 0 \end{bmatrix} X$, $p_0 = (-1, -3)$, $p_1 = (0, -2)$, $p_2 = (-2, -3)$

24. $Y = \begin{bmatrix} 8 \\ -12 \end{bmatrix} + \begin{bmatrix} -2 & -2 \\ 5 & 4 \end{bmatrix} X$, $p_0 = (0, 4)$, $p_1 = (1, 3)$, $p_2 = (0, 6)$

25. $Y = \begin{bmatrix} -2 \\ 0 \end{bmatrix} + \begin{bmatrix} -1 & -2 \\ 0 & 1 \end{bmatrix} X$, $p_0 = (-1, 0)$, $p_1 = (0, -1)$, $p_2 = (0, 0)$

26. $Y = \begin{bmatrix} 13 \\ -26 \\ -1 \end{bmatrix} + \begin{bmatrix} -1 & -2 & 1 \\ 4 & 5 & -2 \\ 0 & 0 & 2 \end{bmatrix} X$, $p_0 = \begin{bmatrix} 3 \\ 4 \\ 1 \end{bmatrix}$, $p_1 = \begin{bmatrix} 4 \\ 3 \\ 1 \end{bmatrix}$, $p_2 = \begin{bmatrix} 2 \\ 6 \\ 2 \end{bmatrix}$, $p_3 = \begin{bmatrix} 4 \\ 2 \\ 1 \end{bmatrix}$

27. $Y = \begin{bmatrix} 2 & -2 & 1 \\ 2 & -5 & 4 \\ 2 & -8 & 7 \end{bmatrix} X - \begin{bmatrix} 1 \\ 4 \\ 6 \end{bmatrix}$, $p_0 = \begin{bmatrix} 1 \\ 1 \\ 2 \end{bmatrix}$, $p_1 = \begin{bmatrix} 2 \\ 2 \\ 3 \end{bmatrix}$, $p_2 = \begin{bmatrix} 2 \\ 3 \\ 4 \end{bmatrix}$, $p_3 = \begin{bmatrix} 1 \\ 2 \\ 4 \end{bmatrix}$

28. $Y = \begin{bmatrix} -3 & 2 & 0 \\ -4 & 3 & 0 \\ 2 & -2 & -1 \end{bmatrix} X + \begin{bmatrix} 8 \\ 8 \\ 8 \end{bmatrix}$, $p_0 = \begin{bmatrix} 4 \\ 4 \\ 4 \end{bmatrix}$, $p_1 = \begin{bmatrix} 5 \\ 5 \\ 3 \end{bmatrix}$, $p_2 = \begin{bmatrix} 5 \\ 6 \\ 3 \end{bmatrix}$, $p_3 = \begin{bmatrix} 4 \\ 4 \\ 5 \end{bmatrix}$

29. $Y = \begin{bmatrix} 0 & -1 & 0 \\ -1 & 0 & 0 \\ -2 & 2 & -1 \end{bmatrix} X - \begin{bmatrix} 1 \\ 1 \\ 6 \end{bmatrix}$, $p_0 = \begin{bmatrix} -2 \\ 1 \\ 0 \end{bmatrix}$, $p_1 = \begin{bmatrix} -1 \\ 2 \\ 1 \end{bmatrix}$, $p_2 = \begin{bmatrix} -2 \\ 2 \\ 1 \end{bmatrix}$, $p_3 = \begin{bmatrix} -3 \\ 1 \\ 1 \end{bmatrix}$

30. $Y = \begin{bmatrix} -4 \\ 2 \\ 0 \end{bmatrix} + \begin{bmatrix} -3 & -4 & -2 \\ 3 & 5 & 3 \\ -2 & -4 & -3 \end{bmatrix} X$, $p_0 = \begin{bmatrix} -2 \\ 1 \\ 0 \end{bmatrix}$, $p_1 = \begin{bmatrix} -1 \\ 0 \\ 1 \end{bmatrix}$, $p_2 = \begin{bmatrix} -3 \\ 3 \\ 2 \end{bmatrix}$, $p_3 = \begin{bmatrix} -3 \\ 2 \\ 0 \end{bmatrix}$

31. The plane is mapped onto itself by a 45° rotation counterclockwise about the point $(1, 0)$. Write a formula for this transformation.

32. The plane is mapped onto itself by reflection in the line $x + y = 1$. Write a formula for this transformation.

33. The plane is mapped into itself by projecting points parallel to the line $y = 2x$ onto the line $x - y = 1$. Write a formula for this transformation.

34. Space is mapped into itself by a rotation of 120° clockwise with axis of rotation the vertical line through the points $(2, 3, -1)$ and $(2, 3, 1)$. Write a formula for this transformation.

35. Space is mapped into itself by projection onto the plane through $(1, 0, 1)$, $(2, -3, 2)$, and $(0, 4, 0)$ parallel to the line through $(1, 0, 1)$ and $(0, 4, 1)$. Write a formula for this transformation.

36. The plane is mapped into itself as follows. A typical point p is first rotated 90° counterclockwise, then translated two units in the direction of the positive y axis, and finally reflected in the line $x = -1$.

(a) Write a formula for this transformation.

Hint: If $Y = T_1(X)$ is the rotation, $Y = T_2(X)$ the translation, and $Y = T_3(X)$ the reflection, then the transformation is $Y = T_3(T_2(T_1(X))) = T(X)$.

(b) Find all the points mapped to themselves by this circuitous route [i.e., solve $X = T(X)$].

(c) Move the origin to any point such that $T(X) = X$ [part (b)], and write the formula for T in the new coordinate system.

(d) Rotate axes 45° and write a formula for the transformation of part (c) in the new coordinate system.

(e) Give a different geometrical description of the transformation T.

37. Space is mapped into itself as follows. A typical point is first rotated $30°$ counterclockwise about the x axis, then $90°$ clockwise about the y axis, then $150°$ clockwise about the z axis.

(a) Write a formula for this transformation and then show that it is linear.

Hint: See the hint to exercise 36(a).

(b) This transformation is in fact a rotation in space. The axis is the set of points such that $X = T(X)$. Find the axis. From a sketch, through what angle are the points rotated about this axis?

38. Let p and q be two points in the plane. Show that a rotation of the plane about q is the same as a rotation about p, through the same angle, followed by a translation.

Hint: Without loss of generality you may assume that p is the origin. Now imitate Example 4.10.

Projective Representation

The remaining exercises introduce a representation of the affine transformation $T(X) = b + AX$ that gives a coordinate-change formula more convenient for machine work than the formula of Proposition 4.9. Such representations are heavily used in computer-graphics applications.

39. Let A, B be n-by-n matrices and let a, b be column vectors in $\mathbf{R}^n$.

(a) Show that

$$\left[\begin{array}{c|c} 1 & 0 \\ \hline a & A \end{array}\right]\left[\begin{array}{c|c} 1 & 0 \\ \hline b & B \end{array}\right] = \left[\begin{array}{c|c} 1 & 0 \\ \hline a + Ab & AB \end{array}\right]$$

where $0 = (0, 0, \ldots, 0)$ in $\mathbf{R}^n$.

(b) Show that if A is invertible, then

$$\underline{A} = \left[\begin{array}{c|c} 1 & 0 \\ \hline a & A \end{array}\right] \quad \text{is invertible}$$

and

$$\underline{A}^{-1} = \left[\begin{array}{c|c} 1 & 0 \\ \hline -A^{-1}a & A^{-1} \end{array}\right]$$

40. Given a vector X in $\mathbf{R}^k$, let $\underline{X}$ denote the vector $1, X$ in $\mathbf{R}^{k+1}$. Let $T: \mathbf{R}^n \to \mathbf{R}^m$ be the affine function $Y = T(X) = b + AX$, where b is taken to be a single-column matrix. Let $\underline{A}$ be the partitioned matrix

$$\underline{A} = \left[\begin{array}{c|c} 1 & 0 \\ \hline b & A \end{array}\right]$$

Show that if $Y = T(X)$ then $\underline{Y} = \underline{A}\underline{X}$ — that is, $T(X)$ is `1↓A+.×1,X`.

Note: The matrix $\underline{A}$ of problem 40 is a *projective* or *homogeneous* representation of the affine transformation T.

41. Let the affine transformation $T: \mathbf{R}^n \to \mathbf{R}^n$ have a projective representation

$$\underline{Y} = \left[\begin{array}{c|c} 1 & 0 \\ \hline b & A \end{array}\right]\left[\begin{array}{c} 1 \\ \hline X \end{array}\right] = \underline{A}\underline{X}$$

(see exercise 40). Let $p_0, p_1, \ldots, p_n$ $(n = 2, 3)$ define a new coordinate system. Let $P = [p_1 - p_0 \,|\, p_2 - p_0 \,|\, \cdots \,|\, p_n - p_0]$ and set

$$\underline{P} = \left[\begin{array}{c|c} 1 & 0 \\ \hline p_0 & P \end{array}\right]$$

Show that in the new coordinate system T has a projective representation

$$\underline{Y}' = (\underline{P}^{-1}\underline{A}\underline{P})\underline{X}'$$

Hint: To compute $\underline{P}^{-1}$ see exercise 39. Then compare $\underline{P}^{-1}\underline{A}\underline{P}$ to the formula given in Proposition 4.9.

In exercises 42 through 45 a function is described. Code the function in APL.

42. Name: *PREP*
Right argument: A set of points $p_0, p_1, \ldots, p_n$ $(n = 2, 3)$ defining a new affine coordinate system stored as $[p_0 \,|\, p_1 \,|\, \cdots \,|\, p_n]$
Result: The matrix $\underline{P}$ of exercise 41

43. Name: *NEWTOOLD*
Right argument: A matrix with columns consisting of X' coordinates
Left argument: Points defining the X' coordinates stored as $[p_0 \,|\, p_1 \,|\, \cdots \,|\, p_n]$
Result: A matrix with columns consisting of the corresponding X coordinates

Hint: Show $\underline{X} = \underline{P}\underline{X}'$ and assume *PREP* (exercise 42) exists.

44. Name: *OLDTONEW*
Right argument: A matrix consisting of X coordinates
Left argument: $[p_0 \,|\, p_1 \,|\, \cdots \,|\, p_n]$, points defining a new coordinate system
Result: A matrix with columns consisting of the corresponding X' coordinates

Hint: See the hint to exercise 43.

45. Name: *NEWAFFN*
Right argument: $[b \,|\, A]$, where $T(X) = b + AX$ is an affine function and A is square
Left argument: $[p_0 \,|\, p_1 \,|\, \cdots \,|\, p_n]$, points defining X' coordinates
Result: $[b' \,|\, A']$ where $T(X') = b' + A'X'$

Hint: Exercises 41, 42.

46. Computer assignment: Use the function *NEWAFFN* of exercise 45 to check your answers to exercises 20 through 30.

4.3 Quadratic Functions in the Plane

Quadratic functions $f:\mathbf{R}^2 \to \mathbf{R}$ can be completely analyzed by means of coordinate changes in $\mathbf{R}^2$. We shall do this here. The method used here will be extended to quadratic functions $f:\mathbf{R}^n \to \mathbf{R}$ in Chapter 5. The crucial step is the "diagonalization" of a symmetric matrix by a rotation of coordinates. This diagonalization process is quite important in multivariate statistics and in some optimization problems.

Suppose that the quadratic function $f:\mathbf{R}^2 \to \mathbf{R}$ is given by

$$f(x, y) = ax^2 + 2bxy + cy^2 + dx + ey + k$$

or

$$f(x, y) = k + [d \quad e]\begin{bmatrix} x \\ y \end{bmatrix} + [x \quad y]\begin{bmatrix} a & b \\ b & c \end{bmatrix}\begin{bmatrix} x \\ y \end{bmatrix}$$

which we will write

$$f(X) = k + BX + X^TAX$$

where $A = A^T$ and B has a single row.

If an affine coordinate change

$$X = p_0 + PX'$$

is made, then

$$\begin{aligned} f(X') &= k + B(p_0 + PX') + (p_0 + PX')^TA(p_0 + PX') \\ &= k + Bp_0 + BPX' + (p_0^T + X'^TP^T)(Ap_0 + APX') \\ &= k + Bp_0 + p_0{}^TAP_0 + BPX' + p_0^TAPX' + X'^TP^TAp_0 + X'^TP^TAPX' \end{aligned}$$

Now as a matrix $X'^TP^TAp_0$ is 1-by-1 and so is certainly symmetric. Thus

$$f(X') = f(p_0) + (B + 2p_0^TA)PX' + X'^T(P^TAP)X'^T$$

First one chooses P so that there is no $x'y'$ term: $(P^TAP)[1; 2] = 0$.

Proposition 4.10 Let

$$A = \begin{bmatrix} a & b \\ b & c \end{bmatrix}$$

Then

$$R_\theta^TAR_\theta = \begin{bmatrix} \lambda & 0 \\ 0 & \mu \end{bmatrix}$$

where

$$\theta = \begin{cases} \dfrac{\pi}{4} & \text{if } a = c \\ \dfrac{1}{2}\operatorname{Tan}^{-1}\dfrac{2b}{a - c} & \text{if } a \neq c \end{cases}$$

Proof Carrying out the matrix multiplication, we see that

$$\left(\begin{bmatrix} \cos\theta & \sin\theta \\ -\sin\theta & \cos\theta \end{bmatrix}\begin{bmatrix} a & b \\ b & c \end{bmatrix}\begin{bmatrix} \cos\theta & -\sin\theta \\ \sin\theta & \cos\theta \end{bmatrix}\right)[2;\,1]$$

$$= (c - a)\sin\theta\cos\theta + b(\cos^2\theta - \sin^2\theta)$$

$$= \frac{c - a}{2}\sin 2\theta + b\cos 2\theta$$

If $c = a$, we choose $\theta = \pi/4$ and $(R_\theta^{-1}AR_\theta)[2;\,1] = 0$. If $a \neq c$, then we choose θ such that

$$\frac{c - a}{2}\sin 2\theta + b\cos 2\theta = 0$$

$$b\cos 2\theta = \frac{a - c}{2}\sin 2\theta$$

$$\theta = \frac{1}{2}\operatorname{Tan}^{-1}\frac{2b}{a - c} \qquad \blacksquare$$

If $P = R_\theta$ is chosen according to Proposition 4.10, then the second-degree terms of $f(X')$ are

$$X'^T P^T A P X' = [x' \quad y']\begin{bmatrix} \lambda & 0 \\ 0 & \mu \end{bmatrix}\begin{bmatrix} x' \\ y' \end{bmatrix} = \lambda x'^2 + \mu y'^2$$

Thus an axis rotation will always eliminate the term $2bxy$.

Next one can shift the origin in an attempt to simplify the linear term $(B + 2p_0^T A)PX'$ as much as possible. If we try to set the linear term to zero, we have

$$\begin{aligned} (B + 2p_0^T A)P &= 0 \\ B + 2p_0^T A &= 0 \qquad (P \text{ is invertible}) \\ B^T + 2Ap_0 &= 0 \qquad (\text{transpose of both sides}) \\ Ap_0 &= -\tfrac{1}{2}B^T \end{aligned}$$

If A is invertible, we may choose $p_0 = -\frac{1}{2}A^{-1}B^T$, and this origin shift will eliminate the first-degree terms (in elementary courses this is called "completing the square"). If A is not invertible, however, it may or may not be possible to choose p_0 so that the linear terms are eliminated.

Assume now that A is invertible. Then by a proper coordinate change we can put f into the form

$$f(x', y') = \lambda x'^2 + \mu y'^2 + \delta \qquad [\delta = f(p_0)]$$

To gain an understanding of such functions, it is useful to plot the *level curves*. These are the curves

$$f(x', y') = \text{constant}$$

and the set of these curves gives a contour map of the graph of the function $z = f(x, y)$ in $\mathbf{R}^3$.

If we absorb δ into the constant, c, the level curves are given by the equation

$$\lambda x'^2 + \mu y'^2 = c$$

which, depending on the signs of λ, μ, $c \neq 0$, can be put into one of the three canonical forms

$$\frac{x'^2}{\alpha^2} + \frac{y'^2}{\beta^2} = 1, \qquad \frac{x'^2}{\alpha^2} - \frac{y'^2}{\beta^2} = 1, \qquad -\frac{x'^2}{\alpha^2} + \frac{y'^2}{\beta^2} = 1$$

where $\alpha^2 = |c|/|\lambda|$, $\beta^2 = |c|/|\mu|$. The first equation is an ellipse and the second and third are hyperbolas. The curves are sketched in parts (a), (b), and (c) of Figure 4.18. The asymptotes for the hyperbolas are $y = \pm\beta x/\alpha$ and are the level curves for $c = 0$. Since the constant c can take on both positive and negative values, there are only two distinct cases for $f(x', y') = \lambda x'^2 + \mu y'^2 + \gamma = \text{const}$. If λ and μ have the same sign, the level curves are ellipses; if λ and μ have opposite signs, the level curves are hyperbolas. Typical level curves for the two cases are shown in Figure 4.19.

Notice that a square matrix A is invertible if and only if $P^{-1}AP$ is invertible. The matrix

$$\begin{bmatrix} \lambda & 0 \\ 0 & \mu \end{bmatrix}$$

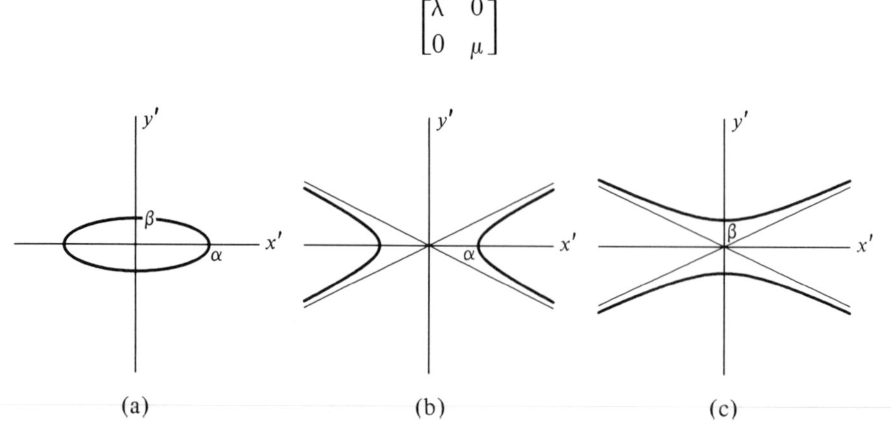

FIGURE 4.18

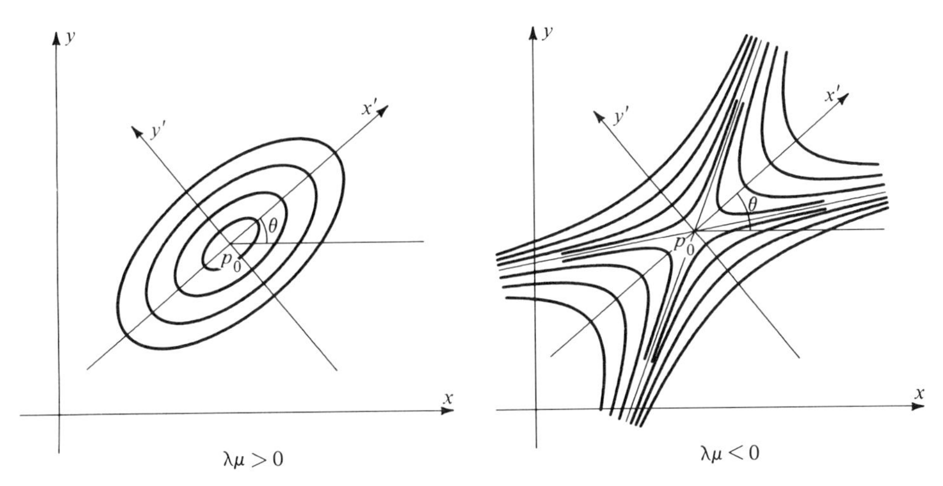

FIGURE 4.19

is invertible if and only if $\lambda \neq 0$ and $\mu \neq 0$. Thus Figure 4.19 applies whenever $\lambda\mu \neq 0$. If λ and μ have the same sign, then f has an absolute minimum at p_0 if λ and μ are both positive [Figure 4.20(a)] and an absolute maximum at p_0 if λ and μ are both negative [Figure 4.20(b)]. If λ and μ have opposite signs, then f has neither maxima nor minima. The graph of $z = f(x, y)$ has a saddle point at p_0 [Figure 4.20(c)].

The x' and y' axes give the directions from p_0 in which the function f changes the slowest and the fastest. If $|\lambda| > |\mu|$, then the x' axis gives the fastest rate of change: if $|\lambda| < |\mu|$, then the y' axis gives the fastest rate of change. If $\lambda = \mu$, the level curves are circles and the rate of change of f is equal in all directions from p_0.

EXAMPLE 4.13 Discuss the quadratic function

$$f(x, y) = 1 + 2x + 3y + x^2 - 4xy + y^2$$

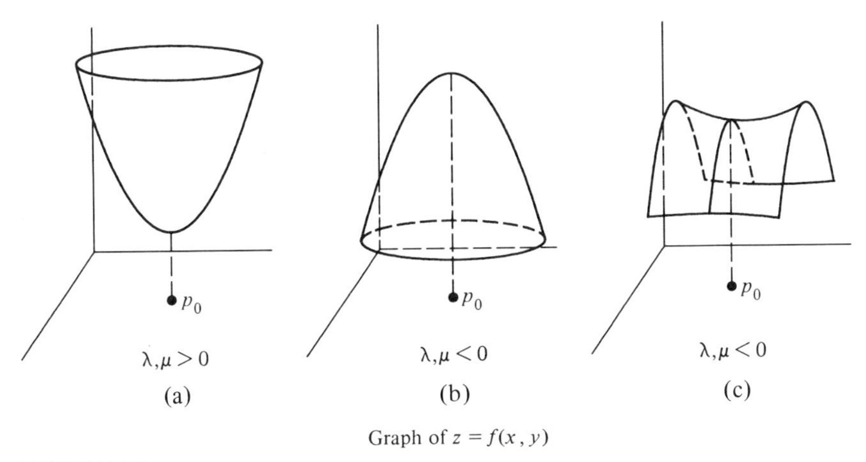

FIGURE 4.20

Solution In matrix form we have

$$f(X) = 1 + [2 \quad 3]X + X^T \begin{bmatrix} 1 & -2 \\ -2 & 1 \end{bmatrix} X = k + BX + X^T AX$$

Since $a = c$, the xy term may be eliminated by a rotation of $\pi/4$, and since

$$A^{-1} = \frac{1}{1-4} \begin{bmatrix} 1 & 2 \\ 2 & 1 \end{bmatrix}$$

translating the origin to

$$p_0 = -\tfrac{1}{2}A^{-1}B^T = \tfrac{1}{6} \begin{bmatrix} 1 & 2 \\ 2 & 1 \end{bmatrix} \begin{bmatrix} 2 \\ 3 \end{bmatrix} = \begin{bmatrix} \frac{4}{3} \\ \frac{7}{6} \end{bmatrix}$$

will eliminate the linear terms. Since

$$R_{\pi/4} = \frac{1}{\sqrt{2}} \begin{bmatrix} 1 & -1 \\ 1 & 1 \end{bmatrix}$$

we have

$$\begin{bmatrix} \lambda & 0 \\ 0 & \mu \end{bmatrix} = \tfrac{1}{2} \begin{bmatrix} 1 & 1 \\ -1 & 1 \end{bmatrix} \begin{bmatrix} 1 & -2 \\ -2 & 1 \end{bmatrix} \begin{bmatrix} 1 & -1 \\ 1 & 1 \end{bmatrix} = \begin{bmatrix} -1 & 0 \\ 0 & 3 \end{bmatrix}$$

Thus the function has no maxima or minima. It has a saddle at $p_0 = (\frac{4}{3}, \frac{7}{6})$.

$$f(p_0) = 1 + [2 \quad 3] \begin{bmatrix} \frac{4}{3} \\ \frac{7}{6} \end{bmatrix} + [4/3 \quad 7/6] \begin{bmatrix} 1 & -2 \\ -2 & 1 \end{bmatrix} \begin{bmatrix} \frac{4}{3} \\ \frac{7}{6} \end{bmatrix} = 91/12$$

So in the new coordinate system

$$f(x', y') = \tfrac{91}{12} + x'^2 - 3y'^2 \qquad \blacksquare$$

EXAMPLE 4.14 Sketch the curve

$$1 + 8x + 12y + 3x^2 + 4xy + 5y^2 = 6$$

Solution This is just one of the level curves of the function

$$f(X) = 1 + [8 \quad 12]X + X^T \begin{bmatrix} 3 & 2 \\ 2 & 5 \end{bmatrix} X$$

The rotation matrix that will eliminate the xy term is

```
      +P←ROT TH←.5×¯3○2×2÷3−5
0.851 0.526
¯0.526  0.851
```

The diagonalization of the matrix A is

```
      +D←(⍉P)+.×(A←2 2⍴3 2 2 5)+.×P
1.76E0        6.943¯18
9.54E¯18      6.24E0
```

Since λ and μ are both positive, the level curves of f are ellipses with centers at

```
      +P0←(8 12⌹A)÷¯2
¯0.727  ¯0.909
```

To convert from radians to degrees divide by $\pi/180$:

```
      TH÷○÷180
¯31.7
```

so the angle of rotation is a bit more than 30° clockwise.

The constant term in the new coordinate system is $f(p_0)$ or

```
      +Δ←1+(8 12 +P0+.×A)+.×P0
¯7.36
```

which is the minimum value attained by f. In the new coordinate system the equation of $f(x, y) = 6$ is

$$-7.36 + 1.76x' + 6.24y' = 6$$

or

$$1.76x'^2 + 6.24y'^2 = 13.4$$

This ellipse cuts the new axes at $x' = \sqrt{13.4/1.76}$ and $y' = \sqrt{13.4/6.24}$, or

```
      +INTERCEPTS←((6−Δ)÷1 1⍉D)*÷2
2.75  1.46
```

To sketch the ellipse, one finds some points on the curve in $x'y'$ coordinates and stores them as columns of a matrix M. The corresponding xy coordinates are obtained by adding p_0 to the columns of PM (see, however, exercise 43 of Exercises 4.2). Notice that the x' axis is the line through the points with $x'y'$ coordinates (± 2.75, 0), and the y' axis is the line through the points with $x'y'$ coordinates (0, ± 1.46):

```
      M
2.750 ¯2.750 0.000  0.000
0.000  0.000 1.460 ¯1.460
```

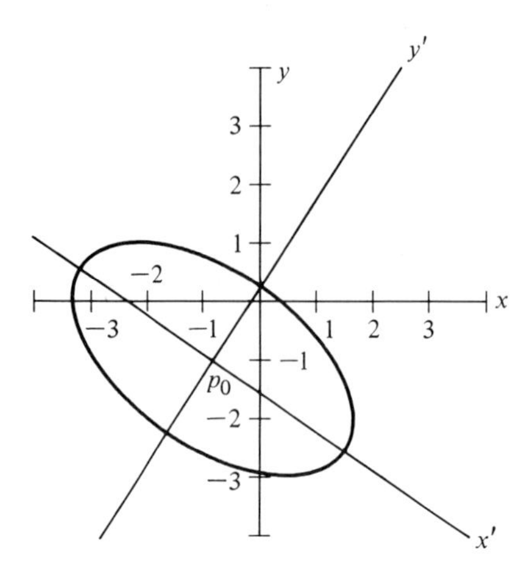

FIGURE 4.21

```
      (⍉4 2⍴P0)+P+.×M
 1.610 ¯3.070    0.042 ¯1.500
¯2.360  0.538    0.336 ¯2.150
```

The curve is sketched using these four points in Figure 4.21. ■

Notice that a given equation may not be satisfied for any pair of points. In the last example the minimum value of f was -7.36; thus the equation

$$-8 = f(x, y) = 1 + 8x + 12y + 3x^2 + 4xy + 5y^2$$

is not satisfied for any pair of real numbers (x, y).

The level surfaces of $f(X) = k + BX + X^TAX$ are ellipses or hyperbolas when A is nonsingular. When A is singular, they are parabolas or pairs of straight lines (degenerate ellipses).

DEGENERATE ELLIPSES

This is the case where A is singular but the equation $Ap_0 = -\frac{1}{2}B^T$ still has solutions. In this case there is a whole line of solutions p_0 and placing the origin of the $x'y'$ coordinate system on this line will eliminate the linear terms.

In this case, since $\lambda = 0$ or $\mu = 0$, f may be put in one of the forms

$$f(x', y') = \delta + \lambda x'^2; \qquad f(x', y') = \delta + \mu y'^2$$

Taking for example the first form, the level curves are of the form $x' = \pm\sqrt{\text{const.}}$, and the graph of $y = f(x, y)$ is a trough or ridge with a horizontal bottom or top (Figure 4.22).

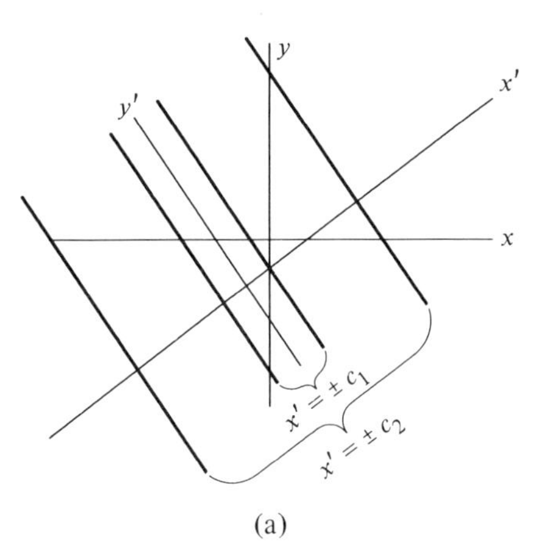

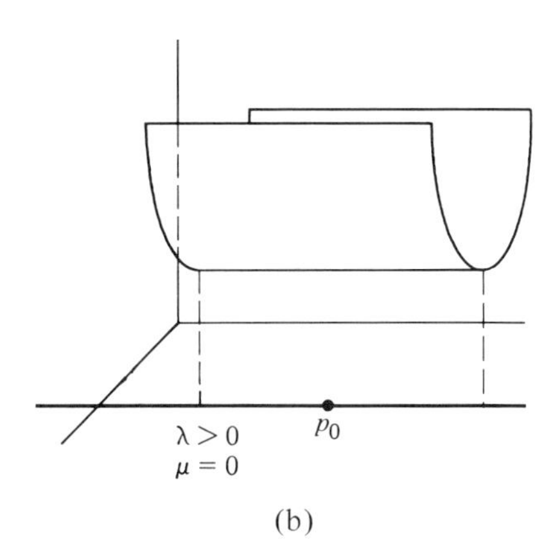

FIGURE 4.22

PARABOLAS

Suppose that $Ap_0 = -\frac{1}{2}B^T$ does not have solutions. If $P^{-1}AP = D$ has a zero column, then the same column of $AP = PD$ is also zero, since $(PD)[;i] = PD[;i]$. Suppose that

$$R_\theta^T A R_\theta = R_\theta^{-1} A R_\theta = \begin{bmatrix} 0 & 0 \\ 0 & \mu \end{bmatrix}$$

Then AR_θ has a zero first column. Using the coordinate change $X = p_0 + R_\theta X'$ gives

$$f(X') = \delta + (BR_\theta + 2p_0^T AR_\theta)X' + X'^T(R_{-\theta} AR_\theta)X'$$

Let

$$p_0 = \begin{bmatrix} x_0 \\ y_0 \end{bmatrix}; \qquad BR_\theta = [b_1 \quad b_2],\ 2AR_\theta = \begin{bmatrix} 0 & a_1 \\ 0 & a_2 \end{bmatrix}$$

Then the X' coefficient is

$$[b_1 \quad b_2] + [x_0 \quad y_0]\begin{bmatrix} 0 & a_1 \\ 0 & a_2 \end{bmatrix} = [b_1 \quad b_2 + a_1 x_0 + a_2 y_0]$$

Since $A \neq 0$, at least one of the numbers a_1, a_2 is nonzero, and so the equation $b_2 + a_1 x_0 + a_2 y_0 = 0$ has an infinity of solutions $p_0 = (x_0, y_0)$. Thus by the proper choice of p_0, f may be put into the form

$$f(x', y') = \delta + \gamma x' + \mu y'^2 \qquad (\gamma = (BR_\theta)[;1])$$

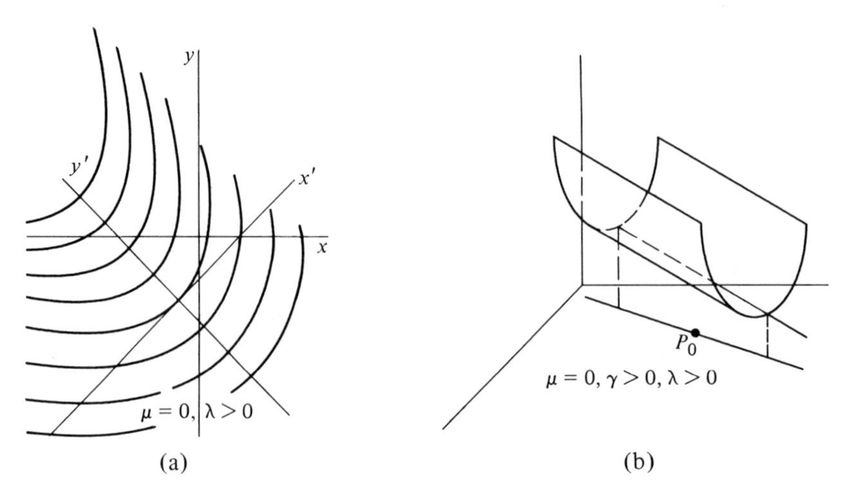

FIGURE 4.23

Similarly, if $\mu = 0$ and $\lambda \neq 0$, f may be put in the form

$$f(x', y') = \delta + \gamma y' + \lambda x'^2 \qquad (\gamma = (BR_\theta)[;2])$$

In either case the level curves $f(X) = 0$ are congruent parabolas and the graph of $z = f(x, y)$ is a trough with a slanted bottom. The function f has neither maxima nor minima (Figure 4.23).

EXAMPLE 4.15 Sketch the curve

$$1 - x + y + x^2 + 2xy + y^2 = 0$$

Solution This is the level curve $f(x, y) = 0$ for

$$f(X) = 1 + [1 \quad 1]X + X^T\begin{bmatrix} 1 & 1 \\ 1 & 1 \end{bmatrix}X$$

The matrix $A = 2\ 2\rho 1$ can be diagonalized by a rotation through $\theta = \pi/4$ degrees (Proposition 4.10). In this case we have

$$R_{-\theta}AR_\theta = \frac{1}{\sqrt{2}}\frac{1}{\sqrt{2}}\begin{bmatrix} 1 & 1 \\ -1 & 1 \end{bmatrix}\begin{bmatrix} 1 & 1 \\ 1 & 1 \end{bmatrix}\begin{bmatrix} 1 & -1 \\ 1 & 1 \end{bmatrix} = \begin{bmatrix} 2 & 0 \\ 0 & 0 \end{bmatrix}$$

$$BR_\theta + 2p_0^TAR_\theta = \frac{1}{\sqrt{2}}\left([-1 \quad 1]\begin{bmatrix} 1 & -1 \\ 1 & 1 \end{bmatrix} + [x_0 \quad y_0]\begin{bmatrix} 2 & 2 \\ 2 & 2 \end{bmatrix}\begin{bmatrix} 1 & -1 \\ 1 & 1 \end{bmatrix}\right)$$

$$= \frac{1}{\sqrt{2}}\left([0 \quad 2] + [x_0 \quad y_0]\begin{bmatrix} 4 & 0 \\ 4 & 0 \end{bmatrix}\right)$$

$$= \sqrt{2}[2(x_0 + y_0) \quad 1]$$

$$= [0 \quad \sqrt{2}] \qquad \text{if we choose } p_0 \text{ on the line } y = -x$$

The new constant term will be $f(p_0)$, and $f(x, -x) = 1 - 2x$, so take $p_0 = (\frac{1}{2}, -\frac{1}{2})$. Then $f(p_0) = 0$, and in the new coordinates

$$f(x', y') = \sqrt{2}y' - 2x'^2$$

and so the level curve $f(X) = 0$ is

$$y' = \sqrt{2}x'^2$$

A table of values for this curve in $x'y'$ coordinates is

```
      +VAL←2 7ρ(6 CHOP ¯3 3),(2*.5)×(6 CHOP ¯3 3)*2
¯3.000  ¯2.000  ¯1.000  0.000  1.000  2.000   3.000
12.700   5.660   1.410  0.000  1.410  5.660  12.700
```

(The function *CHOP* is defined in Section 3.1). Using the formula

$$X = p_0 + PX'$$

the corresponding xy coordinates are

```
      (⍉7 2ρ.5 ¯.5)+(ROT ○÷4)+.×VAL
¯10.600  ¯4.910  ¯1.210   0.500  0.207  ¯2.090  ¯6.380
  6.380   2.090  ¯0.207  ¯0.500  1.210   4.910  10.600
```

Using these points, the curve is sketched in Figure 4.24. ∎

The formulas of Proposition 4.10 are often inconvenient for hand calculation. It is possible to write down a matrix that will diagonalize a given 2-by-2 symmet-

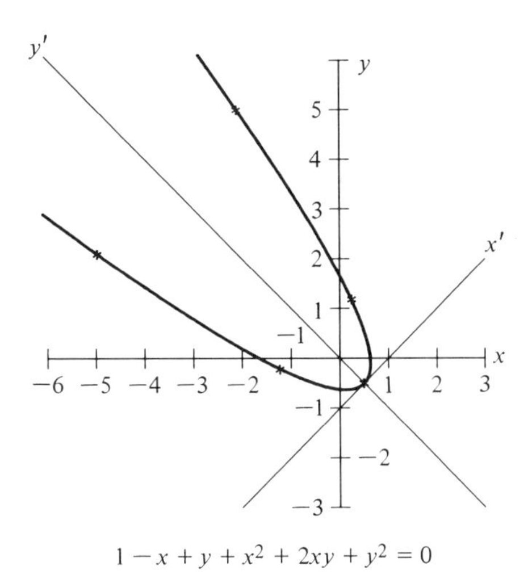

$1 - x + y + x^2 + 2xy + y^2 = 0$

FIGURE 4.24

ric matrix without the use of trigonometric functions. One uses the trigonometric identities

$$\cos\left(\tfrac{1}{2}\operatorname{Tan}^{-1} x\right) = (1 + x^2)^{-1/2}$$
$$\sin\left(\tfrac{1}{2}\operatorname{Tan}^{-1} x\right) = x(1 + x^2)^{-1/2}$$

plus some algebra to discover the next result, which we can check by computing P^TP and $P^{-1}AP$.

PROPOSITION 4.11 Let

$$A = \begin{bmatrix} a & b \\ b & c \end{bmatrix}$$

and set $\alpha = a - c$, $\beta = \sqrt{\alpha^2 + 4b^2}$, $\epsilon = \times b$.†

Then the matrix

$$P = \frac{1}{\sqrt{2\beta}}\begin{bmatrix} \sqrt{\beta + \alpha} & -\epsilon\sqrt{\beta - \alpha} \\ \epsilon\sqrt{\beta - \alpha} & \sqrt{\beta + \alpha} \end{bmatrix}$$

is such that $P^{-1} = P^T$ and

$$P^TAP = \frac{1}{2}\begin{bmatrix} a + c + \beta & 0 \\ 0 & a + c - \beta \end{bmatrix} \quad \blacksquare$$

EXAMPLE 4.16 Briefly describe the function

$$f(x, y) = -2y + 2xy - y^2$$

Solution

$$f(X) = [0 \quad -2]X + X^T\begin{bmatrix} 0 & 1 \\ 1 & -1 \end{bmatrix}X$$

So, using the notation of Proposition 4.11, $\alpha = 1$, $\epsilon = 1$, $\beta = \sqrt{5}$,

$$P = \frac{1}{\sqrt{2\sqrt{5}}}\begin{bmatrix} \sqrt{\sqrt{5} + 1} & -\sqrt{\sqrt{5} - 1} \\ \sqrt{\sqrt{5} - 1} & \sqrt{\sqrt{5} + 1} \end{bmatrix}, \quad P^TAP = \frac{1}{2}\begin{bmatrix} -1 + \sqrt{5} & 0 \\ 0 & -1 - \sqrt{5} \end{bmatrix}$$

and $p_0 = \frac{1}{2}A^{-1}B = (0, -1)$.

Since $f(0, -1) = 1$, the coordinate change $X = p_0 + PX'$ puts f into the form

$$f(x', y') = 1 + \tfrac{1}{2}(\sqrt{5} - 1)x'^2 - \tfrac{1}{2}(1 + \sqrt{5})y'^2$$

†Monadic × gives the sign of its argument.

The level curves are hyperbolas centered at $p_0 = (0, -1)$, and so f has a saddle point at p_0. ■

EXERCISES 4.3

In exercises 1 through 11 use Proposition 4.11 to classify the given curve as a hyperbola, parabola, ellipse, pair of parallel lines (degenerate ellipse), pair of intersecting lines (degenerate hyperbola), point, or empty set (imaginary ellipse).

1. $4y + 5x^2 + 12xy + 8y^2 = -4$
2. $4x^2 + 4y^2 - 4x + 8y + 5 = 0$
3. $2x^2 + 4xy + 2y^2 = -1$
4. $2x^2 + 2xy + 5y^2 + 6y = -1$
5. $5x^2 - 2xy + 11y^2 - 2x + 22y = -12$
6. $-2x - 6y + 2xy = -2$
7. $-3x^2 + 6xy + 5y^2 - 6x - 2y = 1$
8. $8y - 16x + 16x^2 - 16xy + 4y^2 = 0$
9. $4x^2 - 4xy + y^2 + 8x - 4y + 4 = 0$
10. $4x^2 + 8xy + 4y^2 - x - 2y = 3$
11. $2x^2 - 8xy + 8y^2 - 9x + 16y + 10 = 0$
12. Write an APL function called ANG.

    Input: A 2-by-2 symmetric matrix A
    Output: An angle θ in radians such that $R_\theta^T A R_\theta$ is diagonal. Thus `P←ROT ANG A` computes R_θ

In exercises 13 through 17 decide if the given function has a maximum, minimum, saddle point, or none of these. If such a point occurs, where is it?

13. $f(x, y) = 5 + 8y - 4x + 4x^2 + 4y^2$
14. $f(x, y) = 16 + 4y + 5x^2 + 12xy + 8y^2$
15. $f(x, y) = 4 + 8y - 16x + 16x^2 - 16xy + 4y^2$
16. $f(x, y) = 16y - 9x + 2x^2 - 8xy + 8y^2$
17. $f(x, y) = 6 - 6x - 2y - 3x^2 + 6xy + 5y^2$

(Computer assignment) In exercises 18 through 20 perform a rotation and translation of axes to simplify the equation. On an xy coordinate system sketch the x' and y' axes and the curve.

18. $5x^2 - 2xy + 11y^2 - 2x + 22y + 10 = 0$
19. $3 - x - 2y + 4x^2 + 8xy + 4y^2 = 0$
20. $2 - 2y - 6x + 5y^2 + 6xy - 3x^2 = 0$

Projective Representation

The remaining exercises are a continuation of the material developed in exercises 39 through 46 of Exercises 4.2.

21. Let the quadratic function $f: \mathbf{R}^n \to \mathbf{R}$ be given by

$$f(X) = c + BX + X^TAX$$

where $A = A^T$, B is a single-row matrix, and c is a scalar. Let $\underset{\sim}{Q}$ be the symmetric matrix

$$\underset{\sim}{Q} = \left[\begin{array}{c|c} c & B \div 2 \\ \hline B^T \div 2 & A \end{array}\right]$$

Show that the function f is given by

$$f(X) = \underline{X}^T\underset{\sim}{Q}\underline{X}$$

(see exercise 40 of Exercise 4.2 for $\underline{X}$).

Note: The matrix $\underset{\sim}{Q}$ of exercise 21 is a *projective representation* of f.

22. Let $f: \mathbf{R}^n \to \mathbf{R}$ have projective representation $\underset{\sim}{Q}$ (see exercise 21) and let a new coordinate system be defined by the matrix $\underline{P}$ of exercise 41 of Exercises 4.2. Show that a projective representation of f in the new coordinate system is

$$f(X') = \underline{X}'^T(\underline{P}^T\underset{\sim}{Q}\underline{P})\underline{X}'$$

Hint: Compute $\underline{P}^T\underset{\sim}{Q}\underline{P}$ and compare with the computation preceding Proposition 4.10.

23. Write an APL function with

Name: *NEWQUAD*

Right argument: The terms of the function $f(X) = c + BX + X^TAX$ stored as

$$\left[\begin{array}{c|c} c & B \\ \hline 0 & A \end{array}\right]$$

— that is, $(c, B), [1] 0, A$

Left argument: $[p_0\ p_1\ \cdots\ p_n]$ defining X' coordinates

Result: $\left[\begin{array}{c|c} c' & B' \\ \hline 0 & A' \end{array}\right]$ where $f(X') = c' + B'X' + X'^TA'X'$

Hint: Exercises 21, 22.

4.4 Flats and Coordinate Systems in $\mathbf{R}^n$

In this section we formally extend the techniques of the first two sections of this chapter to $\mathbf{R}^n$, the set of vectors with n components. Sections 4.2 and 4.3 deal with cases $n = 2, 3$. The definitions and propositions of this section apply in a rather trivial way to the cases $n = 0$ and $n = 1$ as well as $n \geq 4$.

The first task is to extend the definition of lines and planes to $\mathbf{R}^n$. This is done by defining a *k-dimensional flat in* $\mathbf{R}^n$, $0 \leq k \leq n$. A k-dimensional flat will be called a *point* if $k = 0$, a *line* if $k = 1$, a *plane* if $k = 2$, and a *hyperplane* if $k = n - 1$.

For $n = 2, 3$ we have the definitions

point: $p_0 \in \mathbf{R}^n$

line: $l(t) = p_0 + t(p_1 - p_0)$; $p_0 \neq p_1 \in \mathbf{R}^n$, $-\infty < t < \infty$

plane: $p(t_1, t_2) = p_0 + t_1(p_1 - p_0) + t_2(p_2 - p_0)$; $p_0, p_1, p_2 \in \mathbf{R}^n$, $-\infty < t_1, t_2 < \infty$, and $p_1 - p_0, p_2 - p_0$ linearly independent

The condition $p_0 \neq p_1$ guarantees that $l(t)$ is a line rather than a point. The linear independence of $p_1 - p_0$ guarantees that $p(t_1, t_2)$ sweeps out a plane rather than a line as t_1 and t_2 vary.

DEFINITION 4.2 Let $p_0, p_1, p_2, \ldots, p_k$ be vectors in $\mathbf{R}^n$ with $p_1 - p_0, p_2 - p_0, \ldots, p_k - p_0$ linearly independent. The set of vectors swept out by

$$p(t_1, t_2, \ldots, t_k) = p_0 + t_1(p_1 - p_0) + t_2(p_2 - p_0) + \cdots + t_k(p_k - p_0)$$

as $t_1, t_2, \ldots, t_k$ vary is called a *k-dimensional flat* in $\mathbf{R}^n$.

We should say something about the way points and lines are formally made to fit this definition.

Recall (Proposition 2.11) that the columns of the matrix $P = [p_1 - p_0 \mid p_2 - p_0 \mid \cdots \mid p_k - p_0]$ are linearly independent if and only if the equation $PX = 0$ has only the solution $X = 0$. For a line, $k = 1$, and so P has a single column: $p_1 - p_0$. It follows that $PX = 0$ has only the trivial solution if and only if $p_1 - p_0 \neq 0$ — that is, $p_1 \neq p_0$. Thus a line is a one-dimensional flat by Definition 4.2. For the case of a point in $\mathbf{R}^n$ we take the matrix P to be n by zero ($P \leftarrow (n, 0)\rho 0$) and adopt the convention that an empty set of vectors is linearly independent. Then Definition 4.2 applies, and a point is a zero-dimensional flat.

Taking an empty set of vectors to be linearly independent is not an arbitrary convention. For example, if P is n by zero, then $P^T P$ is 0 by 0 and hence is ID 0. Thus P has a left inverse and the columns of P are linearly independent by Proposition 2.9. Proposition 2.11 leads to the same conclusion (exercise 23).

A more compact notation than that of Definition 4.2 is desirable. We will let T denote the column vector with components $t_1, t_2, \ldots, t_k$ and use the matrix P defined above as $P = [p_1 - p_0 \mid p_2 - p_0 \mid \cdots \mid p_k - p_0]$. Then the k-flat is the set of vectors swept out by

$$p(T) = p_0 + PT \tag{4.2}$$

as the parameter vector T varies in $\mathbf{R}^k$.

Since we no longer have the geometry of $\mathbf{R}^2$ and $\mathbf{R}^3$ to fall back on, there are some basic facts to be verified algebraically. For example, can a set of points be both a k-dimensional flat and an l-dimensional flat with $k \neq l$? The answer is no, but the reason why is not obvious.

To begin with, let us refer to the set of vectors $p(T)$ of the form (4.2) as a *flat* even if the columns of P are not independent. Our first task is to show that a flat is a k-dimensional flat for some k. The next proposition is our main tool.

PROPOSITION 4.12 The flat F_1 defined by

$$q(S) = q_0 + QS, \qquad S \text{ in } \mathbf{R}^l$$

is contained in the flat F_2 defined by

$$p(T) = p_0 + PT, \qquad T \text{ in } \mathbf{R}^k$$

if and only if (1) q_0 is in F_2, (2) $Q = PR$ for some k-by-l matrix R.

Proof First assume that the flat F_1 is contained in the flat F_2. Then, in particular, $q_0 = q(0)$ is a point of F_2, and statement (1) is proved. Let $q_0 = p_0 + PT_0$. Let $q_i = q_0 + Q[;i] = q_0 + QS_i$, where S_i is the ith column of the l-by-l identity matrix. Then q_i is in F_1, hence q_i is in F_2 as well. Let $q_i = p_0 + PT_i$ and define R by $R[;i] = T_i - T_0$. Then

$$Q[;i] = q_i - q_0 = p_0 + PT_i - (p_0 + PT_0) = P(T_i - T_0) = PR[;i]$$

and hence $Q = PR$. This proves statement (2).

Next assume that statements (1) and (2) are true. To show that F_1 is contained in F_2 we must show that for each S in $\mathbf{R}^l$ there is a T in $\mathbf{R}^k$ such that $q_0 + QS = p_0 + PT$. Since (1) is true, we have $q_0 = p_0 + PT_0$, and since (2) is true,

$$q_0 + QS = p_0 + PT_0 + (PR)S = p_0 + P(T_0 + RS)$$

and $T = T_0 + RS$ is in $\mathbf{R}^k$. ■

Now we will show that a flat is indeed a k-dimensional flat for some k. The proof is based on a variation of the formulas of Proposition 3.1. Let P be a matrix and let R be the matrix obtained by dropping the rows of zeros, if any, from the echelon form of P. Let V be the vector of indices of the pivot columns. Then $P = P[;V]R$ (exercise 24).

PROPOSITION 4.13 The flat defined by

$$p(T) = p_0 + PT, \qquad T \text{ in } \mathbf{R}^k$$

is an l-dimensional flat where l is the rank of P.

Proof Using the matrix R and vector V defined above, we have $P = P[;V]R$. Applying Proposition 4.12 with $q_0 = p_0$ and $Q = P[;V]$ shows that the flat $p(T)$ is contained in the flat defined by

$$q(s) = p_0 + P[;V]S, \qquad S \text{ in } \mathbf{R}^l$$

Similarly, the equation $P[;V] = PI[;V]$ shows the reverse inclusion. Thus the two flats are equal. Since the columns of $P[;V]$ are linearly independent, the flat is an l-dimensional flat where l is the number of pivot columns — that is, the rank — of P. ∎

It follows from the next proposition that the dimension of a flat is well defined. A flat cannot be both k-dimensional and l-dimensional with $k \neq l$.

PROPOSITION 4.14 Let the l-dimensional flat F_1 be contained in the k-dimensional flat F_2. Then $l \leq k$, and $l = k$ if and only if $F_1 = F_2$.

Proof Let F_1 be given by $q(s) = q_0 + QS$ and let F_2 be given by $p(T) = p_0 + PT$, where Q is n by l with linearly independent columns and P is n by k with linearly independent columns. By Proposition 4.12 we have $Q = PR$, where R is k by l. Now Q has a left inverse, call it L. Since $I = LQ = (LP)R$, R also has linearly independent columns. This means that every column of R is a pivot column (look at the echelon form); hence there are at least as many rows as columns in R; that is, $k \geq l$.

Now assume that $l = k$. Then R is square and hence invertible. It follows that $P = QR^{-1}$ and so, by Proposition 4.12 again, F_2 will be contained in F_1 provided p_0 is in F_1. But we have great freedom in the choice of p_0. In fact if p_0' is *any* vector in F_2, then the flat given by $p_0' + PT$ is exactly F_2 (exercise 25). Replacing p_0 by q_0, we see that $F_1 = F_2$. ∎

DEFINITION 4.3 Let $p_0, p_1, \ldots, p_k$ be points in $\mathbf{R}^n$. The *flat defined by* $p_0, p_1, \ldots, p_k$ is the flat given by

$$p(T) = p_0 + PT, \qquad T \text{ in } \mathbf{R}^k$$

where, as usual, $P = [p_1 - p_0 | p_2 - p_0 | \cdots | p_k - p_0]$.

The next proposition shows that a more geometric definition of the flat generated by a set of points is possible.

PROPOSITION 4.15 The flat generated by $p_0, p_1, \ldots, p_k$ is the smallest flat containing $p_0, p_1, \ldots, p_k$. That is, any flat that contains $p_0, p_1, \ldots, p_k$ also contains the flat defined by $p_0, p_1, \ldots, p_k$. ∎

The proof is left as an exercise (see exercise 26).

The proof of Proposition 4.13 shows us how to find the dimension of the flat generated by a set of points and also how to eliminate redundant points.

EXAMPLE 4.17 What is the dimension of the flat in $\mathbf{R}^4$ defined by the points $(1, 1, 1, 1)$, $(2, -1, 1, 1)$, $(-1, 6, 2, -1)$, $(-3, 12, 4, -5)$, $(6, -11, 0, 4)$?

Solution Let these points be p_0, p_1, p_2, p_3, p_4, respectively, and set $P = [p_1 - p_0 | p_2 - p_0 | \cdots | p_4 - p_0]$.

```
      P
 1   ¯2   ¯4    5
¯2    5   11  ¯12
 0    1    3   ¯1
 0   ¯2   ¯6    3
```

Following the proof of Proposition 4.13, the echelon form of P is

```
      ECHELON P
1.000  0.000  2.000  0.000
0.000  1.000  3.000  0.000
0.000  0.000  0.000  1.000
0.000  0.000  0.000  0.000
```

Thus p_0, p_1, p_2, p_3, p_4 define the 3-flat in $\mathbf{R}^4$ consisting of all

$$p(T) = p_0 + P[;1 \quad 2 \quad 4]\begin{bmatrix} t_1 \\ t_2 \\ t_3 \end{bmatrix}$$

— that is, the 3-flat is defined by p_0, p_1, p_2, p_4. ∎

Finding the intersection of a k-flat and an l-flat in $\mathbf{R}^n$ is also simple. Suppose that the k-flat is given by

$$p(T) = p_0 + PT, \qquad P[;i] = p_i - p_0$$

as T varies and the l-flat is given by

$$q(T) = q_0 + QT; \qquad Q[;i] = q_i - q_0$$

as T varies. The intersection of these two flats is the set of points expressible both as $p(T_1)$ and as $q(T_2)$ for suitable choices of the parameter vectors T_1 and T_2. This is the set of solutions of

$$p_0 + PT = q_0 + QS$$
$$PT - QS = q_0 - p_0$$

or

$$[P | -Q]\left[\frac{T}{S}\right] = q_0 - p_0$$

where to avoid confusion we have renamed the parameters for the l-flat $S = [s_1, \ldots, s_l]^T$.

EXAMPLE 4.18 Find the intersection in 4-space of the plane through the three points $p_0 = (1, 0, -3, 2)$, $p_1 = (2, -2, -5, 0)$, $p_2 = (2, -1, -5, -2)$ and the plane through the three points $q_0 = (-1, 3, -2, 1)$, $q_1 = (0, 2, -5, -1)$, $q_2 = (-3, 6, 3, 4)$.

Solution Let $P = [p_1 - p_0 | p_2 - p_0]$, $Q = [q_1 - q_0 | q_2 - q_0]$,

```
      P
 1  1
¯2 ¯1
¯2 ¯2
¯2 ¯4
      Q
 1 ¯2
¯1  3
¯3  5
¯2  3
```

and reduce the augmented matrix of the system $[P | -Q]X = q_0 - p_0$ to echelon form:

```
      ECHELON P,(−Q),Q0−P0
 1.00E0       0.00E0      0.00E0      0.00E0      1.20E1
¯347E¯18      1.00E0     ¯3.47E¯18    0.00E0     ¯4.00E0
 0.00E0       1.39E¯17    1.00E0      2.78E¯17   ¯1.60E1
¯1.39E¯17     0.00E0      0.00E0      1.00E0     ¯1.30E1
```

The system has the unique solution

$$X = \begin{bmatrix} t_1 \\ t_2 \\ s_1 \\ s_2 \end{bmatrix} = \begin{bmatrix} 12 \\ -4 \\ -16 \\ -13 \end{bmatrix}$$

so the two planes intersect in the point

```
      P0+P+.×12 ¯4
9  ¯20  ¯19  ¯6
```

which can also be expressed as

```
      Q0+Q+.×−16 13
```

■

In general, two flats may intersect in more than a point. It is convenient to have a procedure for writing down the intersection.

Starting with the equation

$$p_0 + PT = q_0 + QS$$

we get the linear system

$$[P \mid -Q]\left[\frac{T}{S}\right] = q_0 - p_0$$

Row-reducing in the usual way, we get a solution involving arbitrary parameters, say $u_1, u_2, \ldots, u_m$. In fact, the set of solutions will be expressed in the form

$$X = \left[\frac{T}{S}\right] = R_0 + RU, \qquad \text{where} \qquad U = [u_1, u_2, \ldots, u_m]^T$$

Partition R_0 and R in the same way that X is partitioned.

$$R_0 = \left[\frac{v_0}{w_0}\right], \qquad R = \left[\frac{V}{W}\right]$$

Then

$$\left[\frac{T}{S}\right] = \left[\frac{v_0}{w_0}\right] + \left[\frac{v}{w}\right]U = \left[\frac{v_0 + VU}{w_0 + WU}\right]$$

or $T = v_0 + VU$, $S = w_0 + WU$. Substituting in the original equation, we have for any choice of the parameters $U = [u_1, u_2, \ldots, u_m]^T$

$$p_0 + P(v_0 + VU) = q_0 + Q(w_0 + WU)$$

or

$$(p_0 + PU_0) + (PV)U = (q_0 + Qw_0) + (QW)U$$

This means (see exercise 22) that $p_0 + Pv_0 = q_0 + qw_0$ and $PV = QW$.

As the parameter vector U varies,

$$p(U) = (p_0 + Pv_0) + (PV)U$$

sweeps out the flat defined by $p_0', p_1', \ldots, p_m'$, where $p_0' = p_0 + Pv_0$ and $p_i' - p_0 = (PV)[;i]$.

This flat is the intersection of the two given flats.

Incidentally, we have proved

PROPOSITION 4.16 The intersection of two flats is a flat. ■

EXAMPLE 4.19 Find the intersection in $\mathbf{R}^5$ of the flat defined by

$$(-21, -30, -8, -15, -10),$$
$$(-16, -25, -7, -13, -8),$$
$$(-21, -30, -8, -16, -11),$$
$$(-11, -20, -6, -14, -9)$$

and the flat defined by

$$(22, 52, 8, 19, 14),$$
$$(20, 49, 7, 17, 12),$$
$$(2, 32, 4, 16, 11),$$
$$(23, 50, 8, 18, 14)$$

Solution Label the points p_0, p_1, p_2, p_3 and q_0, q_1, q_2, q_3, respectively, and define P and Q as usual:

```
      P
 5    0   10
 5    0   10
 1    0    2
 2   ¯1    1
 2   ¯1    1
      Q
¯2  ¯20    1
¯3  ¯20   ¯2
¯1   ¯4    0
¯2   ¯3   ¯1
¯2   ¯3    0
```

Reducing the augmented matrix of the system $[P \mid -Q]\begin{bmatrix} T \\ S \end{bmatrix} = q_0 - p_0$ gives

```
+E←ECHELON P,(-Q),Q0-P0
 1.00E0      0.00E0     2.00E0      ¯8.67E¯19   4.00E0      ¯4.34E¯18   7.00E0
 5.20E¯18    1.00E0     3.00E0       3.47E¯18   5.00E0       4.34E¯18   8.00E0
¯3.47E¯18    0.00E0    ¯6.94E¯18     1.00E0    ¯1.39E¯17     5.20E¯18   9.00E0
¯1.73E¯18    0.00E0    ¯3.47E¯18     0.00E0    ¯6.94E¯18     1.00E0     1.00E1
 0.00E0      0.00E0     0.00E0       8.67E¯19   0.00E0      ¯2.17E¯18  ¯1.39E¯17
```

The nonpivot columns are $E[;\ 3\ \ 5]$, and so $x_3\ (= t_3)$ and $x_5\ (= s_2)$ may be chosen arbitrarily — say, $x_3 = u_1$, $x_5 = u_2$. Then the solutions are

$$\begin{aligned} x_1 &= 7 - 2u_1 - 4u_2 \\ x_2 &= 8 - 3u_1 - 5u_2 \\ x_3 &= u_1 \\ x_4 &= 9 \\ x_5 &= u_2 \\ x_6 &= 10 \end{aligned}$$

or

$$\begin{bmatrix} x_1 \\ x_2 \\ x_3 \\ x_4 \\ x_5 \\ x_6 \end{bmatrix} = \begin{bmatrix} t_1 \\ t_2 \\ t_3 \\ \hline s_1 \\ s_2 \\ s_3 \end{bmatrix} = \begin{bmatrix} 7 \\ 8 \\ 0 \\ \hline 9 \\ 0 \\ 10 \end{bmatrix} + \begin{bmatrix} -2 & -4 \\ -3 & -5 \\ 1 & 0 \\ \hline 0 & 0 \\ 0 & 1 \\ 0 & 0 \end{bmatrix} \begin{bmatrix} u_1 \\ u_2 \end{bmatrix}$$

Hence

$$T = \begin{bmatrix} 7 \\ 8 \\ 0 \end{bmatrix} + \begin{bmatrix} -2 & -4 \\ -3 & -5 \\ 1 & 0 \end{bmatrix} \begin{bmatrix} u_1 \\ u_2 \end{bmatrix}, \qquad S = \begin{bmatrix} 9 \\ 0 \\ 10 \end{bmatrix} + \begin{bmatrix} 0 & 0 \\ 0 & 1 \\ 0 & 0 \end{bmatrix} \begin{bmatrix} u_1 \\ u_2 \end{bmatrix} = \begin{bmatrix} 9 \\ 0 \\ 10 \end{bmatrix} + \begin{bmatrix} 0 \\ 1 \\ 0 \end{bmatrix} [u_2]$$

Then,

$$p'(U) = \begin{bmatrix} -21 \\ -30 \\ -8 \\ -15 \\ -10 \end{bmatrix} + \begin{bmatrix} 5 & 0 & 10 \\ 5 & 0 & 10 \\ 1 & 0 & 2 \\ 2 & -1 & 1 \\ 2 & -1 & 1 \end{bmatrix} \left(\begin{bmatrix} 7 \\ 8 \\ 0 \end{bmatrix} + \begin{bmatrix} -2 & -4 \\ -3 & -5 \\ 1 & 0 \end{bmatrix} \begin{bmatrix} u_1 \\ u_2 \end{bmatrix} \right)$$

$$p'(u_1, u_2) = \begin{bmatrix} 14 \\ 5 \\ -1 \\ -9 \\ -4 \end{bmatrix} + \begin{bmatrix} 0 & -20 \\ 0 & -20 \\ 0 & -4 \\ 0 & -3 \\ 0 & -3 \end{bmatrix} \begin{bmatrix} u_1 \\ u_2 \end{bmatrix} = \begin{bmatrix} 14 \\ 5 \\ -1 \\ -9 \\ -4 \end{bmatrix} + u_2 \begin{bmatrix} -20 \\ -20 \\ -4 \\ -3 \\ -3 \end{bmatrix}$$

the line through $(14, 5, -1, -9, -4)$ and $(-6, -15, -5, -12, -7)$.

Using S instead of T, we have

$$q'(u_2) = \begin{bmatrix} 22 \\ 52 \\ 8 \\ 19 \\ 14 \end{bmatrix} - \begin{bmatrix} 2 & 20 & -1 \\ 3 & 20 & 2 \\ 1 & 4 & 0 \\ 2 & 3 & 1 \\ 2 & 3 & 0 \end{bmatrix} \left(\begin{bmatrix} 9 \\ 0 \\ 10 \end{bmatrix} + \begin{bmatrix} 0 \\ 1 \\ 0 \end{bmatrix} [u_2] \right)$$

$$q'(u_2) = \begin{bmatrix} 14 \\ 5 \\ -1 \\ -9 \\ -4 \end{bmatrix} + u_2 \begin{bmatrix} -20 \\ -20 \\ -4 \\ -3 \\ -3 \end{bmatrix} \quad \blacksquare$$

The reason for the mysterious appearance and disappearance of u_1 in the last example is that the four points p_0, p_1, p_2, p_3 are coplanar and hence do not define a 3-flat. The point p_3 can be discarded without changing the problem. If this is done, then only one parameter appears in the calculation.

The next two propositions give connections between the theory of flats and previous work on linear equations and affine functions.

PROPOSITION 4.17 The set of solutions of a system of linear equations is a flat. More precisely, the set of solutions of k irredundant linear equations in n variables is an $(n - k)$-dimensional flat in $\mathbf{R}^n$.

Conversely, every flat in $\mathbf{R}^n$ is the set of solutions of a system of linear equations in n variables.

Proof The first statement is just Proposition 3.7 in different terminology. The converse statement may be proved by generalizing the procedure of Example 4.6. $\blacksquare$

Recall that the image of a function $f: \mathbf{R}^n \to \mathbf{R}^m$ is the set of all vectors Y in $\mathbf{R}^m$ such that $Y = f(X)$ for some vector X in $\mathbf{R}^n$. If S is a subset of $\mathbf{R}^n$, we will refer to the set of vectors Y in $\mathbf{R}^m$ such that $Y = f(X)$ for some X *in* S as the *image of* S *under* f. The image of S under f is often written $f(S)$.

PROPOSITION 4.18 Let $T(X) = b + AX$ be an affine function $T: \mathbf{R}^n \to \mathbf{R}^m$. The image of T is an r-dimensional flat in $\mathbf{R}^m$, where r is the rank of A. If π is a flat in $\mathbf{R}^n$, then the image of π under T is a flat in $\mathbf{R}^m$. Further, if $r = n$, then π and $T(\pi)$ have the same dimension.

Proof The image of T is the set of vectors in $\mathbf{R}^m$ swept out by $T(X) = b + AX$ as X varies in $\mathbf{R}^n$ — an r-dimensional flat by Proposition 4.18. Suppose that $p(U) = p_0 + PU$ is an ℓ-dimensional flat in $\mathbf{R}^m$. Then

$$T(p(U)) = b + A(p_0 + PU) = (b + Ap_0) + (AP)U$$

a flat in $\mathbf{R}^m$. If P has linearly independent columns, then it has a left inverse L_1. If $r = n$, then A has a left inverse L_2, and so AP has the left inverse L_1L_2. Thus AP has ℓ columns, and these columns are linearly independent. This shows that $T(p(U))$ is ℓ-dimensional as well. $\blacksquare$

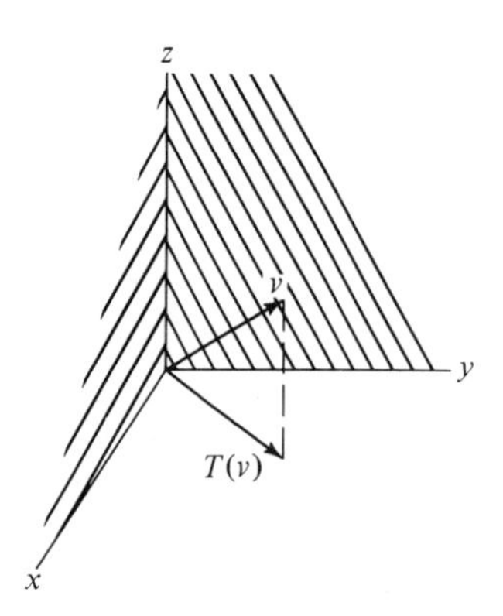

FIGURE 4.25

EXAMPLE 4.20 Let the affine function $T: \mathbf{R}^3 \to \mathbf{R}^3$ be given by

$$T(X) = AX \qquad \text{where } A = \begin{bmatrix} 1 & 0 & 0 \\ 0 & 1 & 0 \\ 0 & 0 & 0 \end{bmatrix}$$

Since the rank of A is two, the image of T is a plane in $\mathbf{R}^3$. In fact, since $T(x, y, z) = (x, y, 0)$, the image of T is the xy plane. T is a projection onto the xy plane. Most planes in $\mathbf{R}^3$ are mapped onto the xy plane by T and so their dimension is preserved. However, planes parallel to the z axis, such as the xz plane and the yz plane, are mapped into lines in the xy plane (Figure 4.25). Similarly, most lines in $\mathbf{R}^3$ project to lines in the xy plane. Lines parallel to the z axis, however, project to points in the xy plane ∎

Proposition 4.18 can be used as the basis of a purely geometric definition of affine map. Affine maps can be characterized as those maps that carry flats to flats and preserve parallelism. (Parallelism in $\mathbf{R}^n$ will be defined in the next section.)

COORDINATES FOR A k-FLAT IN $\mathbf{R}^n$

Geometrically all lines and planes are the same — even if some lie in $\mathbf{R}^3$ and some lie in $\mathbf{R}^{40}$. Every plane is in some sense a copy of $\mathbf{R}^2$, every line a copy of $\mathbf{R}^1$, and every 3-flat a copy of space — $\mathbf{R}^3$. More generally, every k-flat is in some sense a copy of $\mathbf{R}^k$.

Given a k-flat π in $\mathbf{R}^n$, we can make this statement explicit by choosing a coordinate system for π. Such a coordinate system will assign a unique k-tuple of coordinates $(x'_1, x'_i, \ldots, x'_k)$ to each point in π. Although this general statement is interesting and occasionally useful, the most important case arises when $k = n$. In this case $\pi = \mathbf{R}^n$, and we are simply defining a new set of coordinates for $\mathbf{R}^n$. In Section 4.2 this process was carried out for $k = n = 2$ and $k = n = 3$.

If π is a k-flat in $\mathbf{R}^n$, then any set of $k + 1$ points $p_0, p_1, \ldots, p_k$ that define π

can be used to set up an affine coordinate system for π. The origin is p_0 and the x_i' axis is the line (1-flat) through p_0 and p_i. The unit point on the x_i' axis is p_i (see Figures 4.13 and 4.14). If the point p in $\mathbf{R}^n$ has coordinates $(x_1, x_2, \ldots, x_n)$ and p lies in π, then the new coordinates $(x_1', x_2', \ldots, x_k')$ of p are given by

$$X = p_0 + PX'$$

where $P = [p_1 - p_0 | p_2 - p_0 | \cdots | p_k - p_0]$, $X = [x_1, x_2, \ldots, x_n]^T$, and $X' = [x_1', x_2', \ldots, x_k']^T$.

In other words, the new coordinates of p are just the parameters giving p in the parametric representation of the k-flat given by $p_0, p_1, \ldots, p_k$. It is precisely the same formula as we had in Section 4.2 for the special cases $k = n = 2$ and $k = n = 3$.

In the most general case the matrix P is not square (it is n by k), but since the columns of P are linearly independent, P always has left inverses. If L is any one of these left inverses, then

$$X' = L(X - p_0)$$

This latter formula is valid only when the components of X are coordinates of a point in π, since these are the only points with X' coordinates. Convenient choices for L are ⌹P and (k, n)↑GAUSS P. The latter approach can be used for hand computation.

EXAMPLE 4.21 Let π be the plane in $\mathbf{R}^4$ through the three points $p_0 = (1, -1, 0, 1)$, $p_1 = (2, -3, -1, 2)$, $p_2 = (2, -2, -2, 3)$.

(a) Using p_0, p_1, p_2 to define a coordinate system on π, write the coordinate-change formula.

(b) What are original coordinates of the point with new coordinates $(x_1', x_2') = (1, 1)$?

(c) Does the point $(1, 3, 0, -2)$ lie in π?

Solution

(a) The coordinate-change formula is $X = p_0 + [p_1 - p_0 | p_2 - p_0]X'$ or

$$\begin{bmatrix} x_1 \\ x_2 \\ x_3 \\ x_4 \end{bmatrix} = X = \begin{bmatrix} 1 \\ -1 \\ 0 \\ 1 \end{bmatrix} + \begin{bmatrix} 1 & 1 \\ -2 & -1 \\ -1 & -2 \\ 1 & 2 \end{bmatrix} \begin{bmatrix} x_1' \\ x_2' \end{bmatrix}$$

(b)

$$\begin{bmatrix} 1 \\ -1 \\ 0 \\ 1 \end{bmatrix} + \begin{bmatrix} 1 & 1 \\ -2 & -1 \\ -1 & -2 \\ 1 & 2 \end{bmatrix} \begin{bmatrix} 1 \\ 1 \end{bmatrix} = \begin{bmatrix} 3 \\ -4 \\ -3 \\ 4 \end{bmatrix}$$

(c)

$$L = \begin{bmatrix} 2 & 0 & 3 & 2 \\ 2 & 1 & -1 & -1 \end{bmatrix}$$

is a left inverse for P, and so if $(1, 3, 0, -2)$ lies in P, it will have new coordinates:

$$X' = L(X - p_0) = \begin{bmatrix} 2 & 0 & 3 & 2 \\ 2 & 1 & -1 & -1 \end{bmatrix} \left(\begin{bmatrix} 1 \\ 3 \\ 0 \\ -2 \end{bmatrix} - \begin{bmatrix} 1 \\ -1 \\ 0 \\ 1 \end{bmatrix} \right) = \begin{bmatrix} -6 \\ 7 \end{bmatrix}$$

But the only point in π with X' coordinates $(-6, 7)$ is

$$\begin{bmatrix} 1 \\ -1 \\ 0 \\ 1 \end{bmatrix} + \begin{bmatrix} 1 & 1 \\ -2 & -1 \\ -1 & -2 \\ 1 & 2 \end{bmatrix} \begin{bmatrix} -6 \\ 7 \end{bmatrix} = \begin{bmatrix} 2 \\ 4 \\ -8 \\ 9 \end{bmatrix} \neq \begin{bmatrix} 1 \\ 3 \\ 0 \\ -2 \end{bmatrix}$$

Thus the formula $X' = L(X - p_0)$ does not apply to $(1, 3, 0, -2)$; that is, $(1, 3, 0, -2)$ does not lie in π. ■

Notice that part (c) of Example 4.21 is just a typical problem from Section 2.3 dressed up in new terminology (see Examples 2.25 and 2.26).

Now let us restrict our attention to the case $k = n$. The matrix P is then square and invertible, and the coordinate-change formulas

$$X = p_0 + PX'$$
$$X' = P^{-1}(X - p_0)$$

apply to all points in $\mathbf{R}^n$. The basic formulas developed in Section 4.2 for $n = 2, 3$ apply without change for any n. The proofs of the next two propositions are left as exercises.

PROPOSITION 4.19 Let $T: \mathbf{R}^n \to \mathbf{R}^n$ be an affine transformation given by $Y = T(X) = b + AX$. Introduce X' coordinates via the formula $X = p_0 + PX'$. Then in the new coordinate system T is given by

$$Y' = c + BX'$$

where $B = P^{-1}AP$ and $c = P^{-1}(b + (A - I)p_0)$.

In particular if T is linear ($b = 0$) and the coordinate change leaves the origin fixed ($p_0 = 0$), we have

$$Y' = (P^{-1}AP)X' \quad ■$$

PROPOSITION 4.20 Let $f: \mathbf{R}^n \to \mathbf{R}$ be a quadratic function given by

$$f(X) = c + BX + X^TAX$$

with $A = A^T$.

Introduce X' coordinates via the formula $X = p_0 + PX'$. Then in the new coordinate system f is given by

$$f(X') = c' + B'X' + X'^TA'X'$$

where $c' = f(p_0)$, $B' = (B + 2p_0^TA)P$, $A' = P^TAP = A'^T$.

In particular, if f is a quadratic form ($c = 0, B = 0$) and the coordinate change leaves the origin fixed ($p_0 = 0$), then

$$f(X') = X'^T(P^TAP)X' \qquad \blacksquare$$

EXAMPLE 4.22 Let the linear transformation $T: \mathbf{R}^5 \to \mathbf{R}^5$ have matrix

$$A = \begin{bmatrix} 0 & 15 & -6 & -7 & 11 \\ 2 & -9 & 6 & 3 & -10 \\ 1 & -3 & 1 & 7 & 1 \\ 0 & -12 & 4 & 6 & -8 \\ -4 & 12 & -10 & 0 & 17 \end{bmatrix}$$

Choose a new coordinate system for $\mathbf{R}^5$ using

$$p_0 = 0,\ p_1 = (1, 0, -2, 0, -1),$$
$$p_2 = (-2, 1, 2, 1, 0),\ p_3 = (0, -1, 3, 0, 3)$$
$$p_4 = (-3, 2, 1, 2, -2),\ p_5 = (6, -5, 0, -4, 7)$$

What is the matrix of T in the new coordinate system?

Solution Since $p_0 = 0$, the coordinate-change formula is just $X = PX'$, with P given by

P

1	⁻2	0	⁻3	6
0	1	⁻1	2	⁻5
⁻2	2	3	1	0
0	1	0	2	⁻4
⁻1	0	3	⁻2	7

and the matrix of T in the new coordinate system is $P^{-1}AP$ or

```
      (A+.×P)⌹P
 1.00E0      ¯8.75E¯18   ¯1.01E¯17    0.00E0      0.00E0
¯2.72E¯18    2.00E0       0.00E0      0.00E0      0.00E0
¯8.43E¯19   ¯1.25E¯17     3.00E0      0.00E0      0.00E0
 0.00E0      2.51E¯17     4.91E¯17    4.00E0      0.00E0
¯2.99E¯19    1.05E¯17     2.53E¯17    0.00E0      5.00E0
```

Notice that in this new coordinate system T has a simple geometric description. The x_i' axis is stretched by the factor i. ∎

EXAMPLE 4.23 Let $f: \mathbf{R}^5 \to \mathbf{R}$ be

$$\begin{aligned} f(x_1, x_2, x_3, x_4, x_5) = {} & 56x_1^2 + 126x_1x_2 + 64x_1x_3 - 22x_1x_4 - 20x_1x_5 \\ & + 120x_2^2 + 42x_2x_3 - 48x_2x_4 + 36x_2x_5 + 24x_3^2 \\ & - 10x_3x_4 - 32x_3x_5 + 10x_4^2 - 4x_4x_5 + 21x_5^2 \end{aligned}$$

Find the expression for f in the new coordinate system for $\mathbf{R}^5$ defined in Example 4.22.

Solution $f(X) = X^TAX$, where A is given by

```
      A
 56    63    32    11   ¯10
 63   120    21   ¯24    18
 32    21    24    ¯5   ¯16
¯11   ¯24    ¯5    10    ¯2
¯10    18   ¯16    ¯2    21
```

Using the matrix P of Example 4.22, f is given in the new coordinate system by $f(X') = X'^T(P^TAP)X'$. Computing P^TAP, we have

```
     (⍉P)+.×A+.×P
1.000  0.000  0.000  0.000  0.000
0.000  2.000  0.000  0.000  0.000
0.000  0.000  3.000  0.000  0.000
0.000  0.000  0.000  4.000  0.000
0.000  0.000  0.000  0.000  5.000
```

or

$$f(x_1', x_2', x_3', x_4', x_5') = x_1'^2 + 2x_2'^2 + 3x_3'^2 + 4x_4'^2 + 5x_5'^2$$

Clearly f has a global minimum at $X = X' = 0$. ∎

EXERCISES 4.4

In exercises 1 through 5 find the dimension of the flat defined by the given points.

1. $p_0 = (3, -1, -4, -2)$, $p_1 = (4, -3, -3, -2)$, $p_2 = (3, 0, -6, -1)$, $p_3 = (5, -2, -8, 1)$

2. $p_0 = (-7, 1, 5, 9)$, $p_1 = (-6, 1, 5, 9)$, $p_2 = (-7, 2, 5, 10)$, $p_3 = -6, 2, 6, 10)$

3. $p_0 = (-7, -9, 0, 0)$, $p_1 = (-6, -8, -1, 1)$, $p_2 = (-9, -10, 2, -3)$, $p_3 = (-11, -10, 4, -7)$

4. $p_0 = (1, -1, -2, -4)$, $p_1 = (2, -2, -4, -6)$, $p_2 = (2, -1, -6, -6)$, $p_3 = (6, -3, -18, -14)$, $p_4 = (-3, 1, 11, 5)$

5. $p_0 = (1, -3, -7, 2, -6)$, $p_1 = (2, -3, -8, 1, -6)$, $p_2 = (1, -2, -9, 0, -7)$, $p_3 = (3, 0, -15, -6, -9)$, $p_4 = (0, -5, -1, 8, -3)$, $p_5 = (5, 2, -21, -12, -11)$

In exercises 6 through 10 find the flat defined by the set of equations. Use Proposition 4.17 to deduce the number of irredundant equations.

6. $$\begin{aligned} 2x + 2y + 10z &= 18 \\ 3x + 2y + 12z &= 22 \\ x + y + 5z &= 9 \end{aligned}$$

7. $$\begin{aligned} 2x_1 \qquad\quad + 4x_3 + x_4 &= 14 \\ -x_1 + 2x_2 + 4x_3 - x_4 &= 0 \\ x_1 - x_2 - x_3 + x_4 &= 5 \\ x_1 \qquad\quad + 2x_3 \qquad &= 4 \end{aligned}$$

8. $$\begin{aligned} x_1 - 2x_2 - 4x_3 - 6x_4 &= -8 \\ -x_1 + 3x_2 + 7x_3 + 11x_4 &= 15 \\ x_1 - 4x_2 - 10x_3 - 16x_4 &= -22 \\ 2x_2 + 6x_3 + 10x_4 &= 14 \end{aligned}$$

9. $$\begin{aligned} x_1 \qquad\quad - x_3 + x_4 &= 3 \\ x_2 - 2x_3 + x_4 &= 0 \\ x_1 - x_2 + 2x_3 - x_4 &= 2 \\ 2x_3 + 3x_4 &= 7 \end{aligned}$$

10. $$\begin{aligned} x_1 - 2x_2 - 4x_3 + x_4 \qquad\quad &= 0 \\ -2x_1 + 5x_2 + 11x_3 - 3x_4 - x_5 &= -1 \\ -x_1 + x_2 + x_3 + x_4 + 7x_5 &= 10 \\ x_1 - 4x_2 - 10x_3 + 2x_4 - 4x_5 &= -7 \\ 2x_2 + 6x_3 \qquad\quad + 10x_5 &= 16 \end{aligned}$$

In exercises 11 through 15 find the intersection, if any, of the given flats. For questions of parallelism see exercises 21 through 24 of Exercises 4.5.

11. The 3-flat through $(5, -7, -4, 9)$, $(6, -6, -6, 8)$, $(3, -8, 1, 11)$, $(1, -10, 6, 12)$ and the line through $(-15, -20, 42, 35)$ and $(-15, -20, 43, 33)$.

12. The plane through $(-6, -5, -3, -2)$, $(-5, -4, -2, -2)$, $(-7, -5, -4, -1)$ and the plane through $(-18, -9, -9, 6)$, $(-17, -6, -9, 8)$, $(-19, -7, -10, 9)$.

13. The 3-flat through $(4, 4, -8, 0)$, $(5, 2, -9, 6)$, $(3, 7, -9, 5)$, $(1, 11, -6, 3)$ and the line through $(8, -5, -12, 12)$, $(21, -38, -15, 27)$.

14. The plane through $(-2, -2, 5, 2, 6)$, $(-1, -3, 5, 2, 4)$, $(-1, -2, 4, 1, 5)$ and the plane through $(-1, -2, 3, 3, 6)$, $(0, -1, 0, 3, 6)$, $(0, -5, 7, 0, 1)$.

15. The 3-flat through $(3, 4, 0, 9)$, $(4, 3, 1, 10)$, $(4, 4, -1, 9)$, $(-1, 6, 1, 6)$ and the plane through $(-5, 10, -3, 2)$, $(0, 7, -3, 6)$, $(-10, 12, -2, 0)$.

In exercises 16 through 21 a function is given along with a set of points defining a new coordinate system. Find the expression for the function in the new coordinate system.

16. (Computer assignment) $T: \mathbf{R}^4 \to \mathbf{R}^4$ is linear with matrix $\begin{bmatrix} 1 & 4 & -2 & 0 \\ 1 & -3 & 1 & 1 \\ 2 & -4 & 1 & 2 \\ 1 & -8 & 3 & 2 \end{bmatrix}$

$p_0 = 0$, $p_1 = (1, 0, 0, -1)$, $p_2 = (-1, 1, 1, 2)$, $p_3 = (0, 1, 2, 1)$, $p_4 = (-1, 1, 2, 3)$

17. (Computer assignment) $T: \mathbf{R}^5 \to \mathbf{R}^5$ is linear with matrix

$$\begin{bmatrix} 8 & -2 & -3 & -6 & -7 \\ 3 & 1 & -4 & -5 & 3 \\ 3 & -2 & -1 & -5 & 3 \\ -3 & 2 & 5 & 9 & -3 \\ -3 & 2 & 2 & 5 & -2 \end{bmatrix}$$

$p_0 = 0$, $p_1 = (1, 0, 0, 0, -1)$, $p_2 = (1, 1, 1, -1, -1)$, $p_3 = (2, 1, 2, -2, -2)$, $p_4 = (0, -1, -1, 2, 1)$, $p_5 = (2, 1, 1, -1, -1)$

18. (Computer assignment) $T: \mathbf{R}^4 \to \mathbf{R}^4$ by

$$Y = T(X) = \begin{bmatrix} 2 \\ 4 \\ -2 \\ 0 \end{bmatrix} + \begin{bmatrix} 7 & -2 & 4 & 2 \\ 12 & -3 & 8 & 4 \\ -4 & 2 & -1 & 0 \\ -4 & 0 & -4 & -3 \end{bmatrix} X$$

$p_0 = (1, 0, -3, 2)$, $p_1 = (2, 1, -4, 2)$, $p_2 = (2, 2, -4, 2)$, $p_3 = (3, 3, -4, 1)$, $p_4 = (0, -2, -3, 4)$

19. $f: \mathbf{R}^4 \to \mathbf{R}$ is the quadratic form

$$\begin{aligned} f(x_1, x_2, x_3, x_4) = 5x_1^2 + 4x_1x_2 - 6x_1x_3 + 6x_1x_4 - 3x_2^2 + 2x_2x_3 \\ + 2x_2x_4 + x_3^2 - 2x_3x_4 + 2x_4^2 \end{aligned}$$

p_0, p_1, p_2, p_3, p_4 are as in exercise 16

20. $f: \mathbf{R}^5 \to \mathbf{R}$ is the quadratic form

$$\begin{aligned} f(x_1, x_2, x_3, x_4, x_5) = 7x_1^2 - 8x_1x_2 - 14x_1x_3 - 18x_1x_4 + 14x_1x_5 \\ + 11x_2^2 + 2x_2x_3 + 16x_2x_4 - 8x_2x_5 + 15x_3^2 \\ + 26x_3x_4 - 12x_3x_5 + 17x_4^2 - 18x_4x_5 + 8x_5^2 \end{aligned}$$

p_0, p_1, p_2, p_3, p_4, p_5 are as in exercise 17.

21. $f: \mathbf{R}^4 \to \mathbf{R}$ is the quadratic function

$$\begin{aligned} f(x_1, x_2, x_3, x_4) = 19 + 4x_1 + 10x_2 + 22x_3 + 12x_4 + x_1^2 \\ + 4x_1x_2 + 6x_1x_3 + 6x_1x_4 + 6x_2x_3 \\ + 2x_2x_4 + 8x_3^2 + 10x_3x_4 + 3x_4^2 \end{aligned}$$

p_0, p_1, p_2, p_3, p_4 are as in exercise 18.

22. Let p_0, q_0 be points in $\mathbf{R}^n$, P, Q n-by-k matrices, and U a variable vector in $\mathbf{R}^k$.

Suppose that $p_0 + PU = q_0 + QU$ for all U. By choosing special U's, show that $p_0 = q_0$ and $P = Q$.

23. Show that if Proposition 2.11 of Section 2.2 is taken to be the *definition* of a matrix having "linearly independent columns," then an n-by-0 matrix (e.g., `4 3 0 ρ 0`) has "linearly independent columns." Recall that $\iota 0$ is a zero vector (it is the origin in $\mathbf{R}^0$).

24. Let P be a matrix, R the row-echelon form of P with the rows of zeros deleted, and V the vector of indices of the pivot columns. Show that $P = P[;V]R$.

    Hint: Proposition 3.1 and the definition of matrix-vector multiplication.

25. Let π be the flat given by $p(T) = p_0 + PT$ and let q_0 be a point of π. Show that π is the flat π' given by $q(S) = q_0 + PS$.

    Hint: By Proposition 4.12 π' is included in π, and it is sufficient to show that p_0 is a point of π'.

26. Prove Proposition 4.15.

Exercises 27 through 36 refer to Exercises 4.2 and 4.3. Redo the indicated exercises for $\mathbf{R}^n$ with $n > 3$. If no change is necessary for $n > 3$, so state.

27. Exercise 39 of Exercises 4.2
28. Exercise 40 of Exercises 4.2
29. Exercise 41 of Exercises 4.2
30. Exercise 42 of Exercises 4.2
31. Exercise 43 of Exercises 4.2
32. Exercise 44 of Exercises 4.2
33. Exercise 45 of Exercises 4.2
34. Exercise 21 of Exercises 4.2
35. Exercise 22 of Exercises 4.3
36. Exercise 23 of Exercises 4.3

4.5 Subspaces

Suppose that the flat π is defined by the points $p_0, p_1, \ldots, p_k$. Putting $P = [p_1 - p_0 | \cdots | p_k - p_0]$, we have

$$p(T) = p_0 + PT \tag{4.3}$$

Most of the calculations in the previous section involved the matrix P — that is, the points $p_i - p_0$, which do not lie in the flat π. This practice started in Section 4.1, where we noted that although vectors would be drawn at various positions in the plane and space, the tails were, in fact, fixed to the origin. This is illustrated in Figure 4.26, where π is a line in the plane or space. (See also Figures 4.3, 4.6, 4.11, 4.13, and 4.14.) The vectors $p_1 - p_0, \ldots, p_k - p_0$ lie in the flat given by

$$s(T) = PT \tag{4.4}$$

This flat contains the origin, 0.

DEFINITION 4.4 A *subspace* of $\mathbf{R}^n$ is a flat containing the origin.

First a bit of notation. If S is a subset of $\mathbf{R}^n$, the image $T(S)$ of S under the translation $T(X) = a + X$ will be denoted by $a + S$. The set $a + S$ consists of all vectors v of the form $v = a + s$ as s varies in S. We write $S - a$ for $-a + S$. The

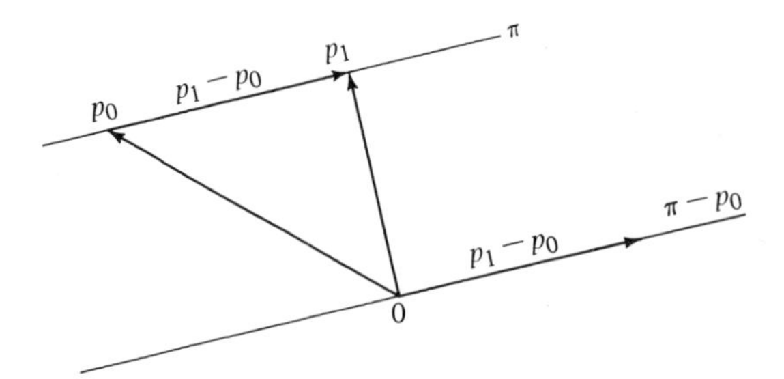

FIGURE 4.26

flat defined in Equation (4.4) is thus $\pi - p_0$. In $\mathbf{R}^2$ and $\mathbf{R}^3$ the flats defined by Equations (4.3) and (4.4) are parallel, and there is only one parallel flat through the origin with the same dimension as π.

We will now generalize these ideas to $\mathbf{R}^n$. The basic fact we will use is given in the next proposition. Notice that if π is a flat, then $a + \pi$ is also a flat. In fact, if π is given by Equation (4.3), then $a + \pi$ is given by $a + p(T) = (a + p_0) + PT$.

PROPOSITION 4.21 Let π be a flat in $\mathbf{R}^n$. If the flats $a + \pi$ and $b + \pi$ have any intersection at all, then they are equal.

Proof First notice that if a flat π is given by $p(T) = p_0 + PT$ as T varies, and if q_0 is a point in π, then the flat π' given by $q(S) = q_0 + PS$ is equal to π. We used this fact in Section 4.4 but left the proof as exercise 23 in Exercises 4.4. To prove it we use Proposition 4.12. Since q_0 is in π and $P = PI$, we have that π' is contained in π. To show that π is contained in π' it is sufficient to show that p_0 is contained in π'. Since q_0 is contained in π, there is a parameter vector T_0 such that $q_0 = p_0 + PT_0$. Hence

$$p_0 = q_0 - PT_0 = q_0 + P(-T_0)$$

and so p_0 is indeed in π'.

Now consider $a + \pi$, given by $(a + p_0) + PT$, and $b + \pi$ given by $(b + p_0) + PT$. If q lies in both these flats, then, by the above reasoning, both are given by $q + PT$. ∎

As an immediate consequence of Proposition 4.21 we have that any two translates of a flat that contain the origin are equal.

DEFINITION 4.5 Let π be a flat in $\mathbf{R}^n$. The *associated subspace* of π is the flat $\pi - p$, where p is any point of π.

If π is given by Equation (4.3), then the associated subspace is given by Equation (4.4).

In $\mathbf{R}^2$ and $\mathbf{R}^3$ if parallel flats of the same dimension intersect, then they are equal. This is in accord with Proposition 4.21. If the flats are of differing dimensions, a line and a plane, say, then if they intersect, one is contained in the other.

DEFINITION 4.6 Two flats in $\mathbf{R}^n$ are said to be *parallel* if one of the associated subspaces is contained in the other.

The associated subspace of π is the only subspace that is parallel to π and has the same dimension as π.

The associated subspace of a point (0-flat) is the origin. Thus a point is parallel to all flats.

In Section 4.4 we saw that affine transformations carry flats to flats. We can now say more.

PROPOSITION 4.22 Affine functions carry parallel flats to parallel flats. Linear functions are precisely those affine functions that carry subspaces to subspaces.

Proof Let π_1 and π_2 be parallel and let S be the associated subspace. Suppose that S is given by Equation (4.4) and that $\pi_1 = a_1 + S$, $\pi_2 = a_2 + S$. If $f(X) = b + AX$ is an affine function $f\colon \mathbf{R}^n \to \mathbf{R}^m$, then

$$f(\pi_i) = b + A(a_i + PT) = (b + Aa_i) + (AP)T$$

as T varies. Thus the associated subspace of $f(\pi_i)$ is given by $s(t) = (AP)T$ for both $i = 1$ and $i = 2$. The flats are thus parallel. If $b = 0$, then $f(S) = (AP)T$, and so if f is linear, it carries subspaces to subspaces. If, on the other hand, $b \neq 0$, then $f(0) = b \neq 0$, so the subspace $\{0\}$ is carried to the 0-flat $\{b\}$, which is not a subspace. ∎

The next proposition gives the connection between subspaces and systems of linear equations. Recall (Proposition 4.17) that flats are simply solution sets for systems of linear equations.

PROPOSITION 4.23 If the flat π is the set of solutions of the system of equations

$$AX = b \tag{4.5}$$

then the associated subspace S is the set of solutions of the associated homogeneous system

$$AX = 0 \tag{4.6}$$

That is, if S is the set of solutions of (4.6) and p is any solution of (4.5), then $p + S$ is the set of solutions of (4.5).

Proof Let S be the set of solutions of (4.6). Since $X = 0$ is a solution of (4.6), the flat S contains 0; that is, S is a subspace. If p is any solution of (4.5) and s is any vector in S, then $A(p + s) = Ap + As = b + 0 = b$, and so $p + S$ consists of solutions of (4.5). On the other hand, if q is any solution of (4.5), then $A(q - p) = b - b = 0$, and so $s = q - p$ is in S and hence $q = p + s$ is in $p + S$. ∎

EXAMPLE 4.24 Consider the flat π defined by the points $p_0 = (3, 3, -9, 5)$, $p_1 = (4, 1, -10, 5)$, $p_2 = (2, 6, -10, 4)$, $p_3 = (0, 10, -7, 2)$, $p_4 = (1, 8, -8, 2)$, and the flat π' defined by $q_0 = (7, -6, -13, 11)$, $q_1 = (20, -39, -16, 26)$, $q_2 = (-6, 27, -10, -4)$.

Letting P have columns $p_i - p_0$ and Q have columns $q_i - q_0$, we have, solving $p_0 + PT = q_0 + QS$,

```
       P
 1.000 ¯1.000 ¯3.000 ¯2.000
¯2.000  3.000  7.000  5.000
¯1.000 ¯1.000  2.000  1.000
 0.000 ¯1.000 ¯3.000 ¯3.000

     Q
 13 ¯13
¯33  33
 ¯3   3
 15 ¯15

+E←ECHELON P,(-Q),Q0-P0
1.00E0   2.78E¯17   0.00E0   1.000E0    2.00E0   ¯2.00E0   0.00E0
0.00E0   1.00E0     0.00E0   ¯2.78E¯17  3.00E0   ¯3.00E0   0.00E0
0.00E0   2.78E¯17   1.00E0   1.00E0     4.00E0   ¯4.00E0   1.11E¯16
0.00E0   1.39E¯17   0.00E0   0.00E0     0.00E0    0.00E0   1.00E0
```

Since the last column is a pivot column, the flats do not intersect. The columns of $-Q$, and hence Q, are linear combinations of the columns of P; in fact $-Q = P[;\ 1\ 2\ 3]E[;\ 5\ 6]$, hence the subspace parallel to π', which is given by $s'(T) = QT$, is contained in the subspace parallel to π, which is given by $s(T) = PT$. Thus the two flats are parallel. ■

Notice that π above is a 3-flat and π' is a line. It can be shown that a 3-flat and a line in $\mathbf{R}^4$ must either intersect or be parallel (exercise 44).

Now let us forget about general flats and concentrate on subspaces.

A flat π is defined by $k + 1$ points $p_0, p_1, \ldots, p_k$. For subspaces, however, we may dispense with p_0. If π is a subspace, then $\pi - p_0$ is another subspace parallel to π, and so $\pi = \pi - p_0$. Thus we can always take π to be defined by $v_0 = 0$, $v_1 = p_1 - p_0, \ldots, v_k = p_k - p_0$. When dealing with subspaces we shall always take $v_0 = 0$ and say that the subspace π is *generated by* $v_1, v_2, \ldots, v_k$. The set of vectors $v_1, \ldots, v_k$ is called a *generating set* for the subspace that they (along with $v_0 = 0$) define. A linearly independent generating set is called a *basis*. Notice that the number of elements in a basis is the dimension of the subspace generated by the basis.

If the subspace S is generated by $v_1, \ldots, v_k$, then a parametric representation of S is all vectors

$$v(T) = PT$$

where $T = [t_1, t_2, \ldots, t_k]^T$ and $P[;i] = v_i$.

EXAMPLE 4.25 Find a basis of the subspace of $\mathbf{R}^4$ consisting of the solutions of the system of linear equations

$$\begin{aligned} x_1 + x_2 + 5x_3 + 9x_4 &= 0 \\ 2x_1 + x_2 + 7x_3 + 13x_4 &= 0 \\ 2x_2 + 6x_3 + 10x_4 &= 0 \\ 2x_1 + 4x_2 + 16x_3 + 28x_4 &= 0 \end{aligned}$$

Solution Since the system $AX = B$ is homogeneous ($B = 0$), we simply row-reduce A rather than $[A|B]$:

```
      A
1   1    5    9
2   1    7   13
0   2    6   10
2   4   16   28

      +E←ECHELON A
1.00E0     0.00E0      2.00E0     4.00E0
0.00E0     1.00E0      3.00E0     5.00E0
0.00E0     1.73E¯18    0.00E0     2.78E¯17
0.00E0     0.00E0      0.00E0     1.39E¯17
```

The nonpivot columns are $E[;\ 3\ 4]$, and so set $x_3 = t_1$, $x_4 = t_2$, and we have

$$\begin{bmatrix} x_1 \\ x_2 \\ x_3 \\ x_4 \end{bmatrix} = t_1 \begin{bmatrix} -2 \\ -3 \\ 1 \\ 0 \end{bmatrix} + t_2 \begin{bmatrix} -4 \\ -5 \\ 0 \\ 1 \end{bmatrix} = \begin{bmatrix} -2 & -4 \\ -3 & -5 \\ 1 & 0 \\ 0 & 1 \end{bmatrix} \begin{bmatrix} t_1 \\ t_2 \end{bmatrix}$$

Thus the subspace is generated by $v_1 = (-2, -3, 1, 0)$, $v_2 = (-4, -5, 0, 1)$. Since v_1, v_2 are linearly independent, they form a basis. The subspace is a plane. ■

EXAMPLE 4.26 What is the dimension of the subspace of $\mathbf{R}^5$ generated by $(1, -1, -2, 1, 0)$, $(0, 1, -1, 0, 1)$, $(2, 1, -7, 2, 3)$, $(4, 1, -13, 4, 5)$?

Solution Since we have a subspace, $v_0 = 0$, and a parametric representation is $s(T) = PT$, where

```
    P
 1   0   2   4
¯1   1   1   1
¯2  ¯1  ¯7 ¯13
 1   0   2   4
 0   1   3   5
```

Row-reducing P, we get

ECHELON P

1.00E0	8.67E⁻19	2.00E0	4.00E0
0.00E0	1.00E0	3.00E0	5.00E0
0.00E0	1.73E⁻18	5.20E⁻18	5.20E⁻18
⁻8.67E⁻19	⁻8.67E⁻19	⁻2.60E⁻18	⁻2.60E⁻18
0.00E0	⁻8.67E⁻19	⁻3.47E⁻18	⁻6.94E⁻18

The first two columns are the pivot columns. Thus (compare Example 4.17) v_1, v_2 is a basis for the subspace generated by v_1, v_2, v_3, v_4, hence the subspace is a plane. ∎

Subspaces are characterized by special properties that often allow them to be identified in contexts somewhat removed from the purely geometric context we have been using. These properties are illustrated in Figure 4.27.

If the points p, q lie in the flat π, then neither $p + q$ nor tp need lie in π [Figure 4.27(a)]. If π is a subspace, however, then $p + q$ and tp will lie in π [Figure 4.27(b)].

PROPOSITION 4.24 Let S be a set of vectors in $\mathbf{R}^n$. The set S is a subspace of $\mathbf{R}^n$ if and only if

1. The zero vector is in S.
2. If u, v are in S, then $u + v$ is in S.
3. If v is in S and t is a scalar, then tv is in S.

Note: (a) Conditions 2 and 3 are equivalent to the statement: If $u_1, u_2, \ldots, u_n$ are in S and $t_1, t_2, \ldots, t_n$ are scalars, then

$$u = t_1u_1 + t_2u_2 + t_3u_3 + \cdots + t_nu_n$$

is in S.

(b) In the presence of condition 3, condition 1 simply assures that the set S contains at least one vector. For if v is any vector in S and

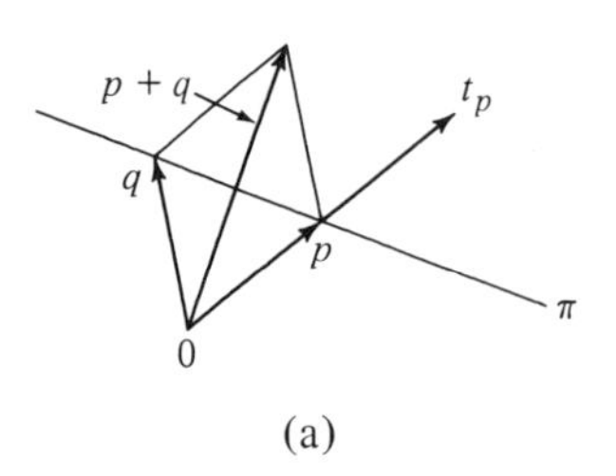

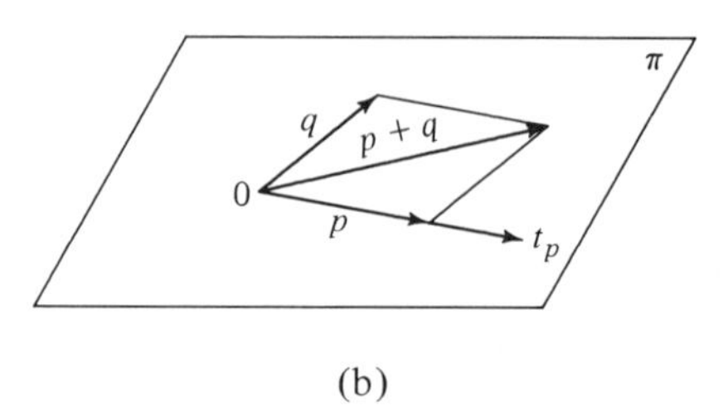

FIGURE 4.27

condition 3 is true, then $0 = 0 \times v$ is also in S. The empty set is not a subspace.

Proof of Proposition 4.24 First assume that S is a subspace. Then condition 1 is true by definition. Suppose that $v_1, \ldots, v_k$ is a basis for S and $P[;i] = v_i$. Then S consists of all vectors $v(T) = PT$ as $T = [t_1, t_2, \ldots, t_k]^T$ varies.

Now, if $u = PT_1$ and $v = PT_2$, then

$$u + v = PT_1 + PT_2 = P(T_1 + T_2)$$
$$tv = tPT_2 = P(tT_2)$$

so statements 2 and 3 are true.

Conversely, assume that statements 1, 2, and 3 are true. The set contains at least one subspace — the 0-flat consisting of the origin. Since S is in $\mathbf{R}^n$, all flats involved have dimension $\leq n$, so there is a k such that S contains a subspace π of dimension k and no subspaces of dimension $k + 1$ $(0 \leq k \leq n)$.

If π is not all of S, then there is a vector v in S but not in π. Let $v_1, \ldots, v_k$ be a basis of π and let π' be the subspace generated by $v_1, v_2, \ldots, v_k, v$. Then the dimension of π' is greater than k (Proposition 4.14), hence π' is not contained in S. But if conditions 2 and 3 hold, then we can show that π' must lie in S, and this contradiction shows that v does not exist — that is, $\pi = S$.

A typical element of π' is

$$\begin{aligned} v(t_1, t_2, \ldots, t_k, t) &= [P\,|\,v]\left[\frac{T}{t}\right] \qquad \text{where } P[;i] = v_i,\ T[i] = t_i \\ &= PT + tv \\ &= u + tv \end{aligned}$$

where u is in π and hence in S. Thus by conditions 2 and 3, $u + tv$ is in S and so π' is in S. ∎

Any matrix A has three important associated subspaces.

DEFINITION 4.7 Let A be an m-by-n matrix.

1. The subspace of $\mathbf{R}^m$ generated by the columns of A is called the *column space* of A.
2. The subspace of $\mathbf{R}^n$ consisting of all vectors v such that $Av = 0$ is called the *null space* of A.
3. The subspace of $\mathbf{R}^n$ generated by the rows of A is called the *row space* of A.

Notice that the row space of A is the column space of A^T. A fourth subspace, the null space of A^T, is sometimes useful but does not have a special name.

The row space and column space are subspaces by definition. The null space is just the set of solutions of $AX = 0$. This is a subspace by Proposition 4.23. It is instructive, however, to use Proposition 4.24 to verify these facts.

EXAMPLE 4.27 Use Proposition 4.24 to verify that the row, column, and null spaces are subspaces.

Solution Since the row space of A is the column space of A^T, it is sufficient to consider null spaces and column spaces.

Null space: Since $A0 = 0$, condition 1 is satisfied. If $Au = 0$ and $Av = 0$, then $A(u + v) = Au + Av = 0 + 0 = 0$, and so condition 2 is satisfied. If t is a scalar and $Av = 0$, then $A(tv) = tAv = t \cdot 0 = 0$, and so condition 3 is satisfied.

Column space: This was done in the Proof of Proposition 4.24. A vector v is in the column space if and only if $v = AT$ for some vector T. Thus, $0 = A0$ shows that condition 1 is satisfied. If $u = AT_1$ and $v = AT_2$, then $u + v = A(T_1 + T_2)$, and so condition 2 is satisfied, and $tu = A(tT_1)$, and so condition 3 is satisfied. ■

In Chapter 3 (Proposition 3.6) it was stated that the matrices A and A^T have the same rank. We can use the concept of row space and column space to prove this. First we prove a preliminary result.

PROPOSITION 4.25 Let A be m-by-n.

1. Let C be n-by-n and invertible. Then A and AC have the same column space.
2. Let B be m-by-m and invertible. Then A and BA have the same row space.

Proof Statement 2 follows from statement 1 by taking transposes (exercise 20). The column space of AC is contained in the column space of A by Proposition 4.12 and, since $A = (AC)C^{-1}$, the reverse inclusion holds as well. ■

Now we are ready for the main result.

PROPOSITION 4.26 Let A be a matrix. Then

$$\text{dim row space } A = \text{rank } A = \text{dim column space } A$$

In particular, rank A = rank A^T.

Proof Let $G = GAUSS\ A$. Then G is invertible and $E = GA$ is the echelon form of A.

Now the dimension of the column space of A is the rank of A (Proposition 4.13) and hence is the number of pivot columns, which is the number of leading 1's in E, which in turn is the number of nonzero rows of E. Since the leading 1's are the only nonzero entries of their column, it follows (see below) that the nonzero rows of E are linearly independent. But since $E = GA$, A and E have the same row space by Proposition 4.25. ■

It is clear that the nonzero rows of an echelon form E are linearly independent from the pattern of zeros and ones involved. We shall need similar observations several times below. The next proposition formalizes the observation.

PROPOSITION 4.27 Let A be a matrix and suppose that there is a vector of row indices V such that $A[V;] = ID \rho V$. Then A has linearly independent columns.

Proof Suppose that $AX = 0$, X a vector. Then

$$\begin{aligned} 0 &= (A + . \times X)[V] \\ &= A[V;] + . \times X \\ &= (ID \rho V) + . \times X \\ &= X \end{aligned}$$

So by Proposition 2.11 A has linearly independent columns. ∎

To apply Proposition 4.27 to the nonzero rows of an echelon form E, suppose that the nonzero rows are $E[W;]$ and the pivot columns are $E[;V]$. Then $E[W; R] = ID \rho V$, so take $A = E[W;]^T$.

The dimensions of the null space and column space are also related.

PROPOSITION 4.28 Let A be a matrix

$$\dim(\text{column space } A) + \dim(\text{null space } A) = \text{number of columns of } A$$

Proof The null space is the set of solutions of $AX = 0$. To prove the proposition, we will just write down a formal description of our solution method. We take $E = ECHELON\ A$ and write down the solutions of $EX = 0$ as follows: Let $E[;V]$ be the *non*pivot columns of E (if there are none, then V is $\iota 0$). The entries of $X[V]$ may be chosen arbitrarily. Say $X[V] = T$. The other coordinates of X may then be written in terms of T. The solutions are then expressed as $X = PT$, where each column of P is derived from a nonpivot column (see exercise 40 and the next example). Since $T = X[V] = (PT)[V] = P[V;]T$ for all choices of T, $P[V] = ID \rho V$ and the columns of P are linearly independent by Proposition 4.27. Thus the dimension of the null space is the number of nonpivot columns.

$$\begin{aligned} &\dim(\text{column space } A) + \dim(\text{null space } A) \\ &\quad = \text{number of pivot columns} + \text{number of nonpivot columns} \\ &\quad = \text{number of columns} \end{aligned}$$ ∎

EXAMPLE 4.28 Find bases of the row space, column space, and null space of

```
      A
 1   ¯1   ¯1    2   11
¯2    3    5   ¯5  ¯23
¯1    0   ¯2    0   ¯4
 1    0    2   ¯1   ¯2
```

Solution

```
      +E←ECHELON A
1.00E0        0.00E0        2.00E0        0.00E0        4.00E0
1.73E¯18      1.00E0        3.00E0        1.73E¯18      5.00E0
0.00E0        0.00E0        0.00E0        1.00E0        6.00E0
0.00E0        1.73E¯18      3.47E¯18      0.00E0        0.00E0
```

The pivot columns are $E[;\ 1\ 2\ 4]$; thus a basis of the column space is $A[;\ 1\ 2\ 4]$ or $(1, -2, -1, 1)$, $(-1, 3, 0, 0)$, $(2, -5, 0, -1)$.

A basis of the row space is $E[\iota 3;]$ or $(1, 0, 2, 0, 4)$, $(0, 1, 3, 0, 5)$, $(0, 0, 0, 1, 6)$. Notice that $E[\iota 3;\ 1\ 2\ 4]$ is $ID\ 3$, as it should be.

To get a basis of the null space we use the nonpivot columns $E[;\ 3\ 5]$. We let $x_3 = t_1$ and $x_5 = t_2$ be arbitrary, and then

$$\begin{aligned} x_1 &= -2t_1 - 4t_2 \\ x_2 &= -3t_1 - 5t_2 \\ x_3 &= t_1 \\ x_4 &= -6t_2 \\ x_5 &= t_2 \end{aligned} \quad \text{or} \quad X = \begin{bmatrix} x_1 \\ x_2 \\ x_3 \\ x_4 \\ x_5 \end{bmatrix} = t_1 \begin{bmatrix} -2 \\ -3 \\ 1 \\ 0 \\ 0 \end{bmatrix} + t_2 \begin{bmatrix} -4 \\ -5 \\ 0 \\ -6 \\ 1 \end{bmatrix} = \begin{bmatrix} -2 & -4 \\ -3 & -5 \\ 1 & 0 \\ 0 & -6 \\ 0 & 1 \end{bmatrix} \begin{bmatrix} t_1 \\ t_2 \end{bmatrix} = PT$$

So a basis of the null space is $(-2, -3, 1, 0, 0)$, $(-4, -5, 0, -6, 1)$. Notice that $P[3\ 5;] = ID\ 2$, as it should. ■

EXERCISES 4.5

In exercises 1 through 5, points $p_0, p_1, \ldots, p_k$ defining a k-flat in $\mathbf{R}^n$ are given. Is the k-flat a subspace? If so, give a basis of the subspace.

1. $p_0 = (-5, 7, 2)$, $p_1 = (-4, 5, 1)$, $p_2 = (-4, 6, 2)$
2. $p_0 = (0, -1, -2)$, $p_1 = (1, -2, -2)$, $p_2 = (1, -1, -1)$
3. $p_0 = (-3, 3, -4, 9)$, $p_1 = (-2, 1, -3, 8)$, $p_2 = (-4, 6, -6, 8)$, $p_3 = (-2, 1, -2, 8)$
4. $p_0 = (1, 1, -2, -1)$, $p_1 = (2, 0, -3, -3)$, $p_2 = (1, 2, -2, -1)$, $p_3 = (-1, 3, 1, 1)$
5. $p_0 = (13, 10, -35, -1, -13)$, $p_1 = (14, 10, -37, 0, -15)$, $p_2 = (14, 11, -37, 0, -15)$, $p_3 = (11, 8, -30, -2, -11)$, $p_4 = (11, 9, -30, -1, -10)$

The sets S given in exercises 6 through 10 are not subspaces, because they are not flats. Which of the conditions of Proposition 4.24 are violated? Give an example of each violation. For example, if condition 3 fails, give a v in S and a scalar t such that tv is not in S.

6. All points in space except the origin.
7. All points (x, y) in the plane such that x and y are integers (positive, negative, or zero).
8. All points in the plane in the first quadrant, axes included.
9. All points in the plane in the first quadrant, axes excluded.
10. All points in the plane in the first and third quadrants, axes included.

11. Let π be the flat defined by $p_0, p_1, \ldots, p_k$.
(a) Show that if there is a single vector v in π and a single scalar $t \neq 1$ such that tv is in π, then π is a subspace.

Hint: From $v = p_0 + PT_1$ and $tv = p_0 + PT_2$, show $0 = p_0 + PT$ has a solution.

(b) Show that if there are two vectors u, v in π such that $u + v$ is in π, then π is a subspace.

In exercises 12 through 19 a matrix A is given. Find bases of the row space of A, the column space of A, and the null space of A. What is the dimension of the null space at A^T?

12. $\begin{bmatrix} 1 & 1 & 5 & 0 \\ -2 & -1 & -7 & 1 \\ -1 & 0 & -2 & 2 \end{bmatrix}$

13. $\begin{bmatrix} -2 & -2 & -10 & 1 \\ 3 & 2 & 12 & -1 \\ -1 & -1 & -5 & 1 \end{bmatrix}$

14. $\begin{bmatrix} -1 & 1 & 1 & 1 \\ 0 & 1 & 3 & 5 \\ 2 & 0 & 4 & 8 \end{bmatrix}$

15. $\begin{bmatrix} 1 & -1 & -1 & -1 \\ 0 & 1 & 3 & 5 \\ -2 & 2 & 2 & 2 \end{bmatrix}$

16. $\begin{bmatrix} 1 & 2 & -2 & 3 \\ -2 & -4 & 5 & -7 \\ 1 & 2 & -4 & 5 \\ -2 & -4 & 2 & -4 \end{bmatrix}$

17. $\begin{bmatrix} 8 & 16 & 2 & 6 \\ 2 & 4 & -1 & 3 \\ 9 & 18 & 2 & 7 \\ 3 & 6 & 0 & 3 \end{bmatrix}$

18. $\begin{bmatrix} 14 & 28 & 7 & 7 & -7 \\ 3 & 6 & 2 & 1 & -1 \\ 14 & 28 & 7 & 7 & -7 \\ 6 & 12 & 3 & 3 & -3 \end{bmatrix}$

19. $\begin{bmatrix} 1 & 2 & -1 & 2 & -2 \\ -1 & -2 & 2 & -3 & 3 \\ -1 & -2 & 0 & -1 & 1 \\ -2 & -4 & 0 & -2 & 2 \end{bmatrix}$

20. Assuming the first statement of Proposition 4.25, prove the second statement. Show that A and BA have the same null space as well.

21. Do the flats of exercise 12 in Exercises 4.4 intersect? Are they parallel?

22. Do the flats of exercise 13 in Exercises 4.4 intersect? Are they parallel?

23. Do the flats of exercise 14 in Exercises 4.4 intersect? Are they parallel?

24. Do the flats of exercise 15 in Exercises 4.4 intersect? Are they parallel?

25. Let A be an m-by-n matrix of rank r.

(a) Let $H = (r, 0) \downarrow GAUSS\ A$. Show that the columns of H^T are a basis of the null space of A^T.

Hint: Show that $HA = 0$, H^T has independent columns, and apply Propositions 4.26 and 4.28.

(b) Show that the columns of $((r, 0) \downarrow GAUSS\ A^T)^T$ are a basis of the null space of A.

Hint: Apply (a) to A^T.

(Computer assignment) In exercises 26 through 33 use the result of exercise 25(b) to complete the null space of the given matrix.

26. The matrix of exercise 12
27. The matrix of exercise 13
28. The matrix of exercise 14
29. The matrix of exercise 15
30. The matrix of exercise 16
31. The matrix of exercise 17
32. The matrix of exercise 18
33. The matrix of exercise 19

34. Let $F: \mathbf{R}^n \rightarrow \mathbf{R}^m$ be a linear transformation. Let S be a subspace of $\mathbf{R}^m$ and let $F^{-1}(S)$ be the set of vectors v in $\mathbf{R}$ such that $F(v)$ lies in S. Show that $F^{-1}(S)$ is a subspace of $\mathbf{R}^n$.

Hint: Use Proposition 4.24.

In exercises 35 through 37 the matrix A of a linear transformation $F: \mathbf{R}^n \rightarrow \mathbf{R}^m$ is given and a subspace S of $\mathbf{R}^m$ is given. Find a basis of $F^{-1}(S)$.

Hint: If S is the set of vectors $v(T) = BT$, then $F^{-1}(S)$ is the set of vectors X for which there exists a vector Y such that $AX = BY$ — that is, for which $[A \mid -B][X \mid Y]^T = 0$.

35. $A = \begin{bmatrix} 1 & 1 & 5 \\ -2 & -1 & -7 \\ -1 & 0 & -2 \end{bmatrix}$, S generated by $(-9, 13, 4)$ and $(-13, 19, 6)$

36. $A = \begin{bmatrix} 1 & -1 & -1 \\ 0 & 1 & 3 \\ -2 & 2 & 2 \end{bmatrix}$, S generated by $(-1, 0, 1)$, $(-5, -5, 4)$

37. $A = \begin{bmatrix} -1 & -2 & -8 \\ 2 & 2 & 10 \\ -1 & -1 & -5 \end{bmatrix}$, S generated by $(-1, 1, -1)$

In exercises 38 through 57 define the APL function described.

38. Name: *NBASE*
Right argument: A matrix E in echelon form
Left argument: The vector V of indices of pivot columns of E
Result: A basis of the null space of E

Suggestion: `1 0 1 0 0 1 \1 2 3` is `1 0 2 0 0 3`. "\" is called *expand*. To expand the number of rows of a matrix A, use ⍀ instead of \.

39. Name: *PIVS*
Right argument: A matrix E in echelon form
Left argument: A scalar M. The component $B = E[i; j]$ is assumed to be zero if `M=M+B` (M sets the scale for fuzzy comparisons.)
Result: The indices of the pivot columns of E

40. Name: *NULLSPACE*
Right argument: A matrix A
Result: A matrix B whose columns are a basis of the null space of A

(Note: You may assume that the functions of exercises 38 and 39 are defined.

41. Name: *COLSPACE*
Right argument: A matrix A
Result: A basis of the column space of A

Note: You may assume that the function *PIVS* of exercise 39 is defined.

42. Name: *CUTSPACE*
Right argument: A matrix B
Left argument: A matrix A with $(\rho B)[1] = (\rho A)[1]$
Result: A basis of the intersection of the column space of A with the column space of B

Note: You may assume that the functions *COLSPACE* of exercise 41 and *NULLSPACE* of exercise 40 are defined.

43. Write a different version of *NULLSPACE* (exercise 40) based on exercise 25.

44. Let H be a hyperplane and π a flat in $\mathbf{R}^n$. Show that if H and π do not intersect, then they must be parallel.

Hint: Let H be given by $p(T) = p_0 + PT$ and let π be given by $q(S) = q_0 + QS$, where P has rank $n - 1$ and the rank of Q equals the dimension of π. Consider the possible placement of pivot columns in $[P \mid -Q \mid q_0 - p_0]$.

CHAPTER FIVE

Orthogonality

In this chapter we introduce the concepts of distance and angle for vectors in $\mathbf{R}^n$ and define orthonormal coordinate systems. These coordinate systems have mutually perpendicular axes and are extremely important in applications.

The concepts of "distance" and "angle" are defined in Section 5.1 via the dot product of two vectors. Distance-preserving mappings and coordinate changes are also discussed.

In Section 5.2 we show that any symmetric matrix may be diagonalized by a distance-preserving coordinate change. This introduces the subject of eigenvalues and eigenvectors for symmetric matrices. Eigenvalues of general matrices are taken up in Chapter 7. The diagonalization of a symmetric matrix is a fundamental technique in optimization (the second-derivative test, Section 5.3*), physics (Section 5.6*), and statistics.

In Section 5.4 we develop formulas for perpendicular projections and reflections and relate them to the least-squares calculations of Chapter 1.

In Section 5.5 the Householder algorithm is developed and used to automate the procedures of Section 5.4. Section 5.5 also contains a function *BACKSUB* that automates the process of writing down the solutions of a linear system, once the echelon form is obtained.

5.1 Distance and Angle

In this section the concepts of "distance" and "angle" in the plane and space will be extended to higher-dimensional spaces. The idea is to express "distance" and "angle" in purely algebraic terms and then use the algebraic formulations as definitions in higher dimensions.

The algebraic formulation is based on the dot product of two vectors.

DEFINITION 5.1 Let v and w be vectors in $\mathbf{R}^n$. The *dot product* of v and w is $v +.\times w$. Alternate notations are $v \cdot w$ and, if v and w are considered to be column vectors, $v^T w = w^T v$.

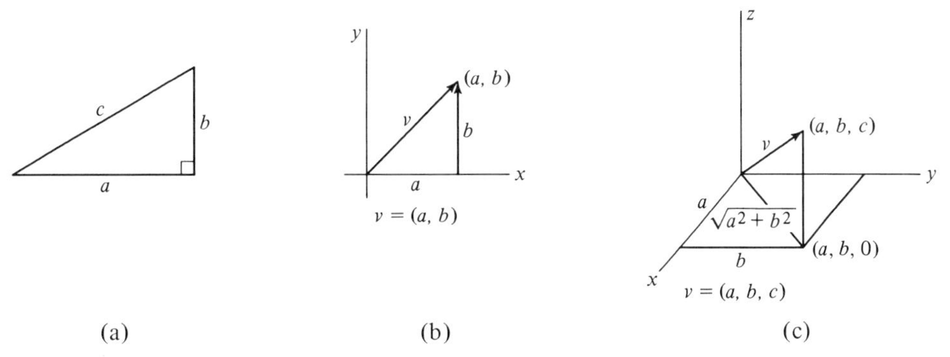

FIGURE 5.1

To write the distance between two points in terms of the dot product we start with the Pythagorean theorem, which states that if the lengths of the two legs of a right triangle are a and b and if the length of the hypotenuse is c, then $a^2 + b^2 = c^2$ [Figure 5.1(a)]. Applying this result to the vector $v = (a, b)$ in the plane, we have $(\text{length } v)^2 = a^2 + b^2$ [Figure 5.1(b)]. Applying the result twice to the vector $v = (a, b, c)$ in space, we have $(\text{length } v)^2 = (\sqrt{a^2 + b^2})^2 + c^2 = a^2 + b^2 + c^2$ [Figure 5.1(c)].

Note that if $v = (v_1, v_2, v_3, \ldots, v_n)$, then $v \cdot v = v_1^2 + v_2^2 + \cdots + v_n^2$.

DEFINITION 5.2 Let v be a vector in $\mathbf{R}^n$. The *length* or *norm* of v is $\|v\| = \sqrt{v \cdot v}$.

To define the distance between two points p and q we note that it should be the length of $q - p$ (Figure 5.2).

DEFINITION 5.3 The *distance* $d(p, q)$ between p and q is $d(p, q) = \|q - p\| = \sqrt{(q - p) \cdot (q - p)}$.

To relate angles to the dot product we use the law of cosines from trigonometry. Recall that the law of cosines extends the Pythagorean theorem to triangles that are not right triangles by adding a correction factor:

$$c^2 = a^2 + b^2 - 2ab \cos \theta$$

where θ is the angle opposite the side of length c (Figure 5.3). If $\theta = \pi/2$, then $\cos \theta = 0$, and we have the Pythagorean theorem. The law of cosines is easily

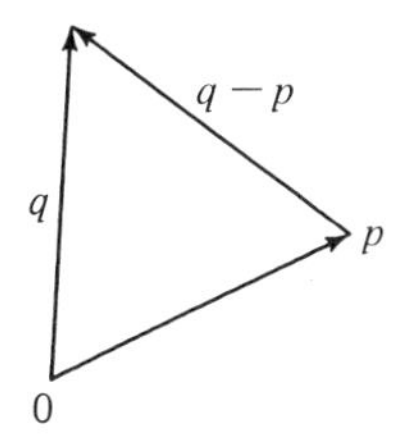

FIGURE 5.2

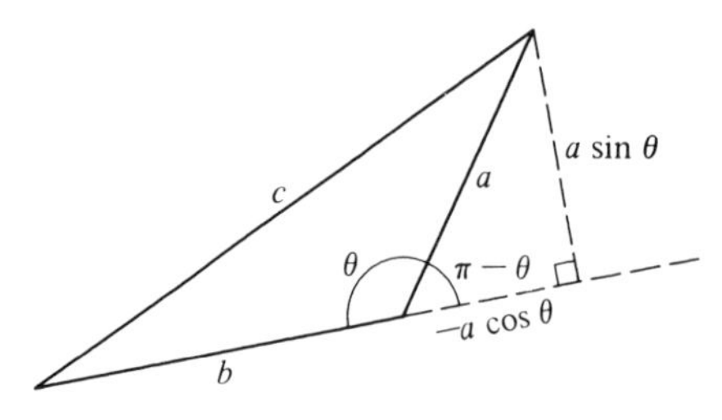

FIGURE 5.3

derived from the Pythagorean theorem. By dropping the perpendicular indicated in Figure 5.3 and using some elementary trigonometry (Exercise 50), one finds

$$(a \sin \theta)^2 + (b - a \cos \theta)^2 = c^2$$

which simplifies to the law of cosines.

Now refer again to Figure 5.2. According to the law of cosines

$$\|q - p\|^2 = \|q\|^2 + \|p\|^2 - 2\|q\|\,\|p\| \cos \theta$$

where θ is the angle between the vectors p and q. When we write p and q as column vectors, this equation becomes

$$\begin{aligned} (q - p)^T(q - p) &= q^Tq + p^Tp - 2\|q\|\,\|p\| \cos \theta \\ q^Tq - q^Tp - p^Tq + p^Tp &= q^Tq + p^Tp - 2\|q\|\,\|p\| \cos \theta \end{aligned}$$

or

$$-2p \cdot q = -2\|p\|\,\|q\| \cos \theta$$

since

$$p^Tq = p \cdot q = q^Tp$$

Thus

$$p \cdot q = \|p\|\,\|q\| \cos \theta$$

If $\|p\|\,\|q\| \neq 0$, this gives

$$\theta = \cos^{-1}\left(\frac{p \cdot q}{\|p\|\,\|q\|}\right) = \cos^{-1}\left[\frac{p \cdot q}{\sqrt{(p \cdot p)(q \cdot q)}}\right]$$

This equation expresses the angle between two vectors in the plane or in space in terms of dot products. In particular *the vectors p and q are perpendicular if and only if $p \cdot q = 0$.*

We would like to use the formula

$$\cos \theta = \frac{p \cdot q}{\|p\|\,\|q\|}$$

to define the angle between the vectors p and q in $\mathbf{R}^n$. There is a problem that must be overcome first. Since $-1 \leq \cos\theta \leq 1$ for any angle θ, the same must be true for $(p \cdot q)/\|p\|\,\|q\|$ or the definition will be nonsense.

The next proposition collects this fact and several other useful results together for reference.

PROPOSITION 5.1 Let v and w be vectors in $\mathbf{R}^n$.

1. $\|v\| \geq 0$ and $\|v\| = 0$ if and only if $v = 0$.
2. $\|\alpha v\| = |\alpha|\,\|v\|$, α any scalar.
3. $|v \cdot w| \leq \|v\|\,\|w\|$ (Schwartz inequality).
4. $\|v + w\| \leq \|v\| + \|w\|$ (triangle inequality).

Proof We leave statements 1 and 2 as Exercises 51 and 52.

To prove the Schwartz inequality we start with the fact that for any scalar x, $0 \leq \|v - xw\|^2$,

$$\begin{aligned} f(x) &= \|v - xw\|^2 \\ &= (v - xw)^T(v - xw) \\ &= v^Tv - 2xv^Tw + x^2w^Tw \\ &= \|w\|^2x^2 - 2(v \cdot w)x + \|v\|^2 \\ &\geq 0 \end{aligned}$$

By the quadratic formula, $f(x) = 0$ when (if $\|w\| = 0$ there is nothing to prove)

$$x = \frac{v \cdot w \pm \sqrt{(v \cdot w)^2 - \|v\|^2\|w\|^2}}{\|w\|^2}$$

Since f has at most one real root, $(v \cdot w)^2 - \|v\|^2\|w\|^2 \leq 0$. Statement 3 follows.

To prove the triangle inequality,

$$\begin{aligned} \|v + w\|^2 &= (v + w) \cdot (v + w) \\ &= \|v\|^2 + 2v \cdot w + \|w\|^2 \\ &\leq \|v\|^2 + 2\|v\|\,\|w\| + \|w\|^2 \qquad \text{by the Schwartz inequality} \\ &= (\|v\| + \|w\|)^2 \end{aligned}$$

Thus $\|v + w\| \leq \|v\| + \|w\|$. ■

The triangle inequality is an algebraic version of the fact that the length of one side of a triangle is always less than or equal to the sum of the lengths of the other two sides (Figure 5.4).

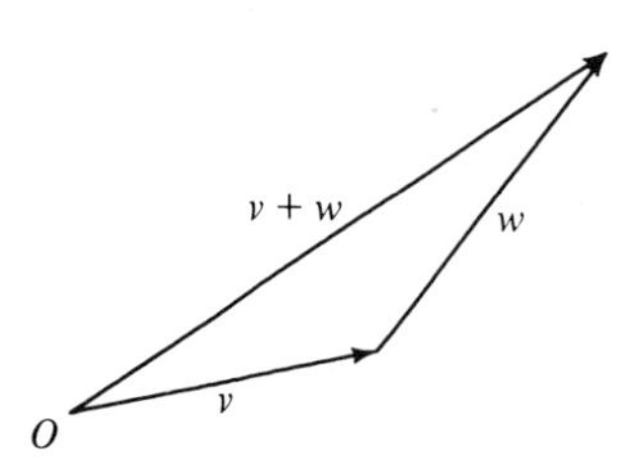

FIGURE 5.4

DEFINITION 5.4 Let v, w be vectors in $\mathbf{R}^n$.

1. The vectors v and w are *perpendicular,* written $v \perp w$, if $v \cdot w = 0$.
2. The angle between v and w is

$$\theta = \begin{cases} \cos^{-1}\left(\dfrac{v \cdot w}{\|v\|\ \|w\|}\right) & \text{if } \|v\|\ \|w\| \neq 0 \\ \text{undefined} & \text{if } \|v\|\ \|w\| = 0 \end{cases}$$

Notice that the zero vector is the only one perpendicular to all vectors or perpendicular to itself.

EXAMPLE 5.1 What is the angle between the line through (1, 2) and (−3, 4) and the line through (−1, 2) and (1, 1)?

Solution Let $p_0 = (1, 2)$, $p_1 = (-3, 4)$, $q_0 = (-1, 2)$, and $q_1 = (1, 1)$. Then the vector $v = p_1 - p_0 = (-4, 2)$ is parallel to the first line and the vector $w = q_1 - q_0 = (2, -1)$ is parallel to the second line. The question is slightly ambiguous, since there are two supplementary angles involved. One angle is obtained using v and w or $-v$ and $-w$ and the other is obtained using $-v$ and w or v and $-w$. The two angles will be

$$\theta = \cos^{-1}\left(\pm\frac{v \cdot w}{\|v\|\ \|w\|}\right)$$

The acute angle is given by $\cos^{-1}(|v \cdot w|/\|v\| \cdot \|w\|)$, and we will take this as the solution.

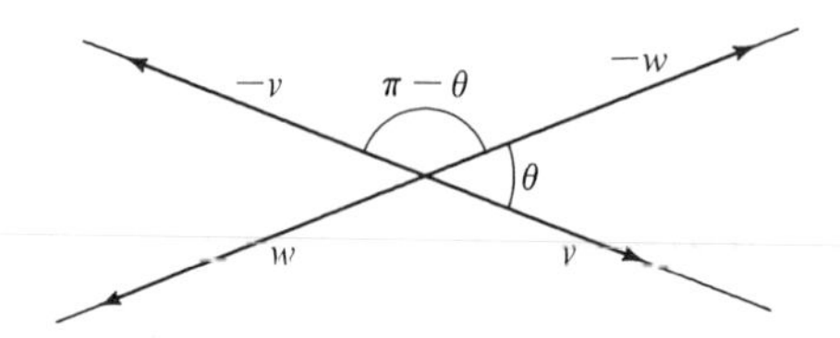

```
        V←¯4 1
        W←2 ¯1
        +RAD←¯2○|V+.×W÷((V+.*2)×W+.*2)*÷2
0.219
        +DEG←RAD÷○÷180
12.5
```

Thus the angle is approximately .22 radians or 12.5 degrees. ■

EXAMPLE 5.2 Find a vector perpendicular to (1, 2, 3) in $\mathbf{R}^3$.

Solution The easy answer, of course, is $v = 0$, since the zero vector is perpendicular to all vectors — even itself. To find a nonzero vector perpendicular to (1, 2, 3) we proceed as follows.

Let $w = (x, y, z)$ be perpendicular to $v = (1, 2, 3)$. Then

$$0 = v \cdot w = x + 2y + 3z$$

Thus the set of vectors perpendicular to (1, 2, 3) is the plane $x + 2y + 3z = 0$. Since $0 \perp v$, this plane is a subspace, the null space of the matrix $[1\ \ 2\ \ 3]$, which is already in echelon form. So

$$w = \begin{bmatrix} x \\ y \\ z \end{bmatrix} = t_1 \begin{bmatrix} -2 \\ 1 \\ 0 \end{bmatrix} + t_2 \begin{bmatrix} -3 \\ 0 \\ 1 \end{bmatrix}$$

and any choice of (t_1, t_2) — for example, (1, 0) or (0, 1) — will give a vector perpendicular to (1, 2, 3). ■

EXAMPLE 5.3 Let the triangle ABC be isosceles with side AB equal to side AC. Show that angle ABC equals angle ACB.

Solution Let $u = A - B$, $v = C - A$. Then $\|u\| = \|v\|$ and $C - B = u + v$. If $\theta_1 = \angle ABC$ and $\theta_2 = \angle ACB$, then

$$\begin{aligned}
\cos\theta_1 &= \frac{u \cdot (u + v)}{\|u\|\,\|u + v\|} = \frac{u \cdot u + u \cdot v}{\|u\|\,\|u + v\|} = \frac{\|u\|^2 + u \cdot v}{\|u\|\,\|u + v\|} \\
&= \frac{\|v\|^2 + u \cdot v}{\|v\|\,\|u + v\|} = \frac{v \cdot v + u \cdot v}{\|v\|\,\|u + v\|} = \frac{v \cdot (u + v)}{\|v\|\,\|u + v\|} \\
&= \cos\theta_2
\end{aligned}$$

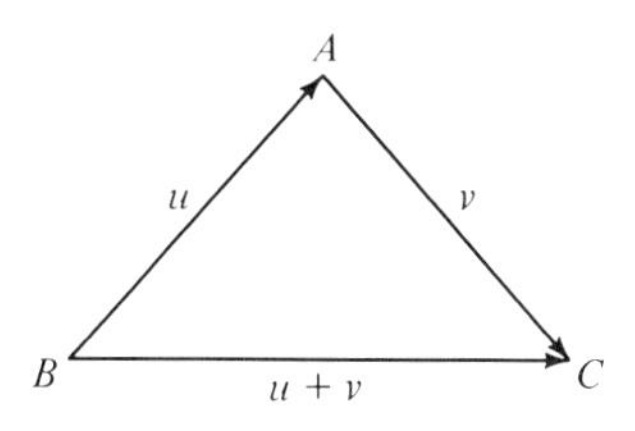

■

EXAMPLE 5.4 Is the triangle in $\mathbf{R}^4$ with vertices $p_1 = (2, 0, 2, 0)$, $p_2 = (5, 3, 5, 3)$, and $p_3 = (9, -1, 9, -1)$ a right triangle?

Solution Let $u = p_1 - p_2 = (-3, -3, -3, -3)$
$v = p_3 - p_1 = (7, -1, 7, -1)$

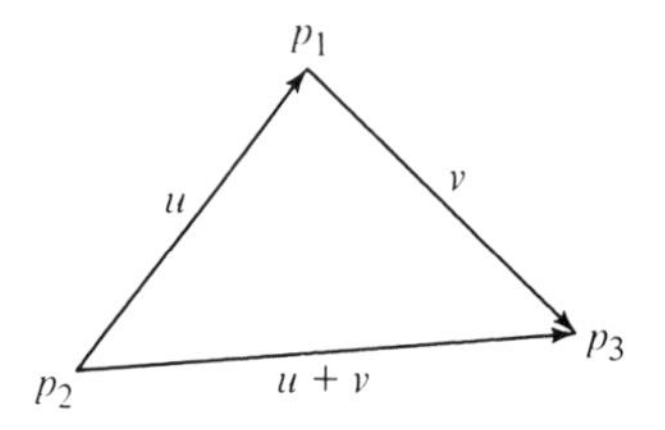

Then $p_3 - p_2 = u + v = (4, -4, 4, -4)$. The dot products among u, v, and $u + v$ may all be calculated at once by forming $A = [u\,|\,v\,|\,u + v]$ and computing $A^T A$. This is because

$$\begin{aligned}(A^T A)[i;j] &= A^T[i;] + . \times A[;j] \\ &= A[;i] + . \times A[;j] \\ &= A[;i] \cdot A[;j]\end{aligned}$$

```
      A
¯3   7   4
¯3  ¯1  ¯4
¯3   7   4
¯3  ¯1  ¯4
      (⍉A)+.×A
 36.00   ¯36.00    0.00
¯36.00   100.00   64.00
  0.00    64.00   64.00
```

Note that $A[;1] \cdot A[;3] = u \cdot (u + v)$ is zero, hence $u \perp u + v$ — that is, side BC is perpendicular to side AC. ■

In the example above, $(A^T A)[i;i] = \|A[;i]\|^2$ and $64 + 36 = 100$; that is, the triangle satisfies the Pythagorean theorem. The next proposition shows that this gives an alternate proof that the triangle is a right triangle.

PROPOSITION 5.2 (*Pythagorean theorem*) Let u, v be vectors in $\mathbf{R}^n$. Then $u \perp v$ if and only if

$$\|u + v\|^2 = \|u\|^2 + \|v\|^2$$

Proof

$$\begin{aligned}\|u + v\|^2 &= (u + v)^T(u + v) = u^Tu + v^Tv + u^Tv + v^Tu \\ &= \|u\|^2 + \|v\|^2 + 2u \cdot v\end{aligned}$$ ■

The trick used in Example 5.4 for computing the dot products is quite important. To get information about the lengths and angles among a set of vectors one stores the vectors as columns of a matrix A and forms the symmetric matrix A^TA. The $i;j$ entry of A^TA is the dot product of $A[;i]$ and $A[;j]$. If this entry is zero, then $A[;i] \perp A[;j]$; if positive, then the angle between $A[;i]$ and $A[;j]$ is acute; and if negative, then the angle between $A[;i]$ and $A[;j]$ is obtuse. The $i;i$ entry of A^TA is $\|A[;i]\|^2$. To obtain the cosine of the angle between $A[;i]$ and $A[;j]$ divide $(A^TA)\ [i;j]$ by $\|A[;i]\|\ \|A[;j]\|$. The quickest way to compute this is

```
I←J←(1 1⍉B←(⍉A)+.×A)*÷2
cos←B÷I∘.×J
```

The angles between vectors in high-dimensional spaces are often used in statistical analyses, although they are not usually thought of as such.

THE CORRELATION COEFFICIENT

Let $x_1, x_2, \ldots, x_n$ be the measurements of one characteristic (e.g., height in inches) of the n individuals in a population and let $y_1, y_2, \ldots, y_n$ be the measurements of another characteristic of the population (e.g., weight in pounds). Let $\bar{x}$ be the average of the x's and $\bar{y}$ the average of the y's. The *correlation coefficient* is usually defined as

$$r_{xy} = \frac{\sum (x_i - \bar{x})(y_i - \bar{y})}{\left[\sum (x_i - \bar{x})^2 \sum (y_i - \bar{y})^2\right]^{1/2}}$$

and is considered to be a measure of the "relatedness" of the two measurements.

The formula for r_{xy} shows that it is $\cos\theta$ for some angle θ in $\mathbf{R}^n$ — but what angle?

Recall (Section 1.3) that, owing to the often arbitrary nature of statistical scores, measurements are often reduced to standard scores or z-scores before being analyzed statistically. The vector of standard scores for the x measurements is $z_x = (X - \bar{x}) \div \sigma_x$ and that for the y measurements is $z_y = (Y - \bar{y}) \div \sigma_y$, where $X[i] = x_i$, $Y[i] = y_i$, σ_x is the standard deviation of X, and σ_y is the standard deviation of Y.

Notice that

$$\sigma_x = \sqrt{\frac{\sum (x_i - x)^2}{n}} = \frac{\|X - \bar{x}\|}{\sqrt{n}}$$

so

$$\begin{aligned} \|z_x\| &= \left\| \frac{(X - \bar{x})}{\sigma_x} \right\| \\ &= \frac{1}{|\sigma_x|} \|X - \bar{x}\| \qquad \text{by Proposition 5.1, statement 2} \\ &= \frac{\sqrt{n}}{\|X - \bar{x}\|} \|X - \bar{x}\| \\ &= \sqrt{n} \end{aligned}$$

Similarly, $\|z_y\| = \sqrt{n}$. In fact, all z-scores have the same length in $\mathbf{R}^n$ and so differ only in direction.

The cosine of the angle between z_x and z_y is

$$\cos\theta = \frac{z_x \cdot z_y}{\|z_x\| \, \|z_y\|} = \frac{1}{n} z_x \cdot z_y = \frac{1}{n} \frac{(X - \bar{x})}{\sigma_x} \cdot \frac{(x - \bar{y})}{\sigma_y}$$

$$\cos\theta = \frac{(X - \bar{x}) \cdot (Y - \bar{y})}{\|X - \bar{x}\| \, \|Y - \bar{y}\|} = r_{xy}$$

Thus the correlation coefficient is the cosine of the angle in $\mathbf{R}^n$ between the z-scores. The measurements X and Y are *perfectly correlated* when $z_x = \pm z_y$ and *uncorrelated* when $z_x \perp z_y$.

When correlations are viewed as angles, certain properties become geometrically evident (see exercise 53).

The next proposition gives some useful results involving matrices and dot products.

PROPOSITION 5.3 Let v and w be vectors in $\mathbf{R}^n$ and let A be n by n.

1. $v \cdot (Aw) = (A^T v) \cdot w$.
2. If $(Av) \cdot w = v \cdot (Aw)$ for all v and w in $\mathbf{R}^n$, then A is symmetric.
3. If $(Av) \cdot (Aw) = v \cdot w$ for all v and w in $\mathbf{R}^n$, then A is invertible and $A^{-1} = A^T$.

Proof

1. Writing v and w as column vectors gives

$$v \cdot (Aw) = v^T A w = (A^T v)^T w$$

2. Notice that if $v = (ID\ n)[;i]$ and $w = (ID\ n)[;j]$, then

$$v \cdot (Aw) = v^T A[;j] = A[i;j]$$

So $A[i;j] = v \cdot (Aw) = (Av) \cdot w = w \cdot (Av) = A[j;i]$; that is, $A = A^T$.

3. $v \cdot w = (Av) \cdot (Aw) = v \cdot (A^T(Aw))$ by statement 1. Now take v and w as in the proof of statement 2 and set $I = ID\ n$. $I[i;j] = v \cdot Iw = v \cdot w = v \cdot (A^T Aw) = (A^T A)[i;j]$. So $A^T A = I$. ∎

DEFINITION 5.5 A matrix is *orthogonal* if A^T is an inverse for A.

For example, the rotation matrix

$$R_\theta = \begin{bmatrix} \cos\theta & -\sin\theta \\ \sin\theta & \cos\theta \end{bmatrix}$$

is orthogonal. Notice that a rotation of the plane does not change the distances between points or alter the angles between lines.

DEFINITION 5.6 Let $f: \mathbf{R}^n \to \mathbf{R}^n$ be a function.

1. We say that f *preserves dot products* if $v \cdot w = f(v) \cdot f(w)$ for all pairs of vectors v, w in $\mathbf{R}^n$.

2. We say that f *preserves distances,* or that it is an *isometry* or a *congruence,* if $d(f(p), f(q)) = d(p, q)$ for all points p, q in $\mathbf{R}^n$.

3. We say that f *preserves angles,* or is a *similarity,* if given three points p_0, p_1, p_2 in $\mathbf{R}$ the angle between $p_1 - p_0$ and $p_2 - p_0$ is the same as the angle between $f(p_1) - f(p_0)$ and $f(p_2) - f(p_0)$ (Figure 5.5).

A similarity need not preserve distances (exercise 44), but it can be shown that an isometry necessarily preserves angles. The next proposition shows this for affine transformations.

PROPOSITION 5.4 Let $Y = f(X) = B + AX$ be an affine transformation from $\mathbf{R}^n$ to $\mathbf{R}^n$.

1. If A is orthogonal, then f preserves distances and angles.
2. If f is an isometry, then A is orthogonal.

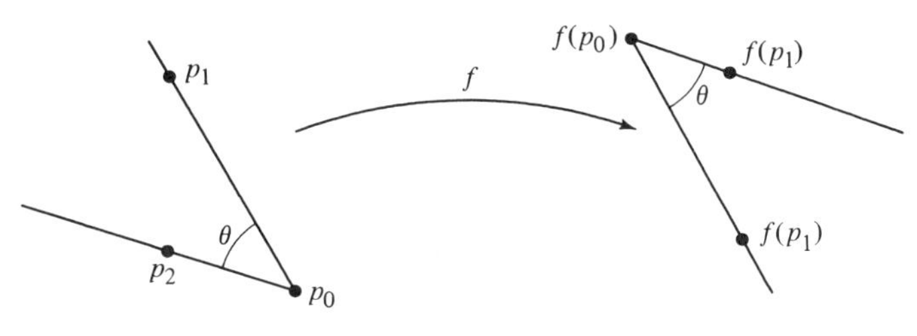

FIGURE 5.5

Proof Statement 1 is exercise 39. We proceed to statement 2. Let p and q be points in $\mathbf{R}^n$.

$$d(f(p), f(q))^2 = \|f(p) - f(q)\|^2 = \|B + Ap - (B + Aq)\|^2$$
$$= \|A(p - q)\|^2 = (A(p - q)) \cdot (A(p - q))$$
$$d(p \cdot q)^2 = \|p - q\|^2 = (p - q) \cdot (p - q)$$

In particular, if we take $p = 0$ and $q = v$, we have

$$v \cdot v = (Av) \cdot (Av)$$

for every vector v in $\mathbf{R}^n$. If w is a second vector in $\mathbf{R}^n$, then

$$[A(v + w)] \cdot [A(v + w)] = (Av) \cdot (Av) + 2(Av) \cdot (Aw) + (Aw) \cdot (Aw)$$

Hence for every pair of vectors v, w in $\mathbf{R}^n$

$$(Av) \cdot (Aw) = \tfrac{1}{2}\{[A(v + w)] \cdot [A(v + w)] - (Av) \cdot (Av) - (Aw) \cdot (Aw)\}$$
$$= \tfrac{1}{2}[(v + w) \cdot (v + w) - v \cdot v - w \cdot w]$$
$$= v \cdot w.$$

Thus A is orthogonal by Proposition 5.3. ■

EXAMPLE 5.5 Which of the following affine functions are isometries?

1. $f\colon \mathbf{R}^n \to \mathbf{R}^n$ is a translation.
2. $f\colon \mathbf{R}^2 \to \mathbf{R}^2$ is a 30-degree rotation clockwise.
3. $f\colon \mathbf{R}^3 \to \mathbf{R}^3$ is a reflection in the xy plane.
4. $f\colon \mathbf{R}^n \to \mathbf{R}^n$ is $Y = f(X) = 3X$.
5. $f\colon \mathbf{R}^4 \to \mathbf{R}^4$ is

$$f(x_1 x_2, x_3, x_4) = \tfrac{1}{2}\begin{bmatrix} 4 + x_1 + x_2 + x_3 + x_4 \\ -2 + x_1 - x_2 + x_3 - x_4 \\ x_1 + x_2 - x_3 - x_4 \\ 6 + x_1 - x_2 - x_3 + x_4 \end{bmatrix}$$

Solution

1. A translation is of the form $f(X) = B + X$ — that is, $A = ID\,n$ — so certainly $A^{-1} = A^T$ and f is an isometry.
2. $Y = f(X) = R_\theta X$ with $\theta = -\pi/6$. Since $R_\theta{}^{-1} = R_\theta{}^T$, this is an isometry.
3. A reflection in the xy plane takes (x, y, z) to $(x, y, -z)$, so

$$Y = f(X) = \begin{bmatrix} 1 & 0 & 0 \\ 0 & 1 & 0 \\ 0 & 0 & -1 \end{bmatrix} X$$

$$\begin{bmatrix} 1 & 0 & 0 \\ 0 & 1 & 0 \\ 0 & 0 & -1 \end{bmatrix}^T \begin{bmatrix} 1 & 0 & 0 \\ 0 & 1 & 0 \\ 0 & 0 & -1 \end{bmatrix} = \begin{bmatrix} 1 & 0 & 0 \\ 0 & 1 & 0 \\ 0 & 0 & -1 \end{bmatrix} \begin{bmatrix} 1 & 0 & 0 \\ 0 & 1 & 0 \\ 0 & 0 & -1 \end{bmatrix}$$

$$= \begin{bmatrix} 1 & 0 & 0 \\ 0 & 1 & 0 \\ 0 & 0 & 1 \end{bmatrix}$$

Hence f is an isometry.

4. In this case $Y = AX$ with

$$A = \begin{bmatrix} 3 & 0 & 0 \\ 0 & 3 & 0 \\ 0 & 0 & 3 \end{bmatrix} = 3I$$

and hence $A^TA = 9I \neq I$. Thus f is not an isometry. (f is a similarity — see Exercise 45.)

5.

$$Y = f(X) = \begin{bmatrix} 2 \\ -1 \\ 0 \\ 3 \end{bmatrix} + \tfrac{1}{2} \begin{bmatrix} 1 & 1 & 1 & 1 \\ 1 & -1 & 1 & -1 \\ 1 & 1 & -1 & -1 \\ 1 & -1 & -1 & 1 \end{bmatrix} X = B + AX$$

and

$$A^TA = \tfrac{1}{4} \begin{bmatrix} 1 & 1 & 1 & 1 \\ 1 & -1 & 1 & -1 \\ 1 & 1 & -1 & -1 \\ 1 & -1 & -1 & 1 \end{bmatrix} \begin{bmatrix} 1 & 1 & 1 & 1 \\ 1 & -1 & 1 & -1 \\ 1 & 1 & -1 & -1 \\ 1 & -1 & -1 & 1 \end{bmatrix}$$

$$= \tfrac{1}{4} \begin{bmatrix} 4 & 0 & 0 & 0 \\ 0 & 4 & 0 & 0 \\ 0 & 0 & 4 & 0 \\ 0 & 0 & 0 & 4 \end{bmatrix} = I$$

Thus f is an isometry. ■

Although distance is defined in terms of the dot product, a function need not preserve dot products in order to preserve distances. In fact any translation is an isometry, but no nontrivial translation will preserve dot products. If

$f(X) = a + X$, a a vector in $\mathbf{R}^n$, then

$$\begin{aligned} f(v)\cdot f(w) &= (a+v)\cdot(a+w) \\ &= a\cdot a + a\cdot(v+w) + v\cdot w \end{aligned}$$

If we restrict our attention to linear transformations, however, the concepts coincide.

PROPOSITION 5.5 Let $Y = f(X) = AX$ be a linear transformation from $\mathbf{R}^n$ to $\mathbf{R}^n$. Then the following statements are equivalent.

1. The function f preserves dot products.
2. The function f is an isometry.
3. The matrix A is orthogonal.

Proof Since distance is defined in terms of the dot product, statement 1 implies statement 2. Statement 2 implies statement 3 by Proposition 5.4. Suppose that the matrix A is orthogonal. Then by Proposition 5.3, statement 1,

$$(Av)\cdot(Aw) = (A^TAv)\cdot w = v\cdot w$$

so statement 3 implies statement 1. ■

COORDINATE CHANGES

Next we wish to discuss the way in which a coordinate change affects the formulas for distance and angle. Because of the way in which translations interfere with the dot product (see the discussion preceding Proposition 5.5 above), we begin by considering coordinate changes in which the origin remains fixed.

DEFINITION 5.7

1. An affine coordinate change $X = p_0 + PX'$ is called *linear* if $p_0 = 0$. That is, the new origin coincides with the old origin.
2. A linear coordinate change is said to *preserve dot products* if $X'\cdot Y' = X\cdot Y$ for all vectors of coordinates X, Y.

The next proposition is immediate from Proposition 5.3, statement 3.

PROPOSITION 5.6 Let $X' = PX$ be a linear coordinate change in $\mathbf{R}^n$. Then $X'^TY' = X^T(P^TP)Y$. In particular the coordinate change preserves dot products if and only if P is orthogonal. ■

The requirement that P be orthogonal restricts the possible coordinate changes a great deal. Suppose that the new coordinate system is defined by the points $p_0, p_1, \ldots, p_n$. We are assuming that $p_0 = 0$, so $P[;i] = p_i - p_0 = p_i$ and the vectors $p_1, p_2, \ldots, p_n$ form a basis of $\mathbf{R}^n$ (Figure 5.6).

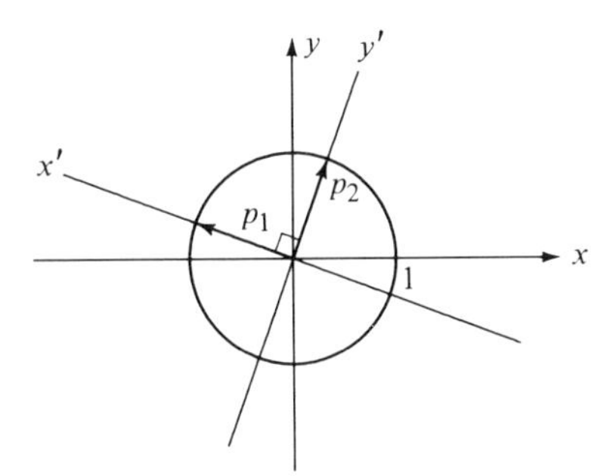

FIGURE 5.6

Now

$$\begin{aligned} p_i p_j &= P[;i] \cdot P[;j] \\ &= (P[;i])^T P[;j] \\ &= P^T[i;]P[;j] \\ &= (\text{ID } n)[i;j] \end{aligned}$$

So $p_i \perp p_j$ when $i \neq j$ and $p_i \cdot p_i = 1$ — that is, $\|p_i\| = 1$. Thus: *The distance and angle formulas remain the same when the new axes are mutually perpendicular and there are no scale changes.*

There is a bit of (somewhat confusing) terminology associated with this situation. When applied to vectors (not matrices!) the term "orthogonal" is a synonym for "perpendicular." A *unit vector* is a vector v with $\|v\| = 1$. It is important to note that $(1/\|v\|)\, v$ is always a unit vector pointing in the same direction as v.

A set of vectors $v_1, \ldots, v_k$ is called *orthonormal* if $v_i \perp v_j$ for $i \neq j$ and $\|v_i\| = 1$. Thus a *matrix* is orthogonal if and only if its columns are an orthonormal set of vectors.

We will refer to a coordinate system as *orthonormal* if the axes are mutually perpendicular and there are no scale changes.

EXAMPLE 5.6 The orthonormal linear coordinate systems in the plane are not hard to describe (see Figure 5.6). First rotate the x and y axes counterclockwise until the positive x axis is rotated into the positive x' axis; say that a rotation through θ is necessary. This rotation gives an intermediate coordinate system

$$\bar{X} = \begin{bmatrix} \bar{x} \\ \bar{y} \end{bmatrix} \qquad \text{where } X = R_\theta \bar{X}$$

Now $\bar{x} = x'$ and, since the x' and y' axes are perpendicular, the $\bar{y}$ axis coincides with the y' axis. The positive directions on the y' and $\bar{y}$ axes may not coincide, however. If they do not, then

$$\bar{X} = \begin{bmatrix} 1 & 0 \\ 0 & -1 \end{bmatrix} X'$$

Thus the linear orthonormal coordinate changes are given by $X = PX'$, where

$$P = \begin{cases} R_\theta & \text{if } x'y' \text{ is a right-handed system}\dagger \\ R_\theta \begin{bmatrix} 1 & 0 \\ 0 & -1 \end{bmatrix} & \text{if } x'y' \text{ is a left-handed system} \end{cases}$$

where

$$0 \leq \theta < 2\pi. \qquad \blacksquare$$

Now suppose we have a coordinate change $X = q + PX'$, where P is orthogonal. If X and Y are vectors of coordinates, then $X \cdot Y \neq X' \cdot Y'$, but this is not important because the formulas for distance and angle are unchanged. That is, we have

$$d(X, Y) = \|X - Y\| = \|X' - Y'\| = d(X', Y')$$

because the constant term q cancels from the difference

$$X - Y = P(X' - Y')$$

and so the calculation reduces to the linear case treated above. The same is true for the formula for $\cos\theta$, since the vectors used in calculating the angle are differences (see Examples 5.1, 5.3, and 5.4).

We close this section with an application to Euclidean plane geometry. We have said that an alternate term for isometry is "congruence." If this terminology is not arbitrary, then two objects in $\mathbf{R}^2$ should be "congruent" if there is a congruence that maps one onto the other.

In secondary school geometry courses, on the other hand, one often says that two figures are congruent if one can be picked up and placed precisely on top of the other. It may be necessary to turn one figure over before the two can be made to match (Figure 5.7).

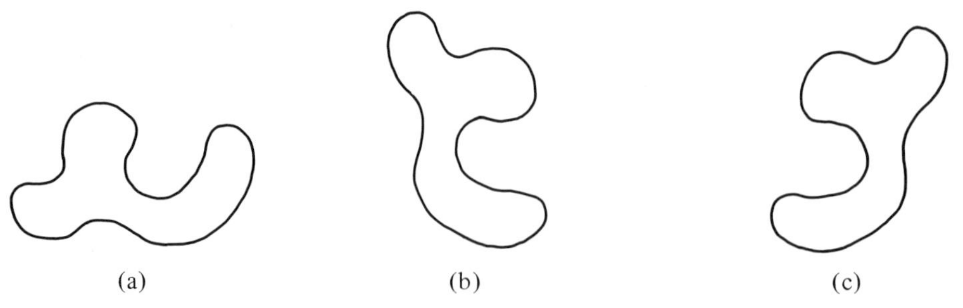

FIGURE 5.7 Three congruent figures in $\mathbf{R}^2$. Figure (b), may be made to coincide with (a) by sliding and rotating; (c) must be flipped over as well.

† That is, if the 90° rotation from the positive x' axis to the positive y' axis is counterclockwise.

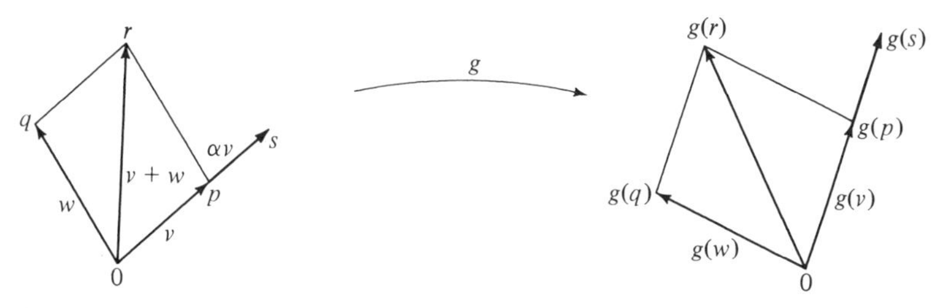

FIGURE 5.8

The next proposition shows, among other things, that the two concepts of congruence coincide. Notice that one way of turning over a figure in the plane is to reflect it in a line — any line (exercise 43).

PROPOSITION 5.7 A congruence in the plane may be factored as a composition of at most three transformations: a reflection in a line, a rotation, and a translation.

Proof Let $f: \mathbf{R}^2 \to \mathbf{R}^2$ be a congruence. Notice that we do not assume that f is affine, so we cannot immediately apply Proposition 5.4. We first show that f is affine.

Let $b = f(0)$ and let $g(x) = f(x) - b$. Since f is a congruence and translation is a congruence, it follows that g is also a congruence (exercise 41). We will use Proposition 2.17 to show that g is linear.

Let v, w be two vectors in $\mathbf{R}^2$. Let us distinguish in this instance between the vectors v, w and the points p, q at their tips. Let r be the tip of $v + w$; see Figure 5.8.

The function g carries the points o, p, q, r to $o = g(o), g(p), g(q), g(r)$. Since $d(o, q) = d(p, r)$, it follows that $d(o, g(q)) = d(g(p), g(r))$, and since $d(o, p) = d(q, r)$, it follows that $d(o, g(p)) = d(g(q), g(r))$. Thus $o, g(p), g(q), g(r)$ form a parallelogram, and hence $g(v + w) = g(v) + g(w)$.

Next we wish to show that $g(\alpha v) = \alpha g(v)$. Let s be the endpoint of αv. Now the points o, p, s are all in a line, and so one of them is between the other two. Three arrangements arise when $\alpha < 0, 0 \leq \alpha \leq 1$, or $\alpha > 1$. Take, for example, the case $\alpha > 1$, which is shown in Figure 5.8. In this case p is between o and s, and so $d(o, s) = d(o, p) + d(p, s)$. Hence $d(o, g(s)) = d(o, g(p)) + d(g(p), g(s))$, which shows that the points lie on a straight line by the triangle inequality. Since $d(o, s) = \alpha d(o, p)$, $d(o, g(s)) = \alpha d(o, g(p))$ and hence $g(\alpha v) = \alpha g(v)$. The other two cases are similar.

By Proposition 2.17 g is a linear map. Say $g(X) = AX$. Now we may apply Proposition 5.4 to show that A is orthogonal.

Any orthogonal matrix A can be used to give a linear coordinate change in $\mathbf{R}^2$, $X = AX'$, which preserves dot products. Thus, by Example 5.6 $A = R_\theta$ or

$$A = R_\theta \begin{bmatrix} 1 & 0 \\ 0 & -1 \end{bmatrix} \qquad \text{where } 0 \leq \theta < 2\pi$$

It follows that

$$f(X) = b + g(X) = \begin{cases} b + R_\theta X \\ \text{or} \\ b + R_\theta \begin{bmatrix} 1 & 0 \\ 0 & -1 \end{bmatrix} \end{cases} \qquad 0 \leq \theta < 2\pi$$

Thus f is (possibly) a reflection (in the x axis) followed by a rotation (about the origin, perhaps through the angle $\theta = 0$) followed by a translation (perhaps $b = 0$). ■

EXAMPLE 5.7 Factor the following congruences of the plane into (possibly) a reflection in the x axis, followed by a counterclockwise rotation about the origin (possibly through $\theta = 0$) followed by a (possibly trivial) translation.

1. Reflection in the line $x + y = 1$.
2. Rotation through the angle α, followed by reflection in the x axis, followed by a rotation through β followed by reflection in the y axis.

Solutions 1. We imitate the proof of Proposition 5.7 and let $b = f(0)$. A sketch (Figure 5.9) shows that $b = (1, 1)$. $g(x) = f(x) - b$ is linear, so $g(X) = AX$, where the columns of A are $g(1, 0)$ and $g(0, 1)$ (Proposition 2.17). Since $f(1, 0) = (1, 0)$, $f(0, 1) = (0, 1)$, and $g(X) = f(X) - (1, 1)$, we have that

$$f(X) = \begin{bmatrix} 1 \\ 1 \end{bmatrix} + \begin{bmatrix} 0 & -1 \\ -1 & 0 \end{bmatrix} X$$

Now

$$A = \begin{bmatrix} 0 & -1 \\ -1 & 0 \end{bmatrix}$$

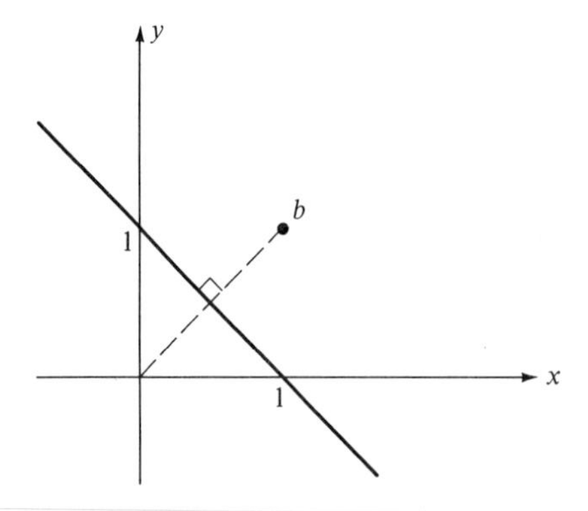

FIGURE 5.9

is not a rotation matrix, since $A[1; 2] \neq -A[2; 1]$, so reflection in the x axis must be involved.

$$f(X) = \begin{bmatrix} 1 \\ 1 \end{bmatrix} + \begin{bmatrix} 0 & 1 \\ -1 & 0 \end{bmatrix}\begin{bmatrix} 1 & 0 \\ 0 & -1 \end{bmatrix}X, \quad \begin{bmatrix} 0 & 1 \\ -1 & 0 \end{bmatrix} = R_{3\pi/2} = R_{-\pi/2}$$

So reflection in $x + y = 1$ is the same as reflection in the x axis, followed by a rotation of 270° (or −90°) followed by the translation that carries the origin to (1, 1).

Here is a second solution. To find a formula for f we pick an orthonormal coordinate system in which the x axis is $x + y = 1$. For example,

$$X = \begin{bmatrix} 1 \\ 0 \end{bmatrix} + R_{-\pi/4}X' \quad \text{or} \quad X' = R_{\pi/4}\left(X - \begin{bmatrix} 1 \\ 0 \end{bmatrix}\right)$$

Then in the $x'y'$ coordinate system f is given by

$$Y' = \begin{bmatrix} 1 & 0 \\ 0 & -1 \end{bmatrix}X'$$

and hence

$$R_{\pi/4}\left(Y - \begin{bmatrix} 1 \\ 0 \end{bmatrix}\right) = \begin{pmatrix} 1 & 0 \\ 0 & -1 \end{pmatrix}R_{\pi/4}\left(X - \begin{bmatrix} 1 \\ 0 \end{bmatrix}\right)$$

which simplifies to

$$f(x) = \begin{bmatrix} 1 \\ 1 \end{bmatrix} + \begin{bmatrix} 0 & -1 \\ -1 & 0 \end{bmatrix}X = \begin{bmatrix} 1 \\ 1 \end{bmatrix} + \begin{bmatrix} 0 & 1 \\ -1 & 0 \end{bmatrix}\begin{bmatrix} 1 & 0 \\ 0 & -1 \end{bmatrix}X$$

2. We have

$$f(X) = \begin{bmatrix} -1 & 0 \\ 0 & 1 \end{bmatrix}R_\beta\begin{bmatrix} 1 & 0 \\ 0 & -1 \end{bmatrix}R_\alpha X$$

Now

$$R_\beta\begin{bmatrix} 1 & 0 \\ 0 & -1 \end{bmatrix} = \begin{bmatrix} 1 & 0 \\ 0 & -1 \end{bmatrix}R_{-\beta}$$

(check it) so

$$\begin{aligned} f(X) &= \begin{bmatrix} -1 & 0 \\ 0 & 1 \end{bmatrix}R_\beta\begin{bmatrix} 1 & 0 \\ 0 & -1 \end{bmatrix}R_\alpha X \\ &= \begin{bmatrix} -1 & 0 \\ 0 & 1 \end{bmatrix}\begin{bmatrix} 1 & 0 \\ 0 & -1 \end{bmatrix}R_{-\beta}R_\alpha X \\ &= -R_{\alpha-\beta}X \\ &= R_{\pi+\alpha-\beta}X \end{aligned}$$

So this composition of rotations and reflections reduces to a single rotation. ■

EXERCISES 5.1

In Exercises 1 through 5 find the distances and angles between the given pair of vectors.

1. (1, 0) and (0, 1).
2. (1, 0) and (1, 1).
3. (1, 0, 0) and (1, 1, 1).
4. (1, 0, 0, 0) and (1, 1, 1, 1).
5. $1 = \iota n$ and $n \rho 1$ as $n \to \infty$.

In exercises 6 through 11 the three vertices p_1, p_2, p_3 of a triangle are given. Identify the triangles as (a) right, (b) isosceles, (c) equilateral, (d) degenerate (i.e., the vertices are collinear), (e) having an obtuse interior angle, or (f) none of these.

Hint: As in Example 5.4 set $u = p_1 - p_2$, $v = p_3 - p_1$, and $A = [u \mid v \mid u + v]$. Then all necessary information is contained in $A^T A$.

6. $p_1 = (\sqrt{3}, 1)$
   $p_2 = (3\sqrt{3}, 3)$
   $p_3 = (\sqrt{3}, 5)$
7. $p_1 = (2, -2, 1)$
   $p_2 = (1, -2, 2)$
   $p_3 = (1, -1, 3)$
8. $p_1 = (2, -2, 1)$
   $p_2 = (1, -2, 2)$
   $p_3 = (1, -1, 1)$
9. $p_1 = (-1, 3, 3, -1)$
   $p_2 = (2, 6, 6, 2)$
   $p_3 = (6, 2, 2, 6)$
10. $p_1 = (2, 6, 6, 2)$
    $p_2 = (5, 1, 1, 5)$
    $p_3 = (-1, 3, 3, -1)$
11. $p_1 = (1, 2, -1)$
    $p_2 = (0, 2, 0)$
    $p_3 = (-3, 2, 3)$

Write the APL function described in exercises 12 through 18.

12. Name: *NORM*
    Right argument: A vector v
    Result: $\|v\|$
13. Name: *DIST*
    Left argument: A vector v in $\mathbf{R}^n$.
    Right argument: A vector w in $\mathbf{R}^n$.
    Result: $d(v, w)$.
14. Name: *SPREAD*
    Left argument: A vector v in $\mathbf{R}^n$.
    Right argument: A vector w in $\mathbf{R}^n$.
    Result: Angle in radians between v and w.
15. Name: *DSPREAD*
    Left argument: A vector v in $\mathbf{R}^n$.
    Right argument: A vector w in $\mathbf{R}^n$.
    Result: The angle in degrees between v and w.
16. Name: *INTANG*
    Right argument: The vertices of a triangle stored as the columns of a matrix.
    Result: The vector consisting of the three interior angles of the triangle in radians.
17. Name: *SIDES*
    Right argument: The vertices of a triangle stored as the columns of a matrix.
    Result: The vector consisting of the lengths of the sides.
18. Name: *AREA4*
    Right argument: The vertices of a triangle stored as the columns of a matrix.
    Result: The area of the triangle.

    Hint: See exercise 14 in Exercises 3.1.

19. Find two linearly independent vectors perpendicular to $u = (1, -1, 1)$.

20. Find a nonzero vector perpendicular to $u = (1, -1, 1)$ and $v = (1, 1, 0)$.

21. Find two nonzero vectors v and w perpendicular to $u = (1, -1, 1)$ and to each other.

22. (a) Find a vector v_2 perpendicular to $v_1 = (1, -1, 1, -1)$ and lying in the subspace generated by v_1 and $u = (1, 0, 1, 0)$.

Hint: Substitute $X = [v_1 \mid u]T$ in $v_1^T X = 0$.]

(b) Find a vector v_3 perpendicular to v_1 and v_2 of part (a) and lying in the subspace generated by v_1, u, and $w = (1, 2, 3, 4)$.

In exercises 23 through 28 an affine function is given. Is the function an isometry?

23. $f(x, y) = \frac{1}{\sqrt{2}}\begin{pmatrix} x - y - \sqrt{2} \\ x + y + \sqrt{2} \end{pmatrix}$

24. $f(x, y) = \begin{pmatrix} x - y - \sqrt{2} \\ x + y + \sqrt{2} \end{pmatrix}$

25. The affine function $f\colon \mathbf{R}^2 \to \mathbf{R}^2$ that carries the triangle with vertices $(0, 0)$, $(1, 0)$, $(0, 1)$ to the triangle with vertices $(3, 4) = f(0, 0)$, $(2, 4) = f(1, 0)$, $(3, 5) = f(0, 1)$.

26. The affine function $f\colon \mathbf{R}^2 \to \mathbf{R}^2$ that carries the triangle with vertices $(0, 0)$, $(1, 0)$, $(0, 1)$ to the triangle with vertices $(3, 5) = f(0, 0)$, $(2, 4) = f(1, 0)$, $(3, 4) = f(0, 1)$.

27. $f\colon \mathbf{R}^6 \to \mathbf{R}^6$ by $f(v) = (\iota 6) + v[1\ 3\ 2\ 6\ 4\ 5]$.

28. $f(x, y, z) = \begin{bmatrix} \frac{x}{\sqrt{3}} + \frac{y}{\sqrt{2}} - \frac{2z}{7} \\ \frac{x}{\sqrt{3}} + \frac{3z}{7} \\ \frac{x}{\sqrt{3}} - \frac{y}{\sqrt{2}} + \frac{6z}{7} \end{bmatrix}$

In exercises 29 through 33 an affine coordinate change is given. Is the new coordinate system orthonormal?

29. $X = \begin{bmatrix} -\sqrt{2} \\ \sqrt{2} \end{bmatrix} + \frac{1}{\sqrt{2}}\begin{bmatrix} 1 & -1 \\ 1 & -1 \end{bmatrix} X'$

30. $X = \begin{bmatrix} -\sqrt{2} \\ \sqrt{2} \end{bmatrix} + \begin{bmatrix} 1 & -1 \\ 1 & 1 \end{bmatrix} X'$

31. The new coordinate system is given by $p_0 = (3, 4)$, $p_1 = (2, 4)$, $p_2 = (3, 5)$.

32. The new coordinate system is given by $p_0 = (3, 5)$, $p_1 = (2, 4)$, $p_2 = (3, 4)$.

33. $X = \begin{bmatrix} \frac{1}{\sqrt{3}} & \frac{1}{\sqrt{2}} & -\frac{2}{7} \\ \frac{1}{\sqrt{3}} & 0 & \frac{3}{7} \\ \frac{1}{\sqrt{3}} & -\frac{1}{\sqrt{2}} & \frac{6}{7} \end{bmatrix} X'$

34. Show that a parallelogram with diagonals of equal length is a rectangle.

35. Show that the law of cosines is true for triangles in $\mathbf{R}^n$.

36. Let $u \perp v$ in $\mathbf{R}^n$ and θ be the angle between u and $u + v$. Show that

$$\cos\theta = \frac{\|u\|}{\|u + v\|} \quad \text{and} \quad \sin\theta = \frac{\|v\|}{\|u + v\|}$$

37. Show that if the Schwartz inequality is an equality, then the two vectors are linearly dependent.

Hint: Let $A = [u \mid v]$ and apply Propositions 2.14 and 2.7.

38. Show that if the triangle inequality $\|u + v\| \leq \|u\| + \|v\|$ is an equality, then u, v, and $u + v$ are parallel. Assume Exercise 37.

39. Show statement 1 of Proposition 5.4.

40. Show that a product of orthogonal matrices is orthogonal.

41. Show that a composition of isometries is an isometry. That is, if f and g are isometries, then $h(X) = f(g(X))$ is an isometry.

42. Show that a composition of similarities is a similarity. That is, if f and g are similarities, then $h(X) = f(g(X))$ is a similarity.

43. Consider $\mathbf{R}^2$ to be the xy plane in $\mathbf{R}^3$. Show that reflection in a line in $\mathbf{R}^2$ can be accomplished by a $180°$ rotation about the line in $\mathbf{R}^3$.

Hint: Since orthonormal coordinate changes preserve the distance and angle formulas, you may assume that the line is the x axis.

44. Show that similarities map lines to lines.

Hint: Similarities preserve $180°$ angles; assume the result of Exercise 37.

45. Show that f: $\mathbf{R}^2 \to \mathbf{R}^2$ given by $Y = f(X) = \alpha X$, α a scalar, is a similarity that is an isometry only when $\alpha = \pm 1$.

46. Assuming the result of Exercise 44:

(a) Show that a similarity maps any triangle to a triangle that is similar.

(b) Let p, q, r be points in $\mathbf{R}^2$, f: $\mathbf{R}^2 \to \mathbf{R}^2$ a similarity, and let $d(f(p)f(q)) = \alpha d(p, q)$ (i.e., $\alpha = d(f(p), f(g)) \div d(p, q)$. Show that $d(f(p), f(r)) = \alpha d(p, r)$ and $d(f(r, f(q)) = \alpha d(r, q)$ also.

Hint: Use part (a).

(c) Show that if f: $\mathbf{R}^2 \to \mathbf{R}^2$ is a similarity, then there is a constant α such that $d(f(p), f(q)) = \alpha d(p, q)$ for all points p and q.

Hint: Given any p_1, q_1 and any p_2, q_2 there are constants α_1, α_2 such that $\alpha_i d(p_i, q_i) = d(f(p_i), f(q_i))$ for $i = 1, 2$. Use part (b) to show that $\alpha_1 = \alpha_2$.

47. Let f: $\mathbf{R}^2 \to \mathbf{R}^2$ be a rotation through the angle θ about the point p. Write f as a rotation about the origin followed by a translation.

Hint: See Example 4.10.

48. Let l_1 and l_2 be two lines intersecting at the origin. Let f: $\mathbf{R}^2 \to \mathbf{R}^2$ be reflection in l_1 followed by reflection in l_2. Show that f is rotation through the angle 2α, where α is the acute angle from l_1 to l_2.

Hint: Formulas for the reflection are given in Example 4.11.

49. Exercise 46 shows that a (not necessarily affine) function f: $\mathbf{R}^2 \to \mathbf{R}^2$ is a similarity if and only if there is a constant $\alpha > 0$ such that $d(f(p), f(q)) = \alpha d(p, q)$ for all p, q in $\mathbf{R}^2$. Assuming this result, show that a similarity is the composition of at most four maps: a

reflection in the x axis, a rotation about the origin, multiplication by a scalar, and a translation.

Hint: Show that $g(X) = (1/\alpha)f(X)$ is a congruence. Use Exercise 33.

50. Derive the law of cosines from the Pythagorean theorem (Figure 5.3).

51. Prove statement 1 of Proposition 5.1.

52. Prove statement 2 of Proposition 5.1.

53. Any three vectors in $\mathbf{R}^n$ lie in a subspace of dimension 3 at most. By putting an orthonormal coordinate system on this subspace, we obtain a geometry-preserving representation of the three vectors in $\mathbf{R}^n$ as three physical vectors in space ($\mathbf{R}^3$). This allows us to analyze the relationships among the vectors geometrically.

(a) Let u, v, w be three vectors in space. Suppose that the angle between u and v is 45° and the angle between v and w is also 45°. Show by a sketch that the angle between u and w can vary from 0° to 90°.

(b) Let X, Y, Z be three measurements on a population. Suppose that the correlations between X and Y and between Y and Z are

$$r_{XY} = r_{YZ} = .707 \simeq \frac{\sqrt{2}}{2}$$

Show that r_{XZ} can vary from 1 to 0 (approximately).

Hint: Use part (a).

(c) Let X, Y, Z can be as in part (b). Suppose that $r_{XY} = .7$ and $r_{YZ} = 5$. Can r_{XY} be negative?

5.2 The Diagonalization of Symmetric Matrices

In this section we will extend the methods used to analyze conics in Section 4.3 to quadratic functions defined on $\mathbf{R}^n$ for $n > 2$. This has several important applications, including the second-derivative test for maxima and minima (Section 5.3*) and moments of inertia (Section 5.6*).

Given a quadratic function

$$f(X) = c + BX + X^TAX$$

where $A = A^T$ is n by n, B is 1 by n, and c is a scalar, we wish to simplify f by an affine coordinate change

$$X = p_0 + PX'$$

Substituting the coordinate-change formula into the expression for f, we obtain

$$f(X') = f(p_0) + (B + 2p_0^TA)PX' + (X')^T(P^TAP)X'$$

Notice that the choice of p_0 does not affect the quadratic term. If the equation $B + 2p_0^TA = 0$ has a solution p_0, then the linear term may be eliminated by choosing p_0 independently of P. In many applications, however, the linear term is automatically zero and the main problem is that of choosing P. One chooses P to eliminate the cross-product terms (i.e., terms of the form $x_i x_j$) from the expression for f. This is equivalent to choosing P so that $P^T AP$ is diagonal — that is, $(P^T AP)[i; j] = 0$ if $i \neq j$.

In Section 4.3 it was shown that if A is 2 by 2, then there is a rotation matrix R_θ such that $R_\theta^T AR_\theta$ is diagonal: this means that $X'^T(R_\theta^T AR_\theta)X'$ has no xy term. In the general case it is possible to find an orthogonal matrix Q such that $Q^T AQ = \Lambda$ is diagonal; however, Q is considerably more difficult to compute.

We will compute Q by the iterative process known as the *Jacobi algorithm*. We begin with a symmetric matrix A that is not diagonal. Suppose that $A[i; j] \neq 0$ with $i \neq j$. Then we use a coordinate change $X = Q_1 X'$ that rotates the $x_i x_j$ plane so that if $A_1 = Q_1^T AQ_1$, then $A_1[i; j] = 0$. Now if A_1 is not diagonal, then there is a component $A[h; k] \neq 0$ with $h \neq k$. This time we use Q_2 to rotate the $x_h x_k$ plane so that if $A_2 = Q_2^T A_1 Q_2$, then $A_2[h; k] = 0$.

Unfortunately this process does not terminate in a finite number of steps, because $A_2[i; j] \neq 0$ in general, even though $A_1[i; j] = 0$. We will describe a way of choosing the coordinate changes Q_i, however, so that the sequence of matrices $A_1, A_2, A_3, \ldots, A_n$ approaches a diagonal matrix as $n \to \infty$. We may then carry out the process until A_n is close enough to a diagonal matrix for our purposes.

If the coordinate-change matrices Q_i are indexed so that

$$A_{i+1} = Q_{i+1}^T A_i Q_{i+1}$$

then

$$\begin{aligned} A_n &= Q_n^T Q_{n-1}^T \cdots Q_2^T Q_1^T AQ_1 Q_2 \cdots Q_n \\ &= (Q_1 Q_2 \cdots Q_n)^T A(Q_1 Q_2 \cdots Q_n) \\ &= Q^T AQ \end{aligned}$$

where $Q = Q_1 Q_2 \cdots Q_n$. The matrices Q_1 will be chosen to be orthogonal, and so Q will be orthogonal (exercise 40 of Exercises 5.1).

The details of the algorithm are contained in the proof sketch of Proposition 5.9 below. First, however, we need some formulas involving matrix multiplication. These formulas involve vectors of row and column indices. We will say that such a vector of indices U is *partitioned by the vectors V and W* if V and W have no components in common and the catenation V, W contains the same entries as U. This means that V, W is just U, possibly with components reordered. For example, $U = (1, 2, 3, 4, 5)$ is partitioned by $V = (2, 4)$ and $W = (5, 3, 1)$.

The next proposition follows easily from Proposition 2.1. The proof is left as exercise 26.

PROPOSITION 5.8 Let A and B be matrices such that AB is defined. Let I, J, U, V, W be appropriate vectors of indices and suppose that U is partitioned by V and W. Then

1. $(AB)[I; J] = A[I;]B[;J]$.
2. $A[I; U]B[U; J] = A[I; V]B[V; J] + A[I; W]B[W; J]$. ■

PROPOSITION 5.9 Let A be an n-by-n matrix. Then statements 1 and 2 are equivalent.

1. A is symmetric.
2. There is an orthogonal matrix Q such that $Q^T A Q$ is diagonal.

Proof We will argue that statement 1 implies statement 2, leaving as exercise 28 the proof that statement 2 implies statement 1.

We need a measure of how "nondiagonal" a matrix is. We define $SS(A)$ to be the sum of the squares of the off-diagonal entries of A. Then A is diagonal if and only if $SS(A) = 0$.

Now suppose that A is not diagonal, say that $A[i; j] \neq 0$ where $i \neq j$. Let $K = (i, j)$ and let L be the rest of the indices from 1 to n so that K and L partition ιn.

Since A is symmetric, so is the 2-by-2 submatrix $A[K; K]$. By Proposition 4.10 there is a rotation matrix R_θ such that $R_\theta^T A[K; K] R_\theta$ is diagonal. Define a matrix Q_1 as follows. Begin with an n-by-n identity matrix ($Q_1 \leftarrow ID\ n$) and then replace the $K;K$-block by $R_\theta (Q_1[K; K] \leftarrow R_\theta)$. Thus $Q_1[K; K] = R_\theta$, $Q_1[L; L] = I$, $Q_1[L; K] = 0$, and $Q_1[K; L] = 0$. The matrix Q_1 is orthogonal (Exercise 27).

Set $A_1 = Q_1^T A Q_1$. Using the formulas of Proposition 5.8, we have

$$\begin{aligned}(AQ_1)[K; K] &= A[K;]Q_1[;K] \\ &= A[K; K]Q_1[K; K] + A[K; L]Q[L; K] \\ &= A[K; K]R_\theta\end{aligned}$$

Similarly $(AQ_1)[L; K] = A[L; K]R_\theta$, $(AQ_1)[K; L] = A[K; L]$, and $(AQ_1)[L; L] = A[L; L]$. Hence

$$\begin{aligned}A_1[K; K] &= Q_1^T[K;](AQ_1)[;K] \\ &= Q_1^T[K; K](AQ_1)[K; K] + Q_1^T[K; L](AQ_1)[L; K] \\ &= R_\theta^T A[K; K]R_\theta + 0 \\ &= R_\theta^T A[K; K]R_\theta\end{aligned}$$

Thus, $A_1[i; j] = 0$. Similarly we have $A_1[L; K] = A[L; K]R_\theta$, $A_1[K; L] = R_\theta^T A[K; L]$, and $A_1[L; L] = A[L; L]$.

We are now in a position to compare $SS(A)$ and $SS(A_1)$. Notice that for any n-by-n matrix B, $SS(B)$ is $SS(B[K; K])$ plus $SS(B[L; L])$ plus the sums of the squares of *all* the entries of $B[L; K]$ and $B[K; L]$.

Since $A_1[L; L] = A[L; L]$, we certainly have $SS(A[L; L]) = SS(A_1[L; L])$.

Notice that the columns of $A_1[K; L]$ are the columns of $A[K; L]$ rotated through the angle $-\theta$ ($R_\theta^T = R_{-\theta}$). Now the sum of the squares of the entries of $A_1[K; L]$ is the sum of the squares of the lengths of its columns. Thus the sums of the squares of the entries of $A_1[K; L]$ and $A[K; L]$ are the same. A similar argument (take transposes) shows that the sums of the squares of the entries of $A_1[L; K]$ and $A[L; K]$ are the same. Finally, $SS(A[K; K]) = 2A[i; j]^2 > 0$ and

$SS(A_1[K;K]) = 0$. Thus

$$SS(A_1) = SS(A) - 2A[i;j]^2 < SS(A)$$

This shows that $SS(A_n)$ decreases with n, but does not show that it decreases to zero. Suppose, however, that $A[i;j]^2$ was the *largest* off-diagonal entry of $A*2$. There are n^2 entries in A, and so there are $n^2 - n$ off-diagonal entries. Since the largest number is always bigger than the average, we have

$$A[i;j]^2 \geq \frac{1}{n^2 - n} SS(A)$$

Therefore

$$\begin{aligned} SS(A_1) &= SS(A) - 2A[i;j]^2 \\ &\leq SS(A) - \frac{2}{n^2 - n} SS(A) \\ &= \left(1 - \frac{2}{n^2 - n}\right) SS(A) \end{aligned}$$

Let

$$q = 1 - \frac{2}{n^2 - n} < 1$$

Then

$$SS(A_n) < q^n\, SS(A) \to 0 \qquad \text{as } n \to \infty \qquad \blacksquare$$

The proof sketch above is not rigorous. It shows that the off-diagonal entries of A_n go to zero, but it does not show that the diagonal of A_n approaches a fixed vector or that the accumulated product $Q_1 Q_2 \ldots Q_n$ approaches a fixed orthogonal matrix Q. It does, however, show how to compute an orthogonal matrix Q so that $Q^T A Q$ is as close to diagonal as desired, and we now develop an APL function to do so. The logical gaps in the development can be filled after eigenvalues are discussed in Chapter 7.

THE FUNCTION *JACOBI*

Given a symmetric matrix A, we will apply the iteration scheme described until we arrive at a matrix A_n whose off-diagonal entries are negligible compared to the original entries of A. We will then take $Q = Q_1 Q_2 \ldots Q_n$ as the coordinate-change matrix.

We will use three auxiliary functions. First we will need a function to find the indices $i;j$ such that $A[i;j]^2$ is maximal. It is sufficient to define a function, call it *JACFIND*, that given a matrix A finds the row and column indices of the largest off-diagonal entry of |A.

We will also use the function *ROT* from Section 4.3 which computes R_θ given θ and a function *ANG* (exercise 12 of Exercises 4.3) which computes θ given $A[K; K]$.

To begin the iteration we set $A_0 = A$ and $Q_0 = ID\ 1{\uparrow}\rho A$. We take B equal to the largest magnitude in A (`B←⌈/,|A`) and stop the iteration when $B = B + A[i;j]$ — that is, when $A[i;j]$ is negligible compared to B (cf. Section 3.4).

```
    ∇ Z←JACOBI A ;B;I;K;Q
[1]    Z←I←ID 1↓ρA
[2]    B←⌈/,|A
[3]   L:K←JACFIND A
[4]    →(B=B+A[K[1];K[2]])/0
[5]    Q←I
[6]    Q[K;K]←ROT ANG A[K;K]
[7]    A←(⍉Q)+.×A+.×Q
[8]    Z←Z+.×Q
[9]    →L
    ∇
```

A function that will do for *JACFIND* is

```
    ∇ Z←JACFIND A ;B
[1]    A←(|A)+(⌈/⍳0)×ID 1↑ρA
[2]    Z←(⌈/A)⍳B←⌈/,A
[3]    Z←Z,A[Z;]⍳B
    ∇
```

The quantity `⌈/⍳0` is the smallest number that can be stored in the machine — "minus infinity" in effect. Adding this number to the diagonal of `|A` ensures that, in cases of practical interest, the diagonal is negative. In fact the diagonal entries will be `⌈/⍳0`, since numbers of practical interest are negligible compared to `⌈/⍳0`.

Line [2] sets B equal to the largest off-diagonal entry of `|A` and computes the index of the first row in which B occurs. Line [3] then adds the column index.

The function *ANG* is left as exercise 1.

Example 5.8 We will follow the operation of *JACOBI* by putting a trace on line [7]. In APL tracing is controlled by the *trace vector* of a function (see Appendix B). The trace vector of the APL function `FCN` is written `T∆FCN`. The trace vector is created and erased with the function and is not listed by the `)VARS` command. To set a trace on lines [6] and [4] of the function `FCN` use the expression

```
T∆FCN←6 4
```

Then every time line [6] of `FCN` is executed the machine prints `FCN[6]` followed by the last value computed on line [6]. Similarly for line [4]. To turn off tracing use the expression

```
T∆FCN←⍳0
```

Tracing is a debugging aid in the APL system, but it is very useful for gaining insight into iterative procedures. We will use it to watch *JACOBI* operate on a 3-by-3 matrix, *A*.

```
      A
 1  2  3
 2  4  5
 3  5  6
      T∆JACOBI←7
      Q←JACOBI A
JACOBI[7]
 1.00E0     ¯3.55E¯1    3.59E0
¯3.55E¯1    ¯9.90E¯2    6.94E¯18
 3.59E0      6.34E¯18   1.01E1
```

The 2;3 entry is set to zero. The 1;3 entry increases, but the 1;2 entry decreases.

```
JACOBI[7]
¯2.45E¯1    ¯3.36E¯1    1.04E¯17
¯3.36E¯1    ¯9.90E¯2   ¯1.16E¯1
 1.18E¯17   ¯1.16E¯1    1.13E1
```

The 1;3 entry is zero. The off-diagonal entries are all an order of magnitude smaller than the off-diagonal entries of *A*.

```
JACOBI[7]
¯5.15E¯1    ¯5.42E¯19  ¯7.31E¯2
¯8.67E¯19    1.72E¯1   ¯9.06E¯2
¯7.31E¯2    ¯9.06E¯2    1.13E1
```

Two orders of magnitude smaller.

```
JACOBI[7]
¯5.15E¯1    ¯5.93E¯4   ¯7.31E¯2
¯5.93E¯4     1.71E¯1   ¯1.09E¯19
¯7.31E¯2    ¯1.31E¯18   1.13E1
JACOBI[7]
¯5.16E¯1    ¯5.93E¯4   ¯1.09E¯18
¯5.93E¯4     1.71E¯1    3.65E¯6
¯2.41E¯19    3.65E¯6    1.13E1
```

Four orders of magnitude smaller

```
JACOBI[7]
¯5.16E¯1    ¯3.44E¯19   3.15E¯9
¯6.56E¯19    1.71E¯1    3.65E¯6
 3.15E¯9     3.65E¯6    1.13E1
```

Six orders of magnitude smaller

```
JACOBI[7]
¯5.16E¯1    ¯1.03E¯15   3.15E¯9
¯1.03E¯15    1.71E¯1   ¯3.40E¯20
 3.15E¯9    ¯1.24E¯18   1.13E1
```

Nine orders of magnitude smaller

```
JACOBI[7]
¯5.16E¯1    ¯1.03E¯15   1.22E¯17
¯1.03E¯15    1.71E¯1   ¯3.40E¯20
 1.31E¯17   ¯1.24E¯18   1.13E1
```

The iteration stops, since ⎕CT is about 3E¯15 for this system.

The coordinate-change matrix Q is

```
      Q
0.737 ¯0.591 0.328
0.328 0.737 0.591
¯0.591 ¯0.328 0.737
```

In the accumulation $Q = Q_1Q_2 \dots Q_n$, little computational error is involved. The computed matrix Q is quite orthogonal (this system carries eighteen digits):

```
      (ID 3)-Q+.×⍉Q
¯1.73E¯18    ¯1.52E¯18    ¯8.67E¯19
¯1.52E¯18    ¯3.47E¯18    ¯1.08E¯18
¯8.67E¯19    ¯1.08E¯18     1.73¯18
```

And Q^TAQ is close to diagonal:

```
      (⍉Q)+.×A+.×Q
¯5.16E¯1     ¯1.03E¯15     2.08E¯17
¯1.03E¯15     1.71E¯1      1.04E¯17
 1.77E¯17     6.72E¯18     1.13E1
```
■

If a quadratic function f: $\mathbf{R}^n \to \mathbf{R}$ can be simplified by a coordinate change to the form

$$f(X') = c + \sum^{n} \lambda_i(x_i')^2$$

then f has a maximum at $X' = 0$ if and only if $\lambda_i \leq 0$ for all i and a minimum at $X' = 0$ if and only if $\lambda_i \geq 0$ for all i. If some λ_i are positive and some negative, then f has neither a maximum nor a minimum at $X' = 0$. In this case f is said to have *saddle* at $X' = 0$.

If the linear term cannot be eliminated by a coordinate change, then the methods of the optional Section 2.7 can be used to show that f has neither a maximum nor a minimum nor a saddle (exercise 23). In this case we say that f has no *critical points*. If the coordinate change $X = p_0 + QX'$ eliminates the linear term, then p_0 is a *critical point* for f.

EXAMPLE 5.9 Discuss the critical points of

$$f(x, y, z) = 4 + 3x - y + 2z + 3x^2 + 2y^2 + 2z^2 + 4xz + 2yz$$

Solution

$$f(x, y, z) = 4 + [3 \quad -1 \quad 2]\begin{bmatrix} x \\ y \\ z \end{bmatrix} + [x \quad y \quad z]\begin{bmatrix} 3 & 0 & 2 \\ 0 & 2 & 1 \\ 2 & 1 & 2 \end{bmatrix}\begin{bmatrix} x \\ y \\ z \end{bmatrix}$$

$$= c + BX + X^TAX$$

To eliminate the linear term we need p_0 such that $B + 2p_0^TA = 0$ or $Ap_0 = -\frac{1}{2}B^T$.

```
      +P0←(3 ¯1 2⌹A)÷¯2
0.5  1  ¯1.5
```

Thus f does have a critical point. To diagonalize A we use

```
      +Q←JACOBI A
0.739 ¯0.427 ¯0.521
0.233 0.888 ¯0.397
0.632 0.172 0.756
```

to obtain

```
      (⍉Q)+.×A+.×Q
4.71E0      8.67E¯19    1.15E¯17
1.73E¯18    2.19E0      2.85E¯18
1.56E¯17    2.17E¯18    9.68E¯2
```

Since $f(p_0)$ is

```
      4+(3 ¯1 2+.×P0)+P0+.×A+.×P0
2.75
```

we see that using the coordinate change $X = p_0 + QX'$ we obtain, to three significant figures,

$$f(x', y', z') = 2.75 + 4.71(x')^2 + 2.19(y')^2 + .0968(z')^2$$

Thus f has an absolute minimum value of 2.75 at the critical point $p_0 = (\frac{1}{2}, 1, -\frac{3}{2})$. ■

Given a symmetric matrix A, Proposition 5.9 says that there is an orthogonal matrix Q such that $Q^TAQ = \Lambda$ is diagonal. How unique is Λ and how unique is Q? We can permute the diagonal entries of Λ by renumbering the axes of the coordinate system that Q defines. Except for order, however, the diagonal of A is unique. The diagonal entries of A are called *eigenvalues* of A. The next proposition gives a very useful characterization of the eigenvalues of A. Although we are assuming that A is symmetric, the concept of an eigenvalue will be extended in Chapter 7 to general matrices in such a way that Proposition 5.10 remains true.

PROPOSITION 5.10 The eigenvalues of A are those scalars λ for which there is a nonzero vector v such that $Av = \lambda v$. That is, the eigenvalues of A are the scalars λ for which the matrix $A - \lambda I$ is singular.

Note: If $Av = \lambda v$ with $v \neq 0$, then v is called an *eigenvector* of A belonging to the eigenvalue λ.

Proof of Proposition 5.10 First assume that $Q^TAQ = \Lambda$, Λ a diagonal matrix. Then, since $Q^T = Q^{-1}$, left multiplication by Q gives

$$AQ = Q\Lambda$$

In particular

$$\begin{aligned} AQ[;i] = (AQ)[;i] = (Q\Lambda)[;i] &= Q\Lambda[;i] \\ &= Q\begin{bmatrix} 0 \\ \vdots \\ \lambda_i \\ \vdots \\ 0 \end{bmatrix} \\ &= \lambda_i Q[;i] \end{aligned}$$

where $\lambda_i = A[i; i]$. Thus the eigenvalues satisfy the condition of the proposition, and the columns of Q are eigenvectors.

Conversely suppose that $Av = \lambda v$. Since Q is invertible with $Q^T = Q^{-1}$, the equation $QX = v$ has solution $w = Q^Tv$. Thus

$$\begin{aligned} Av &= \lambda v \\ AQw &= \lambda Qw \\ Q\Lambda w &= Q\lambda w \\ \Lambda w &= \lambda w \end{aligned}$$

So $\lambda w[i] = \lambda_i w[i]$ for all i. If $w[i] \neq 0$ for some i, then $\lambda = \lambda_i$ and so λ is an eigenvalue. (Further, $w[j] = 0$ for all j's such that $\lambda \neq \lambda_j$.) ∎

Matrix products of the form B^TB often arise. They appeared in the discussion of least-squares approximations in Section 2.3 and in the formula for computing dot products with skewed axes in Proposition 5.6 (see exercises 31 through 34). In statistics they arise in the computation of correlation and covariance matrices. The next proposition gives some alternate characterizations of such products. The converse problem of factoring a symmetric matrix A as B^TB arises in the statistical technique known as a factor analysis.

PROPOSITION 5.11 Let A be a symmetric matrix. Statements (a), (b), and (c) are equivalent:

(a) $A = B^TB$ for some matrix B.

(b) The eigenvalues of A are nonnegative.

(c) $f(X) = X^TAX \geq 0$ for all X.

Statements (d), (e), and (f) also are equivalent:

(d) $A = B^T B$ for some matrix B with linearly independent columns.

(e) The eigenvalues of A are strictly positive.

(f) $f(X) = X^T A X \geq 0$ for all X and $f(X) = 0$ implies $X = 0$.

Proof Let Q be orthogonal such that $Q^T A Q = \Lambda$ is diagonal with 1 1 ⍉ Λ equal to $(\lambda_1, \lambda_2, \ldots, \lambda_n)$. Then in the coordinate system given by $X = QX'$ we have

$$f(X') = X'^T \Lambda X' = \lambda_1 (X_1')^2 + \lambda_2 (X_2')^2 + \cdots + \lambda_n (X_n')^2$$

The equivalence of (b) and (c) and the equivalence of (e) and (f) follow.

If $A = B^T B$, then $f(X) = X^T A X = X^T B^T B X = (BX)^T BX \geq 0$ and $(BX)^T BX = 0$ if and only if $BX = 0$ by Proposition 5.1. Thus (a) implies (c) and, by Proposition 2.11, (d) implies (f).

Last, assume (b) is true. Then $M = \Lambda * \div 2$ is defined and $MM = \Lambda$. Thus $A = Q \Lambda Q^T = QMMQ^T = (QM)\ (QM)^T = B^T B$, where $B = (QM)^T$. Thus (b) implies (a). If (d) is true, then since both M and Q are invertible, the matrix B is invertible. ■

Notice that there is nothing unique about the factorization $A = B^T B$. If C is any matrix such that $C^T C = ID$ 1↑ρB (i.e., a matrix with orthonormal columns), then $(CB)^T CB = B^T C^T CB = B^T B = A$.

A symmetric matrix that satisfies condition (a), (b), or (c) is called *positive*. If A satisfies (d), (e), or (f), A is called *positive definite*.

EXAMPLE 5.10 Is the matrix

$$A = \begin{bmatrix} 2 & -3 & -1 & 2 \\ -3 & 9 & 5 & 2 \\ -1 & 5 & 6 & 2 \\ 2 & 2 & 2 & 8 \end{bmatrix}$$

positive? positive definite? If so, factor $A = B^T B$.

Solution First find an orthogonal matrix Q such that $Q^T A Q = D$ is diagonal.

```
      +D←(⍉Q)+.×A+.×Q←JACOBI A
1.40E¯2     3.53E¯15    2.17E¯18    6.94E¯18
3.54E¯15    1.43E1      3.25E¯18    2.60E¯18
1.73E¯19    3.47E¯18    2.43E0      5.20E¯18
8.56E¯18    6.94E¯18    1.52E¯18    8.21E0
```

To three significant figures the eigenvalues are .014, 14.3, 2.42, and 8.21. Since the eigenvalues are all strictly positive, the matrix A is positive definite.

The proof of Proposition 5.11 shows that one choice of B is $B = (QM)^T$ where $M = D* \div 2$. Although the off-diagonal terms are negligible in D, notice what happens if we take square roots:

```
      D*÷2
1.18E¯1     5.94E¯8     1.47E¯9     2.63E¯9
5.95E¯8     3.79E0      1.80E¯9     1.61E¯9
4.16E¯10    1.86E¯9     1.56E0      2.28E¯9
2.93E¯9     2.63E¯9     1.23E¯9     2.87E0
```

Although this is still diagonal enough for almost all purposes, there is no point in introducing added error when we can avoid it. Since ⎕CT for this system is about 3E¯15, the expression D×10≠10+D will set the negligible terms equal to zero. We use 10, since the largest entry of D (14.3) has magnitude about 10.

```
      D×10≠10+D
0.014    0.000   0.000   0.000
0.000   14.300   0.000   0.000
0.000    0.000   2.430   0.000
0.000    0.000   0.000   8.210

      +M←(D×10≠10+D)*÷2
0.118   0.000   0.000   0.000
0.000   3.790   0.000   0.000
0.000   0.000   1.560   0.000
0.000   0.000   0.000   2.870
```

To three significant figures B is

```
      +B←⍉Q+.×M
 0.102   0.048  ¯0.011  ¯0.034
¯0.632   2.800   2.080   1.340
 0.369  ¯0.712   1.290  ¯0.342
 1.210  ¯0.806  ¯0.137   2.470
```

Always check the accuracy of the factorization by comparing A to $B^T B$.

```
      +E←A-(⍉B)+.×B
¯1.00E¯15   1.98E¯15   1.70E¯15   1.24E¯15
 1.98E¯15   2.21E¯15   5.69E¯16  ¯2.53E¯16
 1.70E¯15   5.69E¯16  ¯3.05E¯16  ¯6.73E¯16
 1.24E¯15  ¯2.53E¯16  ¯6.73E¯16  ¯7.36E¯16
```

Since ⎕CT for this system is about 3E¯15, this error matrix is negligible compared to the original data in the matrix A.

```
      S←⌈/,|A
      S=S+E
1  1  1  1
1  1  1  1
1  1  1  1
1  1  1  1
```

If we do not replace D by D×1≠1+D the error is much worse.

```
      +C←⍉Q+.×D*÷2
 0.102   0.048  ¯0.011  ¯0.034
¯0.632   2.800   2.080   1.340
 0.369  ¯0.712   1.290  ¯0.342
 1.210  ¯0.806  ¯0.137   2.470

      A-(⍉C)+.×C
 6.10E¯8    ¯1.36E¯7    ¯1.20E¯7    ¯8.83E¯8
¯1.36E¯7    ¯1.26E¯7    ¯3.48E¯8     1.14E¯8
¯1.20E¯7    ¯3.48E¯8     1.78E¯8     3.19E¯8
¯8.83E¯8     1.14E¯8     3.19E¯8     4.75E¯8
```
■

If A is any matrix (not necessarily square), then A^TA is a positive matrix and hence has nonnegative eigenvalues.

DEFINITION 5.8 The *singular values* of A are the positive square roots of the eigenvalues of A^TA.

The matrices A and A^TA have the same rank (Proposition 2.14 and Propositions 4.26 and 4.28), and if Q is orthogonal then $\Lambda = Q^T(A^TA)Q = (AQ)^TAQ$ has the same rank as A^TA or A (exercise 25). Suppose that the entries of A are numbers bers derived from experimental measurements. Then the entries will tend to contain statistical errors. Random perturbations of the entries tend to increase the rank. The "true" rank of such a matrix is often estimated using the singular values. The singular values of A are computed and arranged in descending order. A precipitous drop in the size of the $(k + 1)$th singular value is taken as evidence that the true rank of A is k. This is illustrated in the next example.

EXAMPLE 5.11 Consider the matrix

```
      A
¯4   0  ¯4   1   5
 3   0   3   0  ¯3
 7  ¯2   9   1  ¯8
¯4   0  ¯4   1   5
 0  ¯2   2   2   0
```

which has rank equal to 3.

```
      ECHELON A
 1.00000E0     0.00000E0     1.00000E0    ¯4.33681E¯19   ¯1.00000E0
¯1.38778E¯17   1.00000E0    ¯1.00000E0     0.00000E0      1.00000E0
¯1.38778E¯17   0.00000E0     0.00000E0     1.00000E0      1.00000E0
 1.04083E¯17   0.00000E0     3.46945E¯18   8.67362E¯19   ¯6.93889E¯18
 0.00000E0     0.00000E0     0.00000E0     0.00000E0      0.00000E0
```

we can add some "statistical error" to A by using the APL random number generator `?` (Roll). The expression `?N` picks a random integer from 1 to N. Thus, for example, the expression `¯5+?9` picks a random integer between -4 and 4. Roll is a scalar function and operates componentwise on arrays.

Let us add some random noise to A in the fourth decimal place.

```
      +A←A+(¯5+?5 5⍴9)÷1E4
¯4.000300   0.000100  ¯3.999600   1.000400   5.000300
 2.999700  ¯0.000200   3.000000   0.000100  ¯3.000000
 6.999700  ¯1.999900   8.999700   1.000400  ¯7.999700
¯4.000300  ¯0.000400  ¯3.999800   0.999800   4.999800
 0.000300  ¯1.999900   1.999900   2.000400   0.000400
```

Now the rank of A is 5.

```
      ECHELON A
 1.00000E0      9.42972E¯15   ¯1.36453E¯14   ¯5.27356E¯16   ¯1.55709E¯14
¯5.98480E¯14    1.00000E0     ¯1.73264E¯14    3.24914E¯15    2.25514E¯14
 1.13687E¯13    6.77115E¯14    1.00000E0     ¯1.42109E¯14   ¯2.27674E¯13
¯1.13687E¯13   ¯3.94702E¯14    1.13687E¯13    1.00000E0      0.00000E0
 2.27374E¯13    5.16549E¯14   ¯1.13687E¯13    2.84217E¯14    1.00000E0
```

Next compute the singular values of A by diagonalizing $B = A^T A$.

```
      Q←JACOBI B←(⍉A) +.×A
      +D←(⍉Q)+.×B+.×Q
 1.42542E¯7     1.06434E¯13    4.44089E¯16   ¯5.20417E¯18   ¯9.05719E¯17
 1.06414E¯13    3.70445E¯1     3.33067E¯16   ¯1.38778E¯17    4.04194E¯14
 3.462943¯16    3.61690E¯16    3.37513E2      7.28584E¯16    2.16686E¯16
¯1.05993E¯17   ¯2.99240E¯17    7.77156E¯16    1.61025E1     ¯1.37331E¯18
 3.97577E¯17    4.04439E¯14    1.80411E¯16   ¯4.94396E¯17    1.99966E¯10
```

The singular values of A are the square roots of the diagonal entries of D. We would like them sorted in descending order. The APL expression `V[⍒V]` sorts the components of `V` in descending order.

```
      S←(1 1⍉D)*÷2
      S[⍒S]
18.3715   4.0128   0.608642   0.000377547   0.0000141409
```

There is a drop of several orders of magnitude from the third to the fourth singular value. ■

The statements made above about "true" rank are misleading. The question "What is the true rank of A?" is not the sort one wants to ask. A better question is "What is the order of magnitude of the effects that I am neglecting if I assume that A has rank k?" In the example above, if we assume that the rank of A is 3, we are neglecting effects several orders of magnitude below the primary effects. For the sense in which one may "assume that A has rank k" see exercise 30.

We close this section with an application to plane geometry that illustrates the geometric significance of the singular values of a matrix.

First some terminology. We will call an affine transformation $f: \mathbf{R}^2 \to \mathbf{R}^2$ a *stretch* if $f(x, y) = (\alpha x, y)$, $\alpha \geq 0$ in some coordinate system. The map f stretches line segments parallel to a fixed line l by the factor α and does not change the length of line segments perpendicular to l (Figure 5.10). We get the formula $f(x, y) = (\alpha x, y)$ if we choose the x axis parallel to l.

The number $\alpha > 0$ is called the *stretch factor* and the line l is the stretch line. Of course if $\alpha < 1$ then f is really a compression.

PROPOSITION 5.12 Every affine transformation $f(X) = b + AX$ from $\mathbf{R}^2$ to $\mathbf{R}^2$ with A nonsingular can be factored as two stretches with perpendicular stretch lines followed by a congruence. The stretch factors are the singular values of A.

Proof We diagonalize the positive matrix A^TA. There is an orthogonal matrix Q such that $Q^T(A^TA)Q = \Lambda = D^2$ where the matrices Λ and D are diagonal. Since A is nonsingular, A^TA and D are nonsingular. Let $U = AQD^{-1}$. Then U is orthogonal. In fact

$$U^TU = D^{-1}Q^TA^TAQD^{-1} = D^{-1}D^2D^{-1} = I$$

Further,

$$UDQ^T = AQD^{-1}DQ^T = A$$

Now let us introduce the coordinate change

$$X = QX'$$

Since Q is orthogonal, this is an orthonormal coordinate change, and

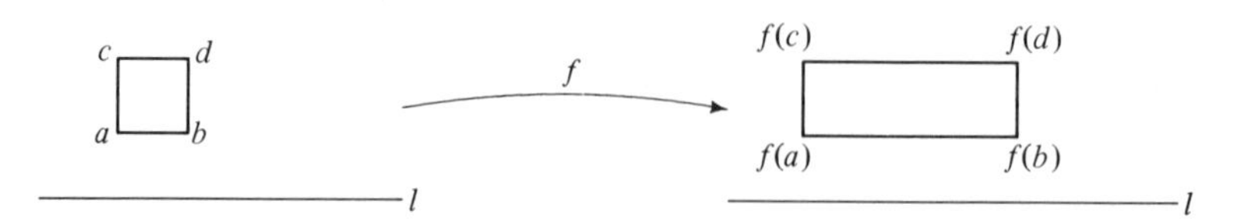

A stretch: $d(f(a), f(c)) = d(a, c)$, $d(f(a), f(b)) = \alpha d(a, b)$

FIGURE 5.10

$$\begin{aligned} Y &= b + AX \\ QY' &= b + AQX' \\ Y' &= Q^T(b + UDQ^TQX') \\ Y' &= (Q^Tb) + UDX' \end{aligned}$$

Thus in the new coordinate system $f(X') = g(h(X'))$, where $h(X') = DX'$ and $g(X') = Q^Tb + UX'$ — a congruence, since U is orthogonal.
Now

$$D = \begin{pmatrix} d_1 & 0 \\ 0 & d_2 \end{pmatrix} = \begin{pmatrix} d_1 & 0 \\ 0 & 1 \end{pmatrix} \begin{pmatrix} 1 & 0 \\ 0 & d_2 \end{pmatrix}$$

is a stretch with stretch factor d_2 and stretch line equal to the x' axis followed by a stretch with stretch factor d_1 and stretch line equal to the y' axis. The stretch factors d_1, d_2 are the singular values of A. ∎

The above proposition easily generalizes to affine maps $f: \mathbf{R}^n \to \mathbf{R}^n$ with non-singular linear part. The only change is that D is then the product of n stretches in mutually perpendicular directions.

The columns of the matrix Q defined in the proof of Proposition 5.12 are vectors parallel to the stretch lines. Such vectors are called *singular vectors* of A.

EXERCISES 5.2

1. Write an APL function *ANG* to be used by *JACOBI*. See Exercise 12 of Exercises 4.3.

In exercises 2 through 7 a symmetric 2-by-2 matrix is given. Use Proposition 4.11 to find an orthogonal matrix Q such that Q^TAQ is diagonal. Write down the eigenvalues of A. Is A positive or positive definite? If A is positive, factor A as $A = B^TB$.

2. $A = \begin{bmatrix} 3 & -1 \\ -1 & 3 \end{bmatrix}$ 3. $A = \begin{bmatrix} 1 & -2 \\ -2 & 4 \end{bmatrix}$ 4. $A = \begin{bmatrix} 1 & -1 \\ -1 & -1 \end{bmatrix}$

5. $A = \begin{bmatrix} 0 & 1 \\ 1 & 0 \end{bmatrix}$ 6. $A = \begin{bmatrix} 0 & -1 \\ -1 & 0 \end{bmatrix}$ 7. $A = \begin{bmatrix} \cos 2\theta & \sin 2\theta \\ \sin 2\theta & -\cos 2\theta \end{bmatrix}$

In exercises 8 through 12 use Proposition 4.11 to find the singular values of the matrix A.

8. $A = \begin{bmatrix} 1 & 2 \\ 3 & 4 \end{bmatrix}$ 9. $A = \begin{bmatrix} 4 & 3 \\ 2 & 1 \end{bmatrix}$ 10. $A = \begin{bmatrix} 4 & -3 \\ 3 & 4 \end{bmatrix}$

11. $A = \begin{bmatrix} 1 & 1 \\ 1 & 0 \\ 1 & -1 \end{bmatrix}$ 12. $A = \begin{bmatrix} 1 & 1 \\ -1 & 2 \\ 1 & 3 \\ -1 & 4 \end{bmatrix}$

(Computer assignment) In exercises 13 through 17 a symmetric matrix A is given. Use the function *JACOBI* to find an orthogonal matrix Q such that Q^TAQ is diagonal. Write down the eigenvalues of A. Is A positive or positive definite? If A is positive, factor A as $A = B^TB$.

13. $A = \begin{bmatrix} 1 & -1 & 1 \\ -1 & 2 & -3 \\ 1 & -3 & -2 \end{bmatrix}$

14. $A = \begin{bmatrix} 1 & 0 & 1 & 0 \\ 0 & 1 & 1 & 0 \\ 1 & 1 & 0 & 1 \\ 0 & 0 & 1 & 1 \end{bmatrix}$

15. $A = \begin{bmatrix} 2 & 1 & 2 \\ 1 & 1 & 1 \\ 2 & 1 & 2 \end{bmatrix}$

16. $A = \begin{bmatrix} 2 & 1 & 1 \\ 1 & 2 & 1 \\ 1 & 1 & 2 \end{bmatrix}$

17. The 6-by-6 Hilbert matrix. The $i;j$ component of this matrix is $1 \div (i + j - 1)$. Note: *JACOBI* converges slowly on Hilbert matrices.

(Computer assignment) Discuss the critical points of the quadratic function defined in Exercises 18 through 22.

18. $f(x, y, z) = 3 + 7x - 2y + 4z + x^2 - 4xy + 4yz - z^2 + y^2$
19. $f(x, y, z) = 2 - 7x + 5z - 2x^2 - y^2 - 2z^2 + 2xz - 2yz$
20. $f(x) = 3 + [1 \quad 0 \quad 7 \quad 2]X + X^T \begin{bmatrix} 1 & 0 & -1 & 0 \\ 0 & 2 & -2 & 0 \\ -1 & -2 & 1 & 1 \\ 0 & 0 & 1 & -1 \end{bmatrix} X$
21. $f(x, y, z) = -1 + x + 2y + 3z + 2x^2 - 2xy + y^2 + 2xz + 5z^2 + 2yz$
22. $f(x, y, z) = -6 + 2x + 4z + 2x^2 - 2xy + y^2 + 2xz + 5z^2 + 2yz$

23. This problem assumes familiarity with the optional Section 2.5*. The *critical points* of a function $f: \mathbf{R}^n \to \mathbf{R}$ are the points at which the derivative $(Df)(X)$ is zero. Using the computation of Example 2.48, show that a quadratic function $f(X) = C + BX + X^TAX$ has critical points if and only if there is a change of variable that eliminates the linear term.

24. Show that the singular values of an orthogonal matrix are all equal to 1.

25. Let A be n by n and P n by n and invertible. Show that A and $P^{-1}AP$ have the same rank.

    Hint: Propositions 4.25 and 4.26.

26. Prove Proposition 5.8.

    Hint: Proposition 2.1.

27. Show that the matrix Q defined on lines [5] and [6] of *JACOBI* is orthogonal.

    Hint: Imitate the calculation in the proof of Proposition 5.9.

28. Show that statement 2 of Proposition 5.9 implies statement 1.

29. (Singular-value decomposition) Let A be a matrix and Q an orthogonal matrix such that $Q^T(A^TA)Q = D^2$, where D is diagonal. Let `(1 1 ⍉ D)[K]` be the nonzero diagonal entries of D (i.e., A, A^TA, and D all have rank `⍴K`). The *singular-value decomposition* of A is

$$A = U\Lambda V$$

where $\Lambda = D[K; K]$ is diagonal, $V = Q[;K]^T$ has orthonormal columns, and $U = AQ[;K]D[;K]^{-1}$ has orthonormal columns. The object of this exercise is to verify the existence of the singular-value decomposition. Let L be the vector of column indices of D that are not components of K.

(a) Show that $Q[;K]^T(A^TA)Q[;K] = D[K;K]^2$.

Hint: $D[K;K]^2 = (Q^T(AA^T)Q)\,[K;K]$. Apply Proposition 5.8.

(b) Show that U has orthonormal columns; that is $U^TU = I$.

(c) Show that $AQ[;L] = 0$.

Hint: $0 = (Q^TA^TAQ)\,[;L] = Q^TA^TAQ[;L]$ and Proposition 2.14.

(d) Show that $QQ^T = Q[;K]Q[;K]^T + Q[;L]Q[;L]^T$.

Hint: $QQ^T = (QQ^T)\,[;] = (QQ^T)\,[\iota N; \iota N]$ where $N = 1 \uparrow \rho Q$. Apply Proposition 5.8.

(e) Show that $A = U\Lambda V$.

30. Apply the singular-value decomposition (exercise 29) to the "perturbed" matrix A of Example 5.11 by assuming that the two smallest eigenvalues are zero. Compute the error matrix $B = A - U\Lambda V$.

Generalized Dot Products

31. Let $\langle X, Y\rangle = X^TY$ denote the dot product in $\mathbf{R}^n$. Let $X = PX'$ be a coordinate change. Show that in the new coordinate system $\langle X', Y'\rangle = (X')^TAY'$, where A is symmetric and positive definite.

Hint: Substitute $X = PX'$ and $Y = PY'$.

32. Define $\langle\ ,\ \rangle$ to be a *generalized dot product* if $\langle X, Y\rangle = X^TAY$ for some symmetric and positive-definite matrix A. Show that if $\langle\ ,\ \rangle$ is a generalized dot product, then there is a coordinate change $X = PX'$ such that in the new coordinate system $\langle X', Y'\rangle = (X')^TY'$. Further, the new axes may be chosen mutually perpendicular (although there will be scale changes).

Hint: Factor $A = P^TP$.

33. Let $\langle\ ,\ \rangle$ be a dyadic function on $\mathbf{R}^n$ such that

(i) $\langle X, Y\rangle = \langle Y, X\rangle$, all X, Y in $\mathbf{R}^n$;

(ii) $\langle X, Y_1 + Y_2\rangle = \langle X, Y_1\rangle + \langle X, Y_2\rangle$, all X, Y_1, Y_2 in $\mathbf{R}^n$;

(iii) $\langle X, \alpha Y\rangle = \alpha\langle X, Y\rangle$ all X, Y in $\mathbf{R}^n$ and scalars α;

(iv) $\langle X, X\rangle \geq 0$ and $\langle X, X\rangle = 0$ if and only if $X = 0$.

(a) Show that we must have

(ii′) $\langle X_1 + X_2, Y\rangle = \langle X_1, Y\rangle + \langle X_2, Y\rangle$, all X_1, X_2, Y in $\mathbf{R}^n$.

(iii′) $\langle \alpha X, Y\rangle = \alpha\langle X, Y\rangle$, all X, Y in $\mathbf{R}^n$ and scalars α.

(b) Show that $\langle\ ,\ \rangle$ is a generalized dot product.

Hint: $A[i;j] = \langle I[;i], I[;j]\rangle$, where I is an identity matrix.

Simultaneous Diagonalization of Two Quadratic Forms

34. Let $f(X) = X^TAX$, $g(X) = X^TBX$ be two quadratic forms on $\mathbf{R}^n$ with f positive definite. Show that there is a coordinate change $X = PX'$ such that $f(X) = (X')^TX'$ and $g(X') = (X')^T\Lambda X'$, where Λ is diagonal.

Hint: Apply the result of exercise 32 to the generalized dot product $\langle X, Y\rangle = X^TAY$ to obtain a new coordinate system in which $\langle X'', Y''\rangle = (X'')^TY''$ is the ordinary dot product. Then apply Proposition 5.11 to the matrix of $g(X'')$.

5.3* (Multivariate Calculus) Optimization — the Second-Derivative Test

This section is a continuation of Section 2.5*. Proposition 2.23, the second-derivative test for max-min, states that the character of a critical point p of a twice continuously differentiable function $f\colon \mathbf{R}^n \to \mathbf{R}$ is the same as that of the quadratic form $Q(X) = X^T D^2 f(p) X$ at the origin.

We will not prove this fact but will give a brief heuristic argument.

The first three terms of the Taylor expansion of f at p are (Proposition 2.25)

$$P(X) = f(p) + Df(p)\,(X - p) + \frac{1}{2!}(X - p)^T D^2 f(p)\,(X - p)$$

If p is a critical point, then by definition $Df(p) = 0$. Changing coordinates so that p becomes the origin ($X = p + X'$) gives

$$P(X') = f(p) + \frac{1}{2!}(X')^T D^2 f(p) X'$$

for this quadratic function. Clearly $P(X')$ has a maximum or minimum at the origin if and only if $Q(X)$ does. Thus Proposition 2.23 states that the character of a critical point p of f is determined by the first three terms of the Taylor expansion at f.

The function *JACOBI* of Section 5.2 makes the second-derivative test practical by making it easy to analyze the function $Y = Q(X)$ at the origin. If $P = JACOBI\ D^2 f(p)$ and $X = PX'$, then in the X' coordinate system we have

$$\begin{aligned} Y' = Q(X') &= (X')^T \begin{bmatrix} \lambda_1 & & 0 \\ & \lambda_2 & \\ 0 & & \ddots\ \lambda_n \end{bmatrix} X' \\ &= \lambda_1 (x_1')^2 + \lambda_2 (x_2')^2 + \lambda_3 (x_3')^2 + \cdots + \lambda_n (x_n')^2 \end{aligned}$$

where the λ_i's are the eigenvalues of $D^2 f(p)$.

If $\lambda_i \geq 0$ for all i, then $Q(0) = 0$ is the minimum value of Q. If $\lambda_i \leq 0$ for all i, then $Q(0) = 0$ is the maximum value of Q.

If $\lambda_i > 0$ for all i, then the minimum value $Q(X) = 0$ is attained only for $X = 0$. If, for example, $\lambda_1 = 0$, then $Q((\alpha, 0, 0, \ldots, 0)) = 0$ for any α, so the minimum is attained on a whole line of points X. Similarly for maxima. In this case $Q = D^2 f(p)$ is singular and the second-derivative test fails.

If some $\lambda_i > 0$ and some $\lambda_j < 0$, then $0 = Q(0)$ is neither a maximum nor a minimum. In fact $Q(i{=}\iota n) = \lambda_i > 0$ and $Q(j{=}\iota n) = \lambda_j < 0$. We shall refer to this third case as a *saddle*. Thus we shall classify critical points into only three types: maxima, minima, and saddles.

EXAMPLE 5.12 The most difficult part of a max-min problem is finding the critical points. The geometry of the following example makes the critical points relatively easy to locate.

Let

$$p_1(x) = \begin{bmatrix} 2 \\ 0 \end{bmatrix} + \begin{bmatrix} \cos x \\ \sin x \end{bmatrix}, \qquad 0 \le x < 2\pi$$

$$p_2(y) = \begin{bmatrix} -2 \\ 0 \end{bmatrix} + \begin{bmatrix} \cos y \\ \sin y \end{bmatrix}, \qquad 0 \le y < 2\pi$$

As x and y vary, the points $p_1(x)$, $p_2(y)$ vary on the circles in Figure 5.11. Let $d(x, y) = \|p_2(y) - p_1(x)\|$. Then clearly d has a minimum at $(x, y) = (\pi, 0)$ and a maximum at $(x, y) = (0, \pi)$. Let us check this with the second-derivative test. As usual with distances we avoid square roots by working with $f(x, y) = d(x, y)^2$:

$$f(x, y) = \|p_2(g) - p_1(x)\|^2 = (4 + \cos x - \cos y)^2 + (\sin x - \sin y)^2$$

$$Df(x, y) = \begin{bmatrix} 2(4 + \cos x - \cos y)(-\sin x) + 2(\sin x - \sin y)\cos x \\ 2(4 + \cos x - \cos y)(\sin y) + 2(\sin x - \sin y)(-\cos y) \end{bmatrix}$$

$$= 2\begin{bmatrix} -4\sin x + \sin(x - y) \\ 4\sin y - \sin(x - y) \end{bmatrix}$$

$$D^2f(x, y) = 2\begin{bmatrix} -4\cos x + \cos(x - y) & -\cos(x - y) \\ -\cos(x - y) & 4\cos y + \cos(x - y) \end{bmatrix}$$

so

$$D^2f(\pi, 0) = 2\begin{bmatrix} 3 & 1 \\ 1 & 3 \end{bmatrix}; \qquad D^2f(0, \pi) = 2\begin{bmatrix} -5 & 1 \\ 1 & -5 \end{bmatrix}$$

Both of these matrices can be diagonalized by a 45° rotation. Taking

$$Q = \frac{1}{\sqrt{2}}\begin{bmatrix} 1 & -1 \\ 1 & 1 \end{bmatrix}$$

we obtain

$$Q^T D^2f(\pi, 0)Q = 2\begin{bmatrix} 4 & 0 \\ 0 & 2 \end{bmatrix}, \qquad Q^T D^2f(0, \pi)Q = 2\begin{bmatrix} -4 & 0 \\ 0 & -6 \end{bmatrix}$$

As expected, the critical point $(\pi, 0)$ is a minimum, since the eigenvalues of $D^2f(\pi, 0)$ are positive, and the critical point $(0,\pi)$ is a maximum, since the eigenvalues of $D^2f(0, \pi)$ are negative.

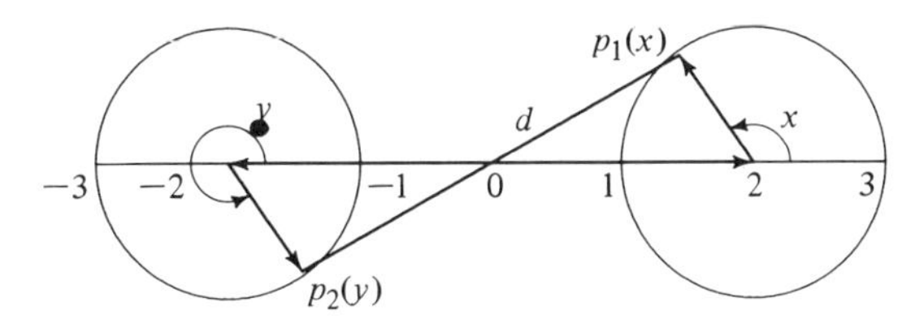

FIGURE 5.11

There are two other critical points for $f(x, y)$: $(0, 0)$ and (π, π).

$$D^2f(0,0) = 2\begin{bmatrix} -3 & -1 \\ -1 & 5 \end{bmatrix}, \qquad D^2f(\pi,\pi) = 2\begin{bmatrix} 5 & -1 \\ -1 & -3 \end{bmatrix}$$

From Figure 5.11, both these critical points have the same character, since they give congruent figures.

```
      +Q←JACOBI A←2 2ρ¯3 ¯1 ¯1 5
0.993 ¯0.122
0.122 0.993
      (⍉Q)+.×A+.×Q
¯3.12E0       ¯1.73E¯18
 0.00E0        5.12E0
```

The eigenvalues differ in sign; thus these two critical points are saddles. ■

The fact that the critical points $(0, 0)$ and (π, π) in the example above are saddles can also be deduced from Figure 5.11.

If we fix p_1 at $(3, 0) = p_1(0)$ and let p_2 vary in a small neighborhood of $(-1, 0) = p_2(0)$, then the length $l(y) = \|p_1(0) - p_2(y)\|$ has a local minimum at $y = 0$. On the other hand if we fix p_2 at $(-1, 0) = p_2(0)$ and let p_1 vary in a small neighborhood of $(3, 0) = p_1(0)$, then $l(x) = \|p_1(x) - p_2(0)\|$ has a local maximum at $x = 0$.

To rephrase: holding x fixed, y has a local minimum at $(0, 0)$, whereas, holding y fixed, x has a local maximum at $(0, 0)$.

This means that the curve $y = 0$ on the surface $z = d(x, y) = \|p_1(x) - p_2(y)\|$ has a maximum at $(0, 0)$, whereas the curve $x = 0$ on this surface has a local minimum. Thus the surface must be a saddle (Figure 5.12).

More generally consider the coordinate curves obtained by letting each x_i vary while the rest of the variables remain constant. If some of these curves have

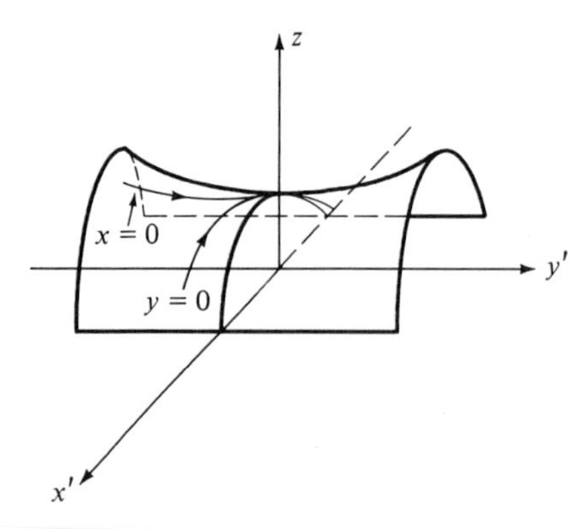

FIGURE 5.12

maxima and others have minima at the critical points, then the critical point must be a saddle. In general, however, this is all that can be said without diagonalizing the second derivative of the critical point. It can happen that these curves all have minima, say, and the point is nonetheless a saddle (exercise 1).

The use of *JACOBI* in Example 5.12 was not necessary. Proposition 4.11 could have been used, since only two variables were involved. *JACOBI* is needed for three or more variables, however.

EXAMPLE 5.13 Verify that the point $(1, 0, -1, 0, 2, 2)$ is a critical point of the function

$$f(x, y, z, w, u, v) = (x-z)^2 + (y-w)^2 + u[(x-2)^2 + y^2 - 1] + v[(z+2)^2 + w^2 - 1]$$

What is the character (i.e., max, min, saddle?) of this critical point?

Solution

$$Df = \begin{bmatrix} 2(x-z) + 2u(x-2) \\ 2(y-w) + 2uy \\ -2(x-z) + 2v(z+2) \\ -2(y-w) + 2vw \\ (x-2)^2 + y^2 - 1 \\ (z+2)^2 + w^2 - 1 \end{bmatrix}$$

and $Df(1, 0, -1, 0, 2, 2) = 0$, so we do have a critical point.

$$D^2f = 2\begin{bmatrix} 1+u & 0 & -1 & 0 & x-2 & 0 \\ 0 & 1+u & 0 & -1 & y & 0 \\ -1 & 0 & 1+v & 0 & 0 & z+2 \\ 0 & -1 & 0 & 1+v & 0 & w \\ x-2 & y & 0 & 0 & 0 & 0 \\ 0 & 0 & z+2 & w & 0 & 0 \end{bmatrix}$$

Setting $A = \frac{1}{2}D^2f(1, 0, -1, 0, 2, 2)$ and $Q = JACOBI\ A$, we obtain

```
        A
 3   0  ¯1   0  ¯1   0
 0   3   0  ¯1   0   0
¯1   0   3   0   0   1
 0  ¯1   0   3   0   0
¯1   0   0   0   0   0
 0   0   1   0   0   0
```

```
      (⍉Q)+.×A+.×Q
2.41E0      0.00E0      ¯2.48E¯16   0.00E0      ¯2.17E¯18   1.19E¯18
0.00E0      2.00E0      0.00E0      ¯3.47E¯18   0.00E0      0.00E0
¯2.45E¯16   0.00E0      4.24E0      0.00E0      2.06E¯18    2.20E¯18
0.00E0      1.73E¯18    0.00E0      4.00E0      0.00E0      0.00E0
¯1.73E¯18   0.00E0      8.67E¯19    0.00E0      ¯4.14E¯1    2.50E¯16
8.67E¯19    0.00E0      2.17E¯18    0.00E0      2.50E¯16    ¯2.36E¯1
```

Since there are both positive and negative eigenvalues, the point is a saddle point. ∎

EXERCISES 5.3*

1. $f(x, y) = x^2 - 3xy + y^2$, being quadratic, has $p = (0, 0)$ as its only critical point.
   (a) Show that the functions $h(x) = f(x, 0)$, $g(y) = f(0, y)$ both have minima at $(0, 0)$.
   (b) Show that $(0, 0)$ is a saddle point of f.
2. Let $p_1(x)$, $p_2(y)$, $p_3(z)$ be the indicated vectors in Figure 5.13, $0 \leq x, y, z < 2\pi$. Let

$$f(x, y, z) = \|p_1(x)\|^2 + \|p_2(y)\|^2 + \|p_3(z)\|^2$$

   (a) Show that

$$f(x, y, z) = 15 + 4(\cos x + \sin y - \cos z)$$

   (b) Compute Df and find the critical points of f.
   (c) Compute D^2f and classify the critical points of f.
3. (Computer assignment) Let f be the function of Example 5.13. Verify that the following points are critical points of f and determine their character.
   (a) $(3, 0, -3, 0, -6, -6)$
   (b) $(1, 0, -3, 0, 4, 4)$
   (c) $(3, 0, -1, 0, -4, -4)$

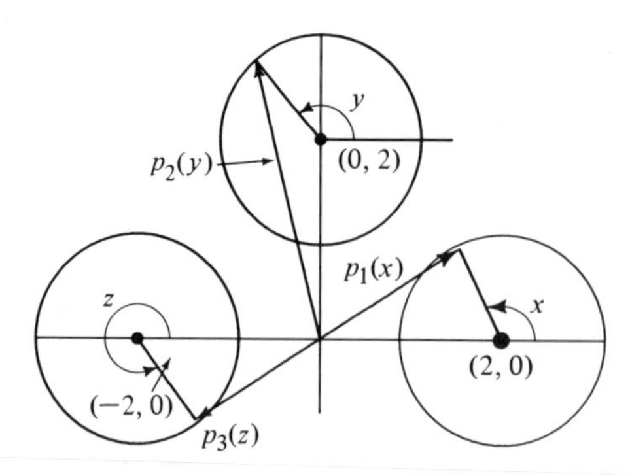

FIGURE 5.13

5.4 Perpendicular Projections and Least Squares

In this section we exhibit the connection between the ideas of perpendicularity and shortest distance. It is an elaboration of the theorem from plane geometry that the shortest distance from a point to a line is attained at the foot of the perpendicular dropped from the point to the line. There is a close connection between these ideas, the least-squares approximations treated in Section 2.3 and the left inverse computed by ⌹.

We begin with the idea of breaking a vector into two components, one parallel to and one perpendicular to a given flat. First some notation.

DEFINITION 5.9 Let S be a set of vectors in $\mathbf{R}^n$. The set of vectors in $\mathbf{R}^n$ perpendicular to S is denoted by $S^\perp$. If S is a subspace of $\mathbf{R}^n$, then $S^\perp$ is called the *orthogonal complement* of S.

We use the notation $\{v_1, v_2, \ldots, v_n\}$ to denote the set of vectors consisting of $v_1, v_2, \ldots, v_n$. If S is the empty set, we take $S^\perp = \mathbf{R}^n$.

PROPOSITION 5.13 The set of vectors $S^\perp$ is always a subspace of $\mathbf{R}^n$. If S is a subspace of $\mathbf{R}^n$ and S is generated by $v_1, v_2, \ldots, v_k$, then $S^\perp = \{v_1, v_2, \ldots, v_k\}^\perp$.

Proof We use Proposition 4.24 to verify that $S^\perp$ is a subspace. The zero vector is in $S^\perp$, since the zero vector is perpendicular to all vectors.

A vector v is in $S^\perp$ if and only if $v \cdot s = 0$ for every s in S. Writing v and s as column vectors, this becomes $v^T s = 0$ and

$$\begin{aligned}(\alpha v)^T s &= \alpha v^T s \\ (v + w)^T s &= (v^T + w^T)s \\ &= v^T s + w^T s\end{aligned}$$

so if v and w are in $S^\perp$, so is αv and $v + w$. Last, suppose that $A[;i] = v_i$. Then S is the column space of A and $v \perp v_i$ for $i = 1, 2, \ldots, k$ if and only if $v^T A = 0$. Suppose that v is in $\{v_1, \ldots, v_k\}^\perp$ and s is in S. Then $s = AT$ for some $T = (t_1, t_2, \ldots, t_k)$, so $v^T s = v^T AT = 0 \cdot T = 0$ and v is in $S^\perp$. Conversely, if v is in $S^\perp$, then $v^T A = 0$, since each v_i is in S. ∎

Suppose that S is the column space of A. Then v is in $S^\perp$ if and only if $v^T A = 0$ or, alternately, $A^T v = 0$. This means that $S^\perp$ is the null space of A^T and hence can be computed by Gaussian reduction.

EXAMPLE 5.14 Compute $S^\perp$ if

$$S = \{(1, 1, 5, 9), (-2, -1, -7, -13), (-1, -3, -11, -19)\}$$

Solution

```
      A
1   ¯2   ¯1
1   ¯1   ¯3
5   ¯7  ¯11
9  ¯13  ¯19
      +E←ECHELON ⍉A
1.000  0.000  2.000  4.000
0.000  1.000  3.000  5.000
0.000  0.000  0.000  0.000
```

The null space is given by

$$X = t_1 \begin{bmatrix} -2 \\ -3 \\ 1 \\ 0 \end{bmatrix} + t_2 \begin{bmatrix} -4 \\ -5 \\ 0 \\ 1 \end{bmatrix}$$

That is, $S^\perp$ is the subspace with basis $(-2, -3, 1, 0)$ and $(-4, -5, 0, 1)$. ∎

If S is the column space of A, then $\dim S = \operatorname{rank} A = \operatorname{rank} A^T = n - \dim(\text{null space } A^T)$ by Propositions 4.26 and 4.28. Thus $\dim(S) + \dim(S^\perp) = n$. Notice that S and $S^\perp$ intersect in a point, the origin. This is because a vector v in both S and $S^\perp$ is perpendicular to itself; that is, $v^T v = 0$ and hence $v = 0$.

The next proposition shows that one may always break a vector into components parallel and perpendicular to a given flat or subspace.

PROPOSITION 5.14 Let S be a subspace of $\mathbf{R}^n$ and let v be a vector in $\mathbf{R}^n$. There are unique vectors $v^\parallel$ in S and $v^\perp$ in $S^\perp$ such that

$$v = v^\parallel + v^\perp$$

Proof Let the columns of A be a basis for S and the columns of B a basis for $S^\perp$. We wish to find $v^\parallel = AX$ and $v^\perp = BY$ such that

$$v = v^\parallel + v^\perp = AX + BY = [A \mid B]\begin{bmatrix} X \\ \hline Y \end{bmatrix}$$

Since $\dim(S) + \dim(S^\perp) = n$, the matrix $[A \mid B]$ is n by n. If $[A \mid B]$ is invertible, then the above equation has a unique solution and the proposition is proved. To show $[A \mid B]$ invertible it is sufficient to show that $[A \mid B]$ has independent columns.

Suppose that $[X_0 \mid Y_0]^T$ is a solution of the corresponding homogeneous equation. Then

$$0 = [A \mid B]\left[\frac{X_0}{Y_0}\right] = AX_0 + BY_0$$

and hence $AX_0 = -BY_0$. This means that the vector $v = AX_0 = B(-Y_0)$ is in both S and $S^\perp$. Hence $v = 0$, which means that

$$AX_0 = 0, \qquad BY_0 = 0$$

But the matrices A and B have independent columns, hence $X_0 = 0$ and $Y_0 = 0$. We have shown that the homogeneous equation has only the trivial solution, so $[A \mid B]$ has independent columns. ∎

DEFINITION 5.10 With the notation of Proposition 5.14, $v^\parallel$ is the *component of* v *parallel to* S; $v^\perp$ is the *component of* v *perpendicular to* S. The vector $v^\parallel$ is also called the *perpendicular projection of* v *onto* S.

In the plane and in space $v^\parallel$ is the foot of the perpendicular dropped from the endpoint of the vector v to the subspace S (Figure 5.14).

PROPOSITION 5.15 Let S be a subspace in $\mathbf{R}^n$, v a vector in $\mathbf{R}^n$. The unique vector in S that is closest to v is $v^\parallel$, the perpendicular projection of v onto S.

Proof Let w be any vector in S. Notice that $v^\parallel - w$ in S implies $(v^\parallel - w) \perp v^\perp$.

$$\begin{aligned} d(v, w)^2 &= \|v - w\|^2 = \|v - v^\parallel + v^\parallel - w\|^2 = \|v^\perp + v^\parallel - w\|^2 \\ &= \|v^\perp\|^2 + \|v^\parallel - w\|^2 \text{ by the Pythagorean theorem (Proposition 5.2)} \\ &\geq \|v^\perp\|^2 \end{aligned}$$

and we get equality only when $v^\parallel - w = 0$. ∎

Suppose that S is the column space of A. To compute $v^\parallel$ we could proceed in the following way. Since $v^\parallel$ lies in S, $v^\parallel = AX$ for some X. Since $v^\perp$ is perpendicu-

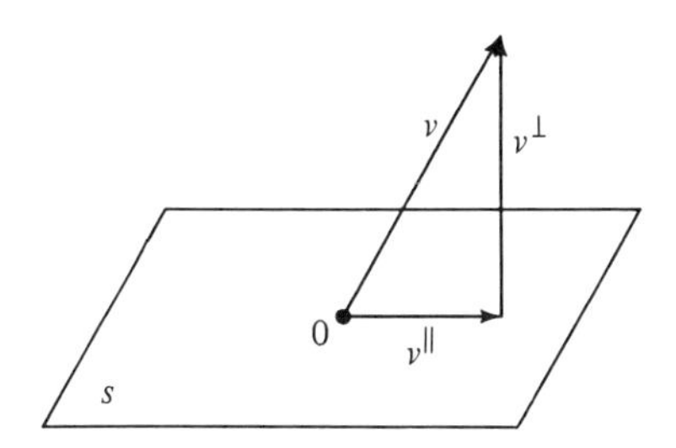

FIGURE 5.14

lar to S, $A^T v^\perp = 0$. Since $v^\perp = v - v^\parallel$, we have that $v^\parallel = AX$ for any solution X of

$$A^T(v - AX) = 0$$

or

$$A^T AX = A^T v$$

This gives an algebraic proof of Proposition 2.15. With this approach we do not need to assume that A has a left inverse. Although $v^\parallel$ is unique given v and S, the equation $v^\parallel = AX$ may have many solutions if A does not have linearly independent columns. Note that the function $(B - AX)$ +.* 2 of Proposition 2.15 is just $\|B - AX\|^2$.

PROPOSITION 5.16 Let A be a matrix, B a vector, and set $f(X) = \|B - AX\|$. The function f has a minimum at any point X satisfying the equation

$$A^T AX = A^T B \qquad \blacksquare$$

In particular if A has linearly independent columns, then $v^\parallel$ is just A +.× v ⌹ A.

EXAMPLE 5.15 Let S be the subspace of $\mathbf{R}^5$ generated by $(1, 1, 0, 2, 3)$ and $(3, 2, -7, 1, 0)$. Let $v = \iota 5$. Write $v = v^\parallel + v^\perp$. What is the distance from v to S?

Solution Let us denote $v^\parallel$ by PROJV and $v^\perp$ by PERPV. Then

```
      +A←⍉2 5⍴1 1 0 2 3 3 2 ¯7 1 0
1  3
1  2
0 ¯7
2  1
3  0
      +PROJV←A+.×(V←⍳5)⌹A
0.795  1.17  2.59  3.44  5.72
      +PERPV←V-PROJV
0.205  0.835  0.406  0.558  ¯0.719
      +DIST←(PERPV+.*2)*÷2
1.32
```

$\blacksquare$

ACCOUNTING FOR VARIATION

Let us take a closer look at how these geometric ideas involving perpendicular projections and shortest distances relate to the least-squares curve-fitting calculations of Section 2.3.

Suppose that some characteristic of a group of individuals is measured, resulting in a vector of values $Y = (y_1, y_2, \ldots, y_n)$. For example, y_i might be the weight of the ith individual. The numbers y_i, of course, vary from person to person, and we want to "account for" this variation. More precisely, we wish to find out "how much" of the variation in weight is "due to" variation in height.

First we need a measure of the "variation." The standard approach is to take the standard deviation or variance. Let $\bar{y}$ be the mean of the y_i's. Then the variance is

$$\sigma_y^2 = \frac{\|Y - \bar{Y}\|^2}{n}$$

where $\bar{Y} = n_\rho \bar{y}$. The n in the denominator is just a scaling factor here and is usually omitted. The measure of variation is the *total sum of squares:*

$$\text{total } SSQ = \|Y - \bar{Y}\|^2 = \Sigma\,(y_i - \bar{y})^2$$

In Section 2.3, recall that we discussed fitting a polynomial of degree k, $y = c_0 + c_1x + \cdots + c_kx^k$, to measured data points (x_i, y_i) (cf. Proposition 2.16). If $k = 0$, the resulting constant polynomial is $y = \bar{y}$. In fact the system $AX = B$ to be solved is

$$\begin{bmatrix} 1 \\ 1 \\ 1 \\ \vdots \\ 1 \end{bmatrix} [c_0] = \begin{bmatrix} y_1 \\ y_2 \\ y_3 \\ \vdots \\ y_n \end{bmatrix}$$

Multiplying through by $A^T = [1 \ldots 1]$, we obtain

$$[1 \quad 1 \ldots 1] \begin{bmatrix} 1 \\ 1 \\ \vdots \\ 1 \end{bmatrix} [c_0] = [1 \quad 1 \ldots 1] \begin{bmatrix} y_1 \\ y_2 \\ \vdots \\ y_n \end{bmatrix}$$

or

$$nc_0 = \sum y_i$$

$$c_0 = \frac{1}{n} \sum y_i = \bar{y}$$

This means that, in $\mathbf{R}^n$, $\bar{Y} = Ac_0$ is the perpendicular projection of Y onto the column space of $A = (n, 1)_\rho 1$ (Figure 5.15).† In particular, $(Y - \bar{Y}) \perp \bar{Y}$.

Now suppose that the weights of the individuals are $X = (x_1, x_2, \ldots, x_n)$. To find the least-squares line $y = c_0 + c_1x$ expressing weight as a function of height

†Next time you see someone calculating averages, be sure to tell him this.

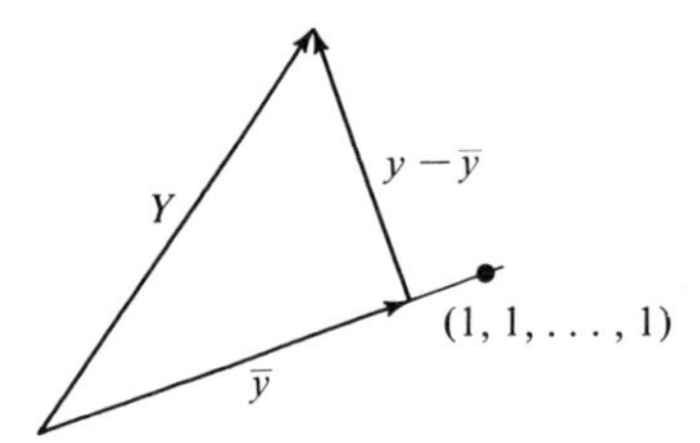

FIGURE 5.15

we find the least-squares solution of $AC = Y$, where $C = (c_0, c_1)$ and the columns of A are $n_\rho 1$ and X. Then $\widehat{Y} = AC$ is the perpendicular projection of Y onto the column space of A (Figure 5.16).

Now $Y\perp = Y - \widehat{Y}$ is perpendicular to all vectors in S. In particular, $(Y - \widehat{Y}) \perp (\widehat{Y} - \overline{Y})$, and hence by the Pythagorean theorem

$$\text{total } SSQ = \|Y - \overline{Y}\|^2 = \|\overline{Y} - \widehat{Y}\|^2 + \|Y - \widehat{Y}\|^2$$

The quantity $\|Y - \overline{Y}\|^2$ is the portion of the total sum of squares "due to" X and is called the *regression sum of squares*. The remaining portion of the total sum of squares, $\|Y - \widehat{Y}\|^2$ is the "unexplained" portion and is called the *residual sum of squares*.

$$\text{Total } SSQ = \text{regression } SSQ + \text{residual } SSQ$$

Dividing through by the total sum of squares gives

$$1 = \frac{\|\overline{Y} - \widehat{Y}\|^2}{\|Y - \overline{Y}\|^2} + \frac{\|Y - \widehat{Y}\|^2}{\|Y - \overline{Y}\|^2}$$

The first term is said to give the fraction of the total SSQ that is attributable to variation in X and the second to give the fraction of the total SSQ left unexplained by X.

Incidentally, reasoning geometrically from Figure 5.16 can be justified. We will see shortly that any flat may be coordinatized with an orthonormal coordinate

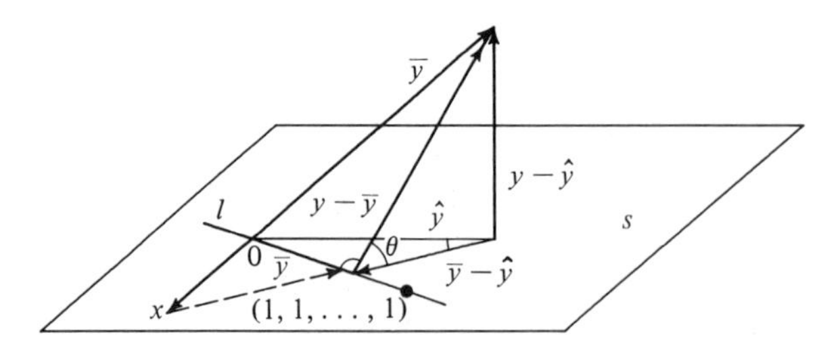

FIGURE 5.16

system. Putting an orthonormal coordinate system on the flat defined by the origin, $n_\rho 1$, Y, and X turns it into a copy of Euclidean space.

EXAMPLE 5.16 Table 5.1 contains the weights (in kilograms) and heights (in inches) of 15 boys aged 9 to 11 engaged in competitive swimming. (The data are extracted from Table 1.3 in Exercises 1.3.) What percentage of the variation in weight in this group is attributable to variation in height?

Solution Let Y be the vector of weights and X the vector of heights. Doggedly applying Proposition 2.16, we have

```
      +TOTSSQ←(Y-Y⌹X∘.*0,⍳0)+.*2
473
      +RESSSQ←(Y-A+.×Y⌹A←X∘.*0,⍳1)+.*2
218
      +REGSSQ←TOTSSQ-RESSSQ
255
      100×(REGSSQ,RESSSQ)÷TOTSSQ
53.9  46.1
```

Thus 54 percent of the variation in weight for this group can be attributed to variation in height, leaving 46 percent "unexplained" by variation in height alone. ■

Figure 5.16 gives more information.

Let l be the line through 0 and $n_\rho 1$. The vector $Y - \overline{Y}$ is perpendicular to l by construction, and, since $Y - \widehat{Y}$ is perpendicular to all of S, it is in particular

TABLE 5.1

Height	*Weight*
50.50	25.50
56.80	35.40
56.50	33.60
54.50	36.90
54.30	28.60
53.00	30.40
53.00	29.10
55.30	30.90
49.50	24.10
56.00	30.90
54.80	26.40
58.80	37.30
51.20	26.80
57.00	47.30
50.60	26.50

perpendicular to l. Thus the plane of the triangle containing these two vectors is perpendicular to l, and in particular

$$(\overline{Y} - \widehat{Y}) \perp l$$

Now if z is a vector in $\mathbf{R}^n$ and $\overline{z}$ is the average of the components of z, then the foot of the perpendicular to l is $n_\rho \overline{z}$. Thus the fact that $\overline{Y} - \widehat{Y}$ is perpendicular to l means that the average of the components of $\widehat{Y}$ is $\overline{y}$.

If we let $\overline{x}$ denote the average of the components of X and set $\overline{X} = n_\rho \overline{x}$, it follows that $X - \overline{X}$ is parallel to $\overline{Y} - \widehat{Y}$, since they lie in the same plane as l and are both perpendicular to l. Hence, if θ is the angle between $X - \overline{X}$ and $Y - \overline{Y}$,

$$\frac{\text{regression } SSQ}{\text{total } SSQ} = \left(\frac{\| Y - \overline{Y} \|}{\| \overline{Y} - \widehat{Y} \|}\right)^2 = \cos^2 \theta = r_{xy}^2$$

where r_{xy} is the xy correlation coefficient (cf. Section 5.1).

Thus, since r_{xy} is symmetric in x and y, we have

$$\begin{aligned} \text{fraction of variation in } y \text{ due to } x &= r_{xy}^2 \\ &= \text{fraction in variation of } x \text{ due to } y \end{aligned}$$

Proposition 5.16 gives one method of computing perpendicular projections. We will now develop an alternate approach that yields a method of constructing orthonormal coordinate systems as well. The method works when $A^T A$ is singular. In this case ⌹ produces a DOMAIN ERROR.

If S is the column space of A and v is a vector in $\mathbf{R}^n$, then $v^{\|}$, the perpendicular projection of v onto S is computed by first getting any solution, w, of

$$A^T A X = A^T v$$

and then setting

$$v^{\|} = Aw$$

Now if the columns of A are orthonormal, then $A^T A = ID\ n$, and so $A^T A X = A^T v$ reduces to $X = A^T v$. Thus $w = A^T v$ and $v^{\|} = A A^T v$.

PROPOSITION 5.17 Let the columns of U form an orthonormal basis of the subspace S of $\mathbf{R}^n$. Let v be a vector in $\mathbf{R}^n$. Then

$$\begin{aligned} \text{the component of } v \text{ parallel to } S \text{ is} \quad & v^{\|} = U U^T v \\ \text{the component of } v \text{ perpendicular to } S \text{ is} \quad & v^{\perp} = (I - U U^T) v \end{aligned}$$

where $I = ID\ n$. ∎

To apply this proposition we need a method of constructing orthonormal bases of a subspace. In concrete terms this means that given a matrix V we wish to construct a matrix U with orthonormal columns and the same column space as V. It is quite useful to be able to do this in such a way that the column space of $V[;\iota k]$, the first k columns of V, is contained in the column space of $U[;\iota k]$, the first k columns of U. We do not demand that these column spaces be equal because we wish to allow V to have linearly dependent columns.

The method of constructing U described below is a variant of the *Gram-Schmidt* process and is intended for hand computation. A method more suited to the requirements of machine computation, the Householder algorithm, is developed in Section 5.5, where it is used to define an APL function *ORTHO* that returns U given V.

THE GRAM-SCHMIDT PROCESS

Given a matrix V, we first construct a matrix W with orthogonal columns such that the column space of $V[;\iota k]$ is contained in the column space of $W[;\iota k]$. We then define the final matrix U by $U[;i] = W[;i]/\|W[;i]\|$. In what follows let $v_i = V[;i]$, $w_i = W[;i]$, $u_i = U[;i]$.

Start with $w_1 = v_1$.

Next we need to choose $w_2 \perp v_1$, such that $[v_1|v_2]$ and $[w_1|w_2]$ have the same column space. Let S_1 be the subspace generated by v_1. Write $v_2 = v_2^{\|} + v_2^{\perp}$ with respect to S_1 and take $w_2 = v_2^{\perp}$ [Figure 5.17(a)]. Then v_1, v_2 and w_1, w_2 span the same subspace and $w_1 \perp w_2$. To get w_3 we let S_2 be the column space of $[w_1|w_2]$ or $[v_1|v_2]$, write $v_3 = v_3^{\|} + v_3^{\perp}$ with respect to S_2 [Figure 5.17(b)], and set $w_3 = v_3^{\perp}$.

In general suppose that $w_1, w_2, \ldots, w_k$ have been constructed such that $[w_1|w_2|\ldots|w_k]$ and $[v_1|v_2|\ldots|v_k]$ have the same column space and $w_i \perp w_j$ for $i \neq j$. Let S_k be the column space of $[w_1|w_2|\ldots|w_k]$ and write $v_{k+1} = v_{k+1}^{\|} + v_{k+1}^{\perp}$ with respect to this subspace. Take $w_{k+1} = v_{k+1}^{\perp}$.

Claim $[v_1|v_2|\ldots|v_k|v_{k+1}]$ and $[w_1|w_2|\ldots|w_k|w_{k+1}]$ have the same column space.

Proof By induction assume that the column spaces of $[v_1|v_2|\ldots|v_k] = V_k$ and $[w_1|w_2|\ldots|w_k] = W_k$ are equal. We must show that v_{k+1} is in the column space of $[W_k|w_{k+1}]$ and that w_{k+1} is in the column space of $[V_k|v_{k+1}]$.

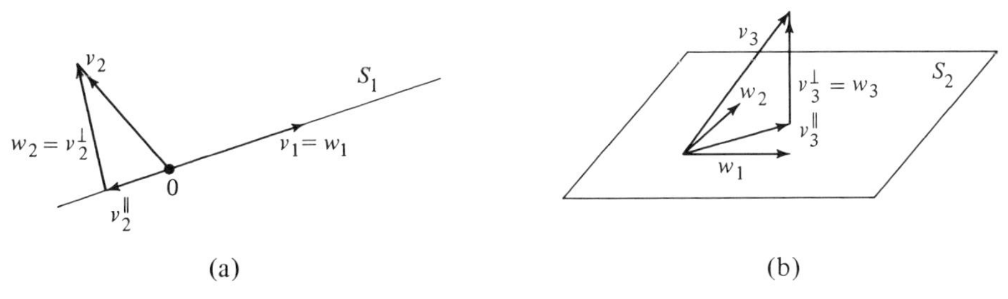

FIGURE 5.17

Since $v_{k+1}^{\|}$ is in S_k, there are vectors T_1, T_2 such that

$$V_k T_1 = v_{k+1}^{\|} = W_k T_2$$

Hence

$$w_{k+1} = v_{k+1} - v_{k+1}^{\|} = [V_k \,|\, v_{k+1}]\left[\frac{-T_1}{1}\right]$$

$$v_{k+1} = v_{k+1}^{\|} + w_{k+1} = [W_k \,|\, w_{k+1}]\left[\frac{T_2}{1}\right] \qquad \blacksquare$$

Suppose that V has n columns. After n steps this process yields a matrix W with orthogonal columns and the same column space as V.

Now suppose that v_3, for example, had been a linear combination of v_1 and v_2. Then [see Figure 5.17(b)] v_3 would lie in S_2 and we would get $v_3 = v_3^{\|}$, $w_3 = v_3^{\perp} = 0$. Thus if V had linearly dependent columns, then W would have some zero columns. The rank of W is the rank of $W^T W$, which is a diagonal matrix with $(W^T W)[i;i] = \|w_i\|^2$. Thus the rank of W is the number of nonzero columns, and deleting the zero columns from W (`W←(0∨.≠W)/W`) gives an orthogonal basis for the column space of V. Hence setting $u_i = w_i/\|w_2\|$ gives an orthonormal basis of the column space of V.

Now consider the details of the computation. Suppose that $w_1, w_2, \ldots, w_k$ have been constructed. Let $W = [w_1 \,|\, w_2 \,|\, \ldots \,|\, w_k]$. Then

$$w_{k+1} = v_{k+1} - v_{k+1}^{\|}$$

where

$$v_{k+1}^{\|} = WT$$

and T is any solution of

$$W^T W X = W^T v_{k+1}$$

The matrix $W^T W$ is diagonal. If we discard zero vectors w_i as they occur, then $W^T W$ is invertible and $(W^T W)^{-1}$ is trivial to calculate, since the inverse of a diagonal matrix is obtained by inverting the diagonal entries. In this case

$$w_{k+1} = v_{k+1} - W(W^T W)^{-1} W^T v_{k+1}$$

Notice, however, that any nonzero multiple of w_{k+1} can be substituted for w_{k+1}. This will change no perpendicularity relations, and w_{k+1} will be normalized in the end anyway. This means that at any stage of the computation we are free to multiply

$$v_{k+1} - W(W^T W)^{-1} W^T v_{k+1}$$

by any number that will eliminate unwanted fractions.

EXAMPLE 5.17 Apply the Gram-Schmidt process to the three vectors $v_1 = (1, 0, 1)$, $v_2 = (2, 1, 3)$, $v_3 = (1, 1, 1)$.

Solution $w_1 = v_1$, so we start with

$$W = \begin{bmatrix} 1 \\ 0 \\ 1 \end{bmatrix}$$

Then

$$\begin{aligned} w_2 &= v_2 - W(W^T W)^{-1} W^T v_2 \\ &= \begin{bmatrix} 2 \\ 1 \\ 3 \end{bmatrix} - \begin{bmatrix} 1 \\ 0 \\ 1 \end{bmatrix} [\tfrac{1}{2}][1 \quad 0 \quad 1] \begin{bmatrix} 2 \\ 1 \\ 3 \end{bmatrix} \\ &= \begin{bmatrix} 2 \\ 1 \\ 3 \end{bmatrix} - \tfrac{1}{2} \begin{bmatrix} 1 \\ 0 \\ 1 \end{bmatrix} [5] \\ &= \begin{bmatrix} 2 \\ 1 \\ 3 \end{bmatrix} - \tfrac{1}{2} \begin{bmatrix} 5 \\ 0 \\ 5 \end{bmatrix} \\ &\parallel \begin{bmatrix} 4 \\ 2 \\ 6 \end{bmatrix} - \begin{bmatrix} 5 \\ 0 \\ 5 \end{bmatrix} \dagger \end{aligned}$$

or

$$w_2 = \begin{bmatrix} -1 \\ 2 \\ 1 \end{bmatrix}$$

Next

$$W = \begin{bmatrix} 1 & -1 \\ 0 & 2 \\ 1 & 1 \end{bmatrix}$$

and

$$\begin{aligned} w_3 &= v_3 - W(W^T W)^{-1} W^T v_3 \\ &= \begin{bmatrix} 1 \\ 1 \\ 1 \end{bmatrix} - \begin{bmatrix} 1 & -1 \\ 0 & 2 \\ 1 & 1 \end{bmatrix} \begin{bmatrix} \frac{1}{2} & 0 \\ 0 & \frac{1}{6} \end{bmatrix} \begin{bmatrix} 1 & 0 & 1 \\ -1 & 2 & 1 \end{bmatrix} \begin{bmatrix} 1 \\ 1 \\ 1 \end{bmatrix} \end{aligned}$$

† For "$\parallel$" read "is parallel to."

$$= \begin{bmatrix}1\\1\\1\end{bmatrix} - \begin{bmatrix}1 & -1\\0 & 2\\1 & 1\end{bmatrix}\begin{bmatrix}\frac{1}{2} & 0\\0 & \frac{1}{6}\end{bmatrix}\begin{bmatrix}2\\2\end{bmatrix} = \begin{bmatrix}1\\1\\1\end{bmatrix} - \begin{bmatrix}1 & -1\\0 & 2\\1 & 1\end{bmatrix}\begin{bmatrix}1\\\frac{1}{3}\end{bmatrix}$$

$$\parallel \begin{bmatrix}3\\3\\3\end{bmatrix} - \begin{bmatrix}1 & -1\\0 & 2\\1 & 1\end{bmatrix}\begin{bmatrix}3\\1\end{bmatrix} = \begin{bmatrix}3\\3\\3\end{bmatrix} - \begin{bmatrix}2\\2\\4\end{bmatrix} = \begin{bmatrix}1\\1\\-1\end{bmatrix}$$

So

$$W = \begin{bmatrix}1 & -1 & 1\\0 & 2 & 1\\1 & 1 & -1\end{bmatrix} \quad \text{and} \quad U = \begin{bmatrix}\frac{1}{\sqrt{2}} & -\frac{1}{\sqrt{6}} & \frac{1}{\sqrt{3}}\\ 0 & \frac{2}{\sqrt{6}} & \frac{1}{\sqrt{3}}\\ \frac{1}{\sqrt{2}} & \frac{1}{\sqrt{6}} & -\frac{1}{\sqrt{3}}\end{bmatrix} \qquad \blacksquare$$

EXAMPLE 5.18 Let S be the subspace of $\mathbf{R}^4$ generated by $(1, -1, 1, -1)$, $(1, 0, 0, 1)$, $(1, 2, 0, 0)$. What is the perpendicular distance from $v = (1, 2, 3, 4)$ to S? What is the vector in S closest to v? Same questions for $(0, 1, 1, 0)$.

Solution Writing $v = v^{\parallel} + v^{\perp}$, the distance is $\|v^{\perp}\|$ and the closest vector is $v^{\parallel}$. We use the Gram-Schmidt process to compute a matrix U whose columns are an orthonormal basis of S.

$$w_1 = (1, -1, 1, -1)$$

$$w_2 = v_2 - W(W^TW)^{-1}W^Tv_2$$

$$w_2 = \begin{bmatrix}1\\0\\0\\1\end{bmatrix} - \begin{bmatrix}1\\-1\\1\\-1\end{bmatrix}\left[\tfrac{1}{4}\right]\begin{bmatrix}1 & -1 & 1 & -1\end{bmatrix}\begin{bmatrix}1\\0\\0\\1\end{bmatrix} = \begin{bmatrix}1\\0\\0\\1\end{bmatrix}$$

$$w_3 = v_3 - W(W^TW)W^Tv_3$$

$$= \begin{bmatrix}1\\2\\0\\0\end{bmatrix} - \begin{bmatrix}1 & 1\\-1 & 0\\1 & 0\\-1 & 1\end{bmatrix}\begin{bmatrix}\frac{1}{4} & 0\\0 & \frac{1}{2}\end{bmatrix}\begin{bmatrix}1 & -1 & 1 & -1\\1 & 0 & 0 & 1\end{bmatrix}\begin{bmatrix}1\\2\\0\\0\end{bmatrix}$$

$$= \begin{bmatrix}1\\2\\0\\0\end{bmatrix} - \begin{bmatrix}1 & 1\\-1 & 0\\1 & 0\\-1 & 1\end{bmatrix}\begin{bmatrix}\frac{1}{4} & 0\\0 & \frac{1}{2}\end{bmatrix}\begin{bmatrix}-1\\1\end{bmatrix} \parallel \begin{bmatrix}4\\8\\0\\0\end{bmatrix} - \begin{bmatrix}1 & 1\\-1 & 0\\1 & 0\\-1 & 1\end{bmatrix}\begin{bmatrix}-1\\2\end{bmatrix}$$

$$= \begin{bmatrix} 4 \\ 8 \\ 0 \\ 0 \end{bmatrix} - \begin{bmatrix} 1 \\ 1 \\ -1 \\ 3 \end{bmatrix} = \begin{bmatrix} 3 \\ 7 \\ 1 \\ -3 \end{bmatrix}$$

so

$$W = \begin{bmatrix} 1 & 1 & 3 \\ -1 & 0 & 7 \\ 1 & 0 & 1 \\ -1 & 1 & -3 \end{bmatrix}$$

Taking

$$U = \begin{bmatrix} \frac{1}{2} & \frac{1}{\sqrt{2}} & \frac{3}{2\sqrt{17}} \\ -\frac{1}{2} & 0 & \frac{7}{2\sqrt{17}} \\ \frac{1}{2} & 0 & \frac{1}{2\sqrt{17}} \\ -\frac{1}{2} & \frac{1}{\sqrt{2}} & -\frac{3}{2\sqrt{17}} \end{bmatrix}$$

we have that $v^{\|} = UU^T v$ for any v in **R** by Proposition 5.17. Computing both projections at once, we have

$$[v_1^{\|} \,|\, v_2^{\|}] = UU^T[v_1 \,|\, v_2]$$

$$= \begin{bmatrix} \frac{1}{2} & \frac{1}{\sqrt{2}} & \frac{3}{2\sqrt{17}} \\ -\frac{1}{2} & 0 & \frac{7}{2\sqrt{17}} \\ \frac{1}{2} & 0 & \frac{1}{2\sqrt{17}} \\ -\frac{1}{2} & \frac{1}{\sqrt{2}} & -\frac{3}{2\sqrt{17}} \end{bmatrix} \begin{bmatrix} \frac{1}{2} & -\frac{1}{2} & \frac{1}{2} & -\frac{1}{2} \\ \frac{1}{\sqrt{2}} & 0 & 0 & \frac{1}{\sqrt{2}} \\ \frac{3}{2\sqrt{17}} & \frac{7}{2\sqrt{17}} & \frac{1}{2\sqrt{17}} & -\frac{3}{2\sqrt{17}} \end{bmatrix} \begin{bmatrix} 1 & 0 \\ 2 & 1 \\ 3 & 1 \\ 4 & 0 \end{bmatrix}$$

$$= \begin{bmatrix} \frac{1}{2} & \frac{1}{\sqrt{2}} & \frac{3}{2\sqrt{17}} \\ -\frac{1}{2} & 0 & \frac{7}{2\sqrt{17}} \\ \frac{1}{2} & 0 & \frac{1}{2\sqrt{17}} \\ -\frac{1}{2} & \frac{1}{\sqrt{2}} & -\frac{3}{2\sqrt{17}} \end{bmatrix} \begin{bmatrix} -1 & 0 \\ \frac{5}{\sqrt{2}} & 0 \\ \frac{4}{\sqrt{17}} & \frac{4}{\sqrt{17}} \end{bmatrix} = \begin{bmatrix} \frac{40}{17} & \frac{6}{17} \\ \frac{45}{34} & \frac{14}{17} \\ -\frac{13}{34} & \frac{2}{17} \\ \frac{5}{34} & -\frac{6}{17} \end{bmatrix}$$

so

$$(1, 2, 3, 4)^{\|} = \tfrac{1}{34}(80, 45, -13, 5), \ \|(1, 2, 3, 4)^{\perp}\| \sim 5.35$$
$$(0, 1, 1, 0)^{\|} = \tfrac{1}{17}(6, 14, 2, -6), \ \|(0, 1, 10)^{\perp}\| \sim 1.03 \qquad \blacksquare$$

From time to time we have encountered reflections in the plane and space. In Section 5.5 we will need the concept of reflection in a subspace of $\mathbf{R}^n$. Notice that the following definition for subspaces coincides with the usual concept when $n = 2, 3$ (Figure 5.18).

DEFINITION 5.11 Let S be a subspace of $\mathbf{R}^n$ and let v be a vector in $\mathbf{R}^n$. The *reflection* $R_s(v)$ *of* v *in* S *is* $R_s(v) = v - 2v^{\perp}$.

PROPOSITION 5.18 Let S be a subspace of $\mathbf{R}^n$ and let the columns of U form an orthonormal basis for S. Then perpendicular projection onto S and reflection in S are linear transformations with matrices

$$P_S = UU^T, \qquad R_S = 2P_S - I$$

respectively. Further

$$P_{S^{\perp}} = I - P_S$$

Proof By Proposition 5.17, $v^{\|} = UU^T v$ and $v^{\perp} = (I - UU^T)v$ for any v in $\mathbf{R}^n$. Thus perpendicular projection, being multiplication by a matrix, is linear. $R_S(v) = v - 2v^{\perp} = [I - 2(I - P_S)]v = (2P_S - I)v$. $\blacksquare$

Notice that there are in general an infinite number of choices for U but only one P_S. The projection and reflection matrices have some special properties that are geometrically evident in $\mathbf{R}^2$ and $\mathbf{R}^3$ (apply the transformation twice).

The projection matrices satisfy

$$P_S = P_S^T, \qquad P_S^2 = P_S$$

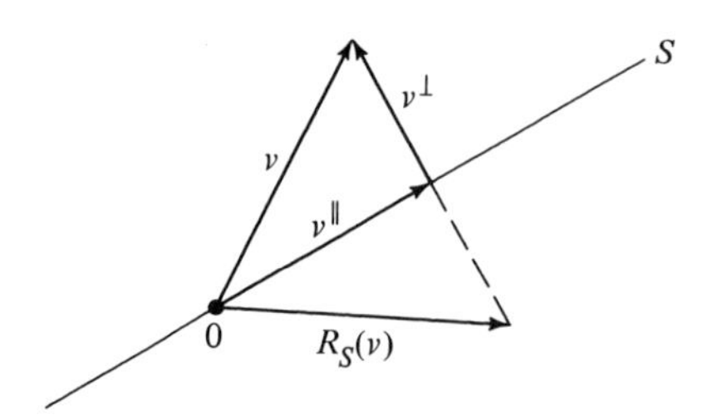

FIGURE 5.18

and the reflection matrices satisfy

$$R_S = R_S^T = R_S^{-1}$$

The verification of these facts is left as exercise 32.

So far we have discussed projections and reflections only for subspaces when clearly the concepts should apply to an arbitrary flat. If we want to discuss projection onto a flat F, we just translate to the origin, perform the projection, and translate back.

To be specific, suppose that p is a point in F. Let S be the unique flat through the origin parallel to F. To project X onto F, we translate F and X by $-p$, perform the projection, and translate back. The resulting transformation is affine. In fact

$$\begin{aligned} Y &= p + P_S(X - p) \\ &= (p - P_S p) + P_S X \\ &= P_{S^\perp} p + P_S X \\ &= p^\perp + P_S X \end{aligned}$$

Notice that if q is another point in F, then $p - q$ is in S, and so

$$\begin{aligned} p^\perp &= P_{S^\perp p} = P_{S^\perp}(q + p - q) \\ &= P_{S^\perp} q + P_{S^\perp}(p - q) \\ &= q^\perp + 0 \end{aligned}$$

So the resulting formula is independent of the choice of p (Figure 5.19).

For reflection in F we have

$$\begin{aligned} Y &= p + R_S(X - p) \\ &= (P - R_S p) + R_S X \\ &= (I - 2P_S - I)p + R_S X \\ &= 2P_{S^\perp} p + R_S X \\ &= 2p^\perp + R_S X \end{aligned}$$

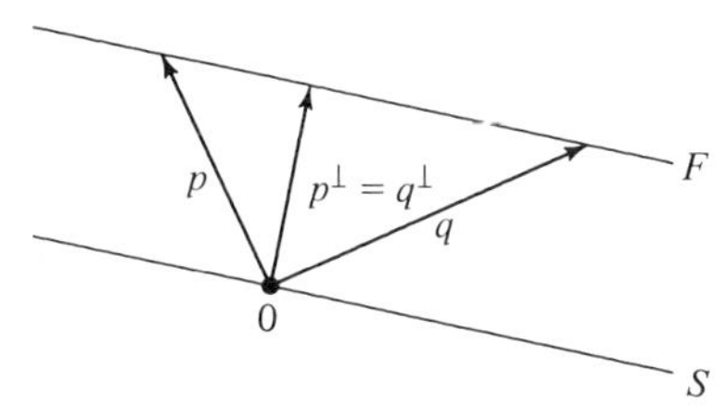

FIGURE 5.19

EXAMPLE 5.19 Let F be the plane in $\mathbf{R}^3$ through the points $p_0 = (1, 0, 0)$, $p_1 = (0, 1, 0)$, and $p_2 = (0, 0, 1)$. Give algebraic expressions for the affine functions:

perpendicular projection onto F, reflection in F

Solution The subspace S parallel to F is generated by $v_1 = p_1 - p_0$ and $v_2 = p_2 - p_0$. We use the Gram-Schmidt process to find an orthornormal basis of S.

$$w_1 = v_1 = (-1, 1, 0), \qquad W = \begin{bmatrix} -1 \\ 1 \\ 0 \end{bmatrix}$$

$$w_2 = v_2 - W(W^TW)^{-1}W^Tv_2$$

$$= \begin{bmatrix} -1 \\ 0 \\ 1 \end{bmatrix} - \begin{bmatrix} -1 \\ 1 \\ 0 \end{bmatrix}[\tfrac{1}{2}][-1 \quad 1 \quad 0]\begin{bmatrix} -1 \\ 0 \\ 1 \end{bmatrix} = \begin{bmatrix} -1 \\ 0 \\ 1 \end{bmatrix} - \tfrac{1}{2}\begin{bmatrix} -1 \\ 1 \\ 0 \end{bmatrix}[1]$$

$$= \begin{bmatrix} -2 \\ 0 \\ 2 \end{bmatrix} - \begin{bmatrix} -1 \\ 1 \\ 0 \end{bmatrix} = \begin{bmatrix} -1 \\ -1 \\ 2 \end{bmatrix}$$

So

$$W = \begin{bmatrix} -1 & -1 \\ 1 & -1 \\ 0 & 2 \end{bmatrix}$$

gives an orthogonal basis of S and

$$U = \begin{bmatrix} -\dfrac{1}{\sqrt{2}} & -\dfrac{1}{\sqrt{6}} \\ \dfrac{1}{\sqrt{2}} & -\dfrac{1}{\sqrt{6}} \\ 0 & \dfrac{2}{\sqrt{6}} \end{bmatrix}$$

gives an orthonormal basis of S. Thus

$$P_S = UU^T = W(W^TW)^{-1}W^T = \begin{bmatrix} -1 & -1 \\ 1 & -1 \\ 0 & 2 \end{bmatrix}\begin{bmatrix} \frac{1}{2} & 0 \\ 0 & \frac{1}{6} \end{bmatrix}\begin{bmatrix} -1 & 1 & 0 \\ -1 & -1 & 2 \end{bmatrix}$$

$$= \tfrac{1}{6}\begin{bmatrix} -1 & -1 \\ 1 & -1 \\ 0 & 2 \end{bmatrix}\begin{bmatrix} 3 & 0 \\ 0 & 1 \end{bmatrix}\begin{bmatrix} -1 & 1 & 0 \\ -1 & -1 & 2 \end{bmatrix}$$

$$= \frac{1}{6}\begin{bmatrix} -1 & -1 \\ 1 & -1 \\ 0 & 2 \end{bmatrix}\begin{bmatrix} -3 & 3 & 0 \\ -1 & -1 & 2 \end{bmatrix}$$

$$= \frac{1}{6}\begin{bmatrix} 4 & -2 & -2 \\ -2 & 4 & -2 \\ -2 & -2 & 4 \end{bmatrix} = \frac{1}{3}\begin{bmatrix} 2 & -1 & -1 \\ -1 & 2 & -1 \\ -1 & -1 & 2 \end{bmatrix}$$

$$p_0^\perp = p_0 - P_S p_0 = \begin{bmatrix} 1 \\ 0 \\ 0 \end{bmatrix} - \frac{1}{3}\begin{bmatrix} 2 \\ -1 \\ -1 \end{bmatrix} = \frac{1}{3}\left(\begin{bmatrix} 3 \\ 0 \\ 0 \end{bmatrix} - \begin{bmatrix} 2 \\ -1 \\ -1 \end{bmatrix}\right) = \frac{1}{3}\begin{bmatrix} 1 \\ 1 \\ 1 \end{bmatrix}$$

and

$$R_S = 2P_S - I = \frac{1}{3}\left(\begin{bmatrix} 4 & -2 & -2 \\ -2 & 4 & -2 \\ -2 & -2 & 4 \end{bmatrix} - \begin{bmatrix} 3 & 0 & 0 \\ 0 & 3 & 0 \\ 0 & 0 & 3 \end{bmatrix}\right)$$

$$R_S = \frac{1}{3}\begin{bmatrix} 1 & -2 & -2 \\ -2 & 1 & -2 \\ -2 & -2 & 1 \end{bmatrix}$$

Thus, $\perp$ projection on F:

$$\frac{1}{3}\left(\begin{bmatrix} 1 \\ 1 \\ 1 \end{bmatrix} + \begin{bmatrix} 2 & -1 & -1 \\ -1 & 2 & -1 \\ -1 & -1 & 2 \end{bmatrix} X\right)$$

Reflection in F:

$$\frac{1}{3}\left(\begin{bmatrix} 2 \\ 2 \\ 2 \end{bmatrix} + \begin{bmatrix} 1 & -2 & -2 \\ -2 & 1 & -2 \\ -2 & -2 & 1 \end{bmatrix} X\right)$$

Notice that $R_S^2 = I$, $P_S^2 = P_S$, as claimed above. ∎

The vector $p^\perp = P_{S^\perp} p$ in Figure 5.19 is the point in F closest to the origin. We record this useful fact for future reference. The proof is left as exercise 33.

PROPOSITION 5.19 Let F be a k-flat in $\mathbf{R}^n$ and let S be the unique k-flat through the origin parallel to F. The point in F closest to the origin is $p^\perp = P_{S^\perp} p$, where p is any point in F. ∎

EXERCISES 5.4

In exercises 1 through 10 a set S is given. Find a basis of $S^{\perp}$.

1. $S = \{(1, 2)\}$ in $\mathbf{R}^2$
2. $S = \{(a, b)\}$ in $\mathbf{R}^2$
3. $S = \mathbf{R}^n$
4. $S = \{0\}$ in $\mathbf{R}^n$
5. $S = \{(1, -2, -4), (1, -1, -1)\}$ in $\mathbf{R}^3$
6. $S = \{(1, -1, -1, -1), (-2, 3, 5, 7), (-2, 0, -4, -8)\}$ in $\mathbf{R}^4$
7. $S = \{(1, 1, 0), (-2, 1, 3), (-1, 0, 1)\}$ in $\mathbf{R}^3$
8. $S = \{(1, -1, -1, 0), (-2, 3, 5, 1), (0, 1, 3, 2)\}$ in $\mathbf{R}^4$
9. $S = \{(1, -1, -2, 0, -1), (1, 0, 0, 1, 8), (-2, 3, 6, 2, 16), (0, -2, -4, -3, -23)\}$ in $\mathbf{R}^5$
10. S is the set of solutions of

$$\begin{aligned} x + y - z &= 0 \\ 2x + 4y - z &= 0 \end{aligned}$$

11. (a) Let S be a subspace of $\mathbf{R}^n$. Show that $(S^{\perp})^{\perp} = S$.
    (b) Use (a) to solve exercise 10.
12. Use exercise 11 to show that if S is the solution space of a system of homogeneous linear equations, then $S^{\perp}$ is the space generated by the coefficient vectors.

In exercises 13 through 17 a matrix A and a vector v are given. Use Proposition 5.16 to find the components of v parallel and perpendicular to the column space of A. What is the distance from v to the column space of A?

13. $v = (1, 1)$, $A = \begin{bmatrix} 1 \\ 0 \end{bmatrix}$
14. $v = (1, 2, 3)$, $A = \begin{bmatrix} 1 & 1 \\ 1 & -1 \\ 1 & 1 \end{bmatrix}$
15. $v = (1, 0, 1)$, $A = \begin{bmatrix} 1 & 1 \\ 0 & 1 \\ 0 & 1 \end{bmatrix}$
16. (Computer assignment) $v = (1, -1, 1, -1)$, $A = \begin{bmatrix} 1 & 2 & 3 \\ 1 & 1 & 0 \\ -1 & 0 & 1 \\ -1 & 3 & 4 \end{bmatrix}$
17. (Computer assignment) $v = (4, 0, 4, 0)$, $A = \begin{bmatrix} 1 & -3 & 2 \\ 0 & 1 & 1 \\ 2 & 0 & 0 \\ 1 & 4 & 6 \end{bmatrix}$

(Computer assignment) Exercises 18, 19, and 20 refer to Table 1.3 of Exercises 1.3.

18. What percentage of the posttest weight variation of the experimental group can be attributed to posttest height variation?

19. What percentage of the pretest weight variation of the control group can be attributed to pretest height variation?

20. What percentage of the posttest weight variation of the control group can be attributed to posttest height variation?

Apply the Gram-Schmidt process to the sets of vectors given in exercises 21 through 25.

21. $\{(1, 0), (1, 1)\}$
22. $\{(1, 1), (2, 2), (2, 3)\}$
23. $\{(1, 1, 0), (1, 0, 1), (0, 1, 1)\}$
24. $\{(1, 0, 1), (1, 1, 0), (0, 1, 1)\}$
25. $\{(1, -1, 1, -1), (1, 0, 1, 0), (0, 1, 0, 1), (1, 1, 1, 1)\}$

In exercises 26 through 31 a subspace S and a vector v are given. Compute the matrices P_S, $P_{S^\perp}$, and R_S. Use these matrices to compute $v^\parallel$, $v^\perp$, and the distance from v to S.

26. $v = (1, 1)$, S generated by $(1, 2)$
27. $v = (1, 1)$, S generated by $(1, -1)$
28. $v = (1, 1)$, S generated by $(-3, -3)$
29. $v = (1, 2, 3)$, S generated by $(1, 0, 0)$, $(1, 1, 1)$
30. $v = (1, 2, 3)$, S generated by $(1, 1, 1)$, $(1, 0, 1)$
31. $v = (1, 0, 1, 0)$, S generated by $(1, -1, 1, -1)$, $(0, 1, 0, 1)$, $(1, 0, 0, 1)$
32. Let S be a subspace of $\mathbf{R}^n$. Show $P_S = P_S^T$, $P_S^2 = P_S$, $R_S = R^T$, and $R_S = R_S^{-1}$.
33. Prove Proposition 5.19.

    Hint: Let q be another point in the flat and show that the Pythagorean theorem applies to the triangle with sides $p^\perp$, $q - p^\perp$, and q.

In exercises 34 through 49 a k-flat F in $\mathbf{R}^n$ is given. Find expressions for the affine functions consisting of perpendicular projection onto F and reflection in F. What is the point in F closest to the origin?

34. The line in $\mathbf{R}^2$ through the points $(-1, 0)$ and $(0, -1)$
35. The line $x + y = 1$ in $\mathbf{R}^2$
36. The line through $(-1, 1)$ and $(0, 3)$
37. The point $(1, 1)$. (A point is a 0-flat.)

    Hint: The subspace $\{(0, 0)\}$ is the column space of $U = 2\ 0\rho 0$. (Just go to the terminal and type `(2 0ρ0)+.×ι0` if you don't believe it.) Further, since $U^T U = ID\ 0$, we take U to have "orthonormal columns."
38. The plane in $\mathbf{R}^3$ through $(1, 0, 0)$, $(0, -1, 0)$, and $(0, 0, -1)$.
39. The plane in $\mathbf{R}^3$ through $(1, 4, 1)$, $(-6, 12, 1)$, and $(7, 7, 1)$
40. The line in $\mathbf{R}^3$ through $(1, 1, 1)$ and $(1, 1, 0)$
41. The line in $\mathbf{R}^3$ through $(1, 0, 1)$ and $(0, -1, 1)$
42. The point $(6, -7, 8)$ in $\mathbf{R}^3$ (see the hint for exercise 37)
43. The 2-flat in $\mathbf{R}^4$ through $(0, 0, 1, 1)$, $(1, -1, 1, -1)$, and $(1, 0, 1, 0)$
44. The 2-flat in $\mathbf{R}^4$ through $(1, 0, 1, 0)$, $(-1, 1, -1, 1)$, and $(1, 0, 0, 1)$
45. The 1-flat in $\mathbf{R}^4$ through $(1, 0, -1, 1)$ and $(1, -1, 1, 0)$

46. The 1-flat in $\mathbf{R}^4$ through (1, 1, 1, 1) and (0, 1, 0, 1)
47. The 3-flat in $\mathbf{R}^4$ through (1, 0, 0, 0), (0, 1, 0, 0), (0, 0, 1, 0), and (0, 0, 0, 1)
48. The 3-flat in $\mathbf{R}^4$ through $(-1, 0, 0, 1)$, $(1, 0, -1, 0)$, $(1, -1, 0, 0)$, and $(0, 0, 1, -1)$
49. The 0-flat $(1, 7, 6, -10)$ in $\mathbf{R}^4$ (see the hint for exercise 37)

5.5 The Householder Algorithm (Automatic Orthonormalization)

The Gram-Schmidt algorithm introduced in Section 5.4 is unreliable in the presence of round-off error (see exercise 1). For this reason another algorithm, due to A. S. Householder, is often used for machine computations. In this section we develop the Householder algorithm and use it to define APL functions for the computation of orthonormal bases and perpendicular projections.

We begin by looking at the Gram-Schmidt process from a different angle. Given a set of vectors, let us store them as the columns of a matrix A. Applying the Gram-Schmidt process to the columns of A produces a matrix U with several nice properties. The column spaces of U and A are the same, and the columns of U form an orthonormal set of vectors: $U^T U = I$. Thus the columns of U provide an orthonormal basis for the column space of A. Often this is all that is needed, but more is true. The kth column of A is in fact a linear combination of no more than the first k columns of U. (Recall that if the columns of A are not independent, then U will have fewer columns than A.) Suppose that

$$A[;k] = U[;\iota k]T_k$$

for some vector of coefficients T_k. Define a matrix R by

$$R[;k] = r\uparrow T_k$$

where r is the number of columns of U ($=$ the rank of A). It follows that

$$A = UR$$

where

$$R = \begin{bmatrix} T_1 & & & \\ & T_2 & & \\ & & T_3 & \\ & 0 & & \ddots \end{bmatrix}$$

The form exhibited for the matrix R is a weak version of a row echelon form, which we will call *upper echelon*.

DEFINITION 5.12 A matrix R is said to be in *upper echelon* form if

1. The zero rows, if any, of R are at the bottom.

   Note: The first nonzero entry of each nonzero row will be called a pivot entry.
2. Each pivot entry has only zeros *below* it in the same column.
3. The pivot entry in a given row is to the right of the pivot entries in the rows above; that is, if $R[i; j]$ and $R[k; l]$ are pivot entries and $i < k$, then $j < l$.

The term "pivot entry" is used to indicate that if one were to row-reduce R to row echelon form, the pivot entries would be used for pivots. An upper echelon form differs from an echelon form in that the pivot entries need not be ones and they may have nonzero entries above them in the same column. A matrix has only one echelon form but may be row-reduced to many different upper echelon forms.

Some special terminology is associated with the equation $A = UR$ above.

DEFINITION 5.13 Let A be a matrix. A *QR factorization* of A is an expression

$$A = QR$$

where Q has orthonormal columns and R is in upper echelon form.

Note that if $A = QR$ is a QR factorization, then multiplication by Q^T gives

$$Q^T A = Q^T QR = R$$

Thus multiplication by Q^T "reduces A to upper echelon form." By our definitions Q need not be invertible, however (it need not be square), and so R need not result from row-reducing A. In the Householder algorithm described below, an invertible Q is used.

EXAMPLE 5.20 In Example 5.17 of the previous section the Gram-Schmidt process was applied to the set of vectors $v_1 = (1, 0, 1)$, $v_2 = (2, 1, 3)$, and $v_3 = (1, 1, 1)$, producing the matrix

$$U = \begin{bmatrix} \frac{1}{\sqrt{2}} & -\frac{1}{\sqrt{6}} & \frac{1}{\sqrt{3}} \\ 0 & \frac{2}{\sqrt{6}} & \frac{1}{\sqrt{3}} \\ \frac{1}{\sqrt{2}} & \frac{1}{\sqrt{6}} & -\frac{1}{\sqrt{3}} \end{bmatrix}$$

In this case we let the columns of A be v_1, v_2, v_3. By the remark above, $R = U^TA$.

$$R = \begin{bmatrix} \frac{1}{\sqrt{2}} & 0 & \frac{1}{\sqrt{2}} \\ -\frac{1}{\sqrt{6}} & \frac{2}{\sqrt{6}} & \frac{1}{\sqrt{6}} \\ \frac{1}{\sqrt{3}} & \frac{1}{\sqrt{3}} & -\frac{1}{\sqrt{3}} \end{bmatrix} \begin{bmatrix} 1 & 2 & 1 \\ 0 & 1 & 1 \\ 1 & 3 & 1 \end{bmatrix} = \begin{bmatrix} \sqrt{2} & \frac{5}{\sqrt{2}} & \sqrt{2} \\ 0 & \frac{1}{\sqrt{6}} & \frac{2}{\sqrt{6}} \\ 0 & 0 & \frac{1}{\sqrt{3}} \end{bmatrix}$$

The QR factorization is

$$\begin{bmatrix} 1 & 2 & 1 \\ 0 & 1 & 1 \\ 1 & 3 & 1 \end{bmatrix} = \begin{bmatrix} \frac{1}{\sqrt{2}} & -\frac{1}{\sqrt{6}} & \frac{1}{\sqrt{3}} \\ 0 & \frac{2}{\sqrt{6}} & \frac{1}{\sqrt{3}} \\ \frac{1}{\sqrt{2}} & \frac{1}{\sqrt{6}} & -\frac{1}{\sqrt{3}} \end{bmatrix} \begin{bmatrix} \sqrt{2} & \frac{5}{\sqrt{2}} & \sqrt{2} \\ 0 & \frac{1}{\sqrt{6}} & \frac{2}{\sqrt{6}} \\ 0 & 0 & \frac{1}{\sqrt{3}} \end{bmatrix} \qquad \blacksquare$$

The Gram-Schmidt process produces QR factorizations. Conversely, any algorithm that produces QR factorizations can be used in place of the Gram-Schmidt process.

The Householder algorithm constructs QR factorizations by a process similar to row reduction. In fact, ⌹ uses the process to solve linear equations. The idea is to use orthogonal matrices instead of elementary matrices to row-reduce a matrix to upper echelon form. To do this we need an orthogonal matrix that can be used in place of pivot matrices to clean up a column.

First we describe how such an orthogonal matrix can be constructed. The construction results from the observation that if v and w are any two vectors in $\mathbf{R}^n$ with $\|v\| = \|w\|$, then there is a reflection matrix R such that $v = Rw$ and $w = Rv$. In fact, R can be taken to be a reflection in the subspace generated by $v + w$ [Figure 5.20(a)].

By Proposition 5.18 we may take $R = 2UU^T - I$, where U is a single-column matrix with $U[;1] = (v + w) \div \|v + w\|$.

The trick is to start with a vector v and set $w = (\|v\|, 0, 0, \ldots, 0)$; then $\|v\| = \|w\|$ and $w = Rv$ has zeros in every component but the first.

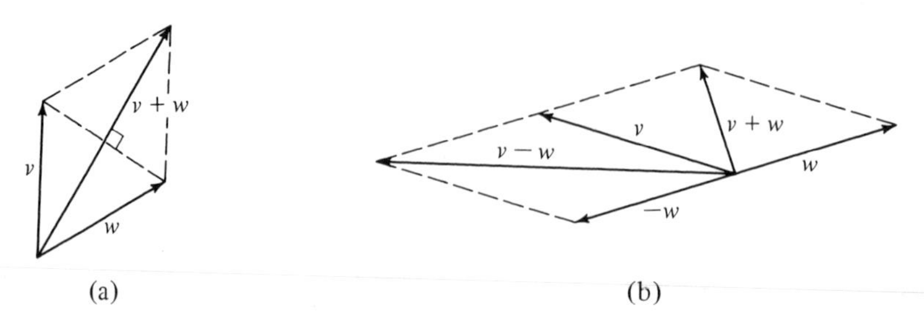

FIGURE 5.20

Given a matrix A, we take $v = A[;1]$ and construct R in such a way that $(\|v\|, 0, \ldots, 0) = Rv$. Then

$$RA = \left[\begin{array}{c|c} \|v\| & \\ 0 & \\ 0 & * \\ \vdots & \\ 0 & \end{array}\right]$$

We will call the reflection R a *special reflection*. Such reflections are members of a general class of reflections called *elementary reflectors* or *Householder transformations*. We will not define the more general class of reflections, since they will not be needed below.

In practice we do not always take $U[;1] = (v + w)/\|v + w\|$. If v is close to $-w$ [Figure 5.20(b)], then $\|v + w\|$ will be small and dividing by $\|v + w\|$ may cause numerical problems. In this case it would be better to reflect in the subspace generated by $v - w$ — that is, to reflect v into $(-\|v\|, 0, \ldots, 0)$.

Because of this problem the APL function *SPRFLCTR* defined below chooses R in such a way that the signs of $1{\uparrow}v$ and $1{\uparrow}w$ are the same unless $0 = 1{\uparrow}v$, in which case we take $0 < 1{\uparrow}w$. If $0 = 1{\downarrow}v$, take $R = \text{ID}\,\rho v$.

```
    ∇ Z←SPRFLCTR V ;L
[1]    Z←IDρV
[2]    L←(V+.*2)*÷2
[3]    →((L=0)∨L∧.=L+1↓V)/0
[4]    V[1]←V[1]+L×(V[1]≥0)-V[1]<0
[5]    Z←(2×V∘.×V÷V+.*2)-Z
    ∇
```

On line [1] the result is provisionally set equal to an identity matrix and on line [2] L is set equal to $\|v\|$. The function stops on line [3] if $\|v\| = 0$ or if $1{\downarrow}v$ is negligible compared to $\|v\|$.

Note that since $0 = 1{\downarrow}w$, the operation $v + w$ affects only $v[1]$ by adding $\pm L$ to $v[1]$. The sign of L is chosen and $v + w$ formed on line [4]. On line [5] the matrix UU^T is formed as an outer product of vectors.

We are now ready to describe the Householder algorithm for reducing a matrix to upper echelon form. As we did with Gaussian reduction, we will use a version of the algorithm that returns a matrix Q, orthogonal this time, such that $Q^TA = R$ is upper echelon. In this case, however, we are more interested in Q than in R.

HOUSEHOLDER ALGORITHM Let the matrix A be m by n.

1. Delete any zero columns from A. If A is a zero matrix, take $Q = ID\,m$ and stop. (Recall our convention that a matrix without rows or without columns is considered to be a zero matrix.)

2. Now let $Q_1 = SPRFLCTR\ A[;1]$. The matrix Q_1 is orthogonal and $Q_1 = Q_1^T$. The matrix $Q_1 A$ has zeros in the first column below the 1;1 position.

3. Let $\underline{A} = 1\ 1 \downarrow Q +.\times A$. By induction on the number of rows we may assume that we have constructed an orthogonal matrix $\underline{Q}$ such that

$$\underline{R} = \underline{Q}^T \underline{A}$$

is an upper echelon form. Define the matrix Q by

$$Q = Q_1 \left[\begin{array}{c|c} 1 & 0 \\ \hline 0 & \underline{Q} \end{array}\right]$$

and stop. ■

To see that this construction works, notice first that $Q^T A = R$ is upper echelon. In fact,

$$\begin{aligned} R = Q^T A &= \left[\begin{array}{c|c} 1 & 0 \\ \hline 0 & \underline{Q} \end{array}\right]^T Q_1 A \\ &= \left[\begin{array}{c|c} 1 & 0 \\ \hline 0 & \underline{Q}^T \end{array}\right] \left[\begin{array}{c|c} * & * \\ \hline 0 & \underline{A} \end{array}\right] \\ &= \left[\begin{array}{c|c} * & * \\ \hline 0 & \underline{R} \end{array}\right] \end{aligned}$$

(The possible presence of columns of zeros in the original A does not substantially affect this computation.) The matrix Q is orthogonal, since

$$\begin{aligned} Q^T Q &= \left[\begin{array}{c|c} 1 & 0 \\ \hline 0 & \underline{Q} \end{array}\right]^T Q_1 Q_1 \left[\begin{array}{c|c} 1 & 0 \\ \hline 0 & \underline{Q} \end{array}\right] \\ &= \left[\begin{array}{c|c} 1 & 0 \\ \hline 0 & \underline{Q}^T \end{array}\right] I \left[\begin{array}{c|c} 1 & 0 \\ \hline 0 & \underline{Q} \end{array}\right] \\ &= \left[\begin{array}{c|c} 1 & 0 \\ \hline 0 & \underline{Q}^T \underline{Q} \end{array}\right] \\ &= I. \end{aligned}$$

The description of the Householder algorithm given above translates immediately into APL notation. In the function $HHLDR$ given below, the scalar B sets the scale for fuzzy comparisons. That is, a number x is taken to be zero if x is negligible compared to $B (B = B + x)$.

```
    ∇   Q←B HHLDR A  ;Q1;Q
[1]     A←(B∨.≠B+A)/A
[2]     Q←ID 1↑ρA
[3]     →(0=1↓ρA)/0
[4]     Q1←SPRFLCTR A[;1]
[5]     Q←B HHLDR 1 1↓Q1+.×A
[6]     Q←Q1+.×((1+1↑ρQ)↑1),[1]0,Q
    ∇
```

Line [1] deletes any columns from A that are negligible compared to B. The expression `((1+1↑ρQ)↑1),[1]0,Q` in line [6] constructs the matrix

$$\left[\begin{array}{c|c} 1 & 0 \\ \hline 0 & \underline{Q} \end{array}\right]$$

For example, if $\underline{Q}$ is

$$\begin{bmatrix} 1 & 2 \\ 3 & 4 \end{bmatrix}$$

then $0, \underline{Q}$ is

$$\begin{bmatrix} 0 & 1 & 2 \\ 0 & 3 & 4 \end{bmatrix}$$

and `(1+1↑ρQ)↑1` is $3 \uparrow 1$ or `1 0 0`. Finally `1 0 0,[1] 0,Q` is

$$\begin{bmatrix} 1 & 0 & 0 \\ 0 & 1 & 2 \\ 0 & 3 & 4 \end{bmatrix}$$

In our applications the number B can be taken to be the largest magnitude in the matrix A. In this case a function such as

```
    ∇ Q←HSHLDR A
[1]    Q←(⌈/,|A)HHLDR A
    ∇
```

is convenient.

EXAMPLE 5.21 Compute a QR factorization of

$$A = \begin{bmatrix} 1 & 6 & 7 \\ 4 & 3 & 7 \\ 2 & -3 & -1 \\ 1 & -2 & -1 \end{bmatrix}$$

Check the accuracy of your answer.

Solution

```
      +Q←HSHLDR A
0.213 0.758 0.515 0.338
0.853 0.162 ¯0.439 ¯0.232
0.426 ¯0.535 0.713 ¯0.155
0.213 ¯0.336 ¯0.184 0.899
```

```
      +R←(⍉Q)+.×A
 4.69E0      2.13E0      6.82E0
 9.54E¯18    7.31E0      7.31E0
 8.67E¯18    3.82E¯17    4.16E¯17
 5.64E¯18    1.39E¯17    2.08E¯17
```

⎕CT for the processor used is about 3E¯15, so R can be taken to be an upper echelon form. To check the orthogonality of Q we see whether Q^T will do for Q^{-1}.

```
      (ID 4)-Q+.×⍉Q
 0.00E0      6.51E¯19    1.19E¯18     1.25E¯18
 6.51E¯19    1.73E¯18    0.00E0       0.00E0
 1.19E¯18    0.00E0      0.00E0       2.17E¯19
 1.25E¯18    0.00E0      2.17E¯19    ¯1.73E¯18
```

Since R was constructed as Q^TA, this check is much the same as comparing A to $QR = QQ^TA$.

```
      A-Q+.×R
 8.67E¯18    0.00E0      6.94E¯18
 0.00E0      0.94E¯18    1.39E¯17
 0.00E0      6.94E¯18    1.04E¯17
 0.00E0      1.39E¯17    1.04E¯17
```
■

We started this section with the problem of constructing an orthonormal basis of the column space of the matrix A. The output of *HHLDR*, however, is a square matrix. The column space of the matrix $Q = HSHLDR\ A$ is all of $\mathbf{R}^n$. It is quite easy to extract an orthonormal basis of the column space of A from Q.

Since Q is invertible, the matrices A and R have the same row space; in particular, they have the same rank (Propositions 4.25 and 4.26). Since R is in upper echelon form, its rank is the number of nonzero rows. This is because it can be put into echelon form by pivoting on the pivot entries, and there is one pivot entry in each nonzero row.

Let us apply these facts to the matrices of Example 5.21. The matrix R has rank 2. The last two rows can be taken to be zero. Thus,

$$A = QR = [Q_1 \mid Q_2]\left[\frac{R_1}{0}\right] = Q_1R_1$$

where $Q_1 = Q[;\iota 2]$. The column space of A is thus contained in the column space of $Q[;\iota 2]$. But these two spaces are both two-dimensional, hence they are equal.

The argument is true in general. The rank of A is the number of nonzero rows of R. If the rank of A is r, then $Q[;\iota r] = U$ is a matrix whose columns form an orthonormal basis of the column space of A. The function *ORTHO* returns U given A.

```
    ∇ Z←ORTHO A ;R;B
[1]    B←⌈/,|A←((ρA),ι1=ρρA)ρA
[2]    R←(⍉Z←B HHLDR A)+.×A
[3]    Z←Z[;ι+/(B+R)∨.≠B]
    ∇
```

The rank of R is taken to be the number of rows that are not negligible compared to B, the largest magnitude in A. The expression `A←((ρA),ι1=ρρA)ρA` leaves A unchanged if A is a matrix but makes A into a single-column matrix if it is a vector. This is because *ORTHO* is sometimes more convenient to use if it accepts vector arguments.

EXAMPLE 5.22 Let S be the column space of the matrix A of Example 5.21. Compute P_S, the perpendicular projection matrix for S, and R_S, the reflection matrix for S. Find $v^{\|}$ and $v^{\perp}$ with respect to S if v is `ι4`.

Solution Using Proposition 5.18 of the previous section,

```
      A
1  6  7
4  3  7
2 ¯3 ¯1
1 ¯2 ¯1

      +U←ORTHO A
0.213 0.758
0.853 0.162
0.426 ¯0.535
0.213 ¯0.336

      +PS←U+.×⍉U
0.621 0.304 ¯0.315 ¯0.209
0.304 0.753 0.277 0.128
¯0.315 0.277 0.468 0.270
¯0.209 0.128 0.270 0.158
```

As a check, we should have $P_S^2 = P_S$.

```
      PS−PS+.×PS
 0.00E0        1.30E¯18      8.67E¯19      2.17E¯19
 1.30E¯18      1.73E¯18      0.00E0        0.00E0
 8.67E¯19      0.00E0       ¯4.34E¯19     ¯4.34E¯19
 2.17E¯19      0.00E0       ¯4.34E¯19      0.00E0
```

For $v^{\|}$ and $v^{\perp}$ we have

```
      +V←ι4
1 2 3 4
      +VPAR←PS+.×V
¯0.551  3.15  2.72  1.49
      +VPERP←V−VPAR
1.55  ¯1.15  0.276  2.51
```

To check $v^{\perp}$, use the fact that the dot product of $v^{\perp}$ with any column of A should be zero.

```
      VPERP+.×A
5.38E¯17  1.67E¯16  2.22E¯16
```

The reflection matrix is

```
      +RS←−(ID 4)−2×PS
 0.241 0.609 ¯0.629 ¯0.418
 0.609 0.507  0.554  0.255
¯0.629 0.554 ¯0.064  0.541
¯0.418 0.255  0.541 ¯0.684
```

As a check, we should have $R_S = R_S^{-1}$.

```
      (ID 4)−RS+.×RS
¯1.73E¯18     4.55E¯18      3.90E¯18      1.52E¯18
 4.55E¯18     5.20E¯18      0.00E0        0.00E0
 3.90E¯18     0.00E0       ¯3.47E¯18     ¯2.60E¯18
 1.52E¯18     0.00E0       ¯2.60E¯18     ¯1.73E¯18
```
■

What about the rows of $Q = HSHLDR\ A$ that are not used by *ORTHO*? They form an orthonormal basis of the orthogonal complement of the column space of A. We collect these facts together for easy reference in the next proposition.

PROPOSITION 5.20 Let A be a matrix of rank r with column space S. Suppose that $A = QR$, where Q is orthogonal and R is in upper echelon form. Then the first r

columns of Q form an orthonormal basis of S and the remaining columns form an orthonormal basis of $S^{\perp}$, the orthogonal complement of S.

Proof Let A be m by n. Then Q is m by m with $Q^T = Q^{-1}$. The rank of Q is m, hence the column space of Q is all of $\mathbf{R}^m$. Since $Q^TQ = I$, the columns of Q form an orthonormal basis of $\mathbf{R}^m$. Choose an arbitrary subset, $Q[;V]$, of the columns of Q. Let the column space of $Q[;V]$ be T. Let T' be the space generated by the remaining columns. Then T' is contained in $T^{\perp}$, since each generator of T' is contained in $T^{\perp}$. But T' and $T^{\perp}$ have the same dimension, since $\dim T + \dim T' = m$, so $T' = T^{\perp}$.

Thus, to finish the proof of the proposition it only remains to identify the subspace S with $Q[;\iota r]$. As indicated above, R and A have the same row space by Proposition 4.26 and hence the same rank r by Proposition 4.27.

By comparing the definitions of echelon form and upper echelon form we see that R can be row-reduced to echelon form by pivoting on each pivot entry and then dividing the row by the pivot entry to obtain a leading 1.

Thus

$$\begin{aligned} r = \text{rank } R &= \text{number of pivot columns} \\ &= \text{number of leading 1's} \\ &= \text{number of pivot entries} \\ &= \text{number of nonzero rows of } R \end{aligned}$$

Since the zero rows of R are all at the bottom,

$$A = QR = [Q[;\iota r] \mid *]\left[\frac{R[\iota r;]}{0}\right] = Q[;\iota r]R[\iota r;]$$

In particular the columns of A are contained in the column space of $Q[;\iota r]$ and hence S is contained in this column space. Since both spaces have dimension r, they are equal. ∎

In the notation of Proposition 5.13, the function $PERP$ below returns a matrix whose columns form an orthonormal basis of $S^{\perp}$. The function differs from $ORTHO$ only in line [3] which drops the first r columns off Q.

```
    ∇ Z←PERP A ;R;B
[1]     B←⌈/,|A←(ρA),ι1=ρρA)ρA
[2]     R←(⍉Z←B HHLDR A)+.×A
[3]     Z←(0,+/(B+R)∨.≠B)↓Z
    ∇
```

The uses of $ORTHO$ and $PERP$ are summarized in the next proposition. Notice that the null space of A is (row space of $A)^{\perp}$.

PROPOSITION 5.21 Let A be a matrix. To compute an orthonormal basis of the given space, use the given expression:

Column space of A	ORTHO A
(Column space of A)$^\perp$	PERP A
Row space of A	ORTHO ⍉A
Null space of A	PERP ⍉A

■

LEAST-SQUARES SOLUTIONS

We close this section with a description of the algorithm used by the dyadic function ⌹ to solve systems of linear equations.

The equation

$$AX = B$$

has solutions if and only if the vector B lies in the column space of the matrix A (Proposition 2.12).

A least-squares solution is a vector V that minimizes

$$f(X) = (B - AX)+.*2 = \|B - AX\|^2$$

(See Proposition 2.15.) That is, the distance from AV to B is minimal. Since the vectors of the form AV make up the column space of A, it follows that a least-squares solution is a solution to

$$AX = B^{\|} = P_S B$$

where S is the column space of A. Let $U = ORTHO\ A$. Then $P_S = UU^T$, so we want a solution of

$$AX = UU^T B$$

Now suppose that $R = U^T A$, $Q = HSHLDR\ A$, $V = PERP\ A$. Then R is upper echelon, $V^T A = 0$, and $Q = [U \mid V]$. Multiplying the equation $AX = UU^T B$ by Q^T will not change the solutions, since Q is invertible (Proposition 2.13).

$$AX = UU^T B$$

$$Q^T AX = Q^T UU^T B$$

$$[U \mid V]^T AX = [U \mid V]^T UU^T B$$

$$\left[\frac{U^T}{V^T}\right] AX = \left[\frac{U^T}{V^T}\right] UU^T B$$

$$\left[\frac{U^T AX}{V^T AX}\right] = \left[\frac{U^T UU^T B}{V^T UU^T B}\right]$$

$$\left[\frac{U^T AX}{0}\right] = \left[\frac{U^T B}{0}\right]$$

or

$$RX = U^T B$$

since $U^T U = I$ and $V^T U = 0$. Thus the least-squares solutions of $AX = B$ are the solutions of $RX = U^T B$, the equation we get if we simply multiply $AX = B$ by U^T.

Domino does not compute U explicitly, however. First the augmented matrix $[A \mid B]$ is formed, and the Householder algorithm is applied to the augmented matrix until the matrix A is reduced to upper echelon form. This is equivalent to multiplying the augmented matrix by Q^T.

$$Q^T[A \mid B] = \left[\begin{array}{c} U^T \\ \hline V^T \end{array}\right] [A \mid B]$$

$$= \left[\begin{array}{c|c} U^T A & U^T B \\ \hline V^T A & V^T B \end{array}\right]$$

$$= \left[\begin{array}{c|c} R & U^T B \\ \hline 0 & V^T B \end{array}\right]$$

The lower portions, the 0 and $V^T B$, can simply be discarded and we have the augmented matrix

$$[R \mid U^T B]$$

It is not difficult to solve $RX = U^T B$ by completing the reduction of R to echelon form. If R is invertible, the system is

$$\begin{aligned} r_{11}x_1 + r_{12}x_2 + \cdots \qquad \cdots + r_{1n}x_n &= b_1 \\ r_{22}x_2 + \cdots \qquad \cdots + r_{2n}x_n &= b_2 \\ &\vdots \\ r_{n-1,n-1}x_{n-1} + r_{n-1,n}x_n &= b_{n-1} \\ r_{n,n}x_n &= b_n \end{aligned}$$

where $r_{ii} \neq 0$ for all i. In this case one can start at the bottom with a pivot on r_{nn}, then a pivot on $r_{n-1,n-1}$, and so on. This process is called back-substitution.

If R is not invertible, then ⌹ assumes the original matrix to be "ill conditioned" and returns a `DOMAIN ERROR`.

Nothing in this procedure forces B to be a vector. The process works exactly the same way if B is a matrix and thus `B⌹A` is defined for matrices B as well.

The monadic function `⌹A` sets `B←ID 1↑⍴A` and then calls the dyadic function ⌹ to compute `B⌹A`.

BACK-SUBSTITUTION

Our treatment of linear equations is so far somewhat incomplete. We have methods of reducing a system of linear equations to echelon form or, now, upper echelon form and APL functions (*ECHELON*, *GAUSS*, *HHLDR*) that implement the procedures that have been written.

The process of writing down the answer once the echelon form has been obtained, however, has not been described with sufficient formality to enable an APL function to be written (but see exercises 40 and 43 of Exercises 4.5).

The function *BACKSUB* described below remedies this situation. The expression

```
A←S BACKSUB R,B
```

solves the equation $RX = B$ where R is upper echelon (in particular, if R is in echelon form). The scalar S sets the scale for comparisons to zero. Numbers are taken to be zero if they are negligible compared to S.

The description of the function *BACKSUB* is somewhat technical and may be omitted without loss of continuity.

To begin, assume that the system $AX = B$ is of the form

$$\begin{array}{rcl}
a_{11}x_1 + a_{12}x_2 + a_{13}x_3 + \cdots + a_{1n}x_n &=& b_1 \\
a_{22}x_2 + a_{23}x_3 + \cdots + a_{2n}x_n &=& b_2 \\
a_{33}x_3 + \cdots + a_{3n}x_n &=& b_3 \\
&\vdots& \\
a_{rn}x_n &=& b_r \\
0 &=& b_{r+1} \\
&\vdots& \\
0 &=& b_{rn}
\end{array}$$

First, if $b_i \neq 0$ for some $i > r$, then the system is inconsistent and need not be considered further. (The actual comparison, of course, is whether b_i is negligible compared to S.) The result returned by *BACKSUB* in this case will be `(N,0)ρ0`, the empty set in $\mathbf{R}^n$.

Otherwise, if $a_{rn} \neq 0$, we take $x_n = b_r \div a_{rn}$ and then move up to the next equation to get a value for x_{n-1}. It is this "moving up" process that must be considered in detail.

Suppose first that we have moved up through the second equation, thus assigning values to $x_2, x_3, \ldots, x_n$. That is, assume that we have solved the system of equations in the box with the solid outline and to this solution we wish to add one more equation (the first) and one more variable (x_1). Once we describe how this is done, we are well on the way to a formal description of a solution process that proceeds by induction on the number of variables — that is, the number of columns of A.

If there is a unique solution, $x_i = c_i$ for $i = 2, 3, \ldots, n$, to the smaller system and of $a_{11} \neq 0$, then there is a unique solution to the original system

$$\begin{bmatrix} x_1 \\ x_2 \\ x_3 \\ \vdots \\ x_n \end{bmatrix} = \begin{bmatrix} (b_1 - a_{12}c_2 - a_{13}c_3 - \cdots - a_{1n}c_n) \div a_{11} \\ c_2 \\ c_3 \\ \vdots \\ c_n \end{bmatrix}$$

If there is a k-flat of solutions ($k > 0$) to the smaller system and if $a_{11} \neq 0$, then there is a k-flat of solutions to the original system. Suppose that the solutions of the smaller system make up the k-flat defined by $p_0, p_1, \ldots, p_k$ in $\mathbf{R}^{n-1}$ (we are labeling our axes $x_2, x_3, \ldots, x_n$ in $\mathbf{R}^{n-1}$ here). Suppose that these points are stored as the columns of a matrix Z. Then the solutions of the original system form the k-flat in $\mathbf{R}^n$ defined by

```
Z←(B[1]-(1↓A[1;])+.×Z÷A[1;1]),[1]Z
```

— that is, the above formula for the case $k = 0$ applied to all the columns of Z.

If there is a k-flat of solutions of the smaller system and $a_{11} = 0$, then a problem arises. If $p = (p_2, p_3, \ldots, p_n)$ is a solution to the smaller system and $a_{11} \neq 0$, then we have the flexibility to choose a scalar $x_1 = p_1$ so that the equation

$$a_{11}p_1 + a_{12}p_2 + a_{13}p_3 + \cdots + a_{1n}p_n = b_1$$

is satisfied. If $a_{11} = 0$, however, we have no room for adjustment, and it can happen that $a_{12}p_2 + a_{13}p_3 + \cdots + a_{1n}p_n \neq b_1$. This means that no first component can be added to p to get a solution of the original system. Thus, before using the first equation to define x_1, we must make sure that we have the set of solutions of the system of equations in the larger box with the dotted outline. This complicates the inductive assumption slightly. If $a_{11} \neq 0$, we assume the solution of the $(n - 1)$-dimensional system in the solid box; if $a_{11} = 0$, we assume the solution of the $(n - 1)$-dimensional system in the larger dotted box.

In the case $a_{11} = 0$, suppose that the solutions of the larger system in the dotted box make up the k-flat defined by the points $p_0, p_1, p_2, \ldots, p_k$ in $\mathbf{R}^{n-1}$. Then the solutions of the original system define a $(k + 1)$-flat in $\mathbf{R}^n$, since the echelon form has one more nonpivot column. This $(k + 1)$-flat can be defined by

$$q_0 = (1, p_0), \quad q_1 = (0, p_0), \quad q_2 = (0, p_1), \quad q_3 = (0, p_2), \quad \ldots, \quad q_{k+1} = (0, p_k)$$

Thus if $p_0, p_1, \ldots, p_k$ are stored as columns of Z, then the points $q_0, q_1, \ldots, q_{k+1}$ are the columns of

$$\left[\begin{array}{c|c} 1 & 0 \\ \hline p_0 & Z \end{array}\right]$$

or `(1,Z[;1]),0,[1]Z`. Of course if the $(n-1)$-dimensional system is inconsistent, then the n-dimensional system is inconsistent as well. So, an $(n-1)$-by-0 matrix Z must be replaced by an n-by-0 matrix `0,[1]Z`.

The problem of assigning a value to x_n — that is, of starting the induction — was passed over too hastily above. The last equations might easily look like this:

$$\begin{aligned} a_{rl}x_l + a_{r,l+1}x_{l+1} + \cdots + a_{rn}x_n &= b_r \\ 0 &= b_{r+1} \\ &\vdots \end{aligned}$$

In this case it is clearly wrong to start the process with $x_n = b_r \div a_{rn}$. We avoid the problem by starting the induction with $n = 0$ rather than $n = 1$.

Consider the equation

$$AX = B$$

where A is m by 0 (`A←(M,0)⍴0`). We take X and B to be vectors. Then, since `(⍴X)=1↓⍴A`, we must have $X = \iota 0$, and hence `(A+.×X)=M⍴0`. This is because a linear combination of an empty set of vectors always results in a zero vector (cf. exercise 23 of Exercises 4.4). It follows that the equation has a solution if and only if B is the zero vector.

The set of solutions of $AX = B$ is a k-flat in $\mathbf{R}^n$, where n is the number of columns of A. Say that the k-flat is given by the $k + 1$ points $p_0, p_1, \ldots, p_k$. Our answer is to be such a set of points stored as the columns of an n-by-$(k+1)$ matrix Z. In the present case $n = 0$. The space $\mathbf{R}^0$ is a single point represented by `⍳0`. So if $B = 0$, then Z is a 0-by-1 matrix, since a single point is a 0-flat. If $B \neq 0$, then Z is 0 by 0, representing the empty set. (The empty set is sometimes said to be a -1-dimensional flat.) To summarize: If A has no columns, then Z is `0 1⍴0` if $B = 0$ and `0 0⍴0` if $B \neq 0$.

This establishes the starting values for the induction, but a detail remains to be explained. In working down the diagonal, it is possible to run out of rows before running out of columns. This brings us to the equation

$$AX = B$$

where A has columns but no rows. In this case any choice of X will do, since $B =$ `⍳0` and `(A+.×X)=⍳0` for any X. Thus $AX = B$ has in this case precisely the same solutions as

$$0 \cdot x_1 + 0 \cdot x_2 + \cdots + 0 \cdot x_n = 0$$

For this reason we treat this case in the same way as the case $a_{11} = 0$ discussed above. Of course we cannot use the expression `A[1;1]=0` for a test, but we can use `A[;1]∧.=0`, since A is always in upper echelon form, or we could use the expression `0=1 1↑A`.

In the function *BACKSUB* below we let A stand for the augmented matrix of the system. Thus `B=A[;1↓ρA]`

```
    ∇ Z←S BACKSUB A
[1]    Z←(0,S∧.=S+A[;1↓ρA])ρ0
[2]    →(1=1↓ρA)/0
[3]    →(S∧.=S+A[;1])/J
[4]    Z←S BACKSUB 1 1↓A
[5]    Z←(((¯1↑A[1;])-(¯1↓1↓A[1;])+.×Z)÷A[1;1]),[1]Z
[6]    →0
[7]   J:Z←0,[1]S BACKSUB 0 1↓A
[8]    →(0=1↓ρZ)/0
[9]    Z←(1,1↓Z[;1]),Z
    ∇
```

Line [1] initializes Z to `0 1ρ0` if `B=A[;1↓ρA]` is negligible compared to S and to `0 0ρ0` otherwise. If the solution is a flat in $\mathbf{R}^0$, then the function stops on line [2]. Line [3] tests if the first column is negligible compared to S. If it is, lines [7], [8], and possibly [9] are executed and the function stops. Otherwise lines [4], [5], and [6] are executed and the function stops. The function cannot reach line [3] unless A has at least two columns. It may be without rows on line [3], however, in which case the function jumps to line [7]. Thus, line [5] cannot be reached unless A contains at least one row and two columns. The expression `¯1↑A[1;]` is used in place of $B[1]$ and `¯1↓1↓A[1;]` is the vector $(a_{12}, a_{13}, \ldots, a_{1n})$.

To solve the equation $AX = B$ one can use

```
(⌈/,|A,B) BACKSUB ECHELON A,B
```

or

```
(⌈/,|A,B) BACKSUB (⍉ HSHLDR A)+.×A,B
```

Least-squares solutions are obtained with

```
(⌈/,|A,B) BACKSUB (⍉ ORTHO A)+.×A,B
```

EXAMPLE 5.23 Consider the system of equations

$$\begin{aligned} 2x_1 - x_2 + x_3 + 3x_4 \quad &= 5 \\ -x_1 + x_2 + x_3 + x_4 \quad &= 1 \\ x_1 - x_2 - x_3 - x_4 + x_5 &= 7 \end{aligned}$$

This is $AX = B$ with

$$A = \begin{bmatrix} 2 & -1 & 1 & 3 & 0 \\ -1 & 1 & 1 & 1 & 0 \\ 1 & -1 & -1 & -1 & 1 \end{bmatrix}, \qquad B = (5, 1, 7)$$

The echelon form of the augmented matrix $[A \mid B]$ is

```
      +E←ECHELON A,B
 1.000  0.000  2.000  4.000  0.000  6.000
 0.000  1.000  3.000  5.000  0.000  7.000
 0.000  0.000  0.000  0.000  1.000  8.000
```

Thus the set of solutions form a plane in $\mathbf{R}^5$ with parametric representation

$$\begin{bmatrix} x_1 \\ x_2 \\ x_3 \\ x_4 \\ x_5 \end{bmatrix} = \begin{bmatrix} 6 \\ 7 \\ 0 \\ 0 \\ 8 \end{bmatrix} + \begin{bmatrix} -2 & -4 \\ -3 & -5 \\ 1 & 0 \\ 0 & 1 \\ 0 & 0 \end{bmatrix} \begin{bmatrix} t_1 \\ t_2 \end{bmatrix}$$

On the other hand *BACKSUB* produces points p_0, p_1, p_2 defining this plane.

```
      S←3
      +ANS1←S BACKSUB E
 0.000  2.000  6.000
¯1.000  2.000  7.000
 1.000  0.000  0.000
 1.000  1.000  0.000
 8.000  8.000  8.000
```

We can check the accuracy by looking at $B - AX$ for each of these points.

```
      BBB←⍉3 3⍴B
      BBB-A+.×ANS1
0.00E0       0.00E0       0.00E0
0.00E0       0.00E0       0.00E0
0.00E0       0.00E0       0.00E0
```

(This extraordinary accuracy arises from the special form of A and B.)

On the other hand we may use *HSHLDR* to reduce *A,B* to upper echelon form.

```
      +R←(⍉HSHLDR A)+.×A,B
 2.45E0     ¯1.63E0     ¯1.73E¯18    1.63E0     4.08E¯1   6.53E0
¯1.73E¯18    5.77E¯1    ¯1.73E0     ¯2.89E0     5.77E¯1   5.77E¯1
 8.67E¯19   ¯1.30E¯18   ¯2.17E¯18   ¯3.04E¯18   7.07E¯1   5.66E0
```

Applying *BACKSUB* to *R* produces

```
      +ANS2←S BACKSUB R
¯1.70E¯17    2.00E0     6.00E0
¯1.00E0      2.00E0     7.00E0
 1.00E0      0.00E0     0.00E0
```

```
1.00E0      1.00E0      0.00E0
8.00E0      8.00E0      8.00E0
```

This time there has been a minimal amount of round-off error (this system carries almost eighteen decimal digits).

```
      BBB-A+.×ANS2
0.00E0       1.39E¯17     1.39E¯17
1.21E¯17     1.04E¯17     6.94E¯18
0.00E0       0.00E0       6.94E¯18
```
■

In spite of the outcome of this example one may in general expect better accuracy from *HSHLDR* than from *ECHELON*.

EXERCISES 5.5

1. Assume ϵ^2 negligible compared to 1 but ϵ not negligible compared to 1 (i.e., $|\epsilon| > \square CT$ but $\epsilon^2 < \square CT$). Let

$$A = \begin{bmatrix} 1 & 1 & 1 \\ \epsilon & 0 & 0 \\ 0 & \epsilon & 0 \\ 0 & 0 & \epsilon \end{bmatrix}$$

(a) Show that applying the Gram-Schmidt process to A in machine arithmetic produces

$$W = \begin{bmatrix} 1 & 0 & 0 \\ \epsilon & -\epsilon & -\epsilon \\ 0 & \epsilon & 0 \\ 0 & 0 & \epsilon \end{bmatrix} \quad \text{and} \quad U = \begin{bmatrix} 1 & 0 & 0 \\ \epsilon & -\dfrac{1}{\sqrt{2}} & -\dfrac{1}{\sqrt{2}} \\ 0 & \dfrac{1}{\sqrt{2}} & 0 \\ 0 & 0 & \dfrac{1}{\sqrt{2}} \end{bmatrix}$$

and the columns of U are *not* orthonormal in machine arithmetic.

(b) Show that if $Q = HSHLDR\ A$, then

$$Q = \begin{bmatrix} 1 & \epsilon & 0 & 0 \\ \epsilon & -1 & 0 & 0 \\ 0 & 0 & -1 & 0 \\ 0 & 0 & 0 & -0 \end{bmatrix} \begin{bmatrix} 1 & 0 & 0 & 0 \\ 0 & \dfrac{1}{\sqrt{2}} & -\dfrac{1}{\sqrt{2}} & 0 \\ 0 & -\dfrac{1}{\sqrt{2}} & -\dfrac{1}{\sqrt{2}} & 0 \\ 0 & 0 & 0 & -1 \end{bmatrix} \begin{bmatrix} 1 & 0 & 0 & 0 \\ 0 & 1 & 0 & 0 \\ 0 & 0 & -\dfrac{1}{\sqrt{3}} & \dfrac{2}{\sqrt{3}} \\ 0 & 0 & \dfrac{2}{\sqrt{3}} & \dfrac{1}{\sqrt{3}} \end{bmatrix}$$

and the columns of Q and hence $U = Q[;\iota 3]$ are orthonormal in machine arithmetic.

In exercises 2 through 6 use the Gram-Schmidt process to construct a QR factorization of the matrix A (cf exercises 21 through 25 of Exercises 5.4).

2. $A = \begin{bmatrix} 1 & 1 \\ 0 & 1 \end{bmatrix}$

3. $A = \begin{bmatrix} 1 & 2 & 2 \\ 1 & 2 & 3 \end{bmatrix}$

4. $A = \begin{bmatrix} 1 & 1 & 0 \\ 1 & 0 & 1 \\ 0 & 1 & 1 \end{bmatrix}$

5. $A = \begin{bmatrix} 1 & 1 & 0 \\ 0 & 1 & 1 \\ 1 & 0 & 1 \end{bmatrix}$

6. $A = \begin{bmatrix} 1 & 1 & 0 & 1 \\ -1 & 0 & 1 & 1 \\ 1 & 1 & 0 & 1 \\ -1 & 0 & 1 & 1 \end{bmatrix}$

In exercises 7 through 14 an APL function is specified. Write an APL function meeting the specifications.

7. Name: *PRJCTR*
   Right argument: A matrix A
   Result: The perpendicular projection matrix P_S where S is the column space of A
   Remarks: Use *ORTHO*. Test your function on exercises 26 through 31 of Exercises 5.4.

8. Name: *COMPS*
   Left argument: A vector v
   Right argument: A matrix A
   Result: A matrix Z with $Z[;1] = v^{\|}$, $Z[;2] = v^{\perp}$, the components of v parallel and perpendicular to the column space of A
   Remarks: *PRJCTR* from exercise 7 can be used. Test the function on exercises 26 through 31 of Exercises 5.4.

9. Name: *RFLCTR*
   Right argument: A matrix A
   Result: The matrix R_S of reflection in the column space S of A
   Remarks: Use *ORTHO* and the function *PRJCTR* of exercise 7. Test the function on exercises 26 through 31 of Exercises 5.4.

10. Name: *SBSP*
    Right argument: A set of points $p_0, p_1, \ldots, p_k$ defining a flat in $\mathbf{R}^n$ stored as the columns of a matrix
    Result: The generators $v_i = p_i - p_0$ of the subspace parallel to the flat, stored as the columns of a matrix

11. Name: *ORDIST*
    Right argument: Points $p_0, p_1, \ldots, p_k$ defining a flat, stored as the columns of a matrix
    Result: The minimum distance from the flat to the origin
    Remarks: Use the results of exercises 10 and 7 or 8. Test the function on exercises 34 through 49 of Exercises 5.4.

12. Name: *PROJ*
    Right argument: Points $p_0, p_1, \ldots, p_k$ defining a flat, stored as the columns of a matrix
    Result: The affine function, $f(X) = b + AX$, of perpendicular projection onto the flat stored as an augmented matrix $[b|A]$
    Remarks: Use the results of exercises 10 and 7. Test the function on exercises 34 through 49 of Exercises 5.4.

13. Name: *REFL*
    Right argument: Points $p_0, p_1, \ldots, p_k$ defining a flat, stored as the columns of a matrix
    Result: The affine function, $f(X) = b + AX$, of reflection in the flat stored as an augmented matrix $[b|A]$
    Remarks: Use the results of exercises 10 and 9. Test the function on exercises 34 through 49 of Exercises 5.4.

14. (Computer assignment) Use *PERP* and Proposition 5.21 to solve exercises 40 through 47 of Exercises 5.4.

(Computer assignment) In exercises 15 through 18 a linear system from Exercises 3.4 is referred to. Solve the system by (a) using *BACKSUB* and *ECHELON*, (b) using *BACKSUB* and *HSHLDR*. If the system is inconsistent, find a least-squares solution using *BACKSUB* and *ORTHO*.

15. Exercise 1.
16. Exercise 3.
17. Exercise 4.
18. Exercise 5.

5.6* Inertia and Principal-Component Analysis (Rayleigh's Principle)

Suppose that we have a problem involving a large number of variables and we wish to simplify things by reducing the number of variables under consideration. We should do this in a way that minimizes the amount of information lost.

To be specific, suppose that two variables called x and y are involved. Perhaps x represents weight and y height for a fixed population of individuals. Suppose that the heights and weights of n individuals are measured giving the data points $p_i = (x_i, y_i)$; $i = 1, 2, \ldots, n$.

If we reduce the number of variables by considering only x or by considering only y, then we are projecting our data points p_i onto the x or y axis and working with the projections instead of the original points p_i. This suggests the following procedure (Figure 5.21). Choose a new coordinate system $x'y'$ such that projection onto the x' axis changes the points p_i as little as possible. Then replace the two variables x and y by the single variable x'. Before we can do this, we need to

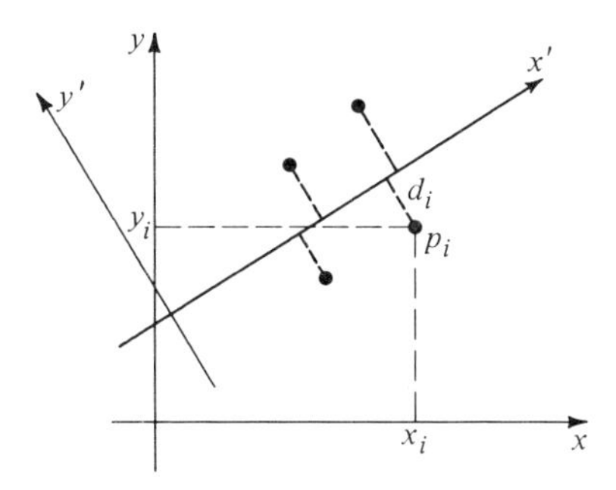

FIGURE 5.21

define what we mean by "change the points p_i as little as possible." In practice this is taken to mean that the quantity $\Sigma_{i=1}^n d_i^2$ the sum of the squares of the perpendicular distances from the points p_i to axis x', is a minimum.

Notice that this is not the same problem as that of fitting a least-squares straight line to the data (x_i, y_i); $i = 1, 2, \ldots, n$. In the case of the least-squares straight line the sum of the squares of the *vertical* distances is minimized (see Example 5.24 below).

The general case of reducing n variables to $k \leq n$ variables reduces to the following geometric problem.

MINIMIZATION PROBLEM Let $p_1, \ldots, p_m$ be the points in $\mathbf{R}^n$, and let F be a k-flat in $\mathbf{R}^n$. Let the affine function π_F be a perpendicular projection onto F. Find F so that the number

$$\sum_{i=1}^{m} \|p_i - \pi_F(p_i)\|^2$$

is a minimum.

This problem arises, for the case $n = 3$, in classical mechanics. Suppose that the points are all particles with the same mass w. For $k = 1$ the flat F is a line and the moment of inertia of the system of masses about this line is

$$\sum w\|p_i - \pi_F(p_i)\|^2$$

Thus the line sought is the one that gives the *smallest moment of inertia*.

For arbitrary k and n the flat F goes through the center of gravity.

PROPOSITION 5.22 The k-flat F of the minimization problem contains the centroid (or mean)

$$\bar{p} = \frac{1}{m}\sum_{i=1}^{m} p_i$$

Proof Notice that the sum of the deviations about the centroid is zero:

$$\sum_{i=1}^{m}(p_i - \bar{p}) = \sum_{i=1}^{m} p_i - \sum_{i=1}^{m}\bar{p} = \sum_{i=1}^{m} p_i - m\bar{p} = 0$$

Now let F be the flat of the minimization problem, S the unique subspace parallel to F, and q any point in F. Let $q = q^{\perp} + q^{\parallel}$ and $\bar{p} = \bar{p}^{\perp} + \bar{p}^{\parallel}$ with respect to S. Let G be the k-flat through $\bar{p}$ parallel to S and F. The perpendicular projection maps onto F and G are then (see Section 5.4)

$$\pi_F(X) = q^{\perp} + P_S X; \qquad \pi_G(X) = \bar{p}^{\perp} + P_S X$$

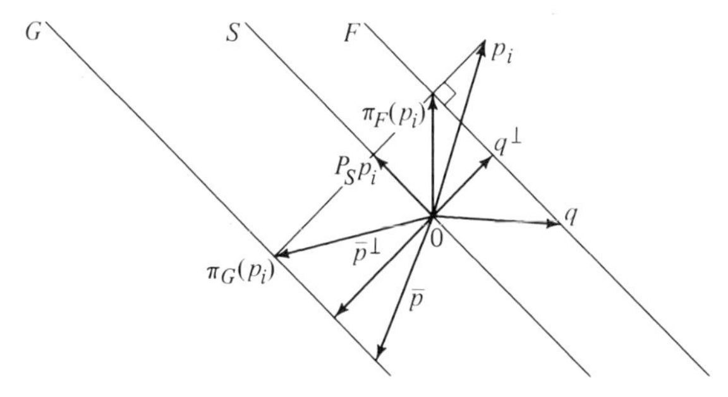

FIGURE 5.22

where P_S is the perpendicular projection matrix for S. Thus

$$\pi_F(X) = \pi_G(X) + q^\perp - \bar{p}^\perp = \pi_G(X) + (q - \bar{p})^\perp$$

(Figure 5.22). Thus

$$\begin{aligned}
&\sum_{i=1}^{m} \|p_i - \pi_F(p_i)\|^2 \\
&\qquad = \sum_{i=1}^{m} \|p_i - \pi_G(p_i) - (q - \bar{p})^\perp\|^2 \\
&\qquad = \sum_{i=1}^{m} [p_i - \pi_G(p_i) + (\bar{p} - q)^\perp] \cdot [p_i - \pi_G(p_i) + (\bar{p} - q)^\perp] \\
&\qquad = \sum_{i=1}^{m} \{\|p_i - \pi_G(p_i)\|^2 + 2[p_i - \pi_G(p_i)] \cdot (\bar{p} - q)^\perp + \|\bar{p}^\perp - q^\perp\|^2\} \\
&\qquad = \sum_{i=1}^{m} \|p_i - \pi_G(p_i)\|^2 + 2\sum_{i=1}^{m} [p_i - \pi_G(p_i)] \cdot (\bar{p} - q)^\perp + m\|\bar{p}^\perp - q^\perp\|^2
\end{aligned}$$

Now

$$\begin{aligned}
&\sum_{i=1}^{m} [p_i - \pi_G(p_i)] \cdot (\bar{p} - q)^\perp \\
&\qquad = \sum_{i=1}^{m} [p_i - \bar{p} + \bar{p} - \pi_G(p_i)] \cdot (\bar{p} - q)^\perp \\
&\qquad = (\bar{p} - q)^\perp \cdot \sum_{i=1}^{m} (p_i - \bar{p}) + (\bar{p} - q)^\perp \cdot \sum_{i=1}^{m} [\bar{p} - \pi_G(p_i)] \\
&\qquad = 0 + 0 = 0
\end{aligned}$$

The first term above is zero, since $\Sigma\,(p_i - \bar{p}) = 0$, and the second term is zero,

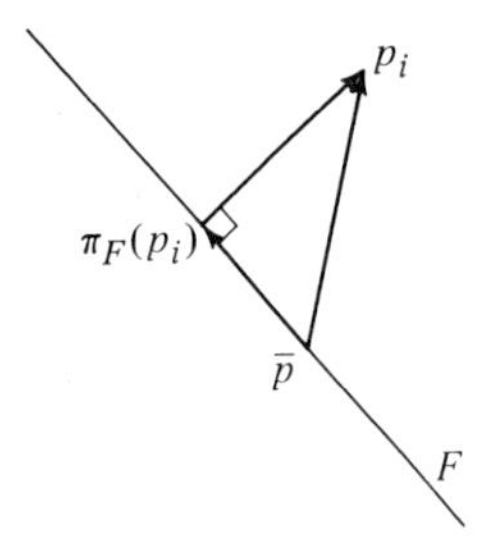

FIGURE 5.23

since each of the terms $\bar{p} - \pi_G(p_i)$ is parallel to G and $(\bar{p} - q)^\perp$ is perpendicular to G. Thus

$$\sum \|p_i - \pi_F(p_i)\|^2 = \sum \|p_i - \pi_G(p_i)\|^2 + m\|\bar{p}^\perp - q^\perp\|^2$$

which is minimal when $\bar{p}^\perp = q^\perp$ — that is, when $F = G$. ■

Since $\bar{p}$ lies in F, we have, by the Pythagorean theorem (Figure 5.23),

$$\sum \|p_i - \bar{p}\|^2 = \sum (\|\pi_F(p_i) - \bar{p}\|^2 + \|p_i - \pi_F(p_i)\|^2)$$

$$= \sum \|\pi_F(p_i) - \bar{p}\|^2 + \sum \|p_i - \pi_F(p_i)\|^2$$

Now the quantity $\Sigma \|p_i - \bar{p}\|^2$ is a constant. It follows that $\Sigma \|p_i - \pi_F(p_i)\|^2$ is a minimum precisely when $\Sigma \|\pi_F(p_i) - \bar{p}\|^2$ is a maximum. We have turned the minimization problem into a

MAXIMIZATION PROBLEM Let $\bar{p}, p_1, p_2, \ldots, p_m$ be points in $\mathbf{R}^n$ ($\bar{p}$ need not be the centroid of $p_1, p_i, \ldots, p_m$ here). Let F be a flat containing $\bar{p}$ and let π_F be a perpendicular projection onto F. Find F so that the number

$$\sum_{i=1}^{m} \|\pi_F(p_i) - \bar{p}\|^2$$

is a maximum.

To discuss the maximum problem, change to new coordinates by translating the origin to $\bar{p}$. This simplifies the formulas. The flat F is now a subspace, call it S as in the proof of Proposition 5.22, and π_F is multiplication by P_S, the perpendicular projection matrix for S.

We will now derive a formula for

$$\sum_{i=1}^{m} \|\pi_F(p_i) - \bar{p}\|^2$$

in terms of the coordinates of the points p_i in the new coordinate system. Let u_1, $u_2, \ldots, u_k$ be an orthonormal basis of S and let $U = [u_1 | u_2 | \ldots | u_k]$. Then $P_S = UU^T$ and, since $\overline{p} = 0$,

$$\sum \|\pi_F(p_i) - \overline{p}\|^2 = \sum \|P_S p_i\|^2$$

Now for each i we have

$$\begin{aligned}\|P_S p_i\|^2 &= (UU^T p_i)^T (UU^T p_i) = p_i^T U (U^T U) U^T p_i = p_i^T U U^T p_i \\ &= (U^T p_i)^T (U^T p_i) = (u_1^T p_i)^2 + (u_2^T p_i)^2 + \cdots + (u_k^T p_i)^2\end{aligned}$$

Now

$$(u_l^T p_i)^2 = (u_l^T p_i)(u_l^T p_i) = (u_l^T p_i)(p_i^T u_l) = u_l^T (p_i p_i^T) u_l$$

and so

$$\sum_{i=1}^{m} (u_l^T p_i)^2 = \sum_i u_l^T (p_i p_i^T) u_l = u_l^T \left(\sum_i p_i p_i^T \right) u_l$$

Now let $A = [p_1 | p_2 | \ldots | p_m]$. Then

$$A^T = \begin{bmatrix} p_1^T \\ \hline p_2^T \\ \hline \vdots \\ \hline p_m^T \end{bmatrix}$$

and

$$AA^T = \sum_i p_i p_i^T$$

The next proposition summarizes this calculation. If the flat F contains $\overline{p}$ and S is the unique subspace parallel to F, then $\pi_F(p_i) - \overline{p} = P_S p_i - P_S \overline{p}$ (Figure 5.22 with G replacing F).

PROPOSITION 5.23 Let the points $\overline{p}, p_1, p_2, \ldots, p_m$ in $\mathbf{R}^n$ be given. Let S be a subspace of $\mathbf{R}^n$ with orthonormal bases $u_1, u_2, \ldots, u_k$ and let P_S be the perpendicular projection matrix for S. Let $A = [p_1 - \overline{p} | p_2 - \overline{p} | \ldots | p_m - \overline{p}]$. Then

$$\sum_{i=1}^{m} \|P_S(p_i - \overline{p})\|^2 = \sum_{l=1}^{k} u_l^T A A^T u_l \qquad \blacksquare$$

The matrix AA^T of Proposition 5.23 is symmetric. Thus the next proposition solves the maximization problem.

PROPOSITION 5.24 Let A be an n-by-n symmetric matrix with eigenvalues $\lambda_1, \lambda_2, \ldots, \lambda_n$, where $\lambda_1 \geq \lambda_2 \geq \lambda_3 \geq \cdots \geq \lambda_n$. If $u_1, u_2, \ldots, u_k$ is an orthonormal set of vectors, define

$$f(u_1, u_2, \ldots, u_k) = \sum u_l^T A u_l$$

Then

$$\lambda_1 + \lambda_2 + \cdots + \lambda_k \geq f(u_1, u_2, \ldots, u_k) \geq \lambda_n + \lambda_{n-1} + \cdots + \lambda_{n-k+1}$$

Further

$$f(u_1, u_2, \ldots, u_k) = \lambda_1 + \lambda_2 + \cdots + \lambda_k$$

when u_i is a unit eigenvector for λ_i, $1 \leq i \leq k$,

$$f(u_1, u_2, \ldots, u_k) = \lambda_n + \lambda_{n-1} + \cdots + \lambda_{n-k+1}$$

when u_i is a unit eigenvector for λ_{n-i+1}, $1 \leq i \leq k$.

Proof Since A is symmetric, it can be diagonalized by an orthonormal change of coordinates. Assume that the coordinate system has been chosen so that

$$A = \begin{bmatrix} \lambda_1 & & & 0 \\ & \lambda_2 & & \\ & & \ddots & \\ 0 & & & \lambda_n \end{bmatrix}$$

We will show that $f(u_1, u_2, \ldots, u_k) \leq \lambda_1 + \lambda_2 + \cdots + \lambda_k$ by showing that $f(u_1, u_2, \ldots, u_k) - (\lambda_1 + \lambda_2 + \cdots + \lambda_k) \leq 0$.

Let $U = [u_1 | u_2 | \ldots | u_k]$. Then $\|u_i\|^2 = 1$ and, since $u_i = U[;i]$,

$$k = \sum_{i=1}^{k} \|U[;i]\|^2 = \sum_{i=1}^{k} \sum_{j=1}^{n} U[j;i]^2 = \sum_{i=1}^{n} \|U[j;]\|^2$$

Now, since A is diagonal,

$$f(u_1, u_2, \ldots, u_k) - \sum_{i=1}^{k} \lambda_i$$

$$= u_1^T A u_1 + u_2^T A u_2 + \cdots + u_k^T A u_k - \sum_{i=1}^{k} \lambda_i$$

$$= \sum_{i=1}^{n} \lambda_i U[i;1]^2 + \sum_{i=1}^{n} \lambda_i U[i;2]^2 + \cdots + \sum_{i=1}^{n} \lambda_i U[i;k]^2 - \sum_{i=1}^{k} \lambda_i$$

$$= \sum_{l=1}^{k} \sum_{i=1}^{n} \lambda_i U[i;l]^2 - \sum_{i=1}^{k} \lambda_i$$

$$= \sum_{i=1}^{n} \sum_{l=1}^{k} \lambda_i U[i; l]^2 - \sum_{i=1}^{k} \lambda_i$$

$$= \sum_{i=1}^{k} \lambda_i \left(\sum_{l=1}^{k} U[i; l]^2 \right) + \sum_{i=k+1}^{n} \lambda_i \left(\sum_{l=1}^{k} U[i; l]^2 \right) - \sum_{i=1}^{k} \lambda_i$$

$$= \sum_{i=1}^{k} \lambda_i (\|U[i;]\|^2 - 1) + \sum_{i=k+1}^{n} \lambda_i \|U[i;]\|^2$$

$$= \sum_{i=1}^{n} (-\lambda_i)(1 - \|U[i;]\|^2) + \sum_{i=k+1}^{n} \lambda_i \|U[i;]\|^2$$

Now for $i \geq k$ we have $\lambda_i \geq \lambda_k$ and hence $-\lambda_i \leq -\lambda_k$. It can be shown (exercise 6) that $1 - \|U[i;]\|^2$ is nonnegative and hence

$$(-\lambda_i)(1 - \|U[i;]\|^2) \leq (-\lambda_k)(1 - \|U[i;]\|^2), \qquad i \leq k$$

Further, since $\lambda_k \geq \lambda_i$ for $i > k$, we have

$$\lambda_i \|U[i;]\|^2 \leq \lambda_k \|U[i;]\|^2, \qquad i > k$$

Hence

$$f(u_1, u_2, \ldots, u_k) - \sum_{i=1}^{n} \lambda_i$$

$$\leq (-\lambda_k) \sum_{i=1}^{k} (1 - \|U[i;]\|^2) + \lambda_k \sum_{i=k+1}^{n} \|U[i;]\|^2$$

$$= \lambda_k \left(-k + \sum_{i=1}^{k} \|U[i;]\|^2 + \sum_{i=k+1}^{n} \|U[i;]\|^2 \right)$$

$$= \lambda_k(-k + k)$$

$$= 0$$

Similarly (exercise 7) one may show that

$$f(u_1, \ldots, u_k) - \sum_{i=1}^{k} \lambda_{n-k+i} \geq 0$$

Notice that we get

$$f(u_1, u_2, \ldots, u_k) = \sum_{i=1}^{k} \lambda_i$$

if we take $U = \pm(n, k)\uparrow ID\ n$ and

$$f(u_1, u_2, \ldots, u_k) = \sum_{i=1}^{k} \lambda_{n-k+i}$$

when $U = \pm(n, -k)\uparrow ID\ n$. By our choice of coordinate system the columns of $ID\ n$ are unit eigenvectors for A. ∎

In the case $k = 1$ Proposition 5.24 is known as Rayleigh's principle.

PROPOSITION 5.25 (*Rayleigh's principle*) Let A be a symmetric matrix with maximum eigenvalue λ_1 and minimum eigenvalue λ_n. Let X be any vector in $\mathbf{R}^n$ and let $f(X)$ be the *Rayleigh quotient*:

$$f(X) = \frac{X^T A X}{X^T X}$$

Then $\lambda_1 \geq f(X) \geq \lambda_n$. ∎

SOLUTION TO THE MINIMIZATION PROBLEM

Let

$$\overline{p} = \frac{1}{m} \sum_{i=1}^{m} p_i$$

and define the matrix A by $A[;i] = p_i - \overline{p}$. The flat F that solves the minimization problem is defined by the points $\overline{p}, \overline{p} + u_1, \ldots, \overline{p} + u_k$ where u_i is an eigenvector belonging to the kth largest eigenvalue of AA^T.

DEFINITION 5.14 The solution to the minimization problem is called the *principal k-flat*. The eigenvectors $u_1, \ldots, u_k$ above are the first k *principal axes*.

Now we will write an APL function to compute the principal k-flats of a set of points $p_1, p_2, \ldots, p_m$ in $\mathbf{R}^n$. We will use the function *JACOBI* from Section 5.2 to diagonalize AA^T. The output of *JACOBI*, however, does not give the eigenvalues of AA^T sorted in descending order. To sort the eigenvalues we use the APL function *downgrade*, denoted by ⍒. If V is a vector, then ⍒V is the set of indices of V — that is, ⍳⍴V, arranged so that V[⍒V] is in descending order.

```
      V←3 1 7 ¯2 6.4 ¯8 1
      ⍒V
3 5 1 2 7 4 6
      V[⍒V]
7  6.4  3  1  1  ¯2  ¯8
```

For computational purposes we start with a matrix $P = [p_1 | p_2 | \cdots | p]$ of data points. The centroid can then be computed as

```
P ← AVE ⍉P
```

where *AVE*, defined in Section 3.1, computes the averages of the columns of a data matrix. Alternately, one can use the expression $p+.\div m$. This takes care of $\overline{p}$. Next we need a function to compute the principal axes from the data matrix P.

```
    ∇ Z←PRAXES P
[1]     P←P−(P+.÷1↑⍴P)∘.+(⍴P)[2]⍴0
[2]     Z←JACOBI P←P+.×⍉P
[3]     Z←Z[;⍒1 1⍉(⍉Z)+.×P+.×Z]
    ∇
```

On line [1] the centroid is computed as $P+.\div m$ and P is replaced by $[p_2 - \overline{p} | p_2 - \overline{p} | \cdots | p_m - \overline{p}]$. On line two the principal axes of PP^T are computed and stored as columns of Z. On line three the columns of Z are arranged in descending order according to the size of the corresponding eigenvalues.

EXAMPLE 5.24 Consider the isosceles triangle with vertices at $(-a, 0)$, $(0, 1)$, and $(a, 0)$ in the xy plane (Figure 5.24). Let us calculate the principal line (1-flat) for the vertices. We have

$$P = \begin{bmatrix} -a & 0 & a \\ 0 & 1 & 0 \end{bmatrix}, \qquad \overline{p} = \tfrac{1}{3}\begin{bmatrix} 0 \\ 1 \end{bmatrix}$$

Hence

$$A = \tfrac{1}{3}\begin{bmatrix} -3a & 0 & 3a \\ -a & 2 & 1 \end{bmatrix}, \qquad AA^T = \begin{bmatrix} a^2 & 0 \\ 0 & \frac{2}{3} \end{bmatrix}$$

Since AA^T is diagonal, the principal axes are $(1, 0)$ and $(0, 1)$. If $a^2 > \frac{2}{3}$, then

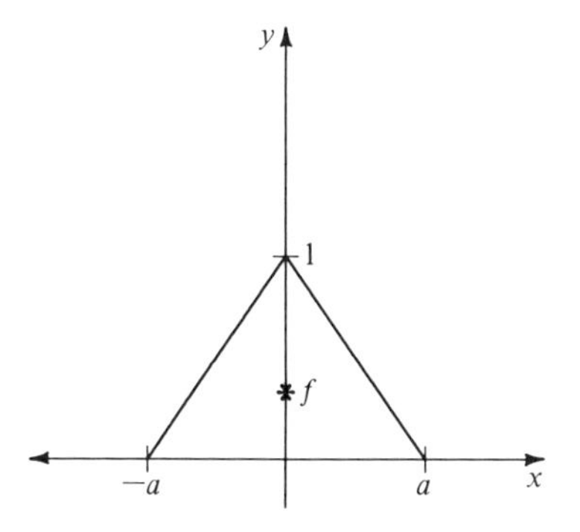

FIGURE 5.24

$u_1 = (1, 0)$ and the principal line is the horizontal line through $p = (0, \frac{1}{3})$. If $a^2 < \frac{2}{3}$, then $u_1 = (0, 1)$ and the principal line is the y axis. If $a^2 = \frac{2}{3}$, then the principal line is undefined. In this case any line through p yields a minimal value of $\Sigma\, d_i^2$.

The case $a^2 = \frac{2}{3}$ occurs when the triangle is equilateral. Thus the principal line is horizontal when the third side of the isosceles triangle is longer than the equal sides, and the principal line is vertical when the third side is the shorter.

Now let us compute the least-squares straight line fitting these points that minimizes the sum of the squares of the *vertical* distances. The methods of Section 2.3 (see Example 2.28) give

$$\begin{bmatrix} 1 & 1 & 1 \\ -a & 0 & a \end{bmatrix} \begin{bmatrix} 1 & -a \\ 1 & 0 \\ 1 & a \end{bmatrix} X = \begin{bmatrix} 1 & 1 & 1 \\ -a & 0 & a \end{bmatrix} \begin{bmatrix} 0 \\ 1 \\ 0 \end{bmatrix}$$

or

$$X = \tfrac{1}{3}\begin{bmatrix} 1 \\ 0 \end{bmatrix}$$

Thus the least-squares straight line is $y = \frac{1}{3}$ and, in contrast to the principal line, is independent of a. ∎

The next example is from physics.

EXAMPLE 5.25 Consider the set of points in Figure 5.25. There are ten points spaced uniformly along the x axis from -3 to $-.3$ and eight points spaced uniformly around the circle with radius 1 and center at 1 on the axis.

We will refer to this configuration as a "tennis racket."

The tennis racket lies in $\mathbf{R}^3$. The z axis is perpendicular to the page. The coordinates of the eighteen points are stored as the columns of the matrix P.

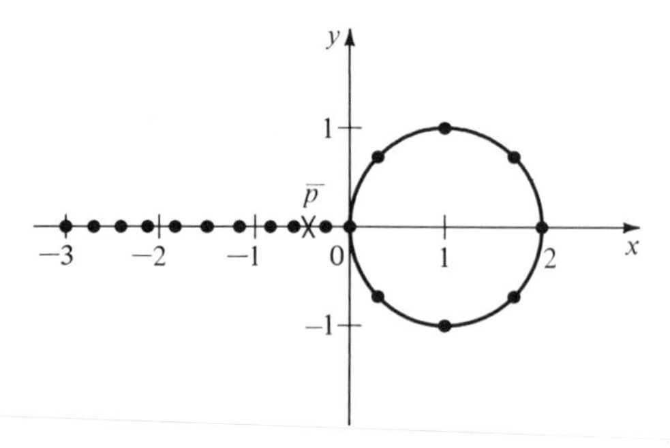

FIGURE 5.25

```
      ⍉P
¯3.00E0      0.00E0      0.00E0
¯2.70E0      0.00E0      0.00E0
¯2.40E0      0.00E0      0.00E0
¯2.10E0      0.00E0      0.00E0
¯1.80E0      0.00E0      0.00E0
¯1.50E0      0.00E0      0.00E0
¯1.20E0      0.00E0      0.00E0
¯9.00E¯1     0.00E0      0.00E0
¯6.00E¯1     0.00E0      0.00E0
¯3.00E¯1     0.00E0      0.00E0
 2.00E0      0.00E0      0.00E0
 1.71E0      7.07E¯1     0.00E0
 1.00E0      1.00E0      0.00E0
 2.93E¯1     7.07E¯1     0.00E0
 0.00E0      1.03E¯18    0.00E0
 2.93E¯1    ¯7.07E¯1     0.00E0
 1.00E0     ¯1.00E0      0.00E0
 1.71E0     ¯7.07E¯1     0.00E0
```

The center of gravity of the tennis racket is on the handle near the strings.

```
      +P←AVE⍉P
¯0.472 ¯0  0
```

The direction of the principal axes, in descending order, are

```
      PRAXES P
1  0  0
0  1  0
0  0  1
```

The first principal axis is along the handle. The second principal axis is perpendicular to the handle and in the plane of the racket. The third principal axis is through the centroid and perpendicular to the plane of the racket.

Recall that the kinetic energy of a mass m moving with velocity v is $\frac{1}{2}mv^2$. In theoretical mechanics it is shown that the energy of a mass rotating about a line l with angular velocity w is $\frac{1}{2}I_l w^2$, where I_l is the moment of inertia about the line l. If we hold w constant and let l vary, then the energy is a minimum when l is the first principal axis. It follows from Rayleigh's principle that the energy is a maximum if the racket is spinning about the third principal axis (exercise 8).

If a tennis racket is tossed into the air, its motion may be analyzed as follows. The centroid moves along some path. At any given moment the racket is rotating about a line l through the centroid. The direction of the line l through the centroid changes in relation to the racket with time. Neglecting air resistance, the energy of rotation about l will be constant while the racket is in the air.

If the racket is launched with l parallel to the third principal axis, the direction of l will remain fixed; it cannot change without the system losing rotational energy. Thus the racket flies with a steady rotation about the third principal axis. This is easy to check using a real racket. Similarly, if the racket is launched with l along the first principal axis, then the direction of l cannot change unless the racket gains energy. This is harder to check with a real racket, since the shape makes it hard to launch a racket rotating only about this axis. These two rotational modes are *stable.* Rotation about the second principal axis, however, is unstable. The direction of the line l may change without changing the rotational energy of the system. It is essentially impossible to make a tennis racket rotate steadily about the second principal axis. ■

PRINCIPAL-COMPONENT ANALYSIS

Principal-component analysis is a statistical technique for reducing the number of variables in a problem. Let us return to the height and weight example that began this section. A sample of m individuals is chosen from a population and their heights and weights are measured, giving m points $p_i = (x_i, y_i)$ — a two-variable problem. One then chooses $x'y'$ coordinates as follows. The origin O' is the centroid $\overline{p}$, the x' axis is along the first principal axis, and the y' axis is perpendicular. The y' coordinate is discarded and one works with the x' coordinate.

In this way we lose the minimal amount of information. But how much information is lost? Let F be the principal k-flat and let π_F be the perpendicular projection function. If $\overline{p}$ is the centroid, then by the Pythagorean theorem we have

$$\sum_{i=1}^{m} \|p_i - \overline{p}\|^2 = \sum_{i=1}^{m} \|\pi_F(p_i) - \overline{p}\|^2 + \sum_{i=1}^{m} \|p_i - \pi_F(p_i)\|^2$$

and F is chosen to minimize the second term on the right-hand side. If we divide this equation by m ($m - 1$ for a sample), we have a statement about variances: Upon division by the left-hand side the m or $m - 1$ in the denominator will cancel and we obtain

$$\frac{\sum \|\pi_F(p_i) - \overline{p}\|^2}{\sum \|p_i - \overline{p}\|^2} = \text{fraction of variance retained}$$

$$\frac{\sum \|p_i - \pi_F(p_i)\|^2}{\sum \|p_i - \overline{p}\|^2} = \text{fraction of variance lost}$$

The proof of the next proposition is left as exercise 9.

PROPOSITION 5.26 Let $p_1, \ldots, p_m$ in $\mathbf{R}^n$ be given. Let F be the principal k-flat for $p_1, p_2, \ldots, p_m$. Let $\overline{p}$ be the centroid and set $A = [p_1 - \overline{p} | p_2 - \overline{p} | \ldots | p_m - \overline{p}]$. If the eigenvalues of AA^T are $\lambda_1 \geq \lambda_2 \geq \cdots \lambda_n$, then

$$\sum_{i=1}^{m} \|\pi_F(p_i) - \bar{p}\|^2 = \lambda_1 + \lambda_2 + \cdots + \lambda_k$$

$$\sum_{i=1}^{m} \|p_i - \pi_F(p_i)\|^2 = \lambda_{k+1} + \lambda_{k+2} + \cdots + \lambda_n \qquad \blacksquare$$

EXERCISES 5.6*

1. (a) How does the principal line of the four points $(a, 1)$, $(a, -1)$, $(-a, 1)$, and $(-a, -1)$ vary with $a > 0$?

   (b) Show that $\Sigma\|(p_i - p)^{\perp}\|^2 = 4$ if $p_1 = (1, 1)$, $p_2 = (1, -1)$, $p_3 = (-1, 1)$, $p_4 = (-1, -1)$ and the flat F is the subspace with orthonormal basis $\{(\cos\theta, \sin\theta)\}$.

2. Find the principal 1-flat for the set of points $(0, 4)$, $(1, 1)$, $(3, 3)$, and $(4, 0)$. Use Proposition 4.10 or 4.11 to diagonalize AA^T.

3. The set of sixteen points in $\mathbf{R}^3$ given in the accompanying table can be thought of as a discrete approximation to a book. Find the stable and unstable axes of rotation of this "book." Test your conclusions with a real book.

   Note: This is *not* a computer exercise.

x	−1	−1	−1	0	0	1	1	1	−1	−1	−1	0	0	1	1	1
y	2	0	−2	2	−2	2	0	−2	2	0	−2	2	−2	2	0	−2
z	.5	.5	.5	.5	.5	.5	.5	.5	−.5	−.5	−.5	−.5	−.5	−.5	−.5	−.5

   Hint: Tape the book shut before you throw it in the air.

4. Distance and perpendicularity are Euclidean concepts. This means that the principal k-flat of a set of points is a Euclidean invariant of the set of points; that is, any congruence that carries set S_1 to set S_2 also carries the principal k-flat for S_1 to the principal k-flat for S_2. In particular, if the set S is mapped into itself by a rotation or reflection, then the principal k-flat will be mapped into itself under the same rotation or reflection. Use this fact to find the principal 2-flat for the "book" of exercise 3 without computation.

5. (Computer assignment) Approximate a boomerang by placing particles at the points $(i, 0, 0)$ for $i = -7, -6, -5, -4, -3, -2, -1$ and at the points $(i, i, 0)$ for $i = 0, 1, 2, 3, 4, 5$. Find the stable and unstable axes of rotation for the boomerang.

6. (a) Show that if Q is a square orthogonal matrix, then Q^T is an orthogonal matrix.

   (b) From (a) show that if Q is n by n and the columns of Q form an orthonormal set of vectors, then the rows of Q also form an orthonormal set of vectors.

   (c) As in the proof of Proposition 5.24, let the columns of U be orthonormal. Let S be the column space of U and let $S^{\perp}$ be the orthogonal complement of S. Let the columns of V form an orthonormal basis of $S^{\perp}$. By applying part (b) to the matrix $Q = [U \mid V]$ show that $\|U[i;]\| \leq 1$ for all valid row indices i.

7. Finish the proof of Proposition 5.24 by showing that

$$f(u_1, u_2, \ldots, u_k) \geq \sum_{i=1}^{k} \lambda_{n-k+i}$$

8. Verify the statement made in Example 5.25 that the energy of the tennis racket is maximal when it is spinning about the third principal axis.

   Hint: Assuming that F contains $\overline{p}$, show that $\Sigma \, \|p_i - \pi_F(p)\|^2$ is maximum when $\Sigma \, \|\pi_F(p_i) - \overline{p}\|^2$ is a minimum. To do this, imitate the argument in the text leading from the minimization problem to the maximization problem.

9. Prove Proposition 5.26

   Hint: Let u_i be a unit eigenvector belonging to λ_i, $1 \leq i \leq n$, and change coordinates by the formula $X = \overline{p} + [u_1 | u_2 | \ldots | u_n]X'$.

CHAPTER SIX

Linear Programming

Linear programming is a technique widely used in industry as a way of helping management make decisions. It is also used in large numerical economic models. Linear programming is an optimization technique. It is concerned with finding maxima and minima of linear functions f: $\mathbf{R}^n \to \mathbf{R}$. Of course such a function $f(X)$ does not usually have maxima or minima if X is allowed to range over all of $\mathbf{R}^n$. In linear programming problems $f(X)$ has a maximum or a minimum because X is not allowed to vary freely but is constrained to be in a specified region of $\mathbf{R}^n$. The constraints imposed are linear inequalities, and this is what makes the subject part of linear algebra.

Section 6.1 is devoted to exhibiting examples of linear programming problems and the technique used to put such problems into standard forms.

In Section 6.2 the geometry of linear programming problems is discussed. The constraints of a linear programming problem restrict the domain of the function $f(X)$ to the region enclosed by a polyhedron in $\mathbf{R}^n$, and we gain insight into the problem by looking at the hyperplane $f(X) = \text{const.}$ near a vertex of this polyhedron.

Section 6.3 describes the simplex algorithm for solving linear programming problems. This is the algorithm upon which many large computational packages for solving linear programming problems are based.

In Section 6.4 we offer some theoretical as opposed to computational applications of linear programming and the closely related topic of game theory.

6.1 Examples of Linear Programming Problems

In this section we concentrate on phrasing problems in linear programming form. Solution methods are deferred to later sections.

We begin with a typical problem involving the allocation of limited resources.

EXAMPLE 6.1 A farmer grows two crops, say corn and soybeans. Suppose that each acre planted with corn requires 1 man-day of labor, \$3 investment (seed and so on), and yields a profit of \$35 at harvest. Each acre of soybeans, on the other hand, requires 2 man-days of labor, \$1 investment, and returns a profit of \$20 at harvest. The farmer has a total of 70 acres available for these two crops. He has \$140 he can invest and 80 man-days of labor available. How many acres of each crop should he plant for maximum profit?

Set up Let x_1 be the number of acres of corn planted and x_2 the number of acres of soybeans. Then

$$\text{profit} = z = 35x_1 + 20x_2$$

This is the function we wish to maximize. This function is unbounded in the whole x_1x_2 plane, but the possible values of x_1 and x_2 are quite restricted.

Since the farmer cannot plant a negative amount of land, we must have $x_1 \geq 0$ and $x_2 \geq 0$.

Further, since only 70 acres are available,

$$x_1 + x_2 \leq 70$$

The limit on available labor imposes the constraint

$$x_1 + 2x_2 \leq 80$$

and the limited amount of capital available forces

$$3x_1 + x_2 \leq 140$$

The problem can be summarized as

$$\begin{array}{ll} \text{maximize} & z = 35x_1 + 20x_2 \\ \text{subject to} & x_1 + x_2 \leq 70 \\ & x_1 + 2x_2 \leq 80 \\ & 3x_1 + x_2 \leq 140 \\ & x_1, x_2 \geq 0 \end{array}$$

We will see in Section 6.3 that the maximum profit is obtained with 40 acres of corn and 20 acres of soybeans. This uses all the labor and all the capital and leaves 10 acres unplanted for a profit of \$1800. ∎

The system of inequalities can be rewritten in matrix form. If v and w are vectors, we will write $v \leq w$ if $v[i] \leq w[i]$ for every i. Then, setting

$$A = \begin{bmatrix} 1 & 1 \\ 1 & 2 \\ 3 & 1 \end{bmatrix}, \quad b = \begin{bmatrix} 70 \\ 80 \\ 140 \end{bmatrix}, \quad c = \begin{bmatrix} 35 \\ 20 \end{bmatrix}, \quad X = \begin{bmatrix} x_1 \\ x_2 \end{bmatrix}$$

the system of example 6.1 becomes

$$\begin{array}{ll} \text{maximize} & z = c^T X \\ \text{subject to} & AX \le b \\ & X \ge 0 \end{array} \tag{6.1}$$

DEFINITION 6.1 The system (6.1) is called a *standard maximum problem*. The function z is called the *objective function* and the condition $X \ge 0$ is called the *positivity condition*.

Many authors refer to a standard maximum problem as a *primal* problem.

EXAMPLE 6.2 A nutritionist is preparing a supplement mixture to add to a commercial preparation of creamed chipped beef on toast. The nutritionist has three mixtures available that can be combined to give the desired mixture. Each gram of the first mixture contains 1 unit of calcium, 1 unit of iron, and costs 70 cents. Each gram of the second mixture contains 1 unit of calcium, 2 units of iron, and costs 80 cents. Each gram of the third mixure contains 3 units of calcium, 1 unit of iron, and costs $1.40. Each batch of the product needs 35 units of calcium and 20 units of iron. How much of each mixture should be added to each batch of the product to meet these requirements at minimum cost?

Set up Let y_i be the amount of the ith mix to be added. Then the cost of the additives is, in cents per batch,

$$w = 70y_1 + 80y_2 + 140y_3$$

Again we have the restrictions $y_i \ge 0$. Further, since 35 units of calcium are needed, we must have

$$y_1 + y_2 + 3y_3 \ge 35$$

and the requirement of 20 units of iron per batch forces

$$y_1 + 2y_2 + y_3 \ge 20$$

The mathematics problem can be summarized as

$$\begin{array}{ll} \text{minimize} & w = 70y_1 + 80y_2 + 140y_3 \\ \text{subject to} & y_1 + y_2 + 3y_3 \ge 35 \\ & y_1 + 2y_2 + y_3 \ge 20 \\ & y_1 \ge 0, \quad y_2 \ge 0, \quad y_3 \ge 0 \end{array}$$

In Section 6.3 we will see that the minimum cost of $18 is attained by adding 5 grams of the second mixture and 10 grams of the third mixture to give precisely 35 units of calcium and 20 units of iron per batch. ■

Taking

$$B = \begin{bmatrix} 1 & 1 & 3 \\ 1 & 2 & 1 \end{bmatrix}, \qquad b = \begin{bmatrix} 70 \\ 80 \\ 140 \end{bmatrix},$$

$$c = \begin{bmatrix} 35 \\ 20 \end{bmatrix}, \qquad Y = \begin{bmatrix} y_1 \\ y_2 \\ y_3 \end{bmatrix}$$

We may rephrase the problem of Example 6.2 as

$$\begin{aligned} &\text{minimize} && w = b^T Y \\ &\text{subject to} && BY \geq c \\ &&& Y \geq 0 \end{aligned} \tag{6.2}$$

DEFINITION 6.2 The system (6.2) is called a *standard minimum problem*. The function w is called the *objective function* and the condition $Y \geq 0$ the *positivity condition*.

We will see in Section 6.2 that there is a close relation between the problems of Examples 6.1 and 6.2 — indeed the solution of one yields the solution of the other.

DEFINITION 6.3 If in Equations (6.1) and (6.2) we have $B = A^T$, then (6.2) is said to be the *dual* of (6.1) and (6.1) is said to be the *dual* of (6.2).

We have not given, and will not give, a formal definition of linear programming except to say that a linear programming problem is one that can be reduced to either a standard maximum problem or a standard minimum problem. In Section 6.3 we will write an APL function *MAX* to simultaneously solve a dual pair of standard problems. Most problems do not automatically come in standard form, however, and a variety of techniques are used to rephrase problems in standard form.

EXAMPLE 6.3 A manufacturer has two plants and three distribution points (warehouses). To meet the local demand for the product, the first distribution point requires 3 carloads of product per week. The second requires 2 carloads per week, and the third requires 3 carloads per week. The first plant can produce 4 carloads per week and the second plant 5 carloads per week. The shipping charges, in hundreds of dollars per carload, are indicated in Figure 6.1.

Determine a shipping schedule that minimizes the shipping cost subject to the restriction that each distribution point receives the required amount.

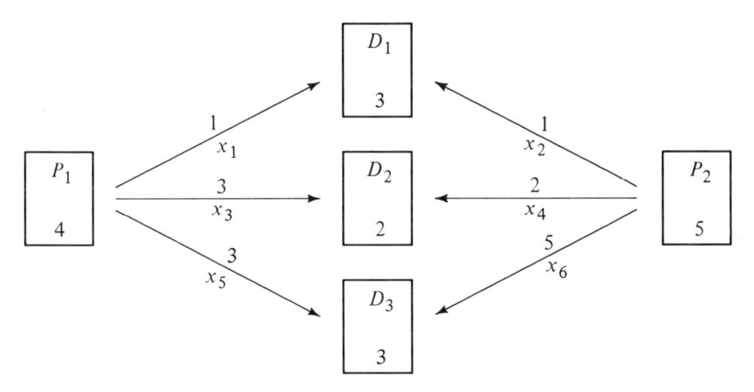

FIGURE 6.1

Set up Let x_i be the number of carloads shipped along the route indicated in Figure 6.1. Then the cost of shipping is

$$w = c^T X = [1 \quad 1 \quad 3 \quad 2 \quad 3 \quad 5]X$$

We wish to minimize w subject to the following restrictions: P_1 cannot ship more than 4 carloads.

$$x_1 + x_3 + x_5 \leq 4$$

P_2 cannot ship more than 5 carloads.

$$x_2 + x_4 + x_6 \leq 5$$

D_1 needs at least 3 carloads.

$$x_1 + x_2 \geq 3$$

D_2 needs at least 2 carloads.

$$x_2 + x_4 \geq 2$$

D_3 needs at least 3 carloads.

$$x_5 + x_6 \geq 3$$

In summary,

$$\begin{array}{llcccccccccccl}
\text{minimize} & w = x_1 & + & x_2 & + & 3x_3 & + & 2x_4 & + & 3x_5 & + & 5x_6 & & \\
\text{subject to} & x_1 & & & + & x_3 & & & + & x_5 & & & \le & 4 \\
& & & x_2 & & & + & x_4 & & & + & x_6 & \le & 5 \\
& x_1 & + & x_2 & & & & & & & & & \ge & 3 \\
& & & & & x_3 & + & x_4 & & & & & \ge & 2 \\
& & & & & & & & & x_5 & + & x_6 & \ge & 3 \\
& & & & & & & & & & & x_1, x_2, x_3, x_4, x_5, x_6 & \ge & 0
\end{array}$$

This system is a minimum problem but is not in standard form. We easily remedy this by multiplying the first two inequalities by -1 to obtain

$$\begin{array}{cccccccc}
-x_1 & & -x_3 & & -x_5 & & \ge & -4 \\
& -x_2 & & -x_4 & & -x_6 & \ge & -5
\end{array}$$

The system can also be rewritten directly as a standard maximum problem. We need only notice that w is a minimum precisely when $z = -w$ is a maximum. Thus we have the standard maximum problem:

$$\begin{array}{llcccccccccccl}
\text{maximize} & z = -x_1 & - & x_2 & - & 3x_3 & - & 2x_4 & - & 3x_5 & - & 5x_6 & & \\
\text{subject to} & x_1 & & & + & x_3 & & & + & x_5 & & & \le & 4 \\
& & & x_2 & & & + & x_4 & & & + & x_6 & \le & 5 \\
& -x_1 & - & x_2 & & & & & & & & & \le & -3 \\
& & & & & -x_3 & - & x_4 & & & & & \le & -2 \\
& & & & & & & & - & x_5 & - & x_6 & \le & -3 \\
& & & & & & & & & & & x_1, x_2, x_3, x_4, x_5, x_6 & \ge & 0
\end{array}$$

Using the methods of Section 6.3, we can show that the solution is $X = (1, 2, 0, 2, 3, 0)$, which has the second plant producing 1 carload per week under capacity. ■

Example 6.3 illustrates two techniques for rephrasing problems. Inequalities may be reversed by multiplying through by a negative one. The problem of minimizing z is the same as that of maximizing $-z$, and the problem of maximizing z is the same as that of minimizing $-z$.

EXAMPLE 6.4 A furniture factory produces three types of couches. The first type uses 1 board foot of framing wood and 3 board feet of cabinet wood, the second requires 2 board feet of framing wood and 2 board feet of cabinet wood, and the third uses 2 board feet of framing wood and 1 board foot of cabinet wood. Currently the factory is producing each month 500 couches of the first type, 300 of the second type, and 200 of the third type.

The supplier informs the factory management that there is a shortage of cabinet wood and the supply to the factory will have to be reduced by 600 board feet per month. In partial compensation the supply of framing wood may be increased by 100 board feet per month. If the profit on the three types of couches is \$10, \$8, and \$5, respectively, how should the production of each type of couch be adjusted to minimize the decrease, if any, in profits?

Set up Let x_i be the change in the number of couches of type i produced each month — positive for an increase in production, negative for a cutback.

The change in profits is then

$$z = \Delta p = 10x_1 + 8x_2 + 5x_3$$

Since more framing wood will be available, the change in the amount of framing wood used will be

$$x_1 + 2x_2 + 2x_3 \leq 100$$

whereas the change in the amount of cabinet wood used is

$$3x_1 + 2x_2 + x_3 \leq -600$$

Since the number of each type of couch produced cannot be less than zero, we have

$$x_1 \geq -500, \qquad x_2 \geq -300, \qquad x_3 \geq -200$$

Minimizing the loss means maximizing Δp. Thus the mathematical problem becomes

$$\begin{aligned} \text{maximize} \quad & z = 10x_1 + 8x_2 + 5x_3 \\ \text{subject to} \quad & x_1 + 2x_2 + 2x_3 \leq 100 \\ & 3x_1 + 2x_2 + x_3 \leq -600 \\ & x_1 \geq -500, \quad x_2 \geq -300, \quad x_3 \geq -200 \end{aligned}$$

This problem is not in the form of a standard maximum problem, since we do not have the positivity condition. The problem must be recast, since the methods of Section 6.3 assume $x_i \geq 0$.

Two techniques are available for rewriting the problem as a standard maximum problem.

The first technique depends on the fact that although the x_i's may be negative, they are bounded below. If we translate coordinates by the formula

$$\begin{bmatrix} x_1 \\ x_2 \\ x_3 \end{bmatrix} = X = X' + \begin{bmatrix} -500 \\ -300 \\ -200 \end{bmatrix}$$

then the condition

$$X \geq \begin{bmatrix} -500 \\ -300 \\ -200 \end{bmatrix}$$

becomes $X' \geq 0$. To make the coordinate change we write the system in matrix form:

$$\begin{aligned} \text{maximize} \quad & z = c^T X \\ \text{subject to} \quad & AX \leq b \\ & X \geq d \end{aligned} \tag{6.3}$$

where

$$c = \begin{bmatrix} 10 \\ 8 \\ 5 \end{bmatrix}, \quad b = \begin{bmatrix} 100 \\ -600 \end{bmatrix}, \quad A = \begin{bmatrix} 1 & 2 & 2 \\ 3 & 2 & 1 \end{bmatrix}, \quad d = \begin{bmatrix} -500 \\ -300 \\ -200 \end{bmatrix}$$

and then substitute $X = X' + d$ to get

$$z = c^T X = c^T(X' + d) = c^T X' + c^T d$$

or

$$z = c^T X' - 8400$$
$$AX = A(X' + d) = AX' + Ad \leq b$$

or

$$AX' \leq b - Ad = \begin{bmatrix} 100 \\ -600 \end{bmatrix} - \begin{bmatrix} -1500 \\ -2300 \end{bmatrix}$$
$$= \begin{bmatrix} 1600 \\ 1800 \end{bmatrix} = b'$$

and

$$X = X' + d \geq d$$

or

$$X' \geq 0$$

Since additive constants change the value of maxima but *not* their location, we have the standard maximum problem

$$\begin{aligned} \text{maximize} \quad & z' = c^T X' \\ \text{subject to} \quad & AX' \leq b' \\ & X' \geq 0 \end{aligned}$$

The second method of obtaining a standard maximum problem when x_i may take on negative values is to split x_i into a positive part and a negative part:

$$x_i = x_i^+ - x_i^-, \qquad x_i^+ \geq 0, \qquad x_i^- \geq 0$$

Then, for example, if $x_i = -1$, we can take $x_i^+ = 0$, $x_i^- = 1 \geq 0$, or perhaps $x_i^+ = 103 \geq 0$, $x_i^- = 104 \geq 0$.

We need not know a lower bound on x_i to use this technique.

In the present example we replace X by

$$X = X^+ - X^-, \qquad X^+ \geq 0, \qquad X' \geq 0$$

in Equation (6.3) to obtain the standard maximum problem

$$\begin{aligned} &\text{maximize} \qquad Z = [c^T \ \ -c^T]\begin{bmatrix} X^+ \\ X^- \end{bmatrix} \\ &\text{subject to} \qquad \begin{bmatrix} A & -A \\ -I & I \end{bmatrix}\begin{bmatrix} X^+ \\ X^- \end{bmatrix} \leq \begin{bmatrix} b \\ -d \end{bmatrix} \\ &\qquad\qquad\qquad \begin{bmatrix} X^+ \\ X^- \end{bmatrix} \geq 0 \end{aligned} \tag{6.4}$$

The system (6.3), incidentally, has an infinity of solutions given by

$$X(t) = \begin{bmatrix} -350 \\ 225 \\ 0 \end{bmatrix} + t\begin{bmatrix} 2 \\ -5 \\ 4 \end{bmatrix}, \qquad -50 \leq t \leq 105$$

For this line segment $z(t) = c^T X(t) = -1700$ independent of t. Thus the profits decrease by \$1700 per month (from \$8400 to \$7700), and this is the smallest possible decrease, given the constraints. ■

The infinity of solutions in the last example indicates that more constraints may be imposed without decreasing profits further.

EXAMPLE 6.5 We continue with the problem of Example 6.4. Since the cutback in supplies is short term, the management of the factory wishes to hold the work force constant during the period. If the first type of couch needs 5 man-hours of labor, the second type 7, and the third type 5, how should the production schedule be changed?

Set up All the constraints of Example 6.5 are in force. In addition, we have that the change in labor requirements should be zero:

$$5x_1 + 7x_2 + 5x_3 = 0$$

Thus we have the system

$$\begin{array}{llr}
\text{maximize} & z = 10x_1 + 8x_2 + 5x_3 & \\
\text{subject to} & 5x_1 + 7x_2 + 5x_3 = 0 & \\
& x_1 + 2x_2 + 2x_3 \leq 100 & \qquad (6.5) \\
& 3x_1 + 2x_2 + x_3 \leq -600 & \\
& x_1 \geq -500, \quad x_2 \geq -300, \quad x_3 \geq -200 &
\end{array}$$

which we wish to rewrite as a standard maximum problem. In Example 6.4 we saw how to take care of the fact that the x_i's may be negative, so the problem is the equality in the first constraint.

Of course, the equality constraint implicitly reduces the number of variables by confining the problem to a hyperplane. We can make this explicit by solving for x_1, say, in terms of x_2 and x_3 and then substituting for x_1 in the remaining inequalities, obtaining a system in only two unknowns. In a large system, however, this procedure can involve a fair amount of computation. The common procedure is to increase the number of constraints by one by replacing the equality with a pair of inequalities

$$\begin{aligned} 5x_1 + 7x_2 + 5x_3 &\leq 0 \\ 5x_1 + 7x_2 + 5x_3 &\geq 0 \end{aligned}$$

or

$$\begin{aligned} 5x_1 + 7x_2 + 5x_3 &\leq 0 \\ -5x_1 - 7x_2 - 5x_3 &\leq 0 \end{aligned}$$

The addition of the equality constraint makes the solution unique: $x_1 = -420$, $x_2 = 400$, $x_3 = -140$. The first type is cut back by 420 units, the second type by 140 units, and the third type increased by 400 units (per month). ∎

NONLINEAR OBJECTIVE FUNCTIONS

The scope of linear programming problems is much wider than the examples so far suggest. Consider again, for example, the problem of fitting a curve to measured data (Figure 6.2).

The least-squares straight line is the line that minimizes the sum of the squares of the distances

$$\sum d_i^2, \qquad d_i = |y_i - (z + bx_i)|$$

A more intuitive procedure would be to simply minimize the sum of the distances

$$w = \sum d_i$$

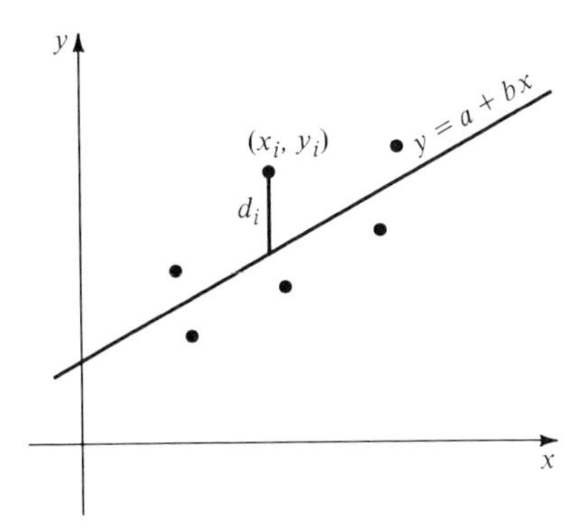

FIGURE 6.2

This method produces a straight line less influenced by outlying points than the least-squares straight line.

The minimization of $w = \Sigma\, d_i$ can be phrased as a linear programming problem. We first prove a preliminary result.

In Example 6.4 we broke variables into positive and negative parts:

$$x = x^+ - x^-, \qquad x^+ \geq 0, \qquad x^- \geq 0$$

There are many choices for x^+ and x^-, however. If $x = -2$, then an obvious choice is $x^+ = 0$, $x^- = 2$ — but $x^+ = 1$, $x^- = 3$, for example, will do as well. The point of the next proposition is that when the proper kind of minimization condition is imposed, then x^+ and x^- are determined by the equations

$$x^+ = \begin{cases} x & \text{if } x \geq 0 \\ 0 & \text{otherwise} \end{cases} \qquad x^- = \begin{cases} 0 & \text{if } x \geq 0 \\ -x & \text{otherwise} \end{cases} \tag{6.6}$$

Notice that when Equations (6.6) hold we have

$$x = x^+ - x^-, \qquad |x| = x^+ + x^- \tag{6.7}$$

PROPOSITION 6.1 Let S be a subset of $\mathbf{R}^n$. Let C_+ and C_- be vectors in $\mathbf{R}^n$ with $(C_+ + C_-)[J] > 0$ for some vector of indices J. Suppose that $X_0^+ \geq 0$, $X_0^- \geq 0$ are vectors in $\mathbf{R}^n$ with $X_0 = X_0^+ - X_0^-$ in S. Suppose further that

$$m = C_+^T X_0^+ + C_-^T X_0^- = \min\,(C_+^T X^+ + C_-^T X^-), \qquad X^+ - X^- \text{ in } S$$

Then $0 = (X_0^+ \llcorner X_0^-)[J]$ and hence the components of $X_0[J]$, $X_0^+[J]$, $X_0^-[J]$ satisfy Equations (6.6) and (6.7).

Proof Recall that $v \llcorner w$ is the component-by-component minimum of v and w. Set $\delta = X_0^+ \llcorner X_0^-$, $Y^+ = X_0^+ - \delta$, $Y^- = X_0^- - \delta$. Then

$$Y^+ - Y^- = (X_0^+ - \delta) - (X_0^- - \delta) = X_0^+ - X_0^- = X_0 \qquad \text{lies in } S$$

Thus

$$\begin{aligned} m &\le C_+^T Y^+ + C_-^T Y^- \\ &= C_+^T X_0^+ + C_-^T X_0^- - (C_+ + C_-)^T \delta \end{aligned}$$

or

$$m \le m - (C_+ + C_-)^T \delta$$

hence

$$0 \le -(C_+ + C_-)^T \delta$$

Since $(C_+ + C_-)[J] > 0$, this implies $\delta[J] \le 0$. But $\delta \ge 0$ as well so $\delta[J] = 0$. ■

Now let us return to the curve-fitting problem. Changing notation a bit, let us take $d_i = y_i - (a + bx_i)$ (d_i now can take on negative values). We wish to minimize $\Sigma\, |d_i|$. Consider the linear programming problem in the unknowns a, b, d_i.

$$\begin{aligned} &\text{minimize} && w = \sum (d_i^+ + d_i^-) \\ &\text{subject to} && a + x_i b + d_i^+ - d_i^- = y_i \quad 1 \le i \le n, \\ &&& d_i^+ \ge 0, \quad d_i^- \ge 0 \end{aligned}$$

We may apply Proposition 6.1 as follows. The set S is the set of all vectors $(a, b, d_1^+, d_1^-, \ldots, d_n^+, d_n^-)$ in $\mathbf{R}^{2(n+1)}$ with $d_i^+, d_i^- \ge 0$. The vectors C_+ and C_- are both $(0, 0, 1, 1, \ldots, 1)$ and $J = (2, 3, \ldots, 2(n + 1))$. It follows from Proposition 6.1 that if $(\alpha, \beta, \delta_1^+, \delta_1^-, \ldots, \delta_n^+, \delta_n^-)$ is a solution of the linear programming problem, then setting $\delta_i = \delta_i^+ - \delta_i^-$ we have

$$w_0 = \sum (\delta_i^+ + \delta_i^-) = \sum |\delta_i|$$

Thus if $w_1 = \min \Sigma\, |d_i|$, we must have $w_1 \le w_0$. On the other hand, given *any* particular value $w = \Sigma\, |d_i|$, we can set $\delta_i^+ = |d_i|$, $\delta_i^- = 0$ and get $w = \Sigma\, (\delta_i^+ + \delta_i^-)$. Hence $w_0 \le w_1$, and so the linear programming problem and the problem of minimizing $\Sigma\, |d_i|$ have the same solution.

This argument immediately generalizes. First, however, a bit of notation.

DEFINITION 6.4

1. The l_1 *norm* on $\mathbf{R}^n$ is the function

$$\|\cdot\|_1 \colon \mathbf{R}^n \to \mathbf{R}$$

given by $\|v\|_1 = \Sigma\, |v_i|$.

2. The l_2 *norm* on $\mathbf{R}^n$ is the function

$$\|\cdot\|_2 \colon \mathbf{R}^n \to \mathbf{R}$$

given by $\|v\|_2 = \sqrt{\Sigma\, v_i^2}$.

3. The l_∞ *norm* on $\mathbf{R}^n$ is the function

$$\|\cdot\|_\infty \colon \mathbf{R}^n \to \mathbf{R}$$

given by $\|v\|_\infty = \max\,\{|v_i|\}$.

The l_2 norm is the norm introduced in Chapter 5, and the least-squares solutions of

$$AX = B$$

are the solutions that minimize $\|B - AX\|_2$. In the discussion above we wished to minimize

$$\sum |d_i| = \sum |y_i - (a + bx_i)| = \|B - AX\|_1$$

where

$$A = \begin{bmatrix} 1 & x_1 \\ 1 & x_2 \\ \vdots & \vdots \\ 1 & x_n \end{bmatrix}, \qquad B = \begin{bmatrix} y_1 \\ y_2 \\ \vdots \\ y_n \end{bmatrix}, \qquad X = \begin{bmatrix} a \\ b \end{bmatrix}$$

The argument given above can be generalized to prove the next proposition.

PROPOSITION 6.2 Let A be a matrix and B a vector. The vectors X that minimize the l_1 norm of the residuals

$$\|B - AX\|_1$$

are the vectors X that appear in the solutions of the linear programming problem

$$\begin{aligned} &\text{minimize} && w = \sum_i (D^+[i] + D^-[i]) \\ &\text{subject to} && Ax + D^+ - D^- = B \\ & && D^+ \geq 0, \qquad D^- \geq 0 \end{aligned}$$ ■

EXAMPLE 6.6 Suppose the three data points $(-1, 0)$, $(0, 1)$, $(1, 0)$ are given. The arrays A, B, X of Proposition 6.2 are

$$A = \begin{bmatrix} 1 & -1 \\ 1 & 0 \\ 1 & 1 \end{bmatrix}, \qquad B = \begin{bmatrix} 0 \\ 1 \\ 0 \end{bmatrix}, \qquad X = \begin{bmatrix} a \\ b \end{bmatrix}$$

The vectors D must be the same size as B — three components. The scalar equation version of the system of Proposition 6.2 is thus

$$\begin{array}{lllllll} \text{minimize} & w = & d_1^+ + d_1^- + d_2^+ + d_2^- + d_3^+ + d_3^- & & & \\ \text{subject to} & a - b + d_1^+ - d_1^- & & & & = 0 \\ & a & + d_2^+ - d_2^- & & & = 1 \\ & a + b & & + d_3^+ - d_3^- & & = 0 \\ & & & d_1^+, d_1^-, d_2^+, d_2^-, d_3^+, d_3^- & & \geq 0 \end{array}$$

To put this system in standard maximum form we can replace w by $z = -w$, split a, b into positive and negative parts, and replace each equality by a pair of inequalities. The result is a system with ten unknowns, six constraints, and positivity conditions that, in matrix form, could be written

$$\text{maximize} \qquad z = -[0 \cdots 0 \,|\, 1 \cdots 1] \begin{bmatrix} X^+ \\ X^- \\ D^+ \\ D^- \end{bmatrix}$$

$$\text{subject to} \qquad \begin{bmatrix} A & -A & I & -I \\ -A & A & -I & I \end{bmatrix} \begin{bmatrix} X^+ \\ X^- \\ D^+ \\ D^- \end{bmatrix} \leq \begin{bmatrix} B \\ -B \end{bmatrix}$$

$$X^+ \geq 0, \quad X^- \geq 0, \quad D^+ \geq 0, \quad D^- \geq 0 \qquad \blacksquare$$

Proposition 6.2 shows that the minimization of $\|B - AX\|_1$ can be rephrased as a linear programming problem. The minimization of $\|B - AX\|_\infty$ can also be treated as a linear programming problem. The trick is to introduce a new variable t constrained by

$$t \geq |d_i|, \qquad \text{all } i$$

where $d_i = B[i] - (AX)[i]$. Then the minimal possible value of t is the largest of the values $|d_i|$. This leads to the next proposition, which we state without proof.

Proposition 6.3 Let A be a matrix and B a vector. The vectors X that minimize the l_∞ norm of the residuals

$$\|B - AX\|_\infty$$

are the vectors X that appear in the solutions of the linear programming problem

$$\begin{aligned} &\text{minimize} && w = t \\ &\text{subject to} && AX + D^+ - D^- = B \\ &&& \begin{bmatrix} t \\ \vdots \\ t \end{bmatrix} - (D^+ + D^-) \geq 0 \\ &&& D^+ \geq 0, \qquad D^- \geq 0 \end{aligned}$$

■

EXAMPLE 6.7 Using the data points $(-1, 0)$, $(0, 1)$, $(1, 0)$ of Example 6.6, the matrices A, B, X, D^+, D^- are the same as for Example 6.6. Thus the scalar equation version of the system of Proposition 6.3 is

$$\begin{aligned} &\text{minimize} && w = t \\ &\text{subject to} && a - b + d_1^+ - d_1^- = 0 \\ &&& a + d_2^+ - d_2^- = 1 \\ &&& a + b + d_3^+ - d_3^- = 0 \\ &&& -d_1^+ + d_1^- + t \geq 0 \\ &&& -d_2^+ + d_2^- + t \geq 0 \\ &&& -d_3^+ + d_3^- + t \geq 0 \\ &&& d_1^+, d_1^-, d_2^+, d_2^-, d_3^+, d_3^- \geq 0 \end{aligned}$$

Notice that the l_1 and l_2 curve-fitting techniques will allow individual deviations $|d_i|$ to be large if the overall sum is minimal. The l_∞ fit, on the other hand, concentrates on making the largest deviation $|d_i|$ as small as possible, even if $\Sigma\ |d_i|$ or $\Sigma\ |d_i|^2$ becomes quite large as a result.

Adding linear constraints to an l_1- or l_∞-minimization problem poses no added problems. A problem such as

$$\begin{aligned} &\text{minimize} && w = \|B - AX\|_1 \\ &\text{subject to} && EX \leq F \end{aligned}$$

can be handled by taking the linear programming problem of Proposition 6.2 and simply adding the constraints $EX \leq F$. ■

EXAMPLE 6.8 A problem need not have an explicit objective function in order to be rephrased as a linear programming problem. Consider the question:

$$\text{Does } AX = B \text{ have solutions with } X \geq 0? \tag{6.8}$$

This question can be answered by solving either of the problems†

†The problem may also be solved by introducing *artificial variables;* see Section 6.3.

$$\begin{cases} \text{minimize} & w = \|B - AX\|_1 \\ \text{subject to} & X \geq 0 \end{cases} \quad \text{or} \quad \begin{cases} \text{minimize} & w = \|B - AX\|_\infty \\ \text{subject to} & X \geq 0 \end{cases}$$

If the solutions of these problems give $w = 0$, then the answer to the question (6.8) is yes, otherwise it is no. ■

The problem (6.8) is much more general than it appears. We will see in Section 6.2 that *every* linear programming problem may be recast in that form!

GOAL PROGRAMMING

Problems using l_1 and l_∞ objective functions fall into a general class of techniques known as "goal programming."

To see the origin of the term consider the set-up of Example 6.3, two plants shipping goods to three distribution points. The firm's management might reasonably have two goals in mind: to keep the plants working at capacity and to give each distribution point precisely what it requires. These goals are in conflict. Indeed, if both plants work at capacity, then $5 + 4 = 9$ carloads are shipped per week, but if the distribution points receive exactly what they need, then $3 + 2 + 3 = 8$ carloads are received each week, and to achieve both these goals is not possible.

If our goals cannot be achieved simultaneously, then what strategy is best? In the above example the goals could be expressed as a matrix equality

$$AX = B$$

that has no solutions. If we cannot meet all our goals, perhaps we should minimize the total deviation from all the goals; that is, perhaps we should minimize

$$w = \|B - AX\|_1$$

On the other hand, it might be reasonable, depending upon the situation, to ensure maximum progress toward all goals — that is, to minimize

$$w = \|B - AX\|_\infty$$

More realistically, some goals may be more important than others. A technique used is to assign "penalties" of various weights to deviations from the various goals. To do this, set $d_i = |B[i] - (AX)[i]|$, the deviation from the ith goal. We then attempt to minimize

$$w = c_1 d_1 + c_2 d_2 + \cdots + c_n d_n$$

where the c_i's are the weights. For example, if the first goal is twice as desirable as the second, we might have $c_1 = 2c_2$.

More generally, there is no reason to assume that falling short of a goal carries the same penalty as overshooting. We can weight the two kinds of deviations differently and try to minimize a function of the form

$$w = c_{+,1}y_1^+ + c_{+,2}y_2^+ + \cdots + c_{+,n}y_n^+ + c_{-,1}y_1^- + c_{-,2}y_2^- + \cdots + c_{-,n}y_n^-$$

Proposition 6.2 shows that such problems can be set up as linear programming problems.

Such specific numerical goals (e.g., ship 5 carloads per week from the second plant) may be mixed with more open-ended goals such as "minimize cost" or "maximize profit." We content ourselves with a single example.

EXAMPLE 6.9 Consider again Figure 6.1. Suppose that oversupply at the first distribution point costs \$15 per carload per week in added storage and undersupply costs \$20 per week in lost revenue. The figures for the other two distribution points are \$10 and \$15 for the second point and \$12 and \$13 for the third. Set

$$d_1 = 3 - x_1 - x_2, \qquad d_2 = 2 - x_3 - x_4, \qquad d_3 = 3 - x_5 - x_6$$

Let us assume that the given bounds on the capacities of the two plants still hold and that there is no penalty for allowing a plant to work under capacity. Minimizing the cost of the operation then becomes the mathematical problem

$$\begin{aligned} \text{minimize} \quad & w = c^TX + c_+^TD^+ + c_-^TD^- \\ \text{subject to} \quad & A_1X + D^+ - D^- = B_1 \\ & A_2X \leq B_2 \end{aligned}$$

where

$$X^T = [x_1 \; x_2 \; x_3 \; x_4 \; x_5 \; x_6], \qquad (D^+)^T = [d_1^+ \; d_2^+ \; d_3^+]$$
$$c^T = 100[1 \; 1 \; 3 \; 2 \; 3 \; 5], \qquad (D^-)^T = [d_1^- \; d_2^- \; d_3^-]$$
$$c_+^T = [15 \; 10 \; 12], \qquad c_-^T = [20 \; 15 \; 13]$$
$$B_1^T = [3 \; 2 \; 3], \qquad B_2^T = [4 \; 5]$$
$$A_1 = \begin{bmatrix} 1 & 1 & 0 & 0 & 0 & 0 \\ 0 & 0 & 1 & 1 & 0 & 0 \\ 0 & 0 & 0 & 0 & 1 & 1 \end{bmatrix}, \qquad A_2 = \begin{bmatrix} 1 & 0 & 1 & 0 & 1 & 0 \\ 0 & 1 & 0 & 1 & 0 & 1 \end{bmatrix} \quad \blacksquare$$

EXERCISES 6.1

In exercises 1 through 10 a linear programming problem is given. Rewrite the problem as a standard maximum problem or a standard minimum problem as indicated.

1. Standard maximum:
$$\begin{aligned} \text{maximize} \quad & z = 2x + 3y \\ \text{subject to} \quad & 5x - 6y \geq 7 \\ & 7x + 8y \leq 9 \\ & x, y \geq 0 \end{aligned}$$

2. Standard minimum:
$$\begin{aligned} \text{minimize} \quad & z = 2x + 3y \\ \text{subject to} \quad & 5x - 6y \geq 7 \\ & 7x + 8y \leq 9 \\ & x, y \geq 0 \end{aligned}$$

3. Standard minimum:
$$\begin{array}{ll} \text{maximize} & z = 2x + 3y \\ \text{subject to} & 5x - 6y \geq 7 \\ & 7x + 8y \leq 9 \\ & x, y \geq 0 \end{array}$$

4. Standard maximum:
$$\begin{array}{ll} \text{minimize} & z = 2x + 3y \\ \text{subject to} & 5x - 6y \geq 7 \\ & 7x + 8y \leq 9 \\ & x, y \geq 0 \end{array}$$

5. Standard maximum:
$$\begin{array}{ll} \text{maximize} & z = 2x + 3y \\ \text{subject to} & 5x - 6y = 7 \\ & 7x + 8y \leq 9 \\ & x, y \geq 0 \end{array}$$

6. Standard minimum:
$$\begin{array}{ll} \text{minimize} & z = 2x + 3y \\ \text{subject to} & 5x - 6y = 7 \\ & 7x + 8y \leq 9 \\ & x, y \geq 0 \end{array}$$

7. Standard maximum:
$$\begin{array}{ll} \text{maximize} & z = 2x + 3y \\ \text{subject to} & 5x - 6y \leq 7 \\ & 7x + 8y \leq 9 \\ & x \geq 0 \end{array}$$

8. Standard minimum:
$$\begin{array}{ll} \text{minimize} & z = 2x + 3y \\ \text{subject to} & 5x - 6y \geq 7 \\ & 7x + 8y \geq 9 \\ & x \geq 0 \end{array}$$

9. Standard maximum:
$$\begin{array}{ll} \text{maximize} & z = 2x + 3y \\ \text{subject to} & 5x - 6y = 7 \\ & 7x + 8y = 9 \end{array}$$

10. Standard minimum:
$$\begin{array}{ll} \text{minimize} & z = 2x + 3y \\ \text{subject to} & 5x - 6y = 7 \\ & 7x + 8y = 9 \end{array}$$

11. Standard maximum:
minimize $z = |2x + 3y - 4|$

12. Standard maximum:
minimize $w = |2x + 3y - 4|$

13. Standard maximum:
minimize $z = |2x + 3y - 4| + |2x + 3y - 5|$

14. Standard minimum:
minimize $w = \max\{|2x - 3|, |3x - 4|\}$

15. Standard maximum:
$$\begin{array}{ll} \text{maximize} & z = 2x + 3y \\ \text{subject to} & -10 \leq x \leq -1 \end{array}$$

16. Standard maximum:
$$\begin{array}{ll} \text{maximize} & z = |2x + 3y| \\ \text{subject to} & x \geq -4,\ y \geq -5 \end{array}$$

17. Standard maximum:
$$\begin{array}{ll} \text{minimize} & w = |4x + 5y| \\ \text{subject to} & x \geq -2,\ y \geq -3 \end{array}$$

18. Standard maximum:
$$\text{minimize} \quad w = \begin{cases} 2x & \text{if } x \geq 0 \\ -3x & \text{if } x \leq 0 \end{cases}$$
subject to $2x + 3y \leq 4$

19. Standard maximum:
$$\text{minimize} \quad w = \begin{cases} -2x & \text{if } x \leq 0 \\ 3x & \text{if } x \geq 0 \end{cases}$$
subject to $2x + 3y = 0$

20. Standard maximum:
$$\begin{array}{ll} \text{minimize} & z = 2|x| - 3|y| \\ \text{subject to} & 2x + 3y = 0 \end{array}$$

21. A furniture factory makes two types of chairs. The first type takes 10 hours of labor to make, uses 2 square yards of fabric and 20 pounds of padding. The second type takes 70 hours of labor, uses 3 square yards of fabric and 10 pounds of padding. The profit on the first type is \$2 per chair and the profit on the second type is \$5 per chair. The resources available are 490 hours of labor, 32 yards of fabric, and 240 pounds of padding per day. Set up a linear programming problem to decide how many chairs of each type should be manufactured per day for maximum profit.

22. A cereal manufacturer wishes to add vitamins A and B to his cereal, which lacks them. He has three vitamin mixtures available. Each gram of the first mixture thrown in the vat would add 1 grain of A and 7 grains of B to each serving at a cost of 4.9 cents per serving. Each gram of the second mixture would add 2 grains of A and 3 grains of B to each serving at a cost of 3.2 cents per serving. Each gram of the third mixture would add 2 grains of A and 1 grain of B to each serving at a cost of 2.4 cents per serving. He wishes to add at least 2 grains of A and 5 grains of B to each serving. Set up a linear programming problem to find out how many grams of each mixture should be tossed into the vat to achieve the minimum requirements at lowest cost.

23. The Natural High Fibre Health Bread Company ships waste sawdust from two sawmills (S_1, S_2) to three bakeries (B_1, B_2, B_3). S_1 produces 10 tons of sawdust per day and S_2 5 tons. Bakeries 1, 2, and 3 need at least 2, 5, and 3 tons of sawdust per day to operate. The accompanying table gives the cost of shipping a ton of sawdust from sawmill S_i to bakery B_j. Set up a linear programming problem to find a shipping schedule that minimizes the cost of shipping sawdust while giving each bakery its minimal requirement. (Let x_{ij} = the amount, in tons, shipped from S_i to B_j.)

	B_1	B_2	B_3
S_1	\$4	\$6	\$8
S_2	\$5	\$7	\$12

24. For the sawmills and bakeries of exercise 23 assume that any unshipped sawdust must be burned at the sawmill at a cost of \$2 per ton in air-pollution fines and any oversupply at the bakeries costs \$1 per ton in labor to throw it over the back fence. Modify the linear programming problem of exercise 23 to take these costs into account.

In exercises 25 through 30 a curve-fitting problem is posed. For each problem find the arrays A and B of Proposition 6.2 or 6.3 as appropriate. How many variables and how many constraints (other than the positivity conditions) would normally be needed to state the problem as a standard maximum problem?

25. Find the straight line that best fits the data in the l_1 sense.

x	0	1	2	4
y	1	0	3	5

26. Find the straight line that best fits the data in the l_∞ sense.

x	0	1	2	4
y	1	0	3	5

27. Find the cubic that best fits the data in the l_∞ sense.

x	1	2	4	-5	6	9
y	1	2	5	4	7	8

28. Find the function of the form

$$z = f(t, y) = \alpha x + \beta y + \gamma$$

x	1	−1	2	−2	0	3
y	2	4	3	−1	0	−2
z	4	2	1	−1	0	−3

that best approximates the data in the l_1 sense.

29. Find the function of the form

$$z = f(x, y) = \alpha x^2 + \beta xy + \gamma y^2$$

x	0	1	3	1	−2	1
y	0	2	2	4	−1	1
z	0	3	1	1	−2	1

that best approximates the data in the l_∞ sense.

30. Find the function of the form

$$w = f(x, y, z) = \alpha x + \beta y + \gamma z + \delta$$

x	0	2	5	−1	1	0
y	0	3	4	1	−1	1
z	0	4	3	−1	1	1
w	1	5	2	1	−1	0

that best approximates the data in the l_1 sense.

In exercises 31 through 35 use the technique of Example 6.8 to set up linear programming problems to answer the (occasionally obvious) given question.

31. (a) Does the system

$$\begin{bmatrix} 1 & 2 \end{bmatrix} \begin{bmatrix} x \\ y \end{bmatrix} = [1]$$

have any solutions with $x \geq 0$, $y \geq 0$?

(b) Does the line

$$x + 2y = 1$$

cut the first quadrant?

32. Does the line $2x + 3y = 4$ go through the rectangle $1 \leq x \leq 2$, $2 \leq y \leq 4$?

33. Does the line $2x + 3y = 6$ go through the square with vertices $(-2, 3)$, $(-1, 3)$, $(-2, 2)$, and $(-1, 2)$?

34. Does the line of intersection of the planes

$$x + 2y - 3z = 1$$
$$2x + 4y - 7z = 0$$

cut through the positive octant?

35. Does the line of intersection of the two planes

$$x + 2y - 3z = 1$$
$$2x + 4y - 7z = 0$$

pierce the cube with corners (3, 3, 3), (3, 4, 3), (4, 3, 3), (4, 4, 4), (3, 3, 2), (3, 4, 2), (4, 3, 2), and (4, 4, 2)?

6.2 The Geometry of Linear Programming

This section develops the basic geometrical ideas pertinent to linear programming problems. First we record a few technical facts that will be needed later.

PROPOSITION 6.4

(a) Let A, B, C be matrices with $B \leq C$. If $A \geq 0$, then $AB \leq AC$ whenever this product is defined. Similarly $BA \leq CA$ whenever the product is defined.

Let A be a matrix, b a vector, and $\pi(T) = p + PT$, T in $\mathbf{R}^n$, a flat.

(b) $A\pi(T) \leq b$ for all T in $\mathbf{R}^n$ implies $AP = 0$.

(c) $A\pi(0) < b$ implies $A\pi(T) < b$ for all T with $\|T\| < \epsilon$ for some $\epsilon > 0$.

(d) $A\pi(T) \leq b$ for all T with $\|T\| < \epsilon$ implies $A\pi(T) \leq b$ for T with $\|T\| = \epsilon$ as well.

Proof

(a) We wish to show $(AB)[i;j] \leq (AC)[i;j]$ for all i, j. By Proposition 2.1 it is sufficient to consider the case where A, B, C are vectors and the product is the dot product. Since $B[k] \leq C[k]$ and $A[k] \geq 0$ for all k, we have $A[k]B[k] \leq A[k]C[k]$ and hence

$$\sum_k A[k]B[k] \leq \sum_k A[k]C[k]$$

(b) From $A\pi(T) \leq b$ we obtain $(AP)T \leq b - Ap$, which reduces us to the case $AT \leq b$ for all T in $\mathbf{R}^n$. If $A[i;j] \neq 0$, let $T[k] = 0$ for all $k \neq j$. Then $(AT)[i] = A[i;j]T[j] \leq b[i]$, which forces either $T[j] \leq b[i] \div A[i;j]$ or $T[j] \geq b[i] \div A[i;j]$, depending on the sign of $A[i;j]$. Neither restriction is allowed.

(c) If v, w are vectors with nonnegative components $\alpha = \lceil/v$ and $\beta = \lceil/w$, then

$$v^T w = \sum_1^n v[k]w[k] \leq \sum_1^n \alpha\beta = n\alpha\beta$$

Now suppose that $b > 0$ has $\beta = \lfloor/b$, A is a matrix with $\alpha = \lceil/,|A$, and T is a vector with $\tau = \lceil/T$ and n components. Then by the above calculation we will have $AT < b$ as long as $\tau < \beta/n\alpha$, and this will be true if $\|T\| < \beta/n\alpha = \epsilon$. The case of $A\pi(T)$ reduces to this case by replacing AP by A and b by $b - Ap$.

(d) Suppose that $\|T_0\| = \epsilon$ but $A\pi(T_0) \nleq b$. Then there is an index i such that $b[i] < (A\pi(T_0))[i] = A[i;]\pi(T_0)$. Let $\pi'(T) = \pi(T + T_0)$. Then $\pi'(0) = \pi(T_0)$, so $(-A[i;])\pi'(0) \leq b[i]$. By (c) there is a $\delta > 0$ such that $(-A[i;])\pi'(T) < -b[i]$ for $\|T\| < \delta$, which means that $b[i] < (A\pi(T_0 + T))[i]$ for $\|T\| < \delta < \epsilon$. But if $T_1 = T_0 - (\delta/\epsilon)T_0$, then $b[i] < (A\pi(T_1))[i]$ with $\|T_1\| = \epsilon - \delta < \epsilon$, which is a contradiction. ∎

In this section we will assume that our linear programming problem has the form

$$\begin{aligned} &\text{maximize} && z = c^T X \\ &\text{subject to} && AX \leq b \end{aligned} \tag{6.9}$$

The techniques described in the last section may be used to put any linear programming problem in this form. Note particularly that the positivity condition $X \geq 0$ does not appear explicitly in (6.9). If the condition is present, then we replace the two conditions

$$AX \leq b, \qquad X \geq 0$$

by the single condition

$$\left[\frac{A}{-I}\right] X \leq \left[\frac{b}{0}\right]$$

The next proposition lies at the root of the geometric approach to linear programming (Figure 6.3).

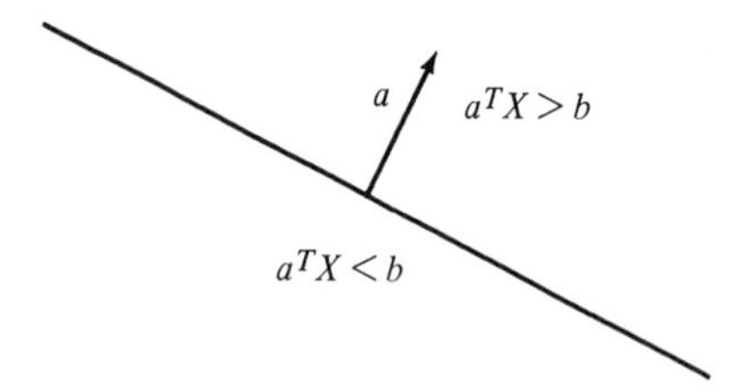

FIGURE 6.3

PROPOSITION 6.5 Let v be a column vector in $\mathbf{R}^n$ and define $f: \mathbf{R}^n \to \mathbf{R}$ by $f(X) = a^T X$. Then

1. a is perpendicular to the hyperplane $f(X) = a^T X = \text{const.}$
2. a points in the direction of increasing $f(X)$.

Proof Fix a constant b. Then $a^T X = b$ is one equation in n unknowns and hence the solutions form a hyperplane [that is, an $(n - 1)$-flat] in $\mathbf{R}^n$ by Proposition 4.17. By Proposition 4.23 the subspace S parallel to this flat is $a^T X = 0$. Thus a is perpendicular to S and hence the component of a parallel to the flat is zero (see Figure 5.19 and the accompanying discussion).

Now suppose that $a^T X_0 = b$ and we move from X_0 in the direction of a to $X_0 + \epsilon a$ for some small ϵ. Then

$$f(X_0 + \epsilon a) = a^T(X_0 + \epsilon a) = b + \epsilon \|a\|^2 > b$$

where $\epsilon > 0$. Thus any movement from X_0 in the direction of a increases $f(X)$. Similarly any movement in the direction $-a$ decreases $f(X)$. ∎

A hyperplane $a^T X = b$ divides $\mathbf{R}^n$ into two *half-spaces* — the half-space $a^T X \geq b$ and the half-space $a^T X \leq b$ (Figure 6.3).

Using Proposition 6.5, we can dispose of most two-dimensional problems with a quick sketch.

EXAMPLE 6.10 Consider the linear programming problem.

$$\begin{aligned} \text{maximize} \quad & z = x + 2y \\ \text{subject to} \quad & x + 3y \leq 18 \\ & x + y \leq 8 \\ & 2x + y \leq 14 \\ & x \geq 0, \quad y \geq 0 \end{aligned}$$

Each inequality defines a half-space (half-plane in this case). For example, $x + 3y \leq 18$ is $f(X) = a^T X \leq 18$, where $a = [1\ \ 3]$. The half-plane defined is on the side of the line $a^T X = 18$ opposite the direction that a points. The inequality $x \geq 0$, on the other hand, is $f(X) = a^T X \geq 0$, where $a = [1\ \ 0]$. In this case the half-plane is the side of the line $a^T X = 0$ *in* the direction that a points.

The set of points satisfying all five inequalities is the shaded set S in Figure 6.4. The arrows on each hyperplane (line) point in the direction of increasing $f(X) = a^T(X)$ — that is, are parallel to a.

The dotted lines in Figure 6.4 are two typical hyperplanes

$$z = c^T X = x + 2y = \text{const.}$$

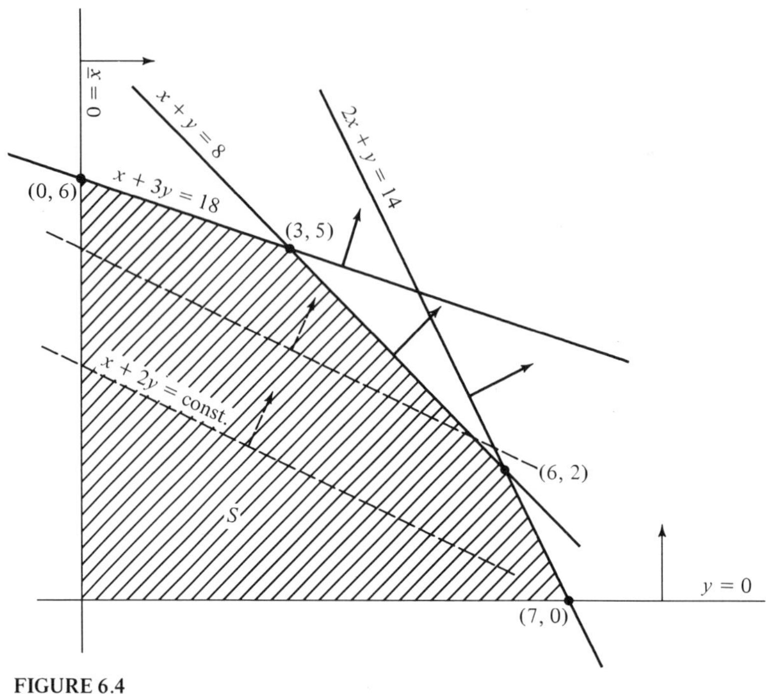

FIGURE 6.4

and the dotted arrows point in the direction of increasing z. From the sketch it is clear that the point of S at which z has its largest value is at the intersection of the lines $x + y = 8$ and $x + 3y = 18$ — that is, at the point (3, 5). ■

Figure 6.4 illustrates several important ideas.

The first thing to notice is that the solution to the problem occurred out on the edge of the set of points S that satisfy the inequalities. The set S is bounded by hyperplanes, and the solution lies on two of them. Algebraically, at the solution point two of the inequalities are equalities.

The second observation we wish to make is a good deal subtler. Why does the maximum of $z = x + 2y = c^TX$ occur at the intersection of $a_1^TX = x + 3y = 18$ and $a_2^TX = x + y = 5$ instead of, say, the intersection of a_2^TX and $a_3^TX = 2x + y = 14$? The answer lies in the slope of the lines $z = c^TX = \text{const}$. If these lines were steeper, $c = [3\ \ 2]$ instead of $[1\ \ 2]$ say, then a sketch shows that the maximum would occur at the intersection of $a_2^TX = 8$ and $a_3^TX = 14$ [Figure 6.5(a)]. There is an intermediate slope $c = [1\ \ 1] = a_2$ where the maximum occurs along the entire line segment at which $a_2^TX = 8$ intersects the set S.

Closer inspection shows that the relationship among the vectors c, a_1, a_2, and a_3 is the crucial factor. Figure 6.6 shows the two intersection points in greater detail with relevant vectors drawn from a common point. The maximum occurs when c lies *between* the vectors a_i at the intersection. For $c = [1\ \ 2]$ as in Figure 6.4 the maximum occurs at the first intersection [Figure 6.6(a)], since c is between a_1

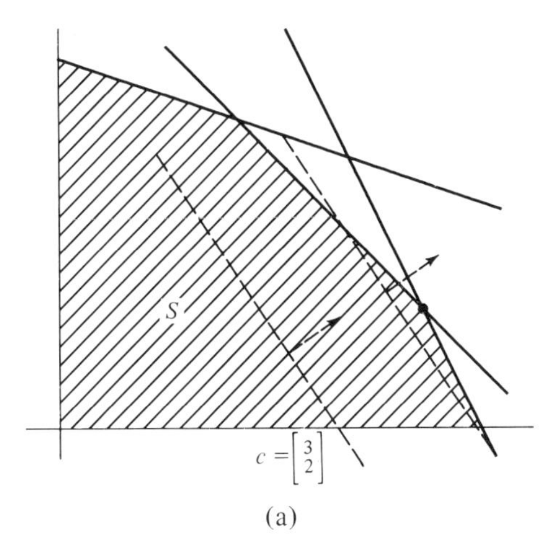

(a)

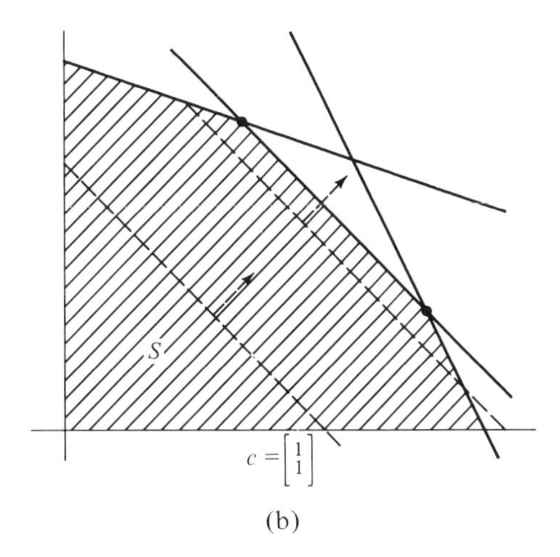

(b)

FIGURE 6.5

and a_2. For $c = [3 \;\; 2]$ on the other hand, the maximum occurs at the second intersection where c is between a_2 and a_3 [Figure 6.6(b)]. For the intermediate case $c = [1 \;\; 1] = a_2$ c is both "between" a_1 and a_2 and "between" a_2 and a_3 as well.

The notion of "between" has a simple algebraic characterization. If a_1, a_2, for example, were used to define a new coordinate system, then c in Figure 6.6(a) would be in the first quadrant. That is, $c = \alpha_1 a_1 + \alpha_2 a_2$ with $\alpha_1, \alpha_2 \geq 0$. In Figure 6.6(b), on the other hand, c is not in the first quadrant of the coordinate system

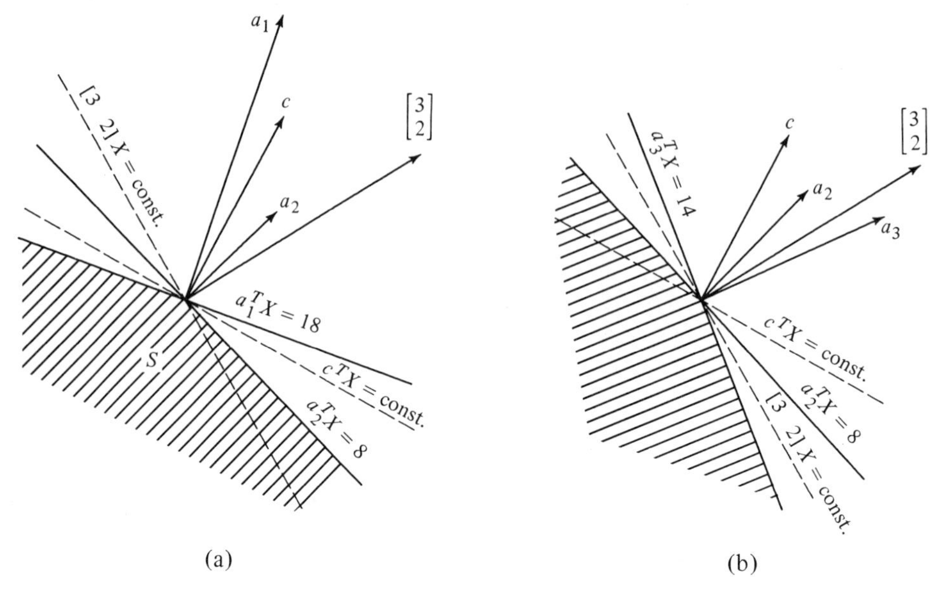

FIGURE 6.6

defined by a_2 and a_3. In fact $c = \alpha_2 a_2 + \alpha_3 a_3$ with $\alpha_3 < 0$. In the plane a vector v is between two vectors w_1 and w_2 if $v = \alpha_1 w_1 + \alpha_2 w_2$ with $\alpha_1 \geq 0$, $\alpha_2 \geq 0$. Setting $A = [w_1 | w_2]$, this means that v is between w_1 and w_2 if the linear system

$$AX = v$$

has a solution with $X \geq 0$.

We proceed to generalize Figure 6.6 to higher dimensions. First a bit of terminology.

DEFINITION 6.5 Let A be a matrix. The *dual cone* of A is the set of vectors X for which $A^TX \leq 0$. If b is a column vector, the dual cone of b is the set of vectors X for which $b^TX \leq 0$.

Translate the coordinate system in Figure 6.6(a) so that the lines intersect at the origin. If $A = [a_1 | a_2]$, then the matrix equation $A^TX \leq 0$ is the pair of vector equations

$$a_1^TX \leq 0, \qquad a_2^TX \leq 0$$

and the set S of Figure 6.6(a) is the dual cone of A. The half-plane below the dotted line is the dual cone of c. The vector c is between a_1 and a_2 if and only if the set S is contained in the half-plane below the dotted line [compare Figure 6.6(b)]. The next proposition generalizes this to higher dimensions. Notice that the problem involved is precisely that of Example 6.8.

PROPOSITION 6.6 (*Farkas' lemma*) Let A be a matrix and c a vector. The equation $AX = c$ has a solution X with $X \geq 0$ if and only if the dual cone of A is contained in the dual cone of c.

Proof First assume that we have a solution $AX = c$ with $X \geq 0$. Let v be any vector in the dual cone of A — that is, $A^T v \leq 0$. Since $X \geq 0$ we have, by Proposition 6.4,

$$0 \geq X^T A^T v = (AX)^T v = c^T v$$

and hence v is in the dual cone of c.

We will prove the converse by induction on the number of columns of A. The proof is quite simple in concept and may be turned into an algorithm for computing positive solutions of $AX = c$.

We begin with the induction step. Assume that the proposition is true for all matrices A with less than n columns and that A has n columns. Let $A' = 0\ 1 \downarrow A$.

Now if the dual cone of A' is contained in the dual cone of c, then we are done. For by the induction hypothesis there is an $X' \geq 0$ such that $A'X' = c$ and hence $[0 \mid X']$ is a solution of $AX = c$ with $X \geq 0$. Thus we may assume that the dual cone of A' is *not* contained in the dual cone of c. Thus there is a vector v such that $A'^T v \leq 0$ but $c^T v > 0$ (see Figure 6.7, where $A = [a_1 \mid a_2]$, $A' = [a_2]$). Let P be any square matrix whose null space is S, the subspace generated by $A[;1]$ ($P = I - P_S$, where P_S is perpendicular projection into S, is an obvious choice, but many other choices are possible). Since the dual cone of A is contained in the dual cone of c, it follows that the dual cone of PA' is contained in the dual cone of Pc. For suppose that $(PA')^T w \leq 0$ and set $a = A[;1]$. Then

$$A^T(P^T w) = [a \mid A']^T P^T w = \left[\frac{a^T P^T w}{A'^T P^T w}\right]$$

$$= \left[\frac{(Pa)^T w}{(PA')^T w}\right] = \left[\frac{0}{(PA')^T w}\right] \leq 0$$

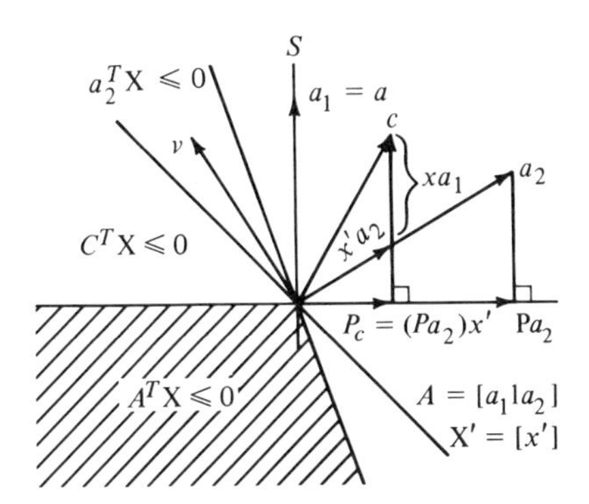

FIGURE 6.7

Thus $P^T w$ is in the dual cone of A and hence the dual cone of c as well. This gives the inequality

$$0 \geq c^T P^T w = (Pc)^T w$$

which shows that w is in the dual cone of Pc. The induction hypothesis now shows, since PA' has $n-1$ columns, that there is an $X' \geq 0$ such that

$$(PA')X' = Pc \quad \text{or} \quad P(c - A'X') = 0$$

Since the null space of P is generated by a, there is a scalar x such that (cf. Figure 6.7) $c - A'X' = xa$ or $[a|A'][x\ X']^T = c$. The vector $[x\ X']^T$ is the solution sought. We must check that $x \geq 0$. This follows from the existence of the vector v above. In fact multiplying $c - A'X' = xa$ by v^T gives

$$v^T c - v^T A'X' = xv^T a$$

or

$$c^T v - (A'^T v)X' = x(a^T v)$$

Now $c^T v > 0$, $A'^T v \leq 0$, and $X' \geq 0$. Thus $x(a^T v) \geq 0$. Now $a^T v > 0$, for if $a^T v \leq 0$ then $A^T v = [a|A']^T v \leq 0$ and, since the dual cone of A is contained in the dual cone of c, we would have $c^T v \leq 0$, which is not true. Thus $a^T v > 0$ and so $x \geq 0$.

It remains to start the induction. Now the dual cone of A is the intersection of the dual cones of the vectors $A[;i]$. If A has no columns $(0 = 1 \downarrow \rho A)$, we take the dual cone of A to be all of $\mathbf{R}^m$. Thus for $n = 0$ the dual cone of c contains the dual cone of A only when $c = 0$. In this case the solution of $AX = c$ exists and is $\iota\, 0$, which we take to be nonnegative. ∎

In Figure 6.4 the set S is the set of points that satisfy the matrix inequality $AX \leq b$ of (6.9). Some special terminology is associated with this set.

Definition 6.6

1. A point satisfying the matrix inequality

$$AX \leq b$$

of (6.9) is called a *feasible point*.

2. The individual inequalities of (6.9)

$$A[i;] \cdot X \leq b[i]$$

(dot products) are called *constraints*.

3. Let v be a feasible point. The constraints

$$A[i;] \cdot v \leq b[i]$$

for which equality holds are called the *active constraints at* v.

4. Let v be a feasible point. The flat obtained by intersecting the hyperplanes

$$A[i;]X = b[i]$$

as i runs through the active constraints at v will be called the *constraint flat* for v and denoted π_v.

Let I be the vector of indices of the active constraints at v. Then π_v is the flat of solutions of $A[I;]X = b[I]$. If the columns of V span the null space of $A[I;]$, then π_v can be parametrized as

$$\pi_v(T) = v + VT$$

(Proposition 4.23).

Referring to Figure 6.4, the set of feasible points is the set S. The points in the interior of S have no active constraints. This means that the matrix $A[I;]$ has no rows ($0 = 1 \uparrow \rho A[I;]$) and so every X in $\mathbf{R}^2$ is a solution: π_v is all of $\mathbf{R}^2$.

For points on the boundary of S there is at least one active constraint, and at the vertices ((0, 0), (0, 6), (3, 5), (6, 2) and (7, 0)) there are two active constraints.

At (3, 5) the active constraints are

$$x + 3y = 18$$
$$x + y = 8$$

and the unique solution of this system is (3, 5). The constraint flat of (3, 5) is just (3, 5) itself. (In this case V has no columns; that is, $0 = 1 \downarrow \rho V$).

In general we define a *vertex* to be a feasible point whose constraint flat is itself — that is, has dimension equal to 0.

Any point on the line segment from (3, 5) to (6, 2), endpoints excluded, has the single active constraint

$$x + y = 8$$

and this line is the constraint flat. It may be parametrized

$$\pi_v(t) = v + \begin{bmatrix} 1 \\ -1 \end{bmatrix} [t]$$

for any v on this line segment.

An important property of the feasible set of a linear programming problem is

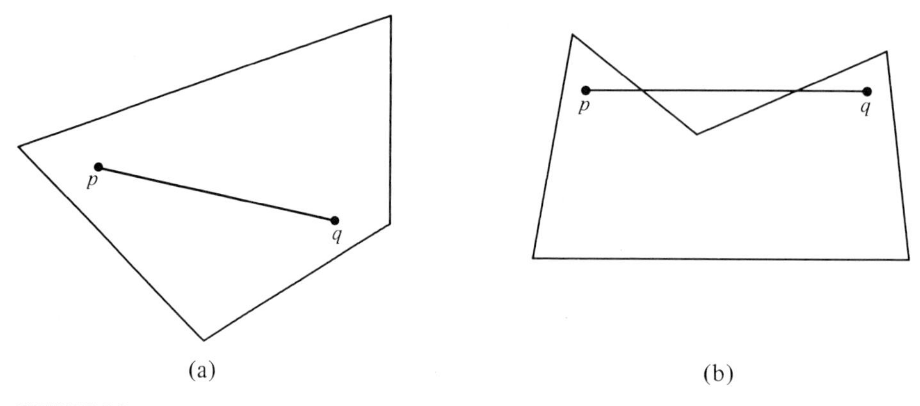

FIGURE 6.8

convexity. Figures 6.8(a) and (b) give examples of sets that are convex and nonconvex, respectively.

DEFINITION 6.7 A set S in $\mathbf{R}^n$ is *convex* if for every pair of points p, q in S the line segment

$$l(t) = p + t(q - p), \qquad 0 \leq t \leq 1$$

lies in S.

PROPOSITION 6.7 The set of solutions of a system of linear inequalities is convex.

Proof Let the system be $AX \leq b$ and suppose p, q are solutions. Let

$$l(t) = p + t(q - p) = (1 - t)p + tq, \qquad 0 \leq t \leq 1$$

Then $1 - t \geq 0$ and $t \geq 0$; thus

$$Al(t) = (1 - t)Ap + tAq \leq (1 - t)b + tb = b \qquad \blacksquare$$

Referring again to Figure 6.4, the points without active constraints are characterized by their being surrounded on all sides by feasible points. The points on the edges, on the other hand, are not completely surrounded by feasible points. If such a point is not a vertex, however, it does lie within a line segment of feasible points. Vertices do not even lie *within* a line segment of feasible points (they do appear as endpoints of such segments). The next proposition is a technical expression of this idea.

PROPOSITION 6.8 Let v be a feasible point of a system of linear inequalities with constraint flat $\pi_v(T) = v + VT$. There is a number $\epsilon > 0$ such that $\pi_v(T)$ is a feasible point with the same active constraints as v for $\|T\| < \epsilon$.

Proof Let the index vectors I and J give the active and inactive constraints, respectively. By the definition of π_v we have $A[I;]\pi_v(T) = b[I]$ for all T, and since $A[J;]\pi(0) < b[J]$, Proposition 6.4(c) shows that $A[J;]\pi(T) < b[J]$ for $\|T\| < \epsilon$ for some $\epsilon > 0$. ■

For the most part the solutions of linear programming problems occur at vertices. A proof of this is complicated by the fact that general linear programming problems need not have solutions and their feasible sets need not have vertices (e.g., maximize $z = y$ subject to $x \leq 0$ in $\mathbf{R}^2$). The next two propositions show the general situation.

PROPOSITION 6.9 Let S be the set of solutions of a system of linear inequalities

$$AX \leq b$$

(a) If S contains an entire flat π, then every constraint flat π_v contains the parallel translate of π through v, and this translate is contained in S.

(b) Let v be a point of S such that the dimension of π_v is minimal among constraint flats. Then π_v lies entirely in S.

Proof Suppose that the flat $\pi(T) = p + PT$ lies entirely in S. Then $A\pi(T) \leq b$ for all T and hence $AP = 0$ by Proposition 6.4(b). Let v be any point of S. The translate of π through v may be parametrized as $\pi'(T) = v + PT$ (Proposition 4.23), and so the computation

$$A(v + PT) = Av + APT = Av + 0 \leq b$$

shows $\pi'(T)$ feasible for all T. Similarly, if the index vector I gives the active constraints at v, then $A[I;]v = b[I]$ and so $A[I;](v + PT) = A[I;]v = b[I;]$ shows that $v + PT$ satisfies the equations defining π_v.

Now assume that the constraint flat $\pi_v(T) = v + PT$ has minimal dimensions for v in S. If there is a parameter vector T_0 such that $p = \pi_v(T_0)$ is not in S, then the set of real numbers r for which the line segment $\pi(t) = v + t(p - v) = v + P(tT_0)$ contains only feasible points for $|t| < r$ is nonempty by Proposition 6.8 and bounded above by any $r > 1$. If r_0 is the least upper bound of this set of real numbers, then by Proposition 6.4(d) the points $\pi(\pm r_0)$ are feasible. Since S is convex, we have, by changing the sign of t if necessary, that $q = \pi(r_0)$ is feasible, but $\pi(r_0 + \delta)$ is not for $\delta > 0$. Now if $\pi_q = \pi_v$, then by Proposition 6.8 $\pi_q(T) = q + VT = v + (r_0T_0 + T)$ is feasible for $\|T\| < \epsilon$, $\epsilon > 0$. Taking $T = \delta T_0$ for $\delta > 0$ sufficiently small, we see that this is impossible. Thus $\pi_q \neq \pi_v$. Since q is contained in π_v, however, we have π_q contained in but not equal to π_v, and hence π_v does not have minimal dimension. This contradiction shows that T_0 does not exist and hence π_v is contained in S. ■

PROPOSITION 6.10 Suppose that the linear programming problem (6.9) has solutions. Then the set of solutions always includes a point v for which the constraint flat π_v has minimal dimension. For any solution point v, π_v is parallel to the hyperplanes $c^T X = \text{const.}$

Proof Suppose that S is the feasible set of (6.9) and that the maximum value attained by $f(X) = c^T X$ is θ. Then the set of solutions is the intersection S' of S with the hyperplane $c^T X = \theta$. Hence S' is the solution set to the system of linear inequalities obtained by adding the constraints $c^T X \leq \theta$ and $-c^T X \leq -\theta$ to $AX \leq b$.

Let the point v in S' have constraint flat π_v when considered to be a point of S. By Proposition 6.8 there is an $\epsilon > 0$ such that $\pi_v(T)$ lies in S for $\|T\| < \epsilon$. In fact, $\pi_v(T)$ must lie in S' for $\|T\| < \epsilon$, for $c^T \pi_v(\pm T) = c^T(v \pm VT) = c^T v \pm (c^T V)T = \theta \pm (c^T V)T$ will be greater than θ for the proper choice of sign unless $c^T V = 0$ — that is, π_v is parallel to $c^T X = \text{const.}$

Thus the constraint flat π_v is the same if v is considered a point of S or a point of S'. Let π_v, v in S', have minimal dimension. Applying Proposition 6.9, we have that π_v lies entirely in S', hence entirely in S, and hence every constraint flat of S contains a translate of π_v. Thus π_v has minimal dimension among the constraint flats of S. ■

Now let us return to the idea (Figure 6.6 and the accompanying discussion) that of a solution of (6.9) the vector c is a positive linear combination of the vectors associated with the active constraints. Assume that v is a solution and the index vector I gives the active constraints at v. We should have a solution, then, of the system

$$A[I;]^T Y = c, \qquad Y \geq 0$$

We can drop reference to I by assuming that the components of Y corresponding to inactive constraints are zero. Then we have a solution of

$$A^T Y = c, \qquad Y \geq 0$$

Note that if w is *any* solution of this latter system and s is *any* feasible point, then

$$f(s) = c^T s = (A^T w)^T s = w^T A s \leq w^T b = b^T w$$

where the inequality is justified by Proposition 6.4(a). Thus $b^T w$ is an upper bound on the values of $f(X) = c^T X$.

This leads to the next proposition — which shows, incidentally, that any algorithm that can find a point of a feasible set can in fact solve linear programming problems.

PROPOSITION 6.11 The vector v is a solution of

$$\begin{array}{ll} \text{maximize} & z = c^T X \\ \text{subject to} & AX \leq b \end{array} \qquad (6.9)$$

if and only if there is a vector w such that $[X \mid Y] = [v \mid w]$ is a solution of the system of linear inequalities

$$\begin{aligned} AX &\leq b \\ A^T Y &= c, \qquad Y \geq 0 \\ c^T X &= b^T Y \end{aligned} \tag{6.10}$$

In this case the components of w corresponding to inactive constraints at v are zero [i.e., $0 = w \times (b - Av)$].

Proof Given $[v \mid w]$ a solution of (6.10), we saw above that $f(v) = c^T v \leq b^T w$, so $f(v)$ is maximal and v is a solution of (6.9).

Since

$$w^T b = b^T w = c^T v = (A^T w)^T v = w^T (Av)$$

we have $w^T(b - Av) = 0$. Since $w \geq 0$ and $b - Av \geq 0$, this dot product is a sum of positive terms, which must then be individually zero — that is, $0 = w \times (b - Av)$.

Conversely, suppose that v is a solution of (6.9). Let the index vector I give the active constraints at v and let the index vector J give the inactive constraints at v. Since we are assuming that a solution exists, it follows from Proposition 6.10 that I is not empty — that is, there are active constraints. Now v is the maximal value of $c^T X$ on the larger set, S_I, of solutions of $A[I;]X \leq b[I]$. For suppose that v' is in S_I and $c^T v' > c^T v$. For the line segment $l(t) = (1 - t)v + tv'$ we have $c^T l(t) = c^T v + t(c^T v' - c^T v) \geq c^T v$ for all $t > 0$. But $l(t)$ lies in S_I by Proposition 6.7. So $A[I;]l(t) \leq b[I]$ for $0 \leq t \leq 1$. Now $A[J;]l(0) < b[J]$ implies that for ϵ sufficiently small, $A[J;]l(\epsilon) < b[J]$. But this means that for some small $\epsilon > 0$, $Al(\epsilon) < b$ and $c^T l(\epsilon) > c^T v$, which means that $c^T v$ is not maximal. This contradiction shows that we cannot have $c^T v' < c^T v$.

Now if v' is any vector such that $A[I;](v' - v) \leq 0$, then $A[I;]v' \leq A[I;]v = b$ and hence $c^T v' \leq c^T v$ — that is, $c^T(v' - v) \leq 0$. This shows that the dual cone of $A[I;]^T = A^T[;I]$ is contained in the dual cone of c and hence by Farkas' lemma (Proposition 6.6) applied to $A^T[;I]$ there is a vector $w_1 \geq 0$ such that $A^T[;I]w_1 = c$. If we define w by $w[I] = w_1$, $w[J] = 0$, then $A^T w = c$ and $w \geq 0$. Further if $e = b - Av$, then $e[I] = 0$ and hence $e^T w = 0$. Thus

$$\begin{aligned} b^T w &= (Av + e)^T w = (Av)^T w + e^T w \\ &= v^T A^T w = v^T c \\ &= c^T v \end{aligned}$$

■

EXAMPLE 6.11 Consider again the system of Example 6.10:

$$\text{maximize} \qquad z = [1 \quad 2]X$$

$$\text{subject to} \quad \begin{bmatrix} 1 & 3 \\ 1 & 1 \\ 2 & 1 \\ -1 & 0 \\ 0 & -1 \end{bmatrix} X \leq \begin{bmatrix} 18 \\ 8 \\ 14 \\ 0 \\ 0 \end{bmatrix}$$

We have seen (Figure 6.4) that the solution to this problem is $X = v = (3, 5)$. The normals to the active constraints at v are just $[1 \;\; 3]$ and $[1 \;\; 1]$ and the vector c lies between these normals [Figure 6.6(a)]. By Farkas' lemma there will be a nonnegative solution of

$$\begin{bmatrix} 1 & 1 \\ 3 & 1 \end{bmatrix} \begin{bmatrix} \alpha \\ \beta \end{bmatrix} = c$$

and indeed

$$\begin{bmatrix} \alpha \\ \beta \end{bmatrix} = \begin{bmatrix} 1 & 1 \\ 3 & 1 \end{bmatrix}^{-1} \begin{bmatrix} 1 \\ 2 \end{bmatrix} = \tfrac{1}{2} \begin{bmatrix} 1 \\ 1 \end{bmatrix}$$

The vector $w = Y$ of Proposition 6.9 is just this vector padded to the proper length with zeros, one zero for each inactive constraint:

$$\begin{bmatrix} 1 & 1 \\ 3 & 1 \end{bmatrix} \begin{bmatrix} \alpha \\ \beta \end{bmatrix} = \begin{bmatrix} 1 & 1 & 2 & -1 & 0 \\ 3 & 1 & 1 & 0 & -1 \end{bmatrix} \begin{bmatrix} \alpha \\ \beta \\ 0 \\ 0 \\ 0 \end{bmatrix} = c$$

For this v and w we have

$$c^T v = [1 \quad 2] \begin{bmatrix} 3 \\ 5 \end{bmatrix} = 13$$

$$b^T w = \tfrac{1}{2}[18 \quad 8 \quad 14 \quad 0 \quad 0] \begin{bmatrix} 1 \\ 1 \\ 0 \\ 0 \\ 0 \end{bmatrix} = \tfrac{26}{2} = 13$$

so indeed $c^T X = b^T Y$. This equation, $c^T X = b^T Y$, serves to rule out such solutions of $A^T Y = c$, $Y \geq 0$, as $Y = \frac{1}{5}[3 \quad 0 \quad 1 \quad 0 \quad 0]$. Geometrically this latter solution comes from the point of intersection of the first constraint $(x + 3y = 18)$ and the third constraint $(2x + y = 14)$ where the dual-cone condition of Farkas' lemma is satisfied but that falls outside the feasible set S. For this value of Y we have

$$b^T Y = [18 \quad 8 \quad 14 \quad 0 \quad 0]\begin{bmatrix}3\\0\\1\\0\\0\end{bmatrix}\tfrac{1}{5} = \tfrac{68}{5} = 13.6 > 13$$

which illustrates the statement from the proof of Proposition 6.9 that

$$AX \leq b \quad \text{and} \quad A^T Y \geq b, \quad Y \geq 0, \quad \text{implies} \quad b^T Y \geq c^T X \qquad \blacksquare$$

EXAMPLE 6.12

$$\begin{aligned} \text{minimize} \quad & u = x + y \\ \text{subject to} \quad & x + 3y \geq -3 \\ & -2x + y \geq -2 \\ & x - y \geq -2 \end{aligned}$$

This is a minimum problem but not a standard one. We rewrite it as a maximum problem:

$$\begin{aligned} \text{maximize} \quad & z = -u = -x - y \\ \text{subject to} \quad & -x - 3y \leq 3 \\ & 2x - y \leq 2 \\ & -x + y \leq 2 \end{aligned}$$

Since

$$\begin{aligned} x + 3y = -3 \quad & \text{goes through} \quad (0, -1) \text{ and } (-3, 0) \\ 2x - y = 2 \quad & \text{goes through} \quad (0, -2) \text{ and } (1, 0) \\ -x + y = 2 \quad & \text{goes through} \quad (0, 2) \text{ and } (-2, 0) \end{aligned}$$

the feasible set is the shaded triangle in Figure 6.9. The direction of increasing z is (Proposition 6.5) $c = (-1, -1)$. Thus the solution would appear to be the intersection of the lines

$$-x - 3y = 3, \qquad -x + y = 2$$

At this point we have

$$\begin{bmatrix} -1 & -1 \\ -3 & 1 \end{bmatrix}\begin{bmatrix} \alpha \\ \beta \end{bmatrix} = \begin{bmatrix} -1 \\ -1 \end{bmatrix}$$

or

$$\begin{bmatrix} \alpha \\ \beta \end{bmatrix} = \begin{bmatrix} -1 & -1 \\ -3 & 1 \end{bmatrix}^{-1}\begin{bmatrix} -1 \\ -1 \end{bmatrix} = \tfrac{1}{2}\begin{bmatrix} 1 \\ 1 \end{bmatrix}$$

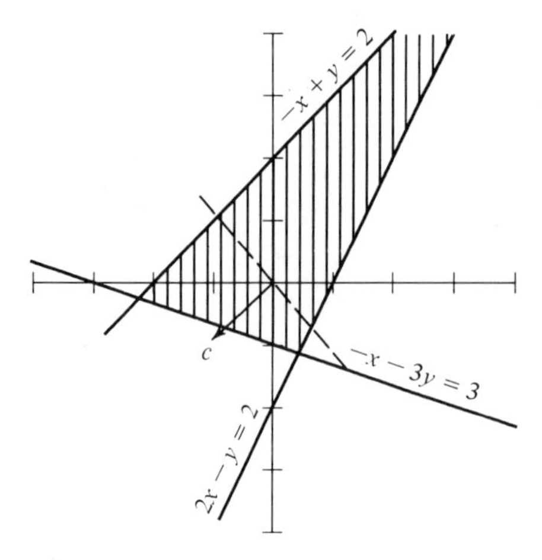

FIGURE 6.9

which is indeed positive. The solution of the problem is the point of intersection of these two lines.

$$v = \begin{bmatrix} x \\ y \end{bmatrix} = \begin{bmatrix} -1 & -3 \\ -1 & 1 \end{bmatrix}^{-1} \begin{bmatrix} 3 \\ 2 \end{bmatrix} = -\tfrac{1}{4}\begin{bmatrix} 9 \\ 1 \end{bmatrix}$$

The vector w of Proposition 6.9 is (α, β) padded with a zero for the inactive constraint $2x - y \leq 2$, so $w = \frac{1}{2}(1, 0, 1)$. The value of the objective function is

$$z = -u = [-1 \quad -1]\begin{bmatrix} -\frac{9}{4} \\ -\frac{1}{4} \end{bmatrix} = \tfrac{5}{2} = [3 \quad 2 \quad 2]\begin{bmatrix} \frac{1}{2} \\ 0 \\ \frac{1}{2} \end{bmatrix}$$

or

$$u = -\tfrac{5}{2}.$$

■

In Section 6.1 we said that there was a close connection between the solutions to dual problems. The relation between the two follows from Proposition 6.11.

PROPOSITION 6.12 Consider the dual problems

$$\begin{cases} \text{maximize} & z = c^T X \\ \text{subject to} & AX \leq b \\ & X \geq 0 \end{cases} \qquad \begin{cases} \text{minimize} & w = b^T Y \\ \text{subject to} & A^T Y \geq c \\ & Y \geq 0 \end{cases}$$

If v is a feasible point for the maximum problem and u is a feasible point for the minimum problem, then

$$c^T v \leq b^T u$$

with equality if and only if v and u are solutions of their respective problems.

In this case the components of w corresponding to the inactive constraints (exclusive of positivity constraints) at v are zero and the components of v corresponding to the inactive constraints (exclusive of positivity constraints) at w are zero.

Proof Suppose that u and v are both feasible. Then $u \geq 0$, $v \geq 0$, $A^T u \geq c$, and $(Av)^T \leq b^T$. Hence, using Proposition 6.4,

$$b^T u \geq (Av)^T u = v^T A^T u \geq v^T c = c^T v$$

Applying Proposition 6.11 to the problem of maximizing $z = c^T X$ subject to

$$\left[\frac{A}{-I}\right] X \leq \left[\frac{b}{0}\right]$$

we find that v is a solution of the maximum problem if and only if there is a vector $Y = [u_1 \ u_2] \geq 0$ such that

$$c = \left[\frac{A}{-I}\right]^T \begin{bmatrix} u_1 \\ u_2 \end{bmatrix} = [A^T - I^T] \begin{bmatrix} u_1 \\ u_2 \end{bmatrix} = A^T u_1 - u_2$$

and $c^T v = [b^T \ 0][u_1^T \ u_2^T]^T = b^T u_1$. Thus suppose that u and v are feasible and that $c^T v = b^T u$. Then set $u_1 = u$ and $u_2 = Au - c \geq 0$, and it follows that v is a solution of the maximum problem. Further, the components of $Y = [u_1^T \ \ u_2^T]^T$ corresponding to the inactive constraints at v are zero. Since the components of Y that are components of u_2 correspond to the positivity constraints, the components of u that correspond to inactive constraints exclusive of positivity constraints are zero.

To prove the corresponding statement, interchanging the roles of u and v, write the minimum problem as the maximum problem:

$$\begin{array}{ll} \text{maximize} & -w = -b^T Y \\ \text{subject to} & -A^T Y \leq -c \\ & Y \geq 0 \end{array}$$

Then the dual problem is

$$\begin{array}{ll} \text{minimize} & -z = -c^T X \\ \text{subject to} & AX \geq -b \\ & X \geq 0 \end{array}$$

and the roles of u and v have been interchanged! ■

EXAMPLE 6.13 Writing the system of Example 6.10 as a standard maximum problem, we have the dual problems:

$$\begin{cases} \text{maximize} & z = [1 \quad 2]X \\ \text{subject to} & \begin{bmatrix} 1 & 3 \\ 1 & 1 \\ 2 & 1 \end{bmatrix} X \leq \begin{bmatrix} 18 \\ 8 \\ 14 \end{bmatrix} \\ & X \geq 0 \end{cases} \qquad \begin{cases} \text{minimize} & w = [18 \quad 8 \quad 14]Y \\ \text{subject to} & \begin{bmatrix} 1 & 1 & 2 \\ 3 & 1 & 1 \end{bmatrix} Y \geq \begin{bmatrix} 1 \\ 2 \end{bmatrix} \\ & Y \geq 0 \end{cases}$$

Here $v = (3, 5)$ and $u = (\frac{1}{2}, \frac{1}{2}, 0)$ — see Example 6.11. The zero component of u corresponds to the inactive constraint $2x + y \leq 14$, and since both components of v are nonzero, both constraints of the minimum problem are active at u.

$$c^T v = [1 \quad 2]\begin{bmatrix} 3 \\ 5 \end{bmatrix} = 13 = [18 \quad 8 \quad 14]\begin{bmatrix} \frac{1}{2} \\ \frac{1}{2} \\ 0 \end{bmatrix} = b^T u \qquad \blacksquare$$

Proposition 6.12 may be rephrased as follows. Solving the dual problems of the proposition is equivalent to finding X and Y such that

$$\begin{aligned} AX &\leq b, \qquad A^T Y \leq c \\ c^T X &= b^T Y \\ X &\geq 0, \qquad Y \geq 0 \end{aligned}$$

We may change the inequalities to equalities by introducing more variables. Define the unknown vectors X' and Y' by

$$AX + X' = b, \qquad A^T Y - Y' = c$$

Then $X' \geq 0$, $Y' \geq 0$. The components of X' are called *slack variables* and the components of Y' *surplus variables* because they measure the slack and surplus, respectively, in the constraints. Using these slack and surplus variables, we may restate Proposition 6.12 as

PROPOSITION 6.13 Solving the dual problems

$$\begin{cases} \text{maximize} & z = c^T X \\ \text{subject to} & AX \leq b \\ & X \geq 0 \end{cases} \qquad \begin{cases} \text{minimize} & w = b^T Y \\ \text{subject to} & A^T Y \geq c \\ & Y \geq 0 \end{cases}$$

is equivalent to finding a solution (X, X', Y, Y') of the system

$$\begin{bmatrix} A & I & 0 & 0 \\ -c^T & 0 & b^T & 0 \\ 0 & 0 & A^T & -I \end{bmatrix} \begin{bmatrix} X \\ X' \\ Y \\ Y' \end{bmatrix} = \begin{bmatrix} b \\ 0 \\ c \end{bmatrix} \tag{6.11}$$

$$(X, X', Y, Y') \geq 0$$

If (X, X', Y, Y') is a solution, then $0 = X \times Y'$, $0 = X' \times Y$. ■

The componentwise products in the conclusion of Proposition 6.13 give the statements in Proposition 6.12 involving zero components and inactive constraints.

Example 6.14 Continuing Example 6.13, we found the solution $X = (3, 5)$, $Y = (\frac{1}{2}, \frac{1}{2}, 0)$. From (6.11) we have $AX + X' = b$ and hence

$$X' = b - Ax = \begin{bmatrix} 18 \\ 8 \\ 14 \end{bmatrix} - \begin{bmatrix} 1 & 3 \\ 1 & 1 \\ 2 & 1 \end{bmatrix} \begin{bmatrix} 3 \\ 5 \end{bmatrix} = \begin{bmatrix} 0 \\ 0 \\ 3 \end{bmatrix}$$

The zero components flag the active constraints. Similarly

$$Y' = A^T Y - c = \begin{bmatrix} 1 & 1 & 2 \\ 3 & 1 & 1 \end{bmatrix} \begin{bmatrix} \frac{1}{2} \\ \frac{1}{2} \\ 0 \end{bmatrix} - \begin{bmatrix} 1 \\ 2 \end{bmatrix} = \begin{bmatrix} 0 \\ 0 \end{bmatrix}$$

and indeed $0 = X \times Y'$, $0 = X' \times Y$. ■

Proposition 6.13 shows that every linear programming problem may be reduced to solving a problem of the form

$$AX = B, \qquad X \geq 0$$

We saw in Example 6.8, on the other hand, that every such problem may be expressed as a linear programming problem. The two types of problems are thus coextensive.

The proof of Farkas' lemma gives one approach to the solution of such problems. Another approach, based on Proposition 6.12, is known as Khachiyan's algorithm. This algorithm, which will find a point in a feasible set in a finite number of steps, caused some excitement in 1979 because theoretical considerations indicated that, for a large number of unknowns and constraints, it should be less costly than current methods. Further investigation showed, however, that the simplex method, discussed in the next section, remains superior for now. The simplex method works with a compact version of Equation (6.11).

EXERCISES 6.2

In exercises 1 through 5 a function $f(X)$ and a system of inequalities is given. Sketch the feasible set of the system of inequalities and identify the vertices. By applying Proposition 6.10 to $f(X)$ and $-f(X)$ find the maximum and minimum values (if they exist) of the function on the set.

1. $f(X) = x + 5y$
   $-x + 3y \leq 5$
   $3x - y \leq 1$
   $x + y \geq -1$

2. $f(X) = 5x + 5y$
   $-x + 4y \leq 9$
   $4x - y \leq 9$
   $x + y \geq 1$

3. $f(X) = 4y$
   $-3x + 2y \leq 10$
   $-x + 3y \leq 8$
   $x + y \leq 4$
   $4x - y \leq 11$

4. $f(X) = 3y$
   $3x + y \geq -8$
   $-3x + y \leq 4$
   $-3x + 4y \geq -2$
   $-x + 4y \geq -6$

5. $f(X) = -2x - 2y$
   $3x + 6y \leq 18$
   $-x + 3y \leq 4$
   $3x - y \leq 4$
   $2x - 3y \leq 5$
   $x + y \geq 0$
   $3x + y \geq -2$

In exercises 6 through 10 a function $z = c^T X$ to be maximized is given and a set of constraints is specified. From a sketch find all vectors v of Proposition 6.11. Then solve the system (6.10) for each v to find all possible vectors w.

6. $z = x + 5y$, the constraints of exercise 1.
7. $z = -x - y$, the constraints of exercise 2.
8. $z = 4y$, the constraints of exercise 3.
9. $z = -3y$, the constraints of exercise 4.
10. $z = 2x + 2y$, the constraints of exercise 5.

In exercises 11 through 14 a minimum problem is given. Restate as a standard minimum problem if necessary and then state the dual problem. Use a sketch to find the solution X of the dual problem and then use Proposition 6.13 to find X', Y, and Y' and thus solve the problem.

11. Minimize $w = 15y_1 + 10y_2 + 2y_3$
    subject to $y_1 + 2y_2 + y_3 \geq 3$
    $3y_1 + y_2 - y_3 \geq 4$
    $y_1, y_2, y_3 \geq 0$

12. Minimize $w = 6y_1 + 11y_2 + y_3$
    subject to $-y_1 + 4y_2 + y_3 \geq 3$
    $2y_1 - y_2 - 2y_3 \geq 1$
    $y_1, y_2, y_3 \geq 0$

13. Minimize $w = 20y_1 + 8y_2 + 12y_3$
    subject to $y_1 + y_2 + 2y_3 \geq 3$
    $4y_1 + y_2 + y_3 \geq 2$
    $y_1, y_2, y_3 \geq 0$

14. Minimize $w = 7y_1 + 5y_2 - 5y_3$
    subject to $y_1 - y_2 + y_3 \geq 0$
    $y_1 + 3y_2 - 3y_3 \geq 4$
    $y_1, y_2, y_3 \geq 0$

15. Minimize $w = y_1 - 2y_2 - y_3$
    subject to $-y_1 + y_2 + y_3 \geq -2$
    $y_1 - y_2 - y_3 \geq 2$

16. (Computer assignment) The APL functions PSLN and POSDIV below are an implementation of the proof of Farkas' lemma. Use them and Proposition 6.13 to check your answers to exercises 11 and 12.

```
    ∇ Z←S PSLN A  ;N;P;SOLN;V;I;X
[1]    Z←((N←¯1+1↓ρA),1)ρ0
[2]    →(S∧.=S+A[;N+1])/0
[3]    Z←0 0ρ0
[4]    →(0=N)/0
[5]    P←P+.×⍉P←PERP A[;1]
[6]    Z←S PSLN P+.×0  1↓A
[7]    →(~SOLN←1=1↓ρZ)/J
[8]    V←-(0 1↓A)+.×Z,[1]¯1
[9]    →((ρA)[1]<I←(S<S+A[;1])⍳1)/J
[10]   →(S≤S+X←V[I;1]÷A[I;1])/END
[11]   Z←S PSLN 0 1↓A
[12]   SOLN←1=1↓ρZ
[13] J:X←0
[14] END:Z←(N,SOLN)ρX,[1]Z
    ∇
```

```
    ∇ Z←C POSDIV A
[1]    Z←,(⌈/,|A,C)PSLN A,C
    ∇
```

The function PERP is from Section 5.5. The variable SOLN is 1 when a solution exists and zero otherwise. To compute a solution of

$$AX = c, \qquad X \geq 0$$

use the expression X←C POSDIV A.

17. The solution of the problem:

$$\begin{array}{ll} \text{minimize} & w = |2x + 1| \\ \text{subject to} & x \geq 0 \end{array}$$

is clearly $x = 0$, $w = 1$

(a) Use Proposition 6.2 to set this problem up as a standard minimum problem in the variables x, d^+, d^-.

(b) Solve the dual problem graphically.

(c) Use Proposition 6.13 to find the value of x, d^+, d^-.

6.3 The Simplex Algorithm

In practice linear programming problems often involve a large number of variables and constraints (several thousand). These problems are run using large programming packages too sophisticated to analyze here. We can, however, describe the algorithm that these packages are based upon.

In this section we will restrict our attention to the dual problems of Proposition 6.13 of the last section and the equivalent equation (6.11).

We begin with the observation that since the feasible points, S, of a standard maximum problem satisfy $X \geq 0$ [or $(-I)X \leq 0$] Proposition 6.4(b) shows that S cannot contain a whole line. Thus if S is not empty, it has vertices, and hence (Proposition 6.10) the solution, if there is one, must occur at a vertex. The simplex algorithm moves from vertex to vertex seeking a solution. To get started, however, the algorithm needs to be given a vertex.

PROPOSITION 6.14 The solutions, if any, of the standard maximum problem

$$\begin{array}{lll} \text{maximize} & z = c^T X & \\ \text{subject to} & AX \leq b, & X \geq 0 \end{array}$$

occur at vertices. If $b \geq 0$, then $X = 0$ is a vertex. ■

The simplex algorithm deals with compact versions of Equation (6.11) called *tableaus* that correspond to vertices. A tableau is a matrix of the form

$$\begin{bmatrix} A_1 & A_2 & Z \\ d_1^T & d_2^T & z \end{bmatrix} \tag{6.12}$$

obtained by applying row operations to the matrix

$$\begin{bmatrix} A & I & b \\ -c^T & 0 & 0 \end{bmatrix} \tag{6.13}$$

extracted from Equation (6.11).

Let v be a feasible point of the standard maximum problem (6.1) and $v' = b - Av$ the vector of values of the slack variables. Notice that the zero components of the vector (v, v') flag the active constraints at v. The zeros of v flag the active constraints of the form $x_i = 0$ and the zeros of v' flag the other active constraints, those for which a slack variable is zero. If the constraint flat at v is $\pi_v(T) = v + PT$, we set $\pi_v'(T) = b - A\pi_v(T)$, and the vector $(\pi_v(T), \pi_v'(T))$ has the same pattern of zeros as the vector (v, v') by Proposition 6.8.

Now let the index vector J give the *nonzero* components of (v, v') and set $p(T) = (\pi_v(T), \pi_v'(T))[J]$. Then $p(T)$ is a flat of solutions of $(A \mid I)[;J]Z = b$. In fact, $p(T)$ gives all solutions of this equation (exercise 23). Since the flats $p(T)$ and $\pi_v(T)$ have the same dimension (the rank of P), it follows that v is a vertex if and only if $p(T)$ is a single point, which is equivalent to $(A \mid I)[;J]$ having independent columns.

Not every set of independent columns of $(A \mid I)$ gives rise to a vertex, however [$(A \mid I)[;J]Z = b$ may have no feasible solutions], and in order to easily recognize vertices we impose some restrictions on the standard maximum problem (6.1).

Let A be m-by-n; then $(A \mid I)$ is m-by-$(n + m)$. Since the feasible set is in $\mathbf{R}^n$, a vertex will in general have n active constraints, hence J will have $n + m - n = m$ components. In this case $(A \mid I)[;J]$ would be an invertible matrix. If there are extra (redundant) active constraints at the vertex, however, then J will have fewer than m components and $(A|I)[;J]$ will not be square. The simplex algorithm proceeds on the assumption that this situation does not arise — and in fact it is not often a problem in practice.

So a vertex gives m independent columns of $(A|I)$. But not every set $B = (A|I)[;J]$ of m independent columns defines a vertex of (6.1). If we set $(v, v')[J] = B^{-1}b$ and set the rest of the components of (v, v') to zero, then v is a vertex if and only if $(v, v') \geq 0$.

Thus: *Vertices of (6.1) without redundant active constraints correspond to sets of m independent columns $B = (A|I)[;J]$ for which $B^{-1}b \geq 0$.*

If v is a vertex without redundant active constraints, then a *tableau for v*, (6.12), is obtained by row-reducing $B = (A|I)[;J]$ in (6.13) to an identity matrix. The product of the elementary matrices for this row reduction is of the form

$$\left[\begin{array}{c|c} B^{-1} & 0 \\ \hline w^T & 1 \end{array}\right] \tag{6.14}$$

where the $(w^T|1)$ appears because (6.13) has one more row than $(A|I)$. Multiplying (6.13) by (6.14) gives

$$\begin{bmatrix} B^{-1}A & B^{-1} & B^{-1}b \\ w^TA - c^T & w^T & w^Tb \end{bmatrix} \tag{6.15}$$

Thus $d_1 = A^Tw - c$, $d_2 = w$, $z = b^Tw$, and $Z = B^{-1}b$. On the other hand,

$$\begin{bmatrix} B^{-1} & 0 \\ w^T & 1 \end{bmatrix}\begin{bmatrix} A & I & b \\ -c^T & 0 & 0 \end{bmatrix}[;J] = \begin{bmatrix} B^{-1} & 0 \\ w^T & 1 \end{bmatrix}\begin{bmatrix} (A|I)[;J] \\ (-c^T|0)[;J] \end{bmatrix}$$

or

$$\begin{bmatrix} I \\ 0 \end{bmatrix} = \begin{bmatrix} I \\ w^TB - (c^T|0)[;J] \end{bmatrix}$$

Thus

$$B^Tw = \begin{bmatrix} c \\ 0 \end{bmatrix}[J;] \quad \text{or} \quad w = (B^T)^{-1}\begin{bmatrix} c \\ 0 \end{bmatrix}[J;]$$

and hence, since inverse commutes with transpose,

$$\begin{aligned} b^Tw &= b^T(B^{-1})^T\begin{bmatrix} c \\ 0 \end{bmatrix}[J;] = (B^{-1}b)^T\begin{bmatrix} c \\ 0 \end{bmatrix}[J;] \\ &= (c^T|0)[;J]\begin{bmatrix} v \\ v' \end{bmatrix}[J;] = c^Tv \end{aligned}$$

Thus, setting $w' = d_1 = A^T w - c$, we have that $(X, X', Y, Y') = (v, v', w, w')$ satisfies all of Equation (6.11) except possibly $(Y, Y') \geq 0$, and the tableau (6.12) is

$$\begin{bmatrix} A_1 & A_2 & \begin{bmatrix} v \\ v' \end{bmatrix}[J;] \\ w' & w & c^T v \end{bmatrix}, \qquad (A_1|A_2)[;J] = I \tag{6.16}$$

Notice that $c^T v = z$ is the value of the objective function at the vertex v. Thus a tableau with $(w, w') \geq 0$ gives a solution of the dual problems of Proposition 6.13.

In practice no row interchanges are used in manipulating tableaus. This does not affect the computations above, since the components of J were not assumed to be in any given order (see Example 6.16 below).

EXAMPLE 6.15 Consider the problem of Example 6.10 (Figure 6.4). The matrix (6.12) is

$$\left[\begin{array}{cc|ccc|c} 1 & 3 & 1 & 0 & 0 & 18 \\ 1 & 1 & 0 & 1 & 0 & 8 \\ 2 & 1 & 0 & 0 & 1 & 14 \\ \hline -1 & -2 & 0 & 0 & 0 & 0 \end{array}\right] \tag{6.17}$$

This matrix is also a tableau (6.13) for the vertex $v = (0, 0)$. The independent columns are $J = 3\ 4\ 5$, so $(v, v')[3\ 4\ 5] = 18\ 8\ 14$ and the other components are zero; that is, $v = (0, 0)$, $v' = (18, 8, 14)$. Taking $w = (0, 0, 0)$, we have $w' = (-1, -2)$ and $c^T v = 0 = b^T w$.

The second, third, and fourth columns are also independent. If we pivot on the 1;2 entry and then set it to 1, we obtain

$$\left[\begin{array}{cc|ccc|c} \frac{1}{3} & 1 & \frac{1}{3} & 0 & 0 & 6 \\ \frac{2}{3} & 0 & -\frac{1}{3} & 1 & 0 & 2 \\ \frac{5}{3} & 0 & -\frac{2}{3} & 0 & 1 & 8 \\ \hline -\frac{1}{3} & 0 & \frac{2}{3} & 0 & 0 & 12 \end{array}\right] \tag{6.18}$$

This time $(A_1|A_2)[;2\ 4\ 5] = I$, so $(v, v') = (0, 6, 0, 2, 8) \geq 0$ and $v = (0, 6)$ is a vertex. $w = (\frac{2}{3}, 0, 0)$, $w' = (-\frac{1}{3}, 0) \ngeq 0$, and $c^T v = 12$.

If we had pivoted instead on the 2;2 or 2;3 entries, then we would not have had $(v, v') \geq 0$, and v would not be feasible. ■

DEFINITION 6.8 In a tableau (6.12) there is a vector J of m indices such that $(A_1|A_2)[;J] = ID\ m$. The vector (v, v') with $(v, v')[J] = z \geq 0$ and other components zero is called a *basic feasible solution*. The columns given by J are called *basic columns* and the corresponding variables *basic variables*.

THE SIMPLEX METHOD

The simplex method is based on the observation that given the tableau for a vertex, it is easy to find an adjacent vertex that is closer to a solution.

Assuming that there are no redundant, active constraints at the vertex v, v is the intersection of n hyperplanes $\pi_1, \ldots, \pi_n$, the (irredundant) active constraints. The intersection of any $n-1$ of these hyperplanes is a line. So, dropping one hyperplane at a time, we have $n-1$ lines intersecting at v. The feasible points along these lines form $n-1$ edges — line segments between vertices (Figure 6.4).

Now suppose that the vertices u and v are connected by an edge. If J_u and J_v are the indices for the basic columns in tableaus for u and v, then since they have $n-1$ hyperplanes defining the edge in common, the vectors J_u and J_v have $n-1$ components in common. Thus we may move from u to v or back by cleaning up one column in the corresponding tableau.

EXAMPLE 6.16 In Example 6.15 we obtained a tableau (6.18) for the vertex $u = (0, 6)$ of the feasible set of Figure 6.4. The vertex $v = (3, 5)$ is connected to $(0, 6)$ by an edge. The vector $J_u = (2, 4, 5)$, and since $v' = b - Av$ we have $(v, v') = (3, 5, 0, 0, 3)$, which shows $J_v = (1, 2, 5)$. Thus to move from u to v we should clean up the first column, leaving the second and fifth columns unchanged. To do this we pivot on the 2;1 entry of (6.18), obtaining

$$\left[\begin{array}{cc|cccc|c} 0 & 1 & -4 & \frac{1}{2} & 0 & 5 \\ 1 & 0 & 13 & \frac{3}{2} & 0 & 3 \\ 0 & 0 & -22 & -\frac{5}{2} & 1 & 3 \\ \hline 0 & 0 & 5 & \frac{1}{2} & 0 & 13 \end{array}\right] \tag{6.19}$$

Strictly construed, this tableau gives $J_v = (2, 1, 5)$ and $(v, v')[2\ \ 1\ \ 5] = (5, 3, 3)$, so we have $(v, v') = (3, 5, 0, 0, 3)$ as we should. [To get $J_v = (1, 2, 5)$ we must interchange rows 1 and 2.] From the last row of (6.19) we see that $w = (5, \frac{1}{2}, 0)$, $w' = (0, 0)$ is a solution of the dual problem, and $b^T w = c^T v = 13$. ∎

In seeking solutions, our situation is somewhat different from that in the above examples. We will know one vertex, v, and we wish to find another vertex closer to a solution. From the above discussion we know that we should be able to reach any of the n adjacent vertices by cleaning up a column of a tableau for v. But which entry should we use for a pivot? Not any entry will do.

For a valid tableau we must have $Z \geq 0$ in (6.12). Suppose we pivot on a_i:

$$\left[\begin{array}{ccc|c} & a_1 & & v_1 \\ & & & \vdots \\ \cdots & a_i & \cdots & v_i \\ & & & \vdots \\ & a_n & & v_n \\ \hline & \delta & & z \end{array}\right] \tag{6.20}$$

Then in the last column we have three cases

v_i is replaced by v_i/a_i
v_j is replaced by $v_j - a_j v_i/a_i$ for $j \neq i$
z is replaced by $z - \delta v_i/a_i$

Since we are starting with a valid tableau, $v_j \geq 0$ for all j. Thus if $a_i < 0$, then $v_i/a_i < 0$ and we do not have a valid tableau. Thus, we need $a_i > 0$. For a valid tableau we also need $v_j - a_j v_i/a_i \geq 0$ or $v_j/a_j \geq v_i/a_i$. So a_i must give the minimum ratio v_j/a_j as j varies.

Last, we would like to move closer to a solution. Since negative entries of $[d_1^T \quad d_2^T]$ in (6.12) show that we do not have a solution, we may as well take a column of (6.20) with $\delta < 0$, since δ will be replaced by 0. If we do this, notice that z, the value of the objective function, becomes $z - \delta v_i/a \geq z$, so we do move closer to a solution.

We will now state the simplex algorithm, in words and APL.

Statement of the Simplex Algorithm

$$T = \begin{bmatrix} A_1 & A_2 & Z \\ d_1^T & d_2^T & z \end{bmatrix} \tag{6.21}$$

is the tableau of a vertex. Suppose that T is m-by-n.

1. Let $T[;j]$ be the column of T containing the most negative entry of $[d_1^T \quad d_2^T]$. This is the pivot column: `J←T[M;]⍳⌊/¯1↓T[N;]`.
If there are no negative entries, stop: `→(T[M;J]≥0)/0`.

2. Let P be the vector of row indices for which $T[P;J]$ is positive: `P←(0<T[;J])/⍳M`.
If there are no such entries, then the problem has no solution, stop: `→(0=⍴P)/0`.

3. Divide the positive entries of column J into the corresponding entries of the last column. Let the ith row give the minimum result: `I←P[S⍳⌊/S←÷/T[P;N,J]]`.

4. Clean up the jth column by pivoting on the $i;j$ entry and return to step 1: `→L,,T←(I,J)LDR(I,J)PIVOT T`. (The functions `LDR` and `PIVOT` are defined in Section 3.2.)

The fact that an empty vector P in step 2 implies that the problem has no solution is left as exercise 24.

EXAMPLE 6.17 We will use the simplex algorithm to solve the standard maximum problem:

$$\begin{aligned} \text{maximize} \quad & z = 35x_1 + 20x_2 \\ \text{subject to:} \quad & x_1 + x_2 \leq 70 \\ & x_1 + 2x_2 \leq 80 \end{aligned}$$

$$3x_1 + x_2 \leq 140$$
$$x_1, x_2 \geq 0$$

of Example 6.1.

Since $b \geq 0$, the origin $(x_1, x_2) = (0, 0)$ is a vertex with tableau (6.13):

$$\left[\begin{array}{cc|ccc|c} 1 & 1 & 1 & 0 & 0 & 90 \\ 1 & 2 & 0 & 1 & 0 & 80 \\ 3 & 1 & 0 & 0 & 1 & 140 \\ \hline -35 & -20 & 0 & 0 & 0 & 0 \end{array}\right]$$

The most negative entry in the last row is -35, and 70 80 140 ÷ 1 1 3 is 70 80 46.7, and so we pivot on the 3. The result is the tableau:

$$\left[\begin{array}{cc|ccc|c} 0 & \frac{2}{3} & 1 & 0 & -\frac{1}{3} & \frac{70}{3} \\ 0 & \frac{5}{3} & 0 & 1 & -\frac{1}{3} & \frac{100}{3} \\ 1 & \frac{1}{3} & 0 & 0 & \frac{1}{3} & \frac{140}{3} \\ \hline 0 & -\frac{25}{3} & 0 & 0 & \frac{35}{3} & \frac{4900}{3} \end{array}\right]$$

The most negative entry in the last row is $-\frac{25}{3}$. Since

$$\tfrac{70}{3}\ \tfrac{100}{3}\ \tfrac{140}{3} \div \tfrac{2}{3}\ \tfrac{5}{3}\ \tfrac{1}{3} \quad \text{is} \quad 35\ 20\ 140$$

we pivot on $\frac{5}{3}$. The result is the tableau:

$$\left[\begin{array}{cc|ccc|c} 0 & 0 & 1 & -\frac{2}{5} & -\frac{1}{15} & 10 \\ 0 & 1 & 0 & \frac{3}{5} & -\frac{1}{5} & 20 \\ 1 & 0 & 0 & -\frac{1}{5} & \frac{7}{15} & 40 \\ \hline 0 & 0 & 0 & 5 & 10 & 1{,}800 \end{array}\right]$$

Thus we have the solution $x_1 = 40$, $x_2 = 20$. The slack variables are $x_1' = 10$, $x_2' = x_3' = 0$. Thus the farmer should plant 40 acres with corn and 20 acres with soybeans. This uses all the available labor, all the available capital, and leaves 10 acres unplanted. The profit is \$1800.

The solution to the dual problem is $y_1 = 0$, $y_2 = 5$, $y_3 = 10$, $y_1' = y_2' = 0$. Thus the nutritionist of Example 6.2 should use 5 grams of the second mixture and 10 grams of the third to give precisely 35 units of calcium ($y_1' = 0$) and 20 units of iron ($y_2' = 0$) per batch for a cost of 1800 cents = \$18. ■

THE FUNCTION MAX

To write an APL function implementing the simplex algorithm is not difficult. The function *MAX* below, which takes a tableau for input and returns a final tableau, incorporates two changes in the above description of the simplex algorithm. First, the tests in steps 1 and 2 are made relative to the largest magnitude in the original tableau (see the line MAX[2] below for the redefinition of the tableau A). Second, the function stops if the maximum has not been found after ten iterations. Although it seems to occur rarely in practice, the simplex algorithm can go into an infinite loop; this is called *cycling*. From our discussion, this can happen only at a vertex with a redundant active constraint. From the discussion that follows (6.20) we see that it can happen only if $v_i = 0$ for some i. That is, the vector Z of (6.20) has a zero component. If MAX returns a tableau with negative components in d_1 or d_2, check Z for zero components. If none are present, give the returned tableau back to MAX for another ten iterations.

```
     ∇ T←MAX  A  ;M;N;I;J;K;P;S
[1]     K←0×(M←(ρA)[1])×N←(ρA)[2]
[2]     A←⌈/,|T←A
[3]   L:J←T[M;]⍳⌊/¯1↓T[M;]
[4]    →(A≤A+T[M;J])/0
[5]    →(0=ρP←(A<A+T[⍳M;J])/⍳M)/0
[6]   I←P[S⍳⌊/S←÷T[P;N,J]
[7]   T←(I,J)LDR(I,J)PIVOT  T
[8]    →(10>K←K+1)/L
     ∇
```

FINDING A VERTEX

In Example 6.17 the origin $(x, y) = (0, 0)$ was a vertex. The reason is that the vector b in Equation (6.1) was positive. If b has negative components, then there may be no obvious vertex to start with.

The way out of this dilemma is to notice that the problem of finding a vertex in such a case can be rephrased as *another linear programming problem — one with an obvious vertex.*

EXAMPLE 6.18

$$\begin{aligned} \text{maximize} \quad & z = 2x_1 + 4x_2 \\ \text{subject to} \quad & -2x_1 + 3x_2 \leq 7 \\ & 3x_1 + x_2 \leq 17 \\ & -x_1 - 4x_2 \leq -13 \\ & x_1, x_2 \geq 0 \end{aligned}$$

Since $b = (7, 17, -13) \ngeq 0$, the origin is not a vertex.

We introduce a new variable x'', called an *artificial variable,* into the last equation and consider the new problem:

$$\begin{array}{ll} \text{maximize} & z = -x'' \\ \text{subject to} & -2x_1 + 3x_2 \le 7 \\ & 3x_1 + x_2 \le 17 \\ & -x'' - x_1 - 4x_2 \le -13 \\ & x_1, x_2, x'' \ge 0 \end{array}$$

Then $(13, 0, 0)$ is a feasible point of the new problem, and since three independent constraints ($x_1 = 0, x_2 = 0, -x_1 - 4x_2 - x'' = -13$) are active at this point, it is a vertex. If we can find a vertex of this latter problem for which $z = -x'' = 0$ (note $-x'' \le 0$ always), we will have a feasible point of the original system. Further, the first column will *not* be a basic column ($x'' = 0$), so the solution to the second problem will give a *vertex* of the first problem.

No solution to the second problem, on the other hand, would imply that the feasible set of the first problem was empty.

We begin with the array (6.13) for the second problem:

```
      T
 0   ¯2    3    1    0    0     7
 0    3    1    0    1    0    17
¯1   ¯1   ¯4    0    0    1   ¯13
 1    0    0    0    0    0     0
```

Here $Z = (13, 0, 0, 7, 17, 0)$, so $J = 1\ 4\ 5$.

To get a tableau for this vertex we must row-reduce $T[;J]$ to the identity — or rather, some permutation of the columns of $T[;J]$ should be row-reduced to the identity augmented by a last row of zeros. It is sufficient to clean up the first column.

```
      +T←3 1 LDR 3 1 PIVOT T
0.000   ¯2.000    3.000    1.000    0.000    0.000     7.000
0.000    3.000    1.000    0.000    1.000    0.000    17.000
1.000    1.000    4.000    0.000    0.000   ¯1.000    13.000
0.000   ¯1.000   ¯4.000    0.000    0.000    1.000   ¯13.000
```

This is a starting tableau for the simplex method. A final tableau is

```
      +F←MAX T
 1.82E¯1    ¯8.67E¯19   1.00E0     9.09E¯2    0.00E0    ¯1.82E¯1    3.00E0
¯1.00E0      0.00E0    ¯5.20E¯18   1.00E0     1.00E0     1.00E0     1.10E1
 2.73E¯1     1.00E0     1.89E¯18  ¯3.64E¯1    0.00E0    ¯2.73E¯1    1.00E0
 1.00E0      0.00E0     0.00E0     0.00E0     0.00E0     0.00E0     0.00E0
```

The maximum value is $-x'' = 0$, so we have a vertex $x_1 = 1$, $x_2 = 3$, $x_1' = 0$, $x_2' = 11$, $x_3' = 0$ — that is, the intersection of the first and third constraints: $-2x_1 + 3x_2 = 7$ and $x_1 + 4x_2 = 13$.

To get the array (6.13) for this vertex, we first discard the first column of F and then insert $[-c^T \ 0 \ 0]$ into the last row.

```
      S←0 1↓F
      S[4;]←-6↑2 4
      S
¯8.673¯19   1.00E0     9.09E¯2   0.00E0   ¯1.82E¯1   3.00E0
 0.00E0    ¯5.20E¯18   1.00E0    1.00E0    1.00E0    1.10E1
 1.00E0     1.89E¯18  ¯3.64E¯1   0.00E0   ¯2.73E¯1   1.00E0
¯2.00E0    ¯4.00E0     0.00E0    0.00E0    0.00E0    0.00E0
```

To get a tableau for this vertex we row-reduce $S[; \ 1 \ 2 \ 4]$ to a permutation of the identity matrix augmented with a last row of zeros.

```
      +S←3 1 LDR 3 1 PIVOT S
0.00E0     1.00E0     9.09E¯2   0.00E0   ¯1.82E¯1   3.00E0
0.00E0    ¯5.20E¯18   1.00E0    1.00E0    1.00E0    1.10E1
1.00E0     1.89E¯18  ¯3.64E¯1   0.00E0   ¯2.73¯1    1.00E0
0.00E0    ¯4.00E0    ¯7.27E¯1   0.00E0   ¯5.45E¯1   2.00E0
      +S←1 2 LDR 1 2 PIVOT 2
 0.000   1.000   0.090   0.000   ¯0.182    3.000
 0.000   0.000   1.000   1.000    1.000   11.000
 1.000   0.000  ¯0.364   0.000   ¯0.273    1.000
 0.000   0.000  ¯0.364   0.000   ¯1.270   14.000
```

This is an initial tableau for the first problem. A final tableau is

```
      +F←MAX S
0.000   1.000   0.273   0.182   0.000    5.000
0.000   0.000   1.000   1.000   1.000   11.000
1.000   0.000  ¯0.090   0.273   0.000    4.000
0.000   0.000   0.909   1.270   0.000   28.000
```

Thus the maximum value of 28 is attained at the vertex $(x_1, x_2) = (4, 5)$. Since $(x_1', x_2', x_3') = (0, 0, 11)$, this vertex is the intersection of the first and second constraints: $-2x_1 + 3x_2 = 7$ and $3x_1 + x_2 = 17$. The solution to the dual problem is $Y = (.909, 1.27, 0)$, $Y' = (0, 0)$. ■

The technique of the last example works in general. For each negative component of the vectors add an artificial variable to the equation. Then maximize the negative sum of the artificial variables.

EXAMPLE 6.19 Find a vertex of the set

$$\begin{aligned} -3x - y &\leq -7 \\ -x - y &\leq -5 \\ -x - 4y &\leq -11 \\ x, y &\geq 0 \end{aligned}$$

Solution The linear programming problem is

$$\begin{array}{ll} \text{maximize} & z = -x_1'' - x_2'' - x_3'' \\ \text{subject to} & \begin{array}{rrrrl} -x_1'' & & & -3x - y \leq & -7 \\ & -x_2'' & & -x - y \leq & -5 \\ & & -x_3'' & - x - 4y \leq & -11 \\ & & & x_1'', x_2'', x_3'', x, y \geq & 0 \end{array} \end{array}$$

The array (6.13) is

$$\left[\begin{array}{rrrrr|rrr|r} -1 & 0 & 0 & -3 & -1 & 1 & 0 & 0 & -7 \\ 0 & -1 & 0 & -1 & -1 & 0 & 1 & 0 & -5 \\ 0 & 0 & -1 & -1 & -4 & 0 & 0 & 1 & -11 \\ \hline 1 & 1 & 1 & 0 & 0 & 0 & 0 & 0 & 0 \end{array}\right]$$

$x_1'' = 7, x_2'' = 5, x_3'' = 11, x = y = x_1' = x_2' = x_3' = 0$ is a vertex. Notice that it is quite easy to obtain the tableau, T, for such a vertex. Just add the sum of the upper rows to the bottom row, then change all the signs in the upper rows.

$$T = \left[\begin{array}{rrrrr|rrr|r} 1 & 0 & 0 & 3 & 1 & -1 & 0 & 0 & 7 \\ 0 & 1 & 0 & 1 & 1 & 0 & -1 & 0 & 5 \\ 0 & 0 & 1 & 1 & 4 & 0 & 0 & -1 & 11 \\ \hline 0 & 0 & 0 & -5 & -6 & 1 & 1 & 1 & -23 \end{array}\right]$$

```
      MAX T

4.3E 19   1.3E0     3.3E 1    1.0E0     0.0E0 | 4.3E 19   1.3E0     3.3E 1  | 3.0E0
1.0E0     3.7E0     6.7E 1    3.2E 18   0.0E0 | 1.0E0     3.7E0     6.7E 1  | 4.0E0
1.1E 19  ¯3.3E 1    3.3E 1    1.4E 19   1.0E0 | 1.1E 19   3.3E 1    3.3E 1  | 2.0E0
-------------------------------------------------------------------------------------
1.0E0     1.0E0     1.0E0     2.6E 18   0.0E0 | 4.3E 19   3.5E 18   1.1E 18 | 1.2E 17
```

Although the value of the objective function is still negative, we take it to be zero, since this machine carries only eighteen digits.

For this final tableau the artificial variables are all zero; $x = 3, y = 2$ is the vertex, which, since $x_2' = x_3' = 0$, is the intersection of the second and third constraints: $x + y = 5$ and $x + 4y = 11$. ■

It is clear from the last example that the artificial-variable technique is a general method for finding a point in the set defined by $AX \leq b$, $X \geq 0$ (a general class of problems that, via Proposition 6.13, can be shown to be coextensive with the set of all linear programming problems). In fact the simplex tableau can do more than find a single point of such a set. It yields a description of the entire set. This is illustrated in the next example for a situation where the geometry is easy to understand.

EXAMPLE 6.20 Does the line $3x + 7y = 29$ intersect the triangle with vertices (2, 1), (4, 1), and (2, 3)? What about the line $3x + 7y = 22$?

A sketch (Figure 6.10) shows the situation. The first line does not intersect the triangle and the second line does. Let us see how the simplex tableau may be used to analyze this situation (and higher-dimensional ones where no sketch is possible).

Equations of the lines through the vertices of the triangle are $x = 2$, $y = 1$, and $x + y = 5$. The triangle is thus the set defined by

$$\begin{aligned} x \quad\quad &\geq 2 \\ y &\geq 1 \\ x + y &\leq 5 \end{aligned}$$

The line $3x + 7y =$ const. intersects the triangle if and only if the set of points defined by the system

$$\begin{cases} 3x + 7y = \text{const.} \\ x \quad\quad \geq 2 \\ \quad\quad y \geq 1 \\ x + \ y \leq 5 \end{cases} \quad \text{or} \quad \begin{cases} 3x + 7y \leq \text{const.} \\ -3x - 7y \leq -\text{const.} \\ -x \quad\quad \leq -2 \\ \quad\quad -y \leq -1 \\ x + \ y \leq 5 \end{cases}$$

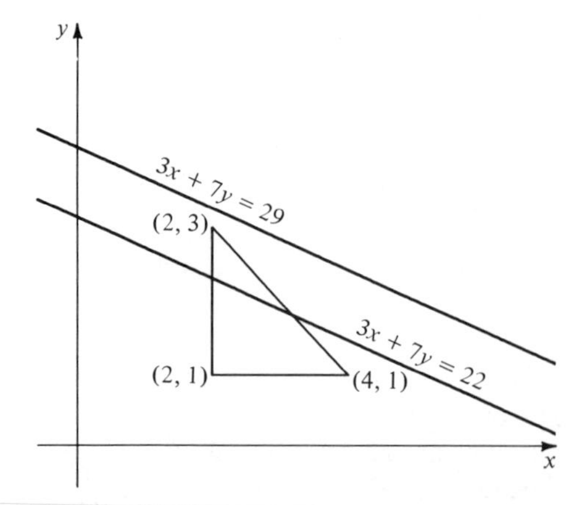

FIGURE 6.10

is nonempty. Since $x \geq 2, y \geq 1$, the positivity conditions $x, y \geq 0$ are automatic. Were this not the case, we would have to break the variables up into positive and negative parts ($x = x^+ - x^-, y = y^+ - y^-$) or move the origin of the coordinate system.

To locate a point of the set we introduce one artificial variable for each negative term on the right-hand side and maximize the negative of their sum. The constant below is either 29 or 22.

$$\begin{array}{llcccrl}
\text{maximize} & z = -x_1'' - x_2'' - x_3'' & & & & & \\
\text{subject to} & & & & 3x + 7y & \leq & \text{const.} \\
 & -x_1'' & & & -3x - 7y & \leq & -\text{const.} \\
 & & -x_2'' & & -x & \leq & -2 \\
 & & & -x_3'' & -y & \leq & -1 \\
 & & & & x + y & \leq & 5
\end{array}$$

The array (6.13) for this system is

$$\left[\begin{array}{rrrrr|rrrrr|r}
0 & 0 & 0 & 3 & 7 & 1 & 0 & 0 & 0 & 0 & \text{const.} \\
-1 & 0 & 0 & -3 & -7 & 0 & 1 & 0 & 0 & 0 & -\text{const.} \\
0 & -1 & 0 & -1 & 0 & 0 & 0 & 1 & 0 & 0 & -2 \\
0 & 0 & -1 & 0 & -1 & 0 & 0 & 0 & 1 & 0 & -1 \\
0 & 0 & 0 & 1 & 1 & 0 & 0 & 0 & 0 & 1 & 5 \\
\hline
1 & 1 & 1 & 0 & 0 & 0 & 0 & 0 & 0 & 0 & 0
\end{array}\right]$$

and a vertex is given by $(x_1'', x_2'', x_3'') = (\text{const.}, 2, 1)$, $(x, y) = (0, 0)$, $(x_1', x_2', x_3', x_4', x_5') = (\text{const.}, 0, 0, 0, 5)$. To obtain a tableau for this vertex we add rows 2, 3, and 4 to the bottom row and then change their sign:

$$T = \left[\begin{array}{rrrrr|rrrrr|c}
0 & 0 & 0 & 3 & 7 & 1 & 0 & 0 & 0 & 0 & \text{const.} \\
1 & 0 & 0 & 3 & 7 & 0 & -1 & 0 & 0 & 0 & \text{const.} \\
0 & 1 & 0 & 1 & 0 & 0 & 0 & -1 & 0 & 0 & 2 \\
0 & 0 & 1 & 0 & 1 & 0 & 0 & 0 & -1 & 0 & 1 \\
0 & 0 & 0 & 1 & 1 & 0 & 0 & 0 & 0 & 1 & 5 \\
\hline
0 & 0 & 0 & -4 & -8 & 0 & 1 & 1 & 1 & 0 & -(3 + \text{const.})
\end{array}\right]$$

Let *T29* and *T22* be the array T with the constant equal to 29 and 22, respectively.

If we apply the simplex algorithm to *T29*, the last row of the final tableau is

```
      (MAX T29)[6;]
0  0  1  ¯1.7E¯18  0  0.75  1  1  3E¯18  1.7  ¯0.5
```

Taking the fourth and ninth components to be zero, we see that the maximum value of the objective function is $-x_1'' - x_2'' - x_3'' = -.5 < 0$. Thus the set of points is empty (Figure 6.10). For T22, however, we have

```
      (F←MAX T22)[6;]
0 1 1 0 0 1 1 0 0 0 0
```

and since $-x_1'' - x_2'' - x_3'' = 0$, we have located a point of the set. Dropping the columns for the artificial variables to give a somewhat more manageable display, we have

```
      0 3↓F
```

0.0E0	0.0E0	1.4E¯1	0.0E0	4.3E¯1	1.0E0	0.0E0	1.3E0
0.0E0	0.0E0	¯1.0E0	¯1.0E0	0.0E0	0.0E0	0.0E0	0.0E0
1.0E0	0.0E0	0.0E0	0.0E0	¯1.0E0	0.0E0	0.0E0	2.0E0
0.0E0	1.0E0	1.4E¯1	0.0E0	4.3E¯1	¯1.7E¯18	0.0E0	2.3E0
0.0E0	0.0E0	¯1.4E¯1	0.0E0	5.7E¯1	1.7E¯18	1.0E0	7.1E¯1
0.0E0	0.0E0	1.0E0	1.0E0	0.0E0	0.0E0	0.0E0	0.0E0

Thus $(x, y) = (2, 2.3)$, the point at which the line $3x + 7y = 22$ cuts the line $x = 2$ (because the slack variable $x_3' = 0$). But what of the other points of intersection?

Recall that if the last row of the pivot column contains a negative number, then as we move to the next vertex the value of the objective function increases. If this entry is positive, however, the value will decrease, and if it is zero, then moving to the next vertex will not change the value of the objective function. The zero entry in the last row and fifth column above indicates that there is another vertex at which the value of $-x_1'' - x_2'' - x_3''$ is zero. Pivoting, we obtain

```
      F[1 4 5;11]÷F[1 4 5;8]
3 5.3 1.3
      0 3↓F←5 8 LDR 5 8 PIVOT F
```

0.0E0	0.0E0	2.5E¯1	0.0E0	4.3E¯19	1.0E0	¯7.5E¯1	7.5E¯1
0.0E0	0.0E0	¯1.0E0	¯1.0E0	0.0E0	0.0E0	0.0E0	0.0E0
1.0E0	0.0E0	¯2.5E¯1	0.0E0	¯1.7E¯18	3.0E¯18	1.7E0	3.3E0
0.0E0	1.0E0	2.5E¯1	0.0E0	4.3E¯19	¯3.0E¯18	¯7.5E¯1	1.7E0
0.0E0	0.0E0	¯2.5E¯1	0.0E0	1.0E0	3.0E¯18	1.7E0	1.3E0
0.0E0	0.0E0	1.0E0	1.0E0	0.0E0	0.0E0	0.0E0	0.0E0

So $(x, y) = (3.3, 1.7)$ is also a point of the set. It is the point at which the line $3x + 7y = 22$ cuts the line $x + y = 5$ (because the slack variable $x_5' = 0$). Since the set of points at which the maximum is attained is convex, the entire edge between this vertex and the previous vertex has $x_1'' + x_2'' + x_3'' = 0$, and so the corresponding values of x and y

$$\begin{bmatrix} x \\ y \end{bmatrix} = (1 - t)\begin{bmatrix} 2 \\ 2.3 \end{bmatrix} + t\begin{bmatrix} 3.3 \\ 1.7 \end{bmatrix}, \qquad 0 \leq t \leq 1$$

lie on the intersection of the line and the triangle (alternately, the intersection of a triangle and a line is convex). ∎

EXERCISES 6.3

In exercises 1 through 10 a standard maximum problem is given. Solve the problem with the simplex algorithm. Give the solution of the dual problem as well.

1. Maximize $z = 44x_1 + 52x_2$
   subject to $x_1 + 3x_2 \leq 14$
   $4x_1 + 2x_2 \leq 18$
   $6x_1 + 3x_2 \leq 35$
   $x_1, x_2 \geq 0$

2. Maximize $z = 15x_1 + 19x_2$
   subject to $2x_1 + 2x_2 \leq 18$
   $3x_1 + 7x_2 \leq 39$
   $4x_1 + 10x_2 \leq 55$
   $x_1, x_2 \geq 0$

3. Maximize $z = 10x_1$
   subject to $5x_1 + 5x_2 \leq 40$
   $2x_1 + 8x_2 \leq 20$
   $9x_1 + 5x_2 \leq 82$
   $x_1, x_2 \geq 0$

4. Maximize $z = 12x_1$
   subject to $6x_1 + 7x_2 \leq 48$
   $9x_1 + 4x_2 \leq 82$
   $5x_1 + 8x_2 \leq 50$
   $x_1, x_2 \geq 0$

5. Maximize $z = 32x_1 + 37x_2$
   subject to $8x_1 + 5x_2 \leq 55$
   $3x_1 + 2x_2 \leq 24$
   $6x_1 + 8x_2 \leq 54$
   $x_1, x_2 \geq 0$

6. Maximize $z = 50x_1 + 14x_2$
   subject to $x_1 + x_2 \leq 16$
   $7x_1 + 2x_2 \leq 67$
   $5x_1 + x_2 \leq 40$
   $x_1, x_2 \geq 0$

7. Maximize $z = 60x_1 + 45x_2$
   subject to $6x_1 + 3x_2 \leq 54$
   $5x_1 + 4x_2 \leq 50$
   $10x_1 + 5x_2 \leq 85$
   $x_1, x_2 \geq 0$

8. Maximize $z = 41x_1 + 53x_2$
   subject to $x_1 + 4x_2 \leq 15$
   $10x_1 + 2x_2 \leq 54$
   $3x_1 + 7x_2 \leq 29$
   $x_1, x_2 \geq 0$

9. Maximize $z = 120x_1 + 140x_2$
   subject to $4x_1 + 10x_2 + 7x_3 \leq 36$
   $9x_1 + 6x_2 + 10x_3 \leq 54$
   $x_1 + 2x_2 + 5x_3 \leq 16$
   $8x_1 + 4x_2 + 4x_3 \leq 40$
   $x_1, x_2, x_3 \geq 0$

10. Maximize $z = 46x_1 + 75x_2 + 87x_3$
    subject to $10x_1 + 2x_2 + x_3 \leq 42$
    $4x_1 + 4x_2 + 2x_3 \leq 52$
    $2x_1 + 7x_2 + 9x_3 \leq 110$
    $7x_1 + 4x_2 + 5x_3 \leq 82$
    $x_1, x_2, x_3 \geq 0$

(Computer assignment) In exercises 11 through 15 use the function *MAX* to solve the standard maximum problem. Give the solution to the dual problem as well.

11. Maximize $z = 162x_1 + 150x_2 + 128x_3$
    subject to $6x_1 + 8x_2 + 7x_3 \leq 126$
    $9x_1 + 8x_2 + 9x_3 \leq 155$
    $10x_1 + 7x_2 + 2x_3 \leq 135$
    $9x_1 + 7x_2 + 5x_3 \leq 132$
    $7x_1 + 2x_2 + 5x_3 \leq 84$
    $x_1, x_2, x_3 \geq 0$

12. Maximize $z = 84x_1 + 34x_2 + 32x_3$
subject to $7x_1 + 2x_2 + x_3 \leq 56$
$6x_1 + x_2 + 2x_3 \leq 45$
$9x_1 + 5x_2 + x_3 \leq 100$
$3x_1 + 8x_2 + 7x_3 \leq 108$
$10x_1 + 10x_2 + 5x_3 \leq 163$
$x_1, x_2, x_3 \geq 0$

13. Maximize $z = 194x_1 + 114x_2 + 150x_3$
subject to $6x_1 + 3x_2 + 3x_3 \leq 74$
$10x_1 + 2x_2 + 6x_3 \leq 90$
$10x_1 + 10x_2 + 10x_3 \leq 210$
$2x_1 + 6x_2 + 5x_3 \leq 121$
$3x_1 + 2x_2 + 2x_3 \leq 43$
$x_1, x_2, x_3 \geq 0$

14. Maximize $z = 39x_1 + 50x_2$
subject to $x_1 + 3x_2 + 8x_3 \leq 29$
$3x_1 + 5x_2 + x_3 \leq 46$
$6x_1 + 7x_2 + 5x_3 \leq 77$
$2x_1 + 6x_2 + 10x_3 \leq 52$
$x_1 + 4x_2 + 6x_3 \leq 30$
$x_1, x_2, x_3 \geq 0$

15. Maximize $z = 162x_1 + 65x_2$
subject to $10x_1 + x_2 + 2x_3 \leq 50$
$8x_1 + 5x_2 + x_3 \leq 90$
$3x_1 + 3x_2 + x_3 \leq 51$
$8x_1 + 10x_2 + 7x_3 \leq 136$
$9x_1 + 7x_2 + 9x_3 \leq 106$
$x_1, x_2, x_3 \geq 0$

For exercises 16 through 22 find *all* the vertices of the set of points (a) by using the technique of artificial variables to find one vertex of the set, (b) by pivoting on columns with zero in the last row to move along edges to other vertices while maintaining the artificial variables at zero. All sets are assumed to be in the first quadrant of $\mathbf{R}^2$ (i.e., $x, y \geq 0$).

16. $x \leq 3,\ y \leq 3,\ x + y \geq 4$
17. $-2x + 3y \leq 2,\ 3x - y \leq 11,\ x + 2y \geq 6$
18. $-x + y \leq 1,\ x + 3y \leq 15,\ x - y \leq 3,\ x + y \geq 5$
19. $2x + y \geq 4,\ x + 2y \geq 5,\ x + 4y \geq 7$
20. $-x + y \leq 1,\ 4x + y \leq 6,\ x + y \geq 3$
21. $2x + y \leq 6,\ x + 3y \leq 8,\ x + y \geq 4$
22. $2x + y \leq 7,\ 2x - y \geq 9$
23. Let v be a feasible point of the standard maximum problem (6.1) and $v' = b - Av$. Let the index vector J give the nonzero component of (v, v') and set $p(T) = (\pi_v(T), b - A\pi_v(T))[J]$, where π_v is the constraint flat for v. Show that $p(T)$ is the flat of solutions of $(A\,|\,I)[;J]z = b$.

Hint: Let $q(T)$ be the solution flat and define

$$\begin{bmatrix} \pi_1(T) \\ \pi_2(T) \end{bmatrix} \quad \text{by} \quad \begin{bmatrix} \pi_1(T) \\ \pi_2(T) \end{bmatrix} [J;] = q(T)$$

the other components being zero. Then $A\pi_1(T) + \pi_2(T) = b$, and one may assume $(\pi_1(0), \pi_2(0)) = (v, v')$. Show that there is an $\epsilon > 0$ such that $(\pi_1(T), \pi_2(T))$ is feasible for $\|T\| < \epsilon$.

24. (a) Show from the definition of d_1, d_2 in (6.12) that

$$c^T X = (\text{const.}) - d^T \begin{bmatrix} X \\ X' \end{bmatrix}$$

(b) Suppose that an entire column of $(A_1 | A_2)$ in 6.12 is negative. Show that the corresponding variable may be increased indefinitely and the point will still be feasible. At the same time $c^T X$ will increase indefinitely. Thus a maximum will not exist.

25. Discuss the problem of Example 6.20 for the case where the line is

(a) $3x + 7y = 13$ (b) $3x + 7y = 27$ (c) $3x + 7y = 19$

26. (Computer assignment) Use the simplex algorithm to verify the answer given for Example 6.3.

27. (Computer assignment) Use the simplex algorithm to verify the answer given for Example 6.5.

6.4* Sociobiology, Game Theory, and Evolution

CASTE IN THE SOCIAL INSECTS

The social insects (ants, termites, bees, and so on) have always posed a special problem for evolutionary biology. Normally we say that the process of natural selection selects those traits that increase an individual's offspring, that spread an individual's genes through the population. If this is strictly the case, then how do sterile workers evolve? The only reasonable answer is that for such insect colonies, natural selection must operate on the colony as if it were, in some sense, a single individual. This answer was originally given by Charles Darwin, but a convincing mechanism for colony-level selection was not proposed until the 1960s (by W. D. Hamilton).

Granted that natural selection operates on an ant colony as a whole, what is the evolutionary advantage of castes? E. O. Wilson has used linear programming techniques to study this question.

In the simplest analysis, if we take evolution to be "survival of the fittest," then species should evolve in such a way as to maximize their "fitness function" subject to the constraints of their environment.

Wilson applies this idea to the social insects in the following way. He takes "fitness" to be success in reproduction. An ant colony can be thought of as a single biological unit. Therefore the reproductive success (= fitness) of a colony is the number of new colonies it founds in a given period of time.

An ant colony typically contains a queen and a number of asexual workers, which may come in various types (soldiers, foragers, nurses, and so on) or *castes*. To found a new colony, a new queen and a number of males are hatched and sent out. Wilson asks: Why do different ant species have different proportions of the various castes in their colonies? That is, given an environment, what ratios of the various castes will maximize the production of virgin queens? He analyzes the question with linear programming techniques.

To actually write down a problem, however, he switches to a dual problem: Minimize the energy cost (= number of workers maintained) while maintaining a given level of virgin queen production. The colony is subject to the constraints of maintaining enough workers to meet the contingencies of the given environment (scarce food, cold weather, anteaters, and so on).

The *cost* of not meeting a particular contingency is measured in terms of the expected lost production of virgin queens per occurrence of the contingency times the expected number of occurrences of the contingency in a given period of time.

Assume two castes and two contingencies. The probability that contingency 1 is not handled successfully is

$$(1 - q_{11})^{\alpha_{11}W_1}(1 - q_{12})^{\alpha_{12}W_2}$$

where q_{ij} is the probability that a worker of caste j is able to cope with contingency i when the worker encounters the contingency (hence $1 - q_{ij}$ is the probability of failure), W_i is the total weight of workers of caste i in an average colony, and α_{ij} is the average number of contacts a worker of caste j can be expected to have with contingency i during the existence of the contingency divided by W_j.

Similarly, the probability that an occurrence of contingency 2 is not handled successfully is

$$(1 - q_{21})^{\alpha_{21}W_1}(1 - q_{22})^{\alpha_{22}W_2}$$

The corresponding costs of not meeting the contingencies are

$$k_1X_1(1 - q_{11})^{\alpha_{11}W_1}(1 - q_{12})^{\alpha_{12}W_2}, \qquad k_2X_2(1 - q_{21})^{\alpha_{21}W_1}(1 - q_{22})^{\alpha_{22}W_2}$$

where k_i is the frequency of contingency i for the given period of time and X_i is the average cost per failure to meet contingency i.

If F_i is the highest tolerable cost due to contingency i, then the constraints are

$$\begin{aligned} k_1X_1(1 - q_{11})^{\alpha_{11}W_1}(1 - q_{12})^{\alpha_{12}W_2} &\leq F_1 \\ k_2X_2(1 - q_{21})^{\alpha_{21}W_1}(1 - q_{22})^{\alpha_{22}W_2} &\leq F_2 \\ W_1, W_2 &\geq 0 \end{aligned}$$

Since the terms in parentheses are less than 1, we must have $F_i < k_iX_i$ or the constraint cannot be active. Although these are nonlinear equations, the boundary curves of the feasible set they define are, in fact, straight lines. The only variables are W_1 and W_2 and, taking logarithms, we obtain the standard minimum problem

Minimize $\qquad u = W_1 + W_2$

Subject to

$$[-\alpha_{11}\ln(1-q_{11})]W_1 + [-\alpha_{12}\ln(1-q_{12})]W_2 \geq -\ln\left(\frac{F_1}{k_1X_1}\right) \qquad (6.22)$$

$$[-\alpha_{21}\ln(1-q_{21})]W_1 + [-\alpha_{22}\ln(1-q_{22})]W_2 \geq -\ln\left(\frac{F_2}{k_2X_2}\right)$$

$$W_1, W_2 \geq 0$$

Since $\ln a < 0$ for $0 < a < 1$, the logarithms in (6.22) are all negative and so the constant terms are all positive.

The feasible set will look something like Figure 6.11.

If we write the system (6.22) in matrix form:

$$\text{minimize } u = [1 \quad 1][W_1 \quad W_2]^T$$

$$\text{subject to} \quad \begin{bmatrix} a_{11} & a_{12} \\ a_{21} & a_{22} \end{bmatrix}\begin{bmatrix} W_1 \\ W_2 \end{bmatrix} \geq \begin{bmatrix} c_1 \\ c_2 \end{bmatrix}$$

Then a_{ij} is a measure of the effectiveness of the ith caste in dealing with contingency j (it increases with both α_{ij} and q_{ij}). Let us number the castes so that caste i is specializing in contingency i. Since caste 1 is specializing in contingency 1 — that is, is better at contingency 1 than contingency 2, $a_{11} > a_{12}$ and so

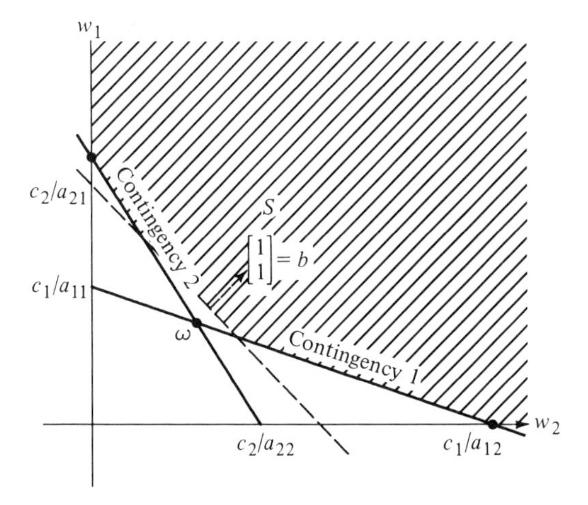

FIGURE 6.11

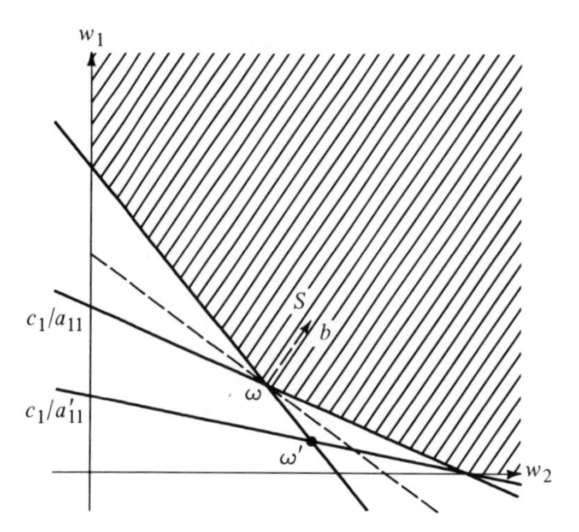

FIGURE 6.12

$c_1/a_{11} < c_1/a_{12}$. This means that the curve for contingency 1 meets the W_2 axis at less than 45°. Similarly, the curve for contingency 2 meets the W_1 axis at less than 45°. Since $b = [1 \;\; 1]^T$, the hyperplanes $u =$ const. meet the axes at 45° and hence (Figure 6.11) the minimum occurs at ω.

Wilson works with Figure 6.11 to derive predictions about how caste ratios can change over evolutionary time. We will give one example.

EXAMPLE 6.21 Suppose that caste 1 becomes more efficient at its task of coping with contingency 1 while all else remains constant. This means that a_{11} increases while a_{12}, a_{22}, a_{21}, c_1, c_2, and b remain fixed. Suppose that $a'_{11} > a_{11}$ is the new coefficient. Then the equilibrium will be shifted to the vertex ω' in Figure 6.12.

Thus the colony will be able to maintain the same level of virgin queen production with fewer workers ($b^T w' < b^T w$), and the relative proportions of the two castes will change. But the more efficient caste has decreased in relative size and the other caste has increased.

This is precisely the opposite of what would be expected if selection operated at the level of the individual rather than the colony.

Thus the theory makes a clear prediction that could conceivably be used to test the hypothesis that selection takes place at the colony level. If selection takes place at the colony level and a caste increases in efficiency, then the caste ratios will shift one way. If, on the other hand, selection takes place on the individual level, then the ratios will shift the other way. ■

GAME THEORY

Let us begin with the children's game "rock–scissors–paper." This is a game for two players. At a signal, each player shows his right hand either flat (paper), with

TABLE 6.1

		Player II		
		1	2	3
	1	0	10	−10
Player I	2	−10	0	10
	3	10	−10	0

two fingers extended (scissors), or clenched into a fist (rock). The winner is determined by the scheme: paper covers rock, rock breaks scissors, scissors cut paper.

The winner usually inflicts some minor physical punishment upon the loser, but let us move the example into the realm of economic theory by assuming that the loser pays the winner some amount, say a dime. Then the various possibilities are summarized in Table 6.1, where 1 = rock, 2 = scissors, 3 = paper.

The entries in Table 6.1 are the amount that player II pays to player I. A negative amount indicates that player I pays player II.

The game of rock–scissors–paper is an example of a *two-person, zero-sum matrix game.* It is a *zero-sum* game because what one player loses, the other wins. Table 6.1 is called the *payoff matrix* of the game.

The question to be answered in such a game is: What strategy will maximize a player's winnings, or at least minimize the losses?

If the first player always shows a fist, say, presumably the second player will figure this out eventually and always choose paper. If he is not to lose in the long run, then, a player must vary his choices and do so in a way that his opponent cannot predict. If, for example, one player decided to cycle through the pattern rock–rock–paper–scissors, presumably his opponent would eventually notice the pattern and begin to play paper–paper–scissors–rock. Therefore, if a player decides that his best strategy is to play rock twice as often as paper and twice as often as scissors, he should devise a scheme for choosing each particular play at random but in such a way that in the long run the proportions of the time that each play is chosen comes out to be rock–rock–scissors–paper. This could be done, for example, by building a spinner on the pattern of Figure 6.13 and making each play by spinning the pointer.

Presumably even with this scheme the opponent would eventually discover the proportions with which the plays were being made. How well can a player do, then, assuming that his opponent has figured out his strategy in the sense that he knows how frequently each play will be made in the long run but cannot tell which play will be made at any particular time.

Here is the general set-up. A matrix A, the *payoff matrix,* is given. It need not be square. For example, the first player might be restricted to scissors and paper but win the ties. The $i;j$ entry of the payoff matrix is the payoff to player I (the *row* player) when player I makes the play i and player II (the *column* player) makes the play j.

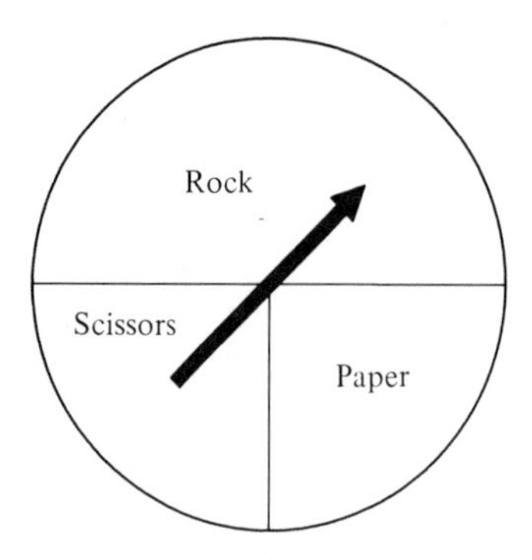

FIGURE 6.13

A strategy for a player is a plan to make each play a fixed proportion of the time in the long run. A strategy then can be represented by a vector p, where $p[i]$ is the fraction of the time that the player will make the ith play in the long run.

For example, if a player wishes to play the proportions of Figure 6.13, then his strategy is given by the vector $p = (\frac{1}{2}, \frac{1}{4}, \frac{1}{4})$.

Notice that the components of such a vector are always nonnegative and sum to one.

DEFINITION 6.9

1. A *probability vector p* is a vector $p \geq 0$ whose components sum to 1.
2. Let A, the payoff matrix for a game, be m-by-n. Then a *strategy for player I* is a probability vector with m components. A *strategy for player II* is a probability vector with n components.
3. A *pure strategy* is a probability vector with one component equal to 1. (It follows that the other components are zero.)
4. The *expected payoff* from strategies p and q is

$$E(p, q) = p^T A q$$

The expected payoff is the average payoff, each play, to player I in the long run. This is because the probability that player I will make the ith play is $p[i]$ and the probability that player II will make the jth play is $q[j]$, hence the probability that the $i;j$th payoff will be made is $p[i]q[j]$. This means that in the long run the payoff $A[i;j]$ is made the fraction $p[i]q[j]$ of the time. It follows that the average payoff will be

$$E(p, q) = \sum_{i,j} A[i; j]p[i]q[j] = p^T A q$$

EXAMPLE 6.22 For the game of rock–scissors–paper the payoff matrix is

$$A = \begin{bmatrix} 0 & 10 & -10 \\ -10 & 0 & 10 \\ 10 & -10 & 0 \end{bmatrix}$$

If player I chooses to play rock all the time and player II chooses to play scissors all the time, then we have the pure strategies $p = (1, 0, 0)$, $q = (0, 1, 0)$, and the expected payoff is

$$E(p, q) = [1 \quad 0 \quad 0] \begin{bmatrix} 0 & 10 & -10 \\ -10 & 0 & 10 \\ 10 & -10 & 0 \end{bmatrix} \begin{bmatrix} 0 \\ 1 \\ 0 \end{bmatrix} = 10$$

Player I can expect to win every game.

If, on the other hand, player II chooses to make each possible play with equal frequency, then $q = (\frac{1}{3}, \frac{1}{3}, \frac{1}{3})$ and

$$E(p, q) = [1 \quad 0 \quad 0] \begin{bmatrix} 0 & 10 & -10 \\ -10 & 0 & 10 \\ 10 & -10 & 0 \end{bmatrix} \begin{bmatrix} 1 \\ 1 \\ 1 \end{bmatrix} \tfrac{1}{3} = 0$$

and in the long run the players should break even. ■

Now let us analyze the situation from player II's point of view. If player II fixes on a strategy q, then player I will figure it out (this is a basic assumption in game theory). Thus, player I will adjust his strategy, p, to the known strategy q. Thus player I will pick a strategy p_2 (the subscript indicates that player II's point of view is being discussed) such that

$$p_2^T A q = \max_p p^T A q \qquad \text{(fixed } q\text{)}$$

Since player II understands this, he must pick the particular $q = q_2$ such that

$$\text{player II:} \qquad p_2^T A q_2 = \min_q \max_p p^T A q$$

This is what the players should do, *provided that* p_2 and q_2 exist. In fact they do exist, as we shall see below.

Player I may go through the same line of reasoning to arrive at the equation

$$\text{player I:} \qquad p_1^T A q_1 = \max_p \min_q p^T A q$$

Now is it possible for each player to settle on a fixed strategy, or will they have to keep shifting about? If $p_1 = p_2$ and $q_1 = q_2$, then they can settle on a fixed strategy. More generally, (the p_i's and q_i's need not be unique) they can settle on a fixed strategy if

$$\max_p \min_q p^T A q = \min_q \max_p p^T A q \tag{6.23}$$

Equation (6.23) is true and is known as the *minimax theorem*. One part of it is easy. Since, for any p, q,

$$\min_q p^T A q \leq p^T A q \leq \max_p p^T A q$$

we do have

$$\max_p \min_q p^T A q \leq \min_q \max_p p^T A q$$

The opposite inequality takes more work. The trick is to set up a closely related linear programming problem — a problem that can be solved to find the optimal strategies of the game given by A.

PROPOSITION 6.15 Let A be the payoff matrix of a two-person zero-sum game. Let v, w be solutions of the pair of dual linear programming problems

$$\begin{cases} \text{maximize} & z = c^T X \\ \text{subject to} & AX \leq b \\ & X \geq 0 \end{cases} \qquad \begin{cases} \text{minimize} & u = b^T y \\ \text{subject to} & A^T y \geq c \\ & y \geq 0 \end{cases} \tag{6.24}$$

where $b = (1, \ldots, 1)$, $c = (1, \ldots, 1)$. Let $\theta = b^T w = c^T v$. Then $p^* = v \div \theta$, $q^* = w \div \theta$ are optimal strategies for the game and the expected payoff is $E(p^*, q^*) = \theta^{-1}$.

Proof Let v and w be solutions; then $b^T w = c^T v$ by Proposition 6.12. Now $b^T w = +/w$ and $c^T v = +/v$. Since $v, w \geq 0$, we have $\theta = 0$ if and only if $v = w = 0$, which violates $A^T w \geq c$. Thus $\theta > 0$ and $p^* = v \div \theta$, $q^* = w \div \theta$ are probability vectors.

Since $Av \leq b$, if p is any probability vector then [Proposition 6.4(a)] $p^T A v \leq p^T b = 1$. Similarly $w^T A q \geq 1$ for any probability vector q. Dividing by θ, we have

$$p^T A q^* \leq \theta^{-1} \leq p^{*T} A q$$

for all probability vectors p, q. Now the pattern of zero components in v and w (Proposition 6.12) implies

$$\theta = b^T w = w^T b = w^T A v = (A^T w)^T v = c^T v = \theta$$

Dividing by θ^2 gives $\theta^{-1} = p^{*T}Aq^*$. Thus,

$$p^{*T}Aq^* \leq p^{*T}Aq \quad \text{shows} \quad p^{*T}Aq^* = \min_q p^{*T}Aq$$
$$p^{*T}Aq^* \geq p^{T}Aq^* \quad \text{shows} \quad p^{*T}Aq^* = \max_p p^{T}Aq^*$$

so

$$p^{*T}Aq^* = \min_q p^{*T}Aq \leq \max_p \min_q p^TAq \leq \min_q \max_p p^TAq$$
$$\leq \max_p p^TAq^* = p^{*T}Aq^*$$

so the minimax theorem holds and we can take

$$p_1 = p_2 = p^*, \qquad q_1 = q_2 = q^* \qquad \blacksquare$$

There is a subtle problem with Proposition 6.15. The proposition states that *if the dual problems have a solution* v, w, then $p^* = v \div \theta$ and $q^* = w \div \theta$ are optimal strategies. How do we know that the pair of problems has solutions? In fact, they need not have solutions (exercise 14), but we can always shift to a pair that does have solutions.

Since $b \geq 0$, the maximum problem has the vertex $X = 0$ (Proposition 6.14). Thus the maximum problem, and hence the minimum problem also, will fail to have a solution only if c^TX is unbounded above. Now if w' is *any* feasible point of the minimum problem, then by Proposition 6.12, $c^TX \leq b^Tw'$ and hence is bounded above. Thus, the problem will have solutions if the feasible set of the minimum problem can be shown to be nonempty. If $A > 0$ (every entry strictly larger than zero), we can pick a w' with sufficiently huge components to guarantee $Aw' \geq c$, $w' \geq 0$.

The trick is the observation that A and $A + \alpha$, α any scalar, define the same game in the sense that

$$p^T(A + \alpha)q = p^TAq + \alpha$$

so that the fixed constant α is added to all the expected payoffs. This means that p^*, q^* will be the same for both A and $A + \alpha$. Thus to apply Proposition 6.15 we first add a large enough α to make all entries of $A + \alpha$ positive.

EXAMPLE 6.23 Consider the game of rock–scissors–paper. The payoff matrix is

$$A = \begin{bmatrix} 0 & 10 & -10 \\ -10 & 0 & 10 \\ 10 & -10 & 0 \end{bmatrix}$$

Adding 11 to each component gives

$$A + 11 = \begin{bmatrix} 11 & 21 & 1 \\ 1 & 11 & 21 \\ 21 & 1 & 11 \end{bmatrix}$$

Since $b = (1, 1, 1) \geq 0$, a starting tableau is

$$T = \left[\begin{array}{ccc|ccc|c} 11 & 21 & 1 & 1 & 0 & 0 & 1 \\ 1 & 11 & 21 & 0 & 1 & 0 & 1 \\ 21 & 1 & 11 & 0 & 0 & 1 & 1 \\ \hline -1 & -1 & -1 & 0 & 0 & 0 & 0 \end{array}\right]$$

and a final tableau is

```
      +F←MAX T
 6.2E¯19     1.0E0      ¯4.3E¯19  │  4.3E¯2    1.0E¯2   ¯2.3E¯2  │ 3.0E¯2
¯2.5E¯19     6.0E¯19     1.0E0    │ ¯2.3E¯2    4.3E¯2    1.0E¯2  │ 3.0E¯2
 1.0E0      ¯2.1E¯19     8.7E¯19  │  1.0E¯2   ¯2.3E¯2    4.3E¯2  │ 3.0E¯2
──────────────────────────────────┼──────────────────────────────┼────────
¯1.3E¯18    ¯4.53¯19    ¯8.7E¯19  │  3.0E¯2    3.0E¯2    3.0E¯2  │ 9.1E¯2
```

Thus the optimal strategies q^* and p^* are

```
      +Q←F[3 1 2;7]÷F[4;7]
0.33  0.33  0.33
      +P←F[4;4 5 6]÷F[4;7]
0.33  0.33  0.33
```

And the expected payoff is, remembering to subtract $\alpha = 11$,

```
      ¯11+P+.×T[ι3;ι3]+.×Q
¯2.8E¯17
```

Thus, as one would expect, the optimal strategies are for each player to choose rock, scissors, and paper with equal frequency and, in the long run, both will break even. ∎

EXAMPLE 6.24 Suppose the first player in rock–scissors–paper is allowed to choose only rock and paper, say, but by way of compensation is allowed to win the ties. What are the best strategies and expected payoff now?

The new payoff matrix is

$$A = \begin{bmatrix} 10 & 10 & -10 \\ 10 & -10 & 10 \end{bmatrix}$$

and, adding 11 to A, starting and final tableaus for the corresponding maximum problem are

```
        T
21     21      1 |  1    0 |  1
21      1     21 |  0    1 |  1
-----------------+---------+----
¯1     ¯1     ¯1 |  0    0 |  0

      +F←MAX T
9.5E¯1    1.0E0       0.0E0     |  4.8E¯2    ¯2.3E¯3 | 4.5E¯2
9.5E¯1   ¯1.7E¯18     1.0E0     | ¯2.3E¯3     4.8E¯2 | 4.5E¯2
--------------------------------+--------------------+-------
9.1E¯1   ¯8.7E¯19    ¯8.7E¯19   |  4.5E¯2     4.5E¯2 | 9.1E¯2
```

The optimal vectors q^* and p^* are

```
      +Q←0,F[1 2;6]÷F[3;6]
0  0.5  0.5
      +P←F[3;4 5]÷F[3;6]
0.5   0.5
```

and the expected payoff is

```
      ¯11+P+.×T[1 2;1 2 3]+.×Q
0
```

Thus, the first player's optimal strategy is to play rock and paper with equal frequency, but the second player's optimal strategy is to play scissors and paper with equal frequency and avoid playing rock altogether (this lessens the probability of ties). In the long run both players will come out even. ■

From Proposition 6.15 it appears that the theory of two-person zero-sum games is a subset of the set of linear programming problems in the sense that each game has a corresponding program (an infinite number in fact) that solves it. In fact, the theories are coextensive in this sense, because every linear programming problem may be rephrased as a two-person, zero-sum game. Thus any general method for solving such games will also solve linear programming problems.

Game theory is not used in solving the everyday problems of industry the way linear programming is, but it has been very influential intellectually, especially in the field of economics. In fact, the term "zero-sum game" has become a stock phrase of newspaper editorial writers.

EVOLUTIONARILY STABLE STRATEGIES

J. Maynard Smith has adapted game theory to the study of the evolution of behavior. His idea is to model contests between animals in game-theoretic terms.

Different individuals of the same species often engage in contests over possession of various kinds of "resources," such as food, territory, and mates. These contests are not invariably duels to the death. A great deal of bluffing can be involved, and the winner is often simply the one that yells the loudest. Gorillas, for example, scream and beat their fists upon their chests but in most cases avoid actual combat.

How does such behavior increase an individual's Darwinian fitness? By "Darwinian fitness" we simply mean reproductive success, ensuring the survival of ones' genes in the population.

When animals confront each other they communicate their intent by behavioral conventions: holding the body in a certain position, showing teeth in a snarl, beating the chest, and so on. These displays will be referred to as *conventional* fighting. Unconventional fighting, on the other hand, is actual fighting. Which gives a better chance of reproductive success, conventional or unconventional fighting? Maynard Smith models the question in terms of a payoff matrix. We will discuss two of his examples.

EXAMPLE 6.25. (*The Game of Hawk and Dove*) Suppose that there are two types of individuals in a population, hawks and doves. The hawk strategy in a contest is to fight unconventionally, to escalate the fighting until victory or serious injury results. The dove strategy is to fight conventionally and to run if the opponent escalates the fighting.

We need some numbers to represent the various possible payoffs. Let us measure the payoff in "Darwinian fitness units" that somehow measure increase in reproductive success — the number of offspring expected to survive to reproductive age, perhaps. Consider a fixed resource. Suppose that possession of the resource is worth, say, $a > 0$ fitness units. Suppose that serious injury is worth $-b$ ($b > 0$) fitness units, and suppose that a long, unresolved contest (dove versus dove) is worth $-c$ ($c > 0$) fitness units with $b > c$.

Now suppose that two individuals enter a contest over the resource. Two pure strategies are available: 1 = hawk and 2 = dove. The payoff matrix is

$$A = \begin{bmatrix} \dfrac{a-b}{2} & a \\ 0 & \dfrac{a}{2} - c \end{bmatrix} \tag{6.25}$$

This is calculated as follows. In a hawk-hawk contest we assume that player I has a 50 percent chance of winning. In hawk-hawk contests, then, the expected payoff to player I would be $\frac{1}{2}a - \frac{1}{2}b$. In a hawk-dove contest the hawk (player I) would win and the payoff would be a. In a dove-hawk contest the dove (player I) would run and thus get 0. In a dove-dove contest we again assume that player I has a 50 percent chance of winning (i.e., of outlasting player II), so the gain to player I is $a/2$ minus the cost of the long contest. ∎

Now if we have a payoff matrix such as (6.25), game theory suggests that we proceed to calculate the optimal strategy for the players. But unless there is some sort of vital principle directing organisms to optimize, it is difficult to see what meaning such a computation would have. We are assuming that player I's behavior, his "chosen strategy" is coded into his genes, and we wish to know what sort of behavior will persist over evolutionary time — that is, will not die out.

Suppose that the population consists entirely of hawks. To be definite let us take $c = 1$, $a = 2$, $b = 4$. Then the payoff matrix is

$$\begin{bmatrix} -1 & 2 \\ 0 & 0 \end{bmatrix}$$

Since there are only hawks in the population, the expected payoff from each contest is -1. Now suppose that, through mutation, a dove strategist arises. To begin with there are many more hawks than doves, hence most hawks will fight with hawks and most doves will also fight with hawks. So, at least when the dove population is small, the expected payoff to a dove strategist is close to $0 > -1$. This means that the dove population will not die out but will persist in the population. This analysis holds as long as $b > a$. Pure hawk is not an evolutionarily stable strategy in a population. (Neither is pure dove; see exercise 15.)

The payoff matrices that arise in this context are all square.

Definition 6.10 Let A be a square payoff matrix. The probability vector p is an *evolutionarily stable strategy* if

1. $E(p, p) \geq E(q, p)$ for all stratcgies q.
2. If $E(p, p) = E(q, p)$, then $E(p, q) > E(q, q)$.

To see what this means, suppose that the population consists entirely of individuals using strategy p and that a mutation arises using strategy q. Suppose that the fraction of the population using p is α and the fraction using q is β ($\alpha + \beta = 1$). If the Darwinian fitness of the individuals is F_0 before a series of contests, then after the series of contests the fitness for the population using strategies p and q is

$$F(p) = F_0 + \alpha E(p, p) + \beta E(p, q)$$
$$F(q) = F_0 + \alpha E(q, p) + \beta E(q, q)$$

This is because the fraction of time that an individual meets a p strategist is α and the fraction of the time he meets a q strategist is β. For evolutionary stability we want $F(p) > F(q)$. Assuming α much larger than β, we get the definition of evolutionarily stable strategy.

Here are two simple propositions on evolutionarily stable strategies.

Proposition 6.16 If the diagonal entry $A[i; i]$ is the strictly largest entry in its column, then pure strategy i is evolutionarily stable. ■

Proposition 6.17 Suppose that p is an evolutionarily stable strategy for A. If $p[i] > 0$, then the ith component of Ap is maximal.

Proof Suppose that $w = Ap$ has maximal component $w[k] = \alpha$ with $k \neq i$. Let q be the kth pure strategy. Then $q \neq p$ and $E(q, p) = q^T Ap = q^T w = w[k] = m$. But

$$\begin{aligned} E(p, p) &= p^T Ap \\ &= p[i]w[i] + \sum_{j \neq i} p[j]w[j] \\ &\leq p[i]w[i] + m \sum_{j \neq i} p[j] \\ &< m\Sigma\, p[i] = m = E(q, p) \qquad \text{if } p[i] > 0 \end{aligned}$$ ■

EXAMPLE 6.26 Let

$$A = \begin{bmatrix} 1 & -1 & -1 \\ 0 & 2 & -2 \\ -1 & 1 & 3 \end{bmatrix}$$

Then all three pure strategies are evolutionarily stable by Proposition 6.16. A population chancing to start off with any of them would persist with them because, although the others are also stable, once one is established, the others cannot arise. ■

The condition of Proposition 6.17 is necessary for p to be evolutionarily stable but it is not sufficient.

EXAMPLE 6.27 (*The Game of Hawk, Dove, and Bourgeois*) A third type of strategy that appears to be important in nature is termed *bourgeois* by Maynard Smith. If you "own" the resource (nest, territory, mate), play hawk; otherwise, play dove.

Assume that in a given contest each contestant has an equal probability of being the owner; for example, the first frog that arrives at the lily pad is the "owner." Then the expected payoff in a hawk-bourgeois contest is the average of the hawk-hawk and hawk-dove payoffs:

$$\frac{1}{2}\left(\frac{a-b}{2} + a\right) = \frac{3a-b}{4}$$

The other entries are computed similarly and the payoff matrix is

$$A = \begin{bmatrix} \dfrac{a-b}{2} & a & \dfrac{3a-b}{4} \\ 0 & \dfrac{a}{2} - c & \dfrac{a-2c}{4} \\ \dfrac{a-b}{4} & \dfrac{3a-2c}{4} & \dfrac{a}{2} \end{bmatrix}$$

and as long as $b > a$ — that is, the penalty for serious injury is greater than the gain from victory; Proposition 6.16 shows that the bourgeois strategy is evolutionarily stable. ■

Notice that in the last example "ownership" is used to settle the contests. Ownership has nothing to do with strength, ferocity, or intelligence. It is the evolutionary equivalent of settling an argument by flipping a coin.

EXERCISES 6.4*

Caste in the Social Insects

1. Show that if there are two castes but only one contingency, then the caste least effective in dealing with the contingency will die out.

2. Suppose there are two castes but more than two contingencies. Show that the caste ratios will be entirely determined by the two most important contingencies.

3. Suppose that contingency 2 increases in frequency or importance. Suppose that caste i is specialized for contingency i. Show that although contingency 1 remains as frequent and important as ever, caste 1 can die out.

4. Show that if there is one caste and two contingencies, then it is to the species' advantage to evolve two castes, one specialized to each contingency, because in the latter case the total weight of workers can be less.

5. Assume two castes and two contingencies. Suppose that the castes are relatively unspecialized (i.e., $a_{11} > a_{12}$ but not much greater, and similarly for the second curve). Show that a relatively small long-term change in the frequency or importance of one of the contingencies can result in a large shift in caste ratios.

6. Assume two castes and two contingencies. Suppose that the castes are quite specialized (e.g., a_{11} quite a bit larger than a_{12} and similarly for the second curve). Show that a relatively large long-term shift in the importance or frequency of one of the contingencies will produce little shift in the caste ratios.

Game Theory

7. Show that the optimal strategies for a given game need not be unique.

   Hint: Examples 6.22 and 6.23.

8. Let A be a payoff matrix. The entry $A[i; j]$ is called a *saddle point* if it is the smallest entry in its row and the largest entry in its column.

   (a) Show that if $\theta = A[i; j]$ is a saddle point, then taking p^* to be the ith pure strategy and q^* to be the jth pure strategy gives a solution to the game.

   Hint: By adding α to A, assume $A > 0$. Show $w = p^*/\theta$, $v = q^*/\theta$ give solutions to the corresponding dual problems.

   (b) Show that all saddle points have the same value.

   Hint: $E(p^*, q^*)$ is the same for all solution points p^*, q^*.

9. Show that

$$E(p^*, q) \leq E(p^*, q^*) \leq E(p, q^*)$$

for all p, q if and only if p^* and q^* are optimal vectors for player I and player II.

Hint: The proof of Proposition 6.15.

10. A matrix is *skew-symmetric* if $A^T = -A$. Suppose that the payoff matrix is skew-symmetric.

(a) Show that $E(p, q) = -E(q, p)$ for all probability vectors p, q.

Hint: Since p^TAq is one by one, $p^TAq = (p^TAq)^T$.

(b) Show that $E(p^*, q^*) = 0$.

Hint: By exercise 9 and part (a) $p^TAq^* \leq p^{*T}Aq^* \leq p^{*T}Aq$ implies $q^{*T}Ap \geq q^{*T}Ap^* \geq q^TAp^*$, so (q^*, p^*) is also a solution. Then, since $E(p^*, q^*) = -E(q^*, p^*)$, $E(p^*, q^*) = 0$.

11. Alter the rock-scissors-paper payoffs (Example 6.22) as follows: rock-scissors, 30 cents to rock; scissors-paper, 20 cents to scissors; paper-rock, 10 cents to paper.

(a) Write the payoff matrix and show that in the long run the players, if they play their optimal strategies, will break even.

Hint: Exercise 10.

(b) (Computer assignment) Use the simplex algorithm to compute a pair of optimal strategies.

12. (Computer assignment) Suppose that the rules of rock-scissors-paper (Example 6.22) are altered so that player I wins 10 cents for all ties but pays 20 cents when he loses. All other payoffs remain the same. What are the optimal strategies and expected payoffs?

13. (Computer assignment) Suppose that the rules of rock-scissors-paper (Example 6.22) are altered so that player I gets 10 cents for rock-rock and paper-paper ties and player II gets 30 cents for scissors-scissors ties. All other payoffs remain the same. What are the optimal strategies and the expected payoff?

14. (Computer assignment) Show that if α is not added to the payoff matrix A of exercise 11, then the dual linear programming problems have no solution.

Evolutionarily Stable Strategies

15. Prove Proposition 6.16.

16. Show that the pure dove strategy is not evolutionarily stable for the hawk-dove game.

17. Verify the payoff matrix of Example 6.27.

18. Show that if, in the payoff matrix A, $A[j; i] > A[i; i]$, then pure strategy i cannot be evolutionarily stable.

Hint: Proposition 6.17.

19. (a) Show that A and $A + \alpha$ have the same evolutionarily stable strategies for any matrix A and scalar α. Given a 2-by-2 matrix A, add α to get $A = \begin{bmatrix} 0 & b \\ a & c \end{bmatrix}$.

(b) Show that if $a < 0$ or $b < c$, then A has an evolutionarily stable strategy.

Hint: Proposition 6.16.

(c) Assume $a \geq 0$, $c \leq b$. If $a + b \neq c$, show that $(p, 1 - p)$ is an evolutionarily stable strategy where $p = (b - c)/(a + b - c)$.

(d) Continuing (c), show that if $a + b = c$, then A has constant columns.

20. Does the matrix $A = \begin{bmatrix} a & b \\ a & b \end{bmatrix}$ have evolutionarily stable strategies?

21. Find an evolutionarily stable strategy for the hawk-dove game.

Hint: Exercise 19.

REFERENCES

1. Wilson, E. O., *The Insect Societies*. Cambridge: Harvard University Press, 1971.
2. Wilson, E. O., *Sociobiology*. Cambridge: Harvard University Press, 1975.
3. Maynard Smith, J. "The Evolution of Behavior," *Scientific American*, September 1978, p. 176.
4. Maynard Smith, J., *J. Theoretical Biology*, **47** (1974), 209.
5. Dawkins, Peter, *The Selfish Gene*. New York: Oxford University Press, 1976.
6. Axelrod, R., and Hamilton, W. D., "The Evolution of Cooperation," *Science* 211 (1981), 1390–1396.

References 1 and 2 are for further discussion of caste in the social insects. References 3, 4, 5 are for the material on evolutionarily stable strategies. Pages 74–94 of reference 5 are relevant, and the mathematics is in reference 4. Reference 6 applies game theory, evolutionarily stable strategies, and computer simulations to investigate the evolution of cooperation between unrelated individuals — for example, the fungus and alga that form a lichen.

CHAPTER SEVEN

Eigenvalues and Eigenvectors

In Section 5.2 we discussed the problem of diagonalizing a symmetric matrix and defined the eigenvalues and eigenvectors of such matrices.

In this chapter we consider the problem of diagonalizing a square, not necessarily symmetric, matrix. The approach is quite different from that used in the treatment of symmetric matrices. In Section 5.2 the eigenvalues were defined as the diagonal entries that appear when the matrix is diagonalized. In this chapter the eigenvalues are defined as the roots of a certain polynomial, the characteristic polynomial of the matrix. With this definition the eigenvalues of a matrix are, in general, complex numbers.

In order to define the characteristic polynomial, we need some facts about determinants. These are developed in Section 7.1. The eigenvalues and eigenvectors of a matrix are then defined in Section 7.2.

Three optional sections are given over to three different applications of the material of Section 7.2. Linear difference and differential equations are discussed in Sections 7.3* and 7.6*, respectively. One of the most interesting aspects here is the way in which complex numbers are used to solve problems that seem at first to involve only real numbers.

In Section 7.4* eigenvectors are used to analyze congruences in three-dimensional space and obtain a geometric description of the nonsingular affine functions $f\colon \mathbf{R}^3 \to \mathbf{R}^3$.

Gerschgorin's theorem, Section 7.5*, gives a useful estimate for the eigenvalues of a matrix.

The approach to eigenvalues and eigenvectors given in Section 7.2 provides theoretical insights but not a practical approach to the computation of eigenvalues. Section 7.7 describes the QR algorithm for computing the eigenvalues of a general matrix. The section ends with some APL functions that are useful for computing eigenvalues and eigenvectors of general matrices.

7.1 Determinants

Determinants are really part of the subject matter of multilinear algebra or tensor analysis. We need them, however, to develop the theory of the eigenvalues of a general (i.e., nonsymmetric) matrix. We will develop only enough of the theory of determinants to enable us to calculate them and use them to discuss eigenvalues.

Determinants are basically volumes — volumes with an algebraic sign attached, and we begin with some plane geometry to explain why one attaches an algebraic sign to a volume. In the plane, of course, the word is area rather than volume.

Let p, q, r be any three points in $\mathbf{R}^2$ and let $A(p, q, r)$ be the area of the triangle with vertices p, q, and r [Figure 7.1(a)]. Ultimately we would like a formula for A in terms of the coordinates of the points p, q, and r. Areas are additive. To be precise, suppose we choose a point, call it o, in the interior of the triangle pqr. This point defines a division of pqr into three smaller triangles: opq, opr, and orq. From Figure 7.1(b) we have the formula

$$A(p, q, r) = A(o, p, q) + A(o, q, r) + A(o, r, p)$$

This formula suggests taking o to be the origin and identifying the points p, q, r with the vectors with their tails at the origin and their heads at p, q, and r, respectively. If we do this, then given two vectors v, w it is sufficient to find a formula for the area $A(o, v, w)$ in order to compute $A(p, q, r)$ for any three points.

There is a problem, however. The formula above works only when o is in the interior. Figure 7.1(c) shows what happens when o is not in the interior of pqr [corresponding triangles in (b) and (c) are shaded similarly]. Clearly the formula above no longer holds in this case. It would hold, however, *if the area of triangle oqr were negative!*

Without further ado we make the convention

$$A(p, q, r) \text{ is } \begin{cases} \text{positive if the circuit } p \to q \to r \to p \text{ is counterclockwise} \\ \text{negative if the circuit } p \to q \to r \to p \text{ is clockwise} \end{cases}$$

This convention makes the formula hold in all cases (Figure 7.2).

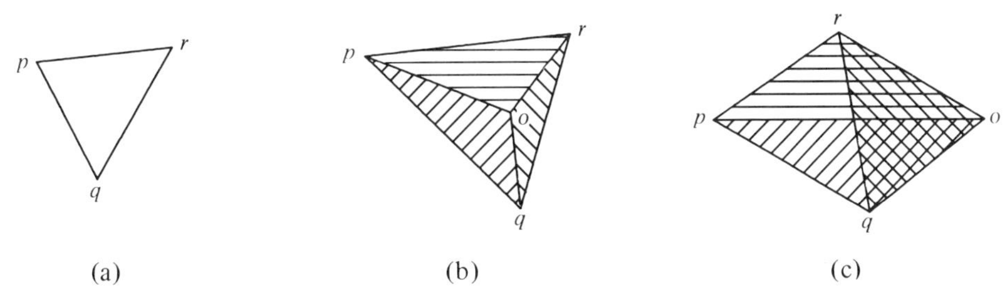

FIGURE 7.1

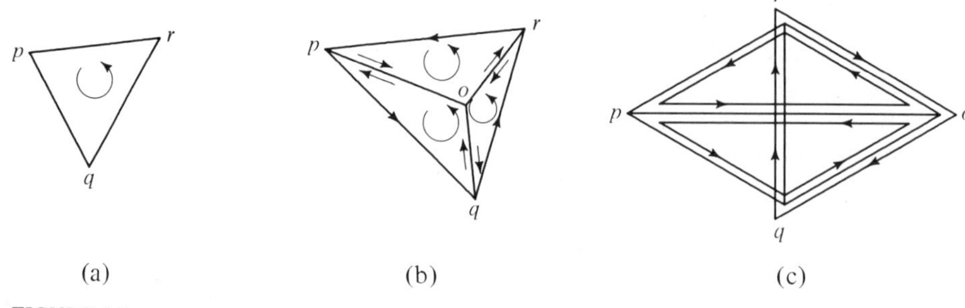

FIGURE 7.2

In fact with this convention we may reduce the calculation of area of an arbitrary polygon to the calculation of areas (with the proper sign) of the form $A(o, u, v)$. Consider, for example, the two polygons in Figure 7.3. The points p_1, p_2, p_3, p_4, and p_5 as well as o are the same in both Figure 7.3(a) and 7.3(b), but different polygons have been formed by connecting the points in different order. By carefully counting up the areas involved, always with the proper sign, the reader will find that, with the obvious notation,

$$A(p_1, p_2, p_3, p_4, p_5) = A(o, p_1, p_2) + A(o, p_2, p_3) + A(o, p_3, p_4) + A(o, p_4, p_5) + A(o, p_5, p_1)$$

$$A(p_1, p_3, p_2, p_4, p_5) = A(o, p_1, p_3) + A(o, p_3, p_2) + A(o, p_2, p_4) + A(o, p_4, p_5) + A(o, p_5, p_1)$$

(The second formula is a bit tricky, but it works.)

The process works for polygons with an arbitrary number of sides, and since a continuous curve can be approximated as closely as desired by a polygonal line, the process can be used to approximate arbitrary areas in the plane. By taking the process to the limit, one obtains a special case of the theorem from the calculus of several variables known as Green's theorem. Such considerations also lead to the mechanical device called the planimeter, which can be used to measure the areas of an irregular shape by tracing the boundary. Nowadays we could also enter the

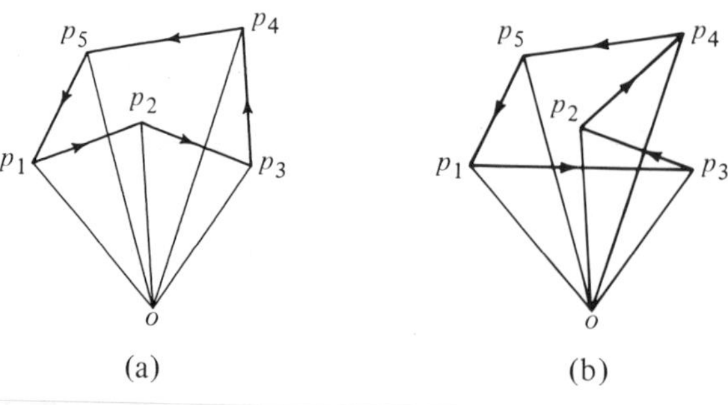

FIGURE 7.3

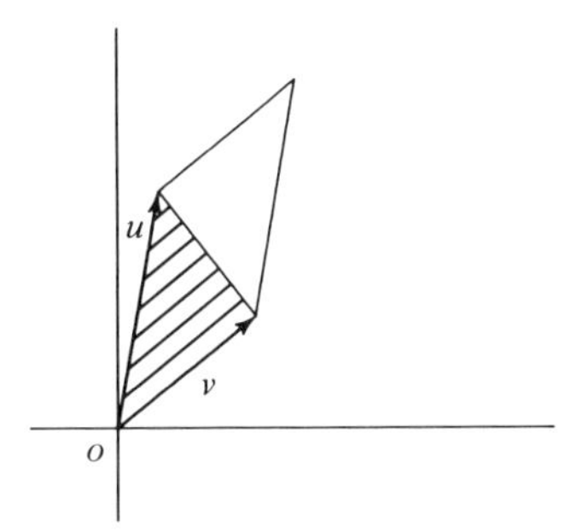

FIGURE 7.4

coordinates of as many boundary points as desired directly into a computer by using an electrical device called a digitizer or graphics tablet.

Let us return to the problem of computing $A(o, u, v)$ for a pair of vectors u, v in $\mathbf{R}^2$ (Figure 7.4). The area of this triangle is half the area of the parallelogram determined by the vectors u and v. We will calculate the area of the parallelogram.

DEFINITION 7.1 Let u, v be vectors in $\mathbf{R}^2$. The *determinant,* $\det(u, v)$, of u and v is the area of the parallelogram determined by u and v. This area is positive if the circuit $o \to u \to v \to o$ is counterclockwise and it is negative if the circuit is clockwise.

The determinant function has four properties that allow us to compute it:

1. $\det(u, v) = -\det(v, u)$
2. $\det(u, v) = \det(u, v + \lambda u)$, λ any scalar
3. $\det(\lambda u, v) = \lambda \det(u, v)$, λ any scalar
4. If $u = (1, 0)$ and $v = (0, 1)$, then $\det(u, v) = 1$

The first property is directly from the definition. The second and third come from the plane geometry formula for the area of a parallelogram:

$$A = bh$$

In Figure 7.5(a) the formula $A = bh$ is illustrated. In Figure 7.5(b) we see that neither b nor h nor the direction of the circuit changes when v is replaced by $v + \lambda u$. In Figure 7.5(c) we see that when u is replaced by λu, h remains un-

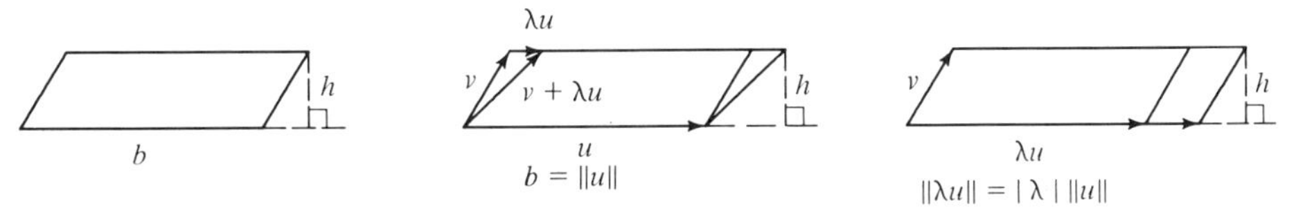

FIGURE 7.5

changed and b is replaced by $|\lambda|b$. If $\lambda > 0$, it follows that property 3 holds. If $\lambda < 0$, then the direction of the circuit $o \to u \to v \to o$ is reversed, so property 3 holds in this case also.

To calculate the determinant of u and v we store them as the *rows* of a matrix. Then properties 1 through 3 can be interpreted as statements about how row-reduction operations change the determinant. Property 1 states that a row interchange changes the sign of the determinant. Property 2 states that a pivot operation does not effect the determinant, and property 3 states that multiplying a row by a scalar multiplies the determinant by the same scalar. This means that we can calculate determinants by row-reducing the matrix with rows u, v.

If A is a 2-by-2 matrix, we define $|A| = \det(A[1;], A[2;])$.

EXAMPLE 7.1 Compute the determinants of the following matrices:

$$\begin{pmatrix} 1 & 3 \\ 2 & 7 \end{pmatrix}, \quad \begin{pmatrix} 6 & 0 \\ 0 & 7 \end{pmatrix}, \quad \begin{pmatrix} 0 & 3 \\ 2 & 1 \end{pmatrix}$$

Solution Row-reducing, we have

1. $\begin{vmatrix} 1 & 3 \\ 2 & 7 \end{vmatrix} = \begin{vmatrix} 1 & 3 \\ 0 & 1 \end{vmatrix}$ property 2 with $u = (1, 3)$, $v = (2, 7)$, $\lambda = -2$

   $= \begin{vmatrix} 1 & 0 \\ 0 & 1 \end{vmatrix}$ (property 2 again)

   $= 1$ (property 4)

2. $\begin{vmatrix} 6 & 0 \\ 0 & 7 \end{vmatrix} = 6 \begin{vmatrix} 1 & 0 \\ 0 & 7 \end{vmatrix}$ (property 3)

   $= 6 \cdot 7 \begin{vmatrix} 1 & 0 \\ 0 & 1 \end{vmatrix}$ (property 3)

   $= 42$ (property 4)

3. $\begin{vmatrix} 0 & 3 \\ 2 & 1 \end{vmatrix} = -\begin{vmatrix} 2 & 1 \\ 0 & 3 \end{vmatrix}$ (property 1)

   $= -2 \begin{vmatrix} 1 & \frac{1}{2} \\ 0 & 3 \end{vmatrix}$ (property 3)

   $= -6 \begin{vmatrix} 1 & \frac{1}{2} \\ 0 & 1 \end{vmatrix}$ (property 3)

   $= -6 \begin{vmatrix} 1 & 0 \\ 0 & 1 \end{vmatrix} = -6$ (property 2 and property 4) ■

In the example above we used two implications of property 1:

2′. $\det(u, v) = \det(u + \lambda v, v)$, λ any scalar

3′. $\det(u, \lambda v) = \lambda \det(u, v)$, λ any scalar

Property 3′ holds, for example, because

$$\det(u, \lambda v) = -\det(\lambda v, u) = -\lambda \det(v, u) = -\lambda\,(-\det(u, v)) = \lambda \det(u, v)$$

We next show that this definition coincides with the definition of the determinant of a 2-by-2 matrix given in Chapter 2.

PROPOSITION 7.1

$$\begin{vmatrix} a & b \\ c & d \end{vmatrix} = ad - bc$$

Proof If $a \neq 0$,

$$\begin{aligned}
\begin{vmatrix} a & b \\ c & d \end{vmatrix} &= a \begin{vmatrix} 1 & \frac{b}{a} \\ c & d \end{vmatrix} && \text{(property 3)} \\
&= a \begin{vmatrix} 1 & \frac{b}{a} \\ 0 & d - \frac{cb}{a} \end{vmatrix} && \text{(property 2)} \\
&= a\left(d - \frac{cb}{a}\right) \begin{vmatrix} 1 & \frac{b}{a} \\ 0 & 1 \end{vmatrix} && \text{(property 3)} \\
&= (ad - cb) \begin{vmatrix} 1 & 0 \\ 0 & 1 \end{vmatrix} && \text{(property 2)} \\
&= ad - cb
\end{aligned}$$

The case $a = 0$ is left as an exercise. ■

EXAMPLE 7.2 What is the area of the polygon with vertices $p_1 = (1, 1)$, $p_2 = (3, 2)$, $p_3 = (2, 3)$, $p_4 = (3, 4)$, $p_5 = (-1, 3)$, in that order?

Solution

$$\begin{aligned}
A(p_1, p_2, p_3, p_4, p_5) = A(o, p_1, p_2) + A(o, p_2, p_3) + A(o, p_3, p_4) \\
+ A(o, p_4, p_5) + A(o, p_5, p_1)
\end{aligned}$$

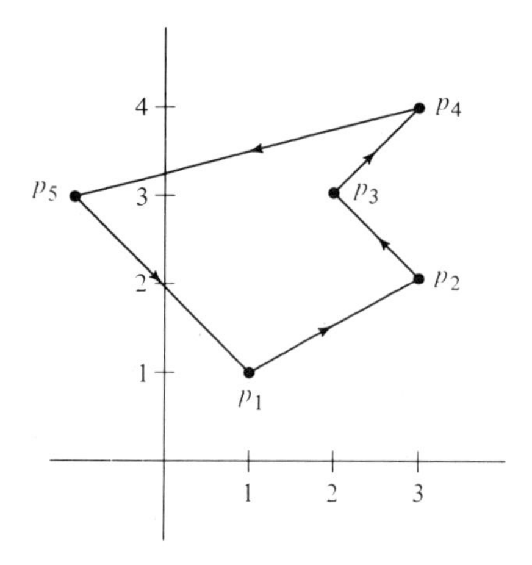

FIGURE 7.6

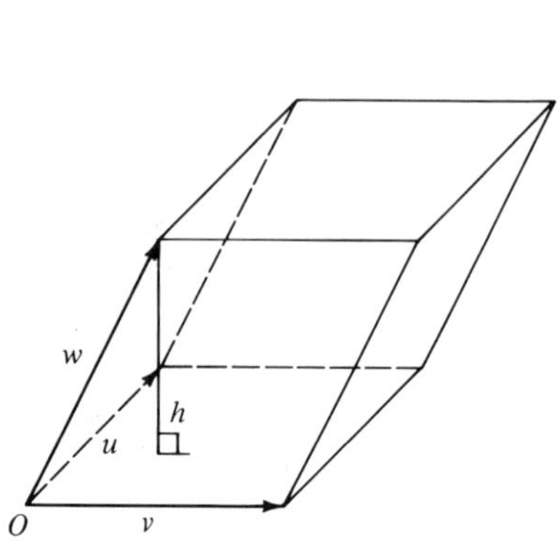

FIGURE 7.7

$$
\begin{aligned}
&= \tfrac{1}{2}\left[\begin{vmatrix} 1 & 1 \\ 3 & 2 \end{vmatrix} + \begin{vmatrix} 3 & 2 \\ 2 & 3 \end{vmatrix} + \begin{vmatrix} 2 & 3 \\ 3 & 4 \end{vmatrix} + \begin{vmatrix} 3 & 4 \\ -1 & 3 \end{vmatrix} + \begin{vmatrix} -1 & 3 \\ 1 & 1 \end{vmatrix}\right] \\
&= \tfrac{1}{2}[(-1) + 5 + (-1) + 13 + (-4)] \\
&= \tfrac{1}{2}(18 - 6) \\
&= 6
\end{aligned}
$$

The polygon is sketched in Figure 7.6. ■

We shall define the determinant of a square matrix by the four properties listed above for the determinant of two vectors in the plane. This shows us how to compute the determinant but leaves logical gaps. It does not prove that the function called "determinant" actually exists, because the four conditions might be too restrictive. If the conditions are not too restrictive, they might not be restrictive enough, and there may be many different functions that satisfy them. We will not, however, pursue these questions further.

DEFINITION 7.2 Let A be a square matrix. The *determinant* of A, written det (A) or $|A|$, is the unique function satisfying conditions 1 through 4.

1. Interchanging two rows changes the sign of the determinant.
2. Adding a multiple of one row to a different row does not change the determinant. In particular, pivoting on any entry of the matrix does not change the determinant.
3. Multiplying a row by a scalar multiplies the determinant by that scalar.
4. The determinant of the n-by-n identity matrix is 1.

For $n = 3$ the three rows of the matrix, considered as vectors in $\mathbf{R}^3$, define a volume called a *parallelopiped* (Figure 7.7). The volume of such a solid is the area of the base (defined by u and v in the figure) times the height, which is the length of the perpendicular dropped from w to the base. The volume is taken to be positive if the circuit around the triangle formed by the tips of the vectors $u = A[1;]$, $v = A[2;]$, $w = A[3;]$ in that order is counterclockwise when viewed from the origin. Otherwise the volume is negative. It is not too difficult to see that the volume of such a parallelopiped satisfies properties 1 through 4.

We now develop some important properties of determinants.

PROPOSITION 7.2 Let A and B be n by n. Then

$$\det(AB) = \det(A)\det(B)$$

Proof First notice that the formula holds when A is an elementary matrix:

A is a switch matrix: A is obtained from $ID\ n$ by interchanging two rows. Hence $\det(A) = -1$. AB is B with two rows interchanged, hence $\det(AB) = -\det(B)$.

A is a multiplier matrix: Suppose A multiplies a row by λ. Then $\det(AB) = \lambda \det(B)$ by property 3. A is obtained by multiplying a row of $ID\ n$ by λ, hence $\det(A) = \lambda$.

A is a pivot matrix: By property 2, $\det(AB) = \det(B)$. A is obtained by multiplying rows of $ID\ n$ by scalars and adding the result to other rows. Hence by property 2 (and 4), $\det(A) = 1$.

The result now follows if A is a product of elementary matrices:

$$A = E_1E_2 \cdots E_n$$

Then

$$\begin{aligned}\det(AB) &= \det(E_1E_2 \cdots E_nB)\\ &= \det(E_1)\det(E_2E_3 \cdots E_nB)\\ &= \det(E_1)\det(E_2) \cdots \det(E_n)\det B\\ &= \det(E_1E_2 \cdots E_n)\det B\\ &= \det(A)\det B\end{aligned}$$

If A is not invertible (see Proposition 3.5) then neither is AB. In this case the determinants are zero. We defer the proof to Proposition 7.5. ∎

PROPOSITION 7.3 Let A, C be square matrices. Then

$$\det\left[\begin{array}{c|c} A & B \\ \hline 0 & C \end{array}\right] = \det(A)\det(C)$$

Proof

$$\left[\begin{array}{c|c} A & B \\ \hline 0 & C \end{array}\right] = \left[\begin{array}{c|c} I & 0 \\ \hline 0 & C \end{array}\right]\left[\begin{array}{c|c} I & B \\ \hline 0 & I \end{array}\right]\left[\begin{array}{c|c} A & 0 \\ \hline 0 & I \end{array}\right]$$

and pivot operations alone will reduce

$$\left[\begin{array}{c|c} I & B \\ \hline 0 & I \end{array}\right]$$

to an identity matrix. Thus by Proposition 7.2

$$\det\left[\begin{array}{c|c} A & B \\ \hline 0 & C \end{array}\right] = \det\left[\begin{array}{c|c} I & 0 \\ \hline 0 & C \end{array}\right]\det\left[\begin{array}{c|c} I & B \\ \hline 0 & I \end{array}\right]\det\left[\begin{array}{c|c} A & 0 \\ \hline 0 & I \end{array}\right] = \det\left[\begin{array}{c|c} I & 0 \\ \hline 0 & C \end{array}\right]\det\left[\begin{array}{c|c} A & 0 \\ \hline 0 & I \end{array}\right]$$

Now precisely the same row reduction that reduces A to echelon form may be used to reduce

$$\left[\begin{array}{c|c} A & 0 \\ \hline 0 & I \end{array}\right]$$

to echelon form, so

$$\det(A) = \det\left[\begin{array}{c|c} A & 0 \\ \hline 0 & I \end{array}\right]$$

Similarly for

$$\left[\begin{array}{c|c} I & 0 \\ \hline 0 & C \end{array}\right] \quad ■$$

PROPOSITION 7.4 Let A be a square matrix. Then $\det(A) = \det(A^T)$.

Proof If $A = E_1E_2 \cdots E_n$, a product of elementary matrices, then $A^T = E_n^TE_{n-1}^T \cdots E_1^T$, so in this case it will be sufficient to prove the proposition for elementary matrices. Switch and multiplier matrices are symmetric, so the question reduces to pivot matrices.

We can factor a pivot matrix into even more elementary matrices. Let $E(i; j, \lambda)$ be the matrix obtained by multiplying the ith row of $ID\ n$ by λ and adding it to the jth row. Then $\det(E(i, j, \lambda)) = 1$, $E(i, j, \lambda)^T = E(j, i, \lambda)$, and a pivot matrix is a product of $E(i, j, \lambda)$'s. If A is singular then so is A^T (Proposition 3.6) and both determinants are zero. See Proposition 7.5. ■

PROPOSITION 7.5 Let A be a square matrix. A is invertible if and only if $\det(A) \neq 0$, and if A is invertible then

$$\det(A^{-1}) = \det(A)^{-1}$$

In particular, $\det(P^{-1}AP) = \det(A)$ for any invertible matrix P.

Proof $A = E_1E_2 \cdots E_nE$, where E_i is elementary and E is the row echelon form of A. Since $\det(E_i) \neq 0$ (see the proof of Proposition 7.2) and $\det(A) = \det(E_1) \det(E_2) \cdots \det(E_n) \det(E)$, we see that $\det(A) \neq 0$ if and only if $\det(E) \neq 0$. A is invertible if and only if E is an identity matrix, and E, being square and echelon, is an identity matrix if and only if it has no zero rows. So the proposition is proved once we establish that a matrix with a zero row has determinant zero.

If A has a row of zeros, multiply the row by 0. Then by property 3, since A remains unchanged,

$$\det(A) = 0 \cdot \det(A) = 0$$

If A^{-1} exists, then

$$\det(A)\det(A^{-1}) = \det(AA^{-1}) = \det(I) = 1$$

So

$$\det(A^{-1}) = \frac{1}{\det(A)}$$

Then

$$\det(P^{-1}AP) = \det(P)^{-1}\det(A)\det(P) = \det(A) \qquad \blacksquare$$

Propositions 7.2 and 7.3 imply a simple algorithm for computing determinants.

Given a matrix A, let R be the special reflection matrix (see Section 5.5) that maps $v = A[;1]$ into $(\|v\|, 0, 0, \ldots, 0)$. Then

$$RA = \left[\begin{array}{c|c} \|v\| & B \\ \hline 0 & C \end{array}\right]$$

and

$$\det(RA) = \det \left[\begin{array}{c|c} \|v\| & B \\ \hline 0 & C \end{array}\right]$$

$$\det(R)\det(A) = \det[\|v\|]\det C = \|v\|\det(C)$$

$$\det(A) = \frac{\|v\|\det(C)}{\det(R)}$$

The matrix C is $(n-1)$ by $(n-1)$, so if $\det(R)$ is reasonably easy to compute, we can invoke mathematical induction for the computation of $\det(C)$ and use the formula to get $\det(A)$.

The matrix R is a reflection in a one-dimensional subspace of $\mathbf{R}^n$.

PROPOSITION 7.6 Let R be an n-by-n special reflection matrix:

$$\det(R) = (-1)^{n-1}$$

Proof Let R be the matrix of reflection in the subspace of $\mathbf{R}^n$ generated by v. Let $X = PX'$ define an orthonormal coordinate system fr which v points along the x_1' axis. Then in X' coordinates we have reflectior in the x_1' axis, so, using Proposition 5.18,

$$P^{-1}RP = \begin{pmatrix} 1 & & & & 0 \\ & -1 & & & \\ & & -1 & & \\ & 0 & & -1 & \\ & & & & -1 \end{pmatrix}$$

So

$$\begin{aligned} \det(R) &= \det(P^{-1}RP), \qquad \text{by Proposition 7.5} \\ &= (-1)^{n-1}, \qquad \text{by property 3} \end{aligned}$$ ∎

If A is n by n, then the formula above becomes

$$\det(A) = (-1)^{n-1}\, \|v\|\, \det(C)$$

We can use the function *SPRFLCTR* from Section 5.5 to compute R, but we must exercise some care. *SPRFLCTR* will return an identity matrix instead of a reflection if the first column of A is already in the proper form. Thus the term $(-1)^{n-1}$ is present only when $R \neq ID\ n$.

```
    ∇ Z←DET A ;R
[1]   A←(R←SPRFLCTR A[;1])+.×A
[2]   Z←A[1;1]×¯1*(¯1+1↑ρA)×∨/,R≠ID 1↑ρA
[3]   →(1=1↑ρA)/0
[4]   Z←Z×DET 1 1↓A
    ∇
```

EXAMPLE 7.3 Choosing a 5-by-5 matrix at random, we have

```
      A
¯94   8  90 ¯13 ¯73
¯11  24 ¯15 ¯77  64
 23 ¯58  88 ¯94  58
 69 ¯25  68  ¯5  63
¯38  75  46 ¯86  84
      DET A
304376154
```

Notice the size of the answer. Each entry has, roughly speaking, an order of magnitude about 10^2 and so the same is true of $\|v\|$ at every level. Since the matrix is 5 by 5, the determinant ought to have an order of magnitude of roughly $(10^2)^5 = 10^{10}$. ∎

The unexpectedly large size of such a determinant complicates using a function such as *DET* to detect invertibility. The determinant must be small compared not to *A* but to A*n, where *A* is *n* by *n*.

For example, the matrix *A* below has rank 4.

```
      A
 ¯21    ¯96    66     61    ¯91
 ¯61    ¯21    17    ¯51     67
  23    ¯25    28    ¯91     42
 425   ¯879   676    230  ¯1009
 ¯87    ¯34    12    ¯12     82
      DET A
¯2.37E¯7
```

Since the expected size of the determinant is about 10^{10}, we can say that 10^{-7} is negligible compared to the expected size of the determinant (for this system ⎕CT is about 1E¯15).

A function such as *DET* is easy to write, and having it around does no harm. Determinants are theoretical tools, however, and are not of much practical use for data processing.

Det (A) is the (signed) volume of the parallelopiped defined by the rows of A. By Proposition 7.3 it is also the volume of the parallelopiped defined by the columns of A.

Now suppose that $y = f(X) = b + AX$ is an affine function from $\mathbf{R}^n$ to $\mathbf{R}^n$. Let C be the "unit parallelopiped" in $\mathbf{R}^n$ — that is, the one determined by the columns of *ID n*. Then f carries C to a translate of the parallelopiped determined by the columns of A. So it carries C, which has a volume of 1, to a parallelopiped with volume det (A). The next proposition, which can be proved by the methods of advanced calculus, generalizes this fact.

PROPOSITION 7.7 Let $f(X) = b + AX$ be an affine map on $\mathbf{R}^n$. Let S be a subset of $\mathbf{R}^n$ with (signed) volume $V[S]$. Then

$$V(f[S]) = \det(A)V(S) \qquad ∎$$

In $\mathbf{R}^2$ and $\mathbf{R}^3$ we have a connection between the volume of the parallelopiped determined by the rows of a matrix A and the determinant of A via a formula for the volume of the parallelopiped. In the case of $\mathbf{R}^2$ the area is the base times the height, where the height is determined by dropping a perpendicular (Figure 7.5). Similarly in $\mathbf{R}^3$ the volume of the parallelopiped is the area of the base times the height, where the height is determined by dropping a perpendicular to the plane of the base (Figure 7.7).

This suggests an inductive definition of the volume of a parallelopiped π whose edges are the vectors $v_1, v_2, \ldots, v_h$.

Suppose we know the volume of the parallelopiped with edges $v_1, v_2, \ldots, v_{h-1}$. Call it A. Let $v_h = v_h^{\|} + v_h^{\perp}$ with respect to the subspace generated by $v_1, v_2, \ldots, v_{h-1}$. Define the (unsigned) volume of π to be $V = \|v_h^{\perp}\|A$. In this case we have $v_1, \ldots, v_h$ vectors in $\mathbf{R}^n$ with $h \leq n$. When $h = n$ we have defined the volume as $\det(A)$, where the rows (or columns) of A are $v_1, \ldots, v_n$. Thus we have two possible definitions for $h = n$. The next proposition shows that the two definitions are consistent.

PROPOSITION 7.8 Let $v_1, \ldots, v_h$ in $\mathbf{R}^n$ be the columns of the matrix A. The square of the volume of the parallelopiped defined by $v_1, \ldots, v_h$ is $\det(A^TA)$. Thus if $h = n$, the square of the volume is $\det(A)^2$.

Proof Let $B = A^T$. Then the rows of B are $v_1, \ldots, v_h$ and $A^TA = BB^T$. We work with B, since elementary row operations are more familiar than elementary column operations. Let $v_h = v^{\|} + v^{\perp}$ with respect to the subspace generated by $v_1, \ldots, v_{h-1}$. Let $B_1 = 0\ {}^{-}1{\downarrow}B$. Then

$$B = \left[\frac{B_1}{v^{\|} + v^{\perp}}\right]$$

and $v^{\|}$ is a linear combination of the rows of B_1. Thus by multiplying the rows of B_1 by scalar and adding them to $v^{\|} + v^{\perp}$ we can row-reduce the matrix B to

$$B_2 = \left[\frac{B_1}{v^{\perp}}\right]$$

since $B_2 = PB$, where $\det(P) = 1$, we have

$$|BB^T| = |PB_2B_2^TP^T| = |P|\,|B_2B_2^T|\,|P| = |B_2B_2^T|$$

Now, since the rows of B_1 are perpendicular to $v^{\perp}$,

$$|B_2B_2^T| = \left|\left[\frac{B_1}{v^{\perp}}\right]\left[\frac{B_1}{v^{\perp}}\right]^T\right| = \left|\left[\frac{B_1}{v^{\perp}}\right][B_1^T|(v^{\perp})^T]\right|$$

$$= \left|\left[\begin{array}{c|c} B_1B_1^T & B_1(v^{\perp})^T \\ \hline v^{\perp}B_1^T & v^{\perp}(v^{\perp})^T \end{array}\right]\right| = \left|\left[\begin{array}{c|c} B_1B_1^T & 0 \\ \hline 0 & \|v^{\perp}\|^2 \end{array}\right]\right|$$

$$= \|v^{\perp}\|^2|B_1B_1^T|$$

By induction we may assume that $\|B_1B_1^T\|$ is the square of the volume determined by $v_1, v_2, \ldots, v_{h-1}$. Thus we need only start the induction.

For $h = 1$, however,

$$|BB^T| = |[v+.\times v]| = \det([\|v\|^2]) = \|v\|^2$$

the square of the length of v. ∎

EXAMPLE 7.4

(a) What is the (unsigned) area of the parallelogram determined by the vectors (1, 1, 1, 1) and (1, −1, 2, 4) in $\mathbf{R}^4$?

(b) What is the (unsigned) volume of the parallelopiped determined by the vectors (1, 5, 6, 1, 5), (1, 0, 1, 0, 2), and (2, 3, −1, −1, −2) in $\mathbf{R}^5$?

Solution

(a) Let

$$A = \begin{bmatrix} 1 & 1 \\ 1 & -1 \\ 1 & 2 \\ 1 & 4 \end{bmatrix}$$

then

$$|A^TA| = \left| \begin{bmatrix} 1 & 1 & 1 & 1 \\ 1 & -1 & 2 & 4 \end{bmatrix} \begin{bmatrix} 1 & 1 \\ 1 & -1 \\ 1 & 2 \\ 1 & 4 \end{bmatrix} \right| = \begin{vmatrix} 4 & 6 \\ 6 & 22 \end{vmatrix} = 88 - 36 = 52$$

So area $= \sqrt{52}$.

(b)

```
        A
 1   1   2
 5   0   3
 6   1  ¯1
 1   0  ¯1
 5   2  ¯2
      (DET (⍉A)+.×A)*÷2
61.22907806
```
∎

Because of the divisions involved, the two methods of evaluating determinants discussed so far are often inconvenient for hand calculation with small matrices. The method of "expansion by minors" is often useful for small matrices or matrices of a special form.

The method is easily derived from the next proposition.

PROPOSITION 7.9 Let the n-by-n matrix $A = [c_1|c_2|\ldots|c_n]$, where $c_i = A[;i]$ is a vector in $\mathbf{R}^n$. Then

$$\det([c_1|c_2|\ldots|c_h + c_h'|\ldots|c_n]) = \det[c_1|c_2|\ldots|c_h|\ldots|c_n] + \det[c_1|c_2|\ldots|c_h'|\ldots|c_n]$$

A similar statement is true if A is partitioned by rows.

Proof We will prove the corresponding statement about rows, since row reduction is more familiar than column reduction. The statement about columns will then follow by taking transposes (Proposition 7.4).

Interchanging rows h and 1 of A will change only the signs of the three determinants involved, so we may as well assume $h = 1$. Then the statement to be proved reads

$$\det\begin{bmatrix} r_1 + r_1' \\ \hline r_2 \\ \hline \vdots \\ \hline r_n \end{bmatrix} = \det\begin{bmatrix} r_1 \\ \hline r_2 \\ \hline \vdots \\ \hline r_n \end{bmatrix} + \det\begin{bmatrix} r_1' \\ \hline r_2 \\ \hline \vdots \\ \hline r_n \end{bmatrix}$$

First notice that if the vectors $\{r_2, r_3, \ldots, r_n\}$ are linearly dependent, then all three matrices are singular and all three determinants are zero. Thus we may assume that $r_2, \ldots, r_n$ are linearly independent.

Next suppose that r_1 is in the space spanned by $r_2, r_3, \ldots, r_n$. Then the first determinant on the right is zero. On the other hand, by multiplying rows r_i with $i \geq 2$ by constants and adding to the first row, we may row-reduce $[r_1 + r_1'|r_2| \cdots |r_n]^T$ to $[r_1'|r_2| \cdots |r_n]^T$ without changing the determinant. Thus the formula holds in this case also.

Thus we may assume that r_1 is not in the subspace spanned by $r_2, \ldots, r_n$. It follows that $r_1, r_2, \ldots, r_n$ are a basis of $\mathbf{R}^n$ (why?) and hence r_1' is a linear combination of $r_1, r_2, \ldots, r_n$. Thus $r_1' = \lambda_1 r_1 + \lambda_2 r_2 + \cdots + \lambda_n r_n$. From this it follows that

$$\begin{bmatrix} r_1 + r_1' \\ \hline r_2 \\ \hline \vdots \\ \hline r_n \end{bmatrix} = \begin{bmatrix} r_1 + \lambda_1 r_1 + \lambda_2 r_2 + \cdots + \lambda_n r_n \\ \hline r_2 \\ \hline \vdots \\ \hline r_n \end{bmatrix}$$

row-reduces to

$$\begin{bmatrix} r_1 + \lambda_1 r_1 \\ \hline r_2 \\ \hline \vdots \\ \hline r_n \end{bmatrix}$$

and

$$\left[\begin{array}{c} r_1' \\ \hline r_2 \\ \hline \vdots \\ \hline r_n \end{array}\right] = \left[\begin{array}{c} \lambda_1 r_1 + \lambda_2 r_2 + \cdots + \lambda_n r_n \\ \hline r_2 \\ \hline \vdots \\ \hline r_n \end{array}\right]$$

row-reduces to

$$\left[\begin{array}{c} \lambda_1 r_1 \\ \hline r_2 \\ \hline \vdots \\ \hline r_n \end{array}\right]$$

Thus

$$\det \left[\begin{array}{c} r_1 + r_1' \\ \hline r_2 \\ \hline \vdots \\ \hline r_n \end{array}\right] = \det \left[\begin{array}{c} r_1 + \lambda_1 r_1 \\ \hline r_2 \\ \hline \vdots \\ \hline r_n \end{array}\right] = (1 + \lambda_1) \det \left[\begin{array}{c} r_1 \\ \hline r_2 \\ \hline \vdots \\ \hline r_n \end{array}\right]$$

$$= \det \left[\begin{array}{c} r_1 \\ \hline r_2 \\ \hline \vdots \\ \hline r_n \end{array}\right] + \lambda_1 \det \left[\begin{array}{c} r_1 \\ \hline r_2 \\ \hline \vdots \\ \hline r_n \end{array}\right] = \det \left[\begin{array}{c} r_1 \\ \hline r_2 \\ \hline \vdots \\ \hline r_n \end{array}\right] + \det \left[\begin{array}{c} r_1' \\ \hline r_2 \\ \hline \vdots \\ \hline r_n \end{array}\right] \qquad \blacksquare$$

EXAMPLE 7.5 Suppose that

$$\begin{vmatrix} a_1 & b_1 & c_1 \\ a_2 & b_2 & c_2 \\ a_3 & b_3 & c_3 \end{vmatrix} = 5, \qquad \begin{vmatrix} a_1 & b_1' & c_1 \\ a_2 & 0 & c_2 \\ a_3 & b_3' & c_3 \end{vmatrix} = -2$$

Evaluate

$$\begin{vmatrix} a_1 & b_1 + 3b_1' & c_1 \\ a_2 & b_2 & c_2 \\ a_3 & b_3 + 3b_3' & c_3 \end{vmatrix}$$

Solution By Proposition 7.9

$$\begin{vmatrix} a_1 & b_1 + 3b_1' & c_1 \\ a_2 & b_2 & c_2 \\ a_3 & b_3 + 3b_3' & c_3 \end{vmatrix} = \begin{vmatrix} a_1 & b_1 & c_1 \\ a_2 & b_2 & c_2 \\ a_3 & b_3 & c_3 \end{vmatrix} + \begin{vmatrix} a_1 & 3b_1' & c_1 \\ a_2 & 0 & c_2 \\ a_3 & 3b_2' & c_3 \end{vmatrix}$$

$$= 5 + 3(-2) = -1 \qquad \blacksquare$$

EXPANSION BY MINORS

Let us apply the last proposition to the evaluation of a 3-by-3 determinant.

$$\begin{vmatrix} a_{11} & a_{12} & a_{13} \\ a_{21} & a_{22} & a_{23} \\ a_{31} & a_{32} & a_{33} \end{vmatrix} = \begin{vmatrix} a_{11} & a_{12} & a_{13} \\ 0 & a_{22} & a_{23} \\ 0 & a_{32} & a_{33} \end{vmatrix} + \begin{vmatrix} 0 & a_{12} & a_{13} \\ a_{21} & a_{22} & a_{23} \\ a_{31} & a_{32} & a_{33} \end{vmatrix}$$

$$= \left|\begin{array}{c|cc} a_{11} & a_{12} & a_{13} \\ \hline 0 & a_{22} & a_{23} \\ 0 & a_{32} & a_{33} \end{array}\right| + \begin{vmatrix} 0 & a_{12} & a_{13} \\ a_{21} & a_{22} & a_{23} \\ 0 & a_{32} & a_{33} \end{vmatrix} + \begin{vmatrix} 0 & a_{12} & a_{13} \\ 0 & a_{22} & a_{23} \\ a_{31} & a_{32} & a_{33} \end{vmatrix}$$

We can apply Proposition 7.3 to the first determinant, and we can switch rows on the other two to bring them into the same form as the first.

$$\begin{vmatrix} a_{11} & a_{12} & a_{13} \\ a_{21} & a_{22} & a_{23} \\ a_{31} & a_{32} & a_{33} \end{vmatrix} = a_{11} \begin{vmatrix} a_{22} & a_{23} \\ a_{32} & a_{33} \end{vmatrix} - \left|\begin{array}{c|cc} a_{21} & a_{22} & a_{23} \\ \hline 0 & a_{12} & a_{13} \\ 0 & a_{32} & a_{33} \end{array}\right| - \left|\begin{array}{c|cc} a_{31} & a_{32} & a_{33} \\ \hline 0 & a_{22} & a_{23} \\ 0 & a_{12} & a_{13} \end{array}\right|$$

$$= a_{11} \begin{vmatrix} a_{22} & a_{23} \\ a_{32} & a_{33} \end{vmatrix} - a_{21} \begin{vmatrix} a_{12} & a_{13} \\ a_{32} & a_{33} \end{vmatrix} + a_{33} \begin{vmatrix} a_{12} & a_{13} \\ a_{22} & a_{23} \end{vmatrix}$$

Notice that in the last term one more row interchange was introduced to put the truncated rows back into their original order. The 2-by-2 determinants are now easily evaluated. The calculation above is an example of expansion by minors on the first column. Any row or column may be used. If we had wanted to use the second column rather than the first, we could have started with

$$\begin{vmatrix} a_{11} & a_{12} & a_{13} \\ a_{21} & a_{22} & a_{23} \\ a_{31} & a_{32} & a_{33} \end{vmatrix} = - \begin{vmatrix} a_{12} & a_{11} & a_{13} \\ a_{22} & a_{21} & a_{23} \\ a_{32} & a_{31} & a_{33} \end{vmatrix}$$

and then proceeded in the same way. The process can be abbreviated as follows.

Let A be a matrix. The ijth *minor* of A, written A_{ij}, is obtained by deleting the ith row and jth column from A, computing the resulting determinant, and then multiplying by $(-1)^{i+j}$.

That is, cross out the ith row and jth column of A and multiply the resulting determinant by the sign assigned to the $i;j$ entry according to the checkerboard pattern

$$\begin{bmatrix} + & - & + & - & \\ - & + & - & + & \cdots \\ + & - & + & - & \\ & & \vdots & & \end{bmatrix}$$

If

$$A = \begin{bmatrix} a_{11} & a_{12} & a_{13} \\ a_{21} & a_{22} & a_{23} \\ a_{31} & a_{32} & a_{33} \end{bmatrix}$$

then the sign pattern is

$$\begin{bmatrix} + & - & + \\ - & + & - \\ + & - & + \end{bmatrix}$$

and

$$A_{11} = \begin{vmatrix} a_{22} & a_{23} \\ a_{32} & a_{33} \end{vmatrix}, \quad A_{12} = -\begin{vmatrix} a_{21} & a_{23} \\ a_{31} & a_{33} \end{vmatrix}, \quad A_{13} = \begin{vmatrix} a_{21} & a_{22} \\ a_{31} & a_{32} \end{vmatrix}$$

and so on. The expansion by minors on the first row of A that was worked out above can now be written

$$\det(A) = a_{11}A_{11} + a_{21}A_{21} + a_{31}A_{31}$$

One should not attempt to remember such formulas, but simply work by crossing out rows and columns and using the sign pattern.

EXAMPLE 7.6 Evaluate

$$\begin{vmatrix} -3 & 8 & -3 & 4 \\ -3 & 8 & -3 & 3 \\ -6 & 12 & -4 & 6 \\ -2 & 2 & 0 & 3 \end{vmatrix}$$

Solution There is a zero in the 4;3 place, so expanding on the fourth row or third column will reduce the problem to one of evaluating three determinants of size 3-by-3. The sign pattern for a 4-by-4 matrix is

$$\begin{bmatrix} + & - & + & - \\ - & + & - & + \\ + & - & + & - \\ - & + & - & + \end{bmatrix}$$

$$\begin{vmatrix} -3 & 8 & -3 & 4 \\ -3 & 8 & -3 & 3 \\ -6 & 12 & -4 & 6 \\ -2 & 2 & 0 & 3 \end{vmatrix} = 2\begin{vmatrix} 8 & -3 & 4 \\ 8 & -3 & 3 \\ 12 & -4 & 6 \end{vmatrix} + 2\begin{vmatrix} -3 & -3 & 4 \\ -3 & -3 & 3 \\ -6 & -4 & 6 \end{vmatrix} + 3\begin{vmatrix} -3 & 8 & -3 \\ -3 & 8 & -3 \\ -6 & 12 & -4 \end{vmatrix}$$

$$= 8\begin{vmatrix} 2 & -3 & 4 \\ 2 & -3 & 3 \\ 3 & -4 & 6 \end{vmatrix} - 6\begin{vmatrix} 1 & -3 & 4 \\ 1 & -3 & 3 \\ 2 & -4 & 6 \end{vmatrix} - 36\begin{vmatrix} 1 & 2 & -3 \\ 1 & 2 & -3 \\ 2 & 3 & -4 \end{vmatrix}$$

Expanding the first 3-by-3 determinant along the bottom row gives

$$\begin{vmatrix} 2 & -3 & 4 \\ 2 & -3 & 3 \\ 3 & -4 & 6 \end{vmatrix} = 3 \begin{vmatrix} -3 & 4 \\ -3 & 3 \end{vmatrix} + 4 \begin{vmatrix} 2 & 4 \\ 2 & 3 \end{vmatrix} + 6 \begin{vmatrix} 2 & -3 \\ 2 & -3 \end{vmatrix}$$

$$= -9 \begin{vmatrix} 1 & 4 \\ 1 & 3 \end{vmatrix} + 8 \begin{vmatrix} 1 & 4 \\ 1 & 3 \end{vmatrix} + 6 \cdot 0$$

$$= 1$$

Multiplying the first column of the second 3-by-3 determinant by 3 and adding to the second gives

$$\begin{vmatrix} 1 & -3 & 4 \\ 1 & -3 & 3 \\ 2 & -4 & 6 \end{vmatrix} = \begin{vmatrix} 1 & 0 & 2 \\ 1 & 0 & 3 \\ 2 & 2 & 6 \end{vmatrix}$$

$$= -0 \begin{vmatrix} 1 & 3 \\ 2 & 6 \end{vmatrix} + 0 \begin{vmatrix} 1 & 4 \\ 2 & 6 \end{vmatrix} - 2 \begin{vmatrix} 1 & 4 \\ 1 & 3 \end{vmatrix}$$

$$= 2$$

Multiplying the first row of the third 3-by-3 determinant by -1 and adding to the second row gives

$$\begin{vmatrix} 1 & 2 & -3 \\ 1 & 2 & -3 \\ 2 & 3 & 4 \end{vmatrix} = \begin{vmatrix} 1 & 2 & -3 \\ 0 & 0 & 0 \\ 2 & 3 & 4 \end{vmatrix} = 0$$

Thus

$$\begin{vmatrix} -3 & 8 & -3 & 4 \\ -3 & 8 & -3 & 3 \\ -6 & 12 & -4 & 6 \\ -2 & 2 & 0 & 3 \end{vmatrix} = 8(1) - 6(2) = -4 \qquad \blacksquare$$

Two results that can be useful in theoretical discussions are not hard to derive using the expansion-by-minors formulas. We will have no use for them but list them for the sake of completeness. They should not be used for computational purposes.

PROPOSITION 7.10 (*Cramer's rule*) Let A be an invertible n-by-n matrix and B a vector in $\mathbf{R}^n$. Define A_i to be A with $A[;i]$ replaced by B. The unique solution of

$$AX = B$$

is given by $x_i = \det(A_i)/\det(A)$. $\blacksquare$

PROPOSITION 7.11 (*The adjoint formula*) Let A be n by n. Define $\bar{A}$ by $\bar{A}[i;j] = A_{ji}$, the $j;i$ cofactor of A. Then

$$\bar{A}A = \det(A)I$$

where $I = ID\ n$. Thus if A is invertible, $A^{-1} = (\det A)^{-1}\bar{A}$. ∎

EXERCISES 7.1

Evaluate the determinants of the matrices in exercises 1 through 27 by any method.

1. $\begin{bmatrix} -1 & -3 & -2 \\ 0 & 2 & 0 \\ 4 & 4 & 5 \end{bmatrix}$

2. $\begin{bmatrix} 1 & 1 & 1 \\ 2 & 1 & -2 \\ -2 & 1 & 4 \end{bmatrix}$

3. $\begin{bmatrix} 1 & 1 & 1 \\ 0 & 2 & -1 \\ 0 & 0 & 3 \end{bmatrix}$

4. $\begin{bmatrix} -2 & -1 & 2 \\ 4 & 3 & -2 \\ -6 & -2 & 5 \end{bmatrix}$

5. $\begin{bmatrix} 3 & -1 & -2 \\ -6 & 8 & 10 \\ 4 & -4 & -5 \end{bmatrix}$

6. $\begin{bmatrix} 12 & -3 & -7 \\ 13 & -2 & -9 \\ 9 & -3 & -4 \end{bmatrix}$

7. $\begin{bmatrix} 2 & 0 & 0 \\ 3 & -1 & -2 \\ -4 & 4 & 5 \end{bmatrix}$

8. $\begin{bmatrix} -9 & -5 & 1 & 3 \\ 14 & 11 & -4 & -2 \\ 4 & 6 & -3 & 2 \\ -10 & -3 & -1 & 6 \end{bmatrix}$

9. $\begin{bmatrix} 4 & 3 & -5 & -3 \\ -3 & 2 & 3 & 1 \\ 3 & 3 & -4 & -3 \\ -4 & 0 & 4 & 3 \end{bmatrix}$

10. $\begin{bmatrix} -1 & 0 & -2 & -2 \\ 4 & 3 & 2 & 2 \\ -3 & 0 & -1 & -3 \\ 3 & 0 & 2 & 4 \end{bmatrix}$

11. $\begin{bmatrix} -4 & 0 & 3 & -3 \\ 1 & 1 & -1 & 1 \\ -7 & 0 & 6 & -3 \\ -1 & 0 & 1 & 2 \end{bmatrix}$

12. $\begin{bmatrix} -5 & -2 & -2 & -2 \\ 10 & 5 & 4 & 2 \\ 5 & 1 & 3 & 1 \\ -3 & -1 & -2 & 2 \end{bmatrix}$

13. $\begin{bmatrix} 5 & 7 & 3 & 3 \\ 2 & 5 & 0 & 2 \\ -2 & -6 & -1 & -2 \\ -6 & -11 & -3 & -4 \end{bmatrix}$

14. $\begin{bmatrix} 0 & 0 & 2 & -1 \\ 0 & 3 & 0 & 0 \\ 0 & 0 & -1 & 0 \\ 2 & 0 & -2 & 3 \end{bmatrix}$

15. $\begin{bmatrix} -3 & -4 & -2 & 2 \\ -1 & 2 & 0 & 1 \\ 0 & -2 & 1 & 0 \\ -6 & -4 & -2 & 5 \end{bmatrix}$

16. $\begin{bmatrix} 9 & 6 & 2 & 0 & -2 \\ -16 & -11 & -4 & 0 & 4 \\ 2 & 2 & 3 & -2 & 0 \\ -4 & -2 & 2 & -3 & 2 \\ -6 & -4 & 0 & -2 & 3 \end{bmatrix}$

17. $\begin{bmatrix} 3 & 4 & 2 & 2 & 2 \\ 0 & 1 & 0 & 0 & 0 \\ -2 & -4 & -1 & -2 & -2 \\ 0 & -2 & 0 & -1 & 0 \\ -2 & -2 & -2 & 0 & -1 \end{bmatrix}$

18. $\begin{bmatrix} 3 & 2 & 2 & -4 & -2 \\ 8 & -7 & -4 & 0 & 0 \\ -8 & 16 & 11 & -8 & -4 \\ 4 & 4 & 4 & -7 & -4 \\ -4 & 4 & 2 & 0 & 1 \end{bmatrix}$

19. $\begin{bmatrix} 1 & 0 & 0 & 0 & 0 \\ 2 & 11 & -4 & -4 & -4 \\ 2 & 12 & -3 & -4 & -6 \\ 2 & 6 & -4 & -3 & 0 \\ 2 & 12 & -4 & -4 & -5 \end{bmatrix}$

20. $\begin{bmatrix} 27 & -12 & -8 & -4 & 2 \\ 40 & -19 & -12 & -4 & 4 \\ 18 & -6 & -5 & -6 & 0 \\ 26 & -12 & -8 & -3 & 2 \\ 0 & 0 & 0 & 0 & 1 \end{bmatrix}$

21. $\begin{bmatrix} -1 & -12 & 8 & -4 & -2 \\ 0 & 9 & -4 & 4 & 0 \\ 0 & 4 & -1 & 2 & 0 \\ 0 & -16 & 8 & -7 & 0 \\ 0 & 0 & 0 & 0 & 1 \end{bmatrix}$

22. $\begin{bmatrix} 3 & 0 & 0 & 0 & -2 \\ 4 & -1 & 0 & 0 & -2 \\ -2 & 2 & 1 & 0 & 0 \\ 4 & 0 & 0 & 1 & -4 \\ 4 & 0 & 0 & 0 & -3 \end{bmatrix}$

23. $\begin{bmatrix} 31 & 18 & -14 & -8 & -8 \\ -12 & -7 & 6 & 4 & 2 \\ 48 & 28 & -21 & -12 & -10 \\ -18 & -10 & 8 & 5 & 4 \\ 36 & 20 & -16 & -8 & -7 \end{bmatrix}$

24. $\begin{bmatrix} -4 & 11 & 22 & -15 & -5 & 9 & -3 \\ -2 & -3 & -10 & 6 & 2 & -2 & 4 \\ -4 & 4 & 7 & -4 & 0 & 4 & 0 \\ -14 & 2 & -2 & 1 & 2 & 6 & 8 \\ 8 & 4 & 14 & -8 & -3 & 0 & -8 \\ -9 & 9 & 18 & -13 & -5 & 10 & 1 \\ -5 & 13 & 26 & -17 & -5 & 11 & -4 \end{bmatrix}$

25. $\begin{bmatrix} 41 & -20 & -4 & -4 & 14 & -6 & 4 \\ 117 & -57 & -15 & -12 & 39 & -18 & 10 \\ 48 & -24 & -5 & -4 & 16 & -8 & 4 \\ -57 & 28 & 11 & 7 & -17 & 10 & -4 \\ 9 & -4 & -3 & 0 & 4 & -2 & 0 \\ -105 & 52 & 15 & 12 & -33 & 17 & -8 \\ -33 & 16 & 7 & 4 & -11 & 6 & -3 \end{bmatrix}$

26. $\begin{bmatrix} 6 & -24 & 10 & 4 & -6 & -10 & -9 \\ 8 & -37 & 16 & 6 & -10 & -16 & -14 \\ 2 & 0 & -1 & -2 & -2 & 0 & 2 \\ -2 & 8 & -4 & 1 & 2 & 4 & 2 \\ 1 & 4 & -2 & -4 & 1 & 2 & 3 \\ -10 & 44 & -20 & -4 & 12 & 19 & 16 \\ -6 & 36 & -16 & -8 & 8 & 16 & 15 \end{bmatrix}$

27. $\begin{bmatrix} 24 & -16 & 6 & -8 & -5 & -1 & -2 \\ 28 & -17 & 6 & -8 & -8 & -4 & -2 \\ -17 & 12 & -3 & 8 & 1 & 1 & 2 \\ 5 & -4 & 4 & -1 & -1 & 3 & 0 \\ 9 & -8 & 0 & -4 & 0 & -1 & -2 \\ 13 & -8 & 6 & -4 & -3 & 2 & 0 \\ -19 & 8 & -8 & 0 & 11 & 3 & -1 \end{bmatrix}$

Let $p_1 = (0, 0)$, $p_2 = (1,1)$, $p_3 = (3, 1)$, $p_4 = (2, 2)$, $p_5 = (1, 3)$, $p_6 = (-1, 2)$. In exercises 28 through 35 a polygon in $\mathbf{R}^2$ is given by specifying the order of the vertices that are taken from the set of points p_1 through p_6. Sketch the polygon and compute its (signed) area.

28. $p_1p_2p_3p_4p_5p_6$ 29. $p_5p_4p_3p_2p_1p_6$ 30. $p_1p_3p_2p_4p_5p_6$

31. $p_1p_3p_4p_2p_5p_6p_2$
32. $p_1p_3p_4p_2p_6p_5p_2$
33. $p_1p_3p_5p_6p_1p_2p_6$
34. $p_1p_2p_4p_2p_3$
35. $p_1p_2p_4p_2p_5p_6p_2$

36. Show that if π is a polygon in the plane and the coordinates of the vertices of π are all integers, then the area of π is either an integer or an integer plus $\frac{1}{2}$.

37. Let R be the matrix of reflection in an h-dimensional subspace of $\mathbf{R}^n$. What is $\det(R)$?

Hint: Modify the proof of Proposition 7.6.

38. (a) Show that the linear transformation with matrix $\begin{bmatrix} a & 0 \\ 0 & b \end{bmatrix}$ carries the unit circle $x^2 + y^2 = 1$ to the ellipse $(x^2/a^2) + (y^2/b^2) = 1$.

Hint: Parametrize the circle as

$$\sigma(t) = \begin{bmatrix} \cos t \\ \sin t \end{bmatrix}$$

apply the mapping and eliminate the parameter t.

(b) Derive a formula for the area of an ellipse with semi-axes a and b.

Hint: Proposition 7.7.

In exercises 39 through 45 vectors defining parallelopipeds in $\mathbf{R}^n$ are given. Use Proposition 7.8 to compute the (unsigned) volume of the given parallelopiped.

39. $(1, 1, 1), (1, 0, 1)$
40. $(1, 0, 1), (-1, 1, 1)$
41. $(-2, 0, 2), (1, 0, 1)$
42. $(1, 1, -1, 0), (1, 0, 1, 0)$
43. $(1, 1, 0, 0), (0, 1, 0, 1), (1, 0, 1, 0)$
44. $(1, 0, 1, 0), (0, 1,.0, 1), (1, 1, 1, 1)$
45. $(1, 1, 1, 1, 1), (2, 1, 3, -2, -3), (-1, 1, 0, 0, 0), (1, 0, -1, 0, 1)$

46. Write an APL function called *VOL* that takes a matrix whose columns define a parallelopiped in $\mathbf{R}^n$ and returns the (unsigned) volume of the parallelopiped.

47. Show that a matrix with two identical columns or rows has a zero determinant.

48. Show that a matrix of integers with determinant equal to ± 1 has an inverse matrix whose entries are integers.

49. Show that if A has zeros below the main diagonal, then the determinant is the product of the diagonal elements.

Hint: Expand by minors down the first column and use induction.

50. Show that an orthogonal matrix has determinant equal to ± 1.

Hint: Consider $\det(A^T A)$.

51. (a) Let A be symmetric. Show that $\det(A)$ is the product of the eigenvalues of A.

Hint: Apply Proposition 5.9 and exercise 49.

(b) For an arbitrary square matrix A, show that $|\det(A)|$ is the product of the singular values of A.

52. Write an APL function:

Name: *PLANIMETER*
Input: Points $p_1, p_2, \ldots, p_n, p_1$ in $\mathbf{R}^2$ stored as the columns of a matrix P
Output: The area of the polygon $p_1p_2p_3 \cdots p_n$

7.2 Eigenvalues and Eigenvectors

In Section 5.2 the problem of diagonalizing a symmetric matrix was solved. In this section we approach the problem for a general matrix A. The solution will be less complete because not every matrix can be diagonalized.

We begin by considering a typical diagonal matrix.

$$D = \begin{bmatrix} d_1 & & & \\ & d_2 & 0 & \\ & & \ddots & \\ & 0 & & d_n \end{bmatrix}$$

This matrix defines a linear transformation $Y = f(X) = DX$ from $\mathbf{R}^n$ to $\mathbf{R}^n$. In terms of coordinates $x_1, x_2, \ldots, x_n$ we have

$$f(x_1, x_2, \ldots, x_n) = (d_1x_1, d_2x_2, d_3x_3, \ldots, d_nx_n)$$

Thus the ith coordinate of X is stretched by the factor d_i. This means that for any vector v lying along the X_i axis

$$f(v) = Dv = d_iv$$

This gives us the approach to diagonalization we shall use. Given a matrix A, we look for vectors v and scalars λ such that

$$Av = \lambda v$$

If we can find a basis of $\mathbf{R}^n$ consisting of such v's, say $v_1, \ldots, v_n$, then we can define a coordinate change $X = PX'$ so that v_i points along the x_i' axis. In X' coordinates the function $f(X) = AX$ will have the form

$$f(x_1', x_2', \ldots, x_n') = (d_1x_1', d_2x_2', \ldots, d_nx_n')$$

and so

$$D = P^{-1}AP$$

will be diagonal.

The problem then is to find both the v's and the λ's. That is, we want all solutions of

$$AX = \lambda X$$

where not only X but λ also is unknown. This is no longer a system of linear equations, the λ makes it nonlinear, so our previously developed techniques do not apply.

An approach is available that is quite valuable for theoretical purposes, though it does not give a reliable computational method. Rewrite the equation above as

$$AX - \lambda X = 0$$

or

$$(A - \lambda I)X = 0$$

where $I = ID\ n$. The one obvious solution to the equation, λ arbitrary and $X = 0$, does us no good whatever. The zero vector cannot be used to define an axis of a coordinate system. Thus we are interested only in solutions with $X \neq 0$. But this means that $A - \lambda I$ must be singular. Thus we wish to find the values of λ for which $A - \lambda I$ is singular. Once such a λ is found, say $\lambda = d$, then the corresponding X's are just the null space of $A - dI$, and we have at least two ways (*ECHELON* and *PERP*) of computing null spaces. To find the λ's for which $A - \lambda I$ is singular we could now reduce the matrix $A - \lambda I$ to find the condition on λ for an infinity of solutions of $(A - \lambda I)X = 0$. This is a tedious procedure, and *ECHELON* cannot be used because of the unknown parameter λ.

Alternately one can use the fact that $A - \lambda I$ is singular if and only if $\det(A - \lambda I) = 0$ and try to find the roots of the nonlinear equation:

$$p(\lambda) = \det(A - \lambda I) = 0$$

This last approach is very useful for theoretical arguments, although it is impractical as a computational method.

EXAMPLE 7.7 Diagonalize the matrix

$$A = \begin{bmatrix} 0 & 3 & -2 \\ -2 & 5 & -2 \\ -2 & 2 & 1 \end{bmatrix}$$

Solution First we compute the function $p(\lambda) = \det(A - \lambda I)$. Notice that $A - \lambda I$ is just A with λ subtracted from each diagonal entry. Expanding by minors down the first column, we have

$$p(\lambda) = \begin{vmatrix} -\lambda & 3 & -2 \\ -2 & 5-\lambda & -2 \\ -2 & 2 & 1-\lambda \end{vmatrix} = -\begin{vmatrix} \lambda & 3 & -2 \\ 2 & 5-\lambda & -2 \\ 2 & 2 & 1-\lambda \end{vmatrix}$$

$$= -\left\{\lambda \begin{vmatrix} 5-\lambda & -2 \\ 2 & 1-\lambda \end{vmatrix} - 2\begin{vmatrix} 3 & -2 \\ 2 & 1-\lambda \end{vmatrix} + 2\begin{vmatrix} 3 & -2 \\ 5-\lambda & -2 \end{vmatrix}\right\}$$

$$= -\{\lambda(\lambda^2 - 6\lambda + 9) - 2(7 - 3\lambda) + 4(2 - \lambda)\}$$

$$= -(\lambda^3 - 6\lambda^2 + 11\lambda - 6)$$

Thus we need to solve the equation

$$-p(\lambda) = q(\lambda) = \lambda^3 - 6\lambda^2 + 11\lambda - 6 = 0$$

This example has been chosen with integer roots. The possible roots are the divisors of 6 — that is, ± 1, ± 2, ± 3, ± 6.

Since $q(1) = 1 - 6 + 11 - 6 = 0$, $\lambda - 1$ is a factor. Long division shows that

$$\begin{aligned} q(\lambda) &= (\lambda - 1)\frac{\lambda^3 - 6\lambda^2 + 11\lambda - 6}{\lambda - 1} \\ &= (\lambda - 1)(\lambda^2 - 5\lambda + 6) \\ &= (\lambda - 1)(\lambda - 2)(\lambda - 3) \end{aligned}$$

Thus the possible values of λ are $\lambda = 1, 2, 3$.

Now we must calculate the null spaces of $A - I$, $A - 2I$, $A - 3I$.

```
        ECHELON A-ID 3
1.000   0.000  ¯1.000
0.000   1.000  ¯1.000
0.000   0.000   0.000
```

so the null space is $t\begin{bmatrix}1\\1\\1\end{bmatrix}$

```
        ECHELON A-2×ID 3
1.000   0.000  ¯0.500
0.000   1.000  ¯1.000
0.000   0.000   0.000
```

so the null space is $t\begin{bmatrix}\frac{1}{2}\\1\\1\end{bmatrix}$

```
        ECHELON A-3×ID 3
1.00E0        ¯1.00E0       0.00E0
3.47E¯18      ¯3.47E¯18     1.00E0
0.00E0         0.00E0       0.00E0
```

so the null space is $t\begin{bmatrix}1\\1\\0\end{bmatrix}$

If we take any matrix of the form

$$P = \begin{bmatrix} \alpha & \frac{\beta}{2} & \gamma \\ \alpha & \beta & \gamma \\ \alpha & \beta & 0 \end{bmatrix}$$

with $\alpha\beta\gamma \neq 0$, then $X = PX'$ defines a new coordinate system in which the matrix of the linear transformation $y = AX$ becomes

$$y' = P^{-1}APX' = \begin{bmatrix} 1 & 0 & 0 \\ 0 & 2 & 0 \\ 0 & 0 & 3 \end{bmatrix} X' \quad ■$$

As the above calculation makes clear, $p(\lambda) = \det(A - \lambda I)$ is a polynomial in λ.

DEFINITION 7.3

1. Let A be an n-by-n matrix. The *characteristic polynomial* of A is $p(\lambda) = \det(A - \lambda I)$.
2. The *eigenvalues* of A are the roots of the characteristic polynomial.
3. Let λ be an eigenvalue of A. The *eigenspace* belonging to λ is the null space of $A - \lambda I$. It is the set of vectors v such that $Av = \lambda v$.
4. Let λ be an eigenvalue of A. An *eigenvector* belonging to λ is any *nonzero* element of the eigenspace of λ.

The zero vector is not considered to be an eigenvector, since it is of no use in defining a new coordinate system.

PROPOSITION 7.12 Let A be an n-by-n matrix. The characteristic polynomial of A is of the form

$$p(\lambda) = (-1)^n\lambda^n + c_{n-1}\lambda^{n-1} + c_{n-2}\lambda^{n-2} + \cdots + c_1\lambda + c_0$$

Further $c_0 = \det(A)$.

Proof Expanding $\det(A - \lambda I)$ by minors down the first column, we have

$$\begin{vmatrix} a_{11}-\lambda & a_{12} & a_{13} & \cdots \\ a_{21} & a_{22}-\lambda & a_{23} & \cdots \\ a_{31} & a_{32} & a_{33}-\lambda & \cdots \\ \vdots & \vdots & \vdots & \end{vmatrix}$$

$$= (a_{11}-\lambda)\begin{vmatrix} a_{22}-\lambda & a_{23} & \cdots \\ a_{32} & a_{33}-\lambda & \cdots \\ \vdots & \vdots & \end{vmatrix} - a_{21}\begin{vmatrix} a_{12} & a_{13} & \cdots \\ a_{22} & a_{33}-\lambda & \cdots \\ \vdots & \vdots & \end{vmatrix} + \cdots$$

The first term is $(a_{11} - \lambda)$ times $\det(B - \lambda I)$, where $B = 1\ 1 \downarrow A$. Thus if we assume by induction that an $(n-1)$-by-$(n-1)$ matrix has characteristic polynomial beginning $(-1)^{n-1}\lambda^{n-1}$, then the highest-degree term in this expansion is $(-1)^n\lambda^n$.

Further, $c_0 = p(0) = \det(A - 0 \cdot I) = \det(A)$. ∎

Note: It may be shown that the coefficient of the $(n-k)$-degree term is, except for sign, the sum of the determinants $\det(A[V; V])$ as the vector V takes on the values $m + \iota k$, $m = 0, 1, \ldots, n-k+1$.

Proposition 7.12 shows that an n-by-n matrix has *at most n* eigenvalues.

PROPOSITION 7.13 Let A be a matrix and let $v_1, \ldots, v_k$ be eigenvectors belonging to the eigenvalues $\lambda_1, \lambda_2, \ldots, \lambda_k$, respectively. Then if the λ_i are all different, the set of vectors $\{v_1, v_2, \ldots, v_k\}$ is linearly independent.

Proof Suppose that the set is dependent. Then there is an $l < k$ such that $v_1, \ldots, v_l$ are independent with $v_{l+1} = \alpha_1 v_1 + \alpha_2 v_2 + \cdots + \alpha_l v_l$. Multiplying through the equation by A, we get

$$\lambda_{l+1} v_{l+1} = \alpha_1 \lambda_1 v_1 + \alpha_2 \lambda_2 v_2 + \cdots + \alpha_l \lambda_l v_l$$

Multiplying the original equation by λ_{l+1} gives

$$\lambda_{l+1} v_{l+1} = \lambda_{l+1}(\alpha_1 v_1 + \alpha_2 v_2 + \cdots + \alpha_l v_l)$$

Subtracting these two equations gives

$$0 = [v_1 | v_2 | \ldots | v_l] \begin{bmatrix} \alpha_1(\lambda_1 - \lambda_{l+1}) \\ \alpha_2(\lambda_2 - \lambda_{l+1}) \\ \vdots \\ \alpha_l(\lambda_l - \lambda_{l+1}) \end{bmatrix}$$

Since $v_1, \ldots, v_l$ are independent, we have

$$\alpha_1(\lambda_1 - \lambda_{l+1}) = \alpha_2(\lambda_2 - \lambda_{l+1}) = \cdots = \alpha_l(\lambda_l - \lambda_{l+1}) = 0$$

Since the eigenvalues are all distinct, it follows that $\alpha_i = 0, i = 1, 2, \ldots, l$ and hence $v_{l+1} = 0$. But then v_{l+1} is *not* an eigenvector. ■

Proposition 7.13 gives us our simplest test for diagonalizability.

PROPOSITION 7.14 Let A be an n-by-n matrix. If A has n distinct eigenvalues, then there is a coordinate change $X = PX'$ such that $P^{-1}AP$ is diagonal.

Proof Let $\lambda_1, \lambda_2, \ldots, \lambda_n$ be the eigenvalues. Since $A - \lambda_i I$ is singular, the dimension of the null space of $A - \lambda_i I$ is at least 1, and so there is a $v_i \neq 0$ such that $Av_i = \lambda_i v_i$. Let $P = [v_1 | v_2 | \ldots | v_n]$. ■

If the matrix A does not have n distinct eigenvalues, then it may not be diagonalizable.

EXAMPLE 7.8 Is

$$A = \begin{bmatrix} 1 & 1 \\ 0 & 1 \end{bmatrix}$$

diagonalizable?

Solution The characteristic polynomial is

$$p(x) = \begin{vmatrix} 1-\lambda & 1 \\ 0 & 1-\lambda \end{vmatrix} = (\lambda - 1)^2$$

So the only eigenvalue is $\lambda = 1$. Since

$$A - I = \begin{bmatrix} 0 & 1 \\ 0 & 0 \end{bmatrix}$$

the eigenspace belonging to $\lambda = 1$ is

$$t\begin{bmatrix} 1 \\ 0 \end{bmatrix}$$

We may choose $v_1 = (1, 0)$ to define the x_1' axis, but we cannot get a second independent eigenvector to define the x_2' axis. ■

DEFINITION 7.4 Let $p(\lambda)$ be the characteristic polynomial of A. The eigenvalue λ_i has *multiplicity* k if $(\lambda - \lambda_i)^k$ divides $p(\lambda)$ but $(\lambda - \lambda_i)^{k+1}$ does not divide $p(\lambda)$.

In Example 7.8 above the eigenvalue $\lambda = 1$ has multiplicity equal to 2. The example immediately generalizes.

PROPOSITION 7.15 Let λ be an eigenvalue of A. If the dimension of the null space of $A - \lambda I$ is less than the multiplicity of λ, then A is not diagonalizable. ■

EXAMPLE 7.9 Is the matrix

$$A = \begin{bmatrix} 3 & 2 & 1 \\ 1 & 3 & 1 \\ -3 & -4 & -1 \end{bmatrix}$$

diagonalizable?

Solution

$$\begin{vmatrix} 3-\lambda & 2 & 1 \\ 1 & 3-\lambda & 1 \\ -3 & -4 & -1-\lambda \end{vmatrix}$$

$$\begin{aligned} &= (3-\lambda)\begin{vmatrix} 3-\lambda & 1 \\ -4 & -1-\lambda \end{vmatrix} - \begin{vmatrix} 2 & 1 \\ -4 & -1-\lambda \end{vmatrix} + (-3)\begin{vmatrix} 2 & 1 \\ 3-\lambda & 1 \end{vmatrix} \\ &= (3-\lambda)(\lambda^2 - 2\lambda + 1) - (2 - 2\lambda) - 3(\lambda - 1) \\ &= (3-\lambda)(\lambda - 1)^2 - (\lambda - 1) = (\lambda - 1)[(3-\lambda)(\lambda - 1) - 1] \\ &= -(\lambda - 1)(\lambda - 2)^2 \end{aligned}$$

Thus the eigenvalues are $\lambda = 1, 2, 2$. The corresponding null spaces are given by

```
      ECHELON A-ID 3
1.00E0        0.00E0        0.00E0
1.73E¯18      1.00E0        5.00E¯1
3.47E¯18      0.00E0        0.00E0

      ECHELON A-2×ID 3
1.00E0        0.00E0        1.00E0
1.73E¯18      1.00E0        1.73E¯18
2.60E¯18      0.00E0        2.60E¯18
```

The null space of $A - 2I$ is one-dimensional and the multiplicity of the eigenvalue $\lambda = 2$ is two. Therefore the matrix is not diagonalizable. ∎

Computing the polynomial $p(\lambda) = \det(A - \lambda I)$ can be tedious. Here is a way to compute it by machine. First we use the definition $p(\lambda) = \det(A - \lambda I)$ to compute some specific values of $p(\lambda)$. The function *POLYAT* takes A as a left argument and a vector $X = (x_1, x_2, \ldots, x_n)$ as a right argument. It returns the vector $(p(x_1), p(x_2), \ldots, p(x_n))$.

```
    ∇ Z←A POLYAT X
[1]     Z←⍳0
[2]     →(0=⍴X)/0
[3]     Z←(DET A-(1↑X)×ID 1↓⍴A),A POLYAT 1↓X
    ∇
```

The function is defined by induction on the number of components of X.

Using this function, we can find a number of points (x_i, y_i) on the graph of $p(\lambda)$. If A is n by n and we have more than n points on the graph, we can then use domino to calculate the coefficients of $p(\lambda)$, because the least-squares polynomial of degree n through the points can only be $p(\lambda)$.

Example 7.10 Is the matrix

$$A = \begin{bmatrix} -10 & -9 & -3 & -6 \\ 6 & 5 & 3 & 3 \\ 6 & 6 & 2 & 3 \\ 6 & 6 & 0 & 5 \end{bmatrix}$$

diagonalizable? (It is known to have integral eigenvalues.)

Solution A polynomial of degree 4 is determined by its values at five points. Of course we can use more than five if we wish; it can't hurt.

```
      Y←A POLYAT X←⍳7
      +C←Y⌹X∘.*4 3 2 1 0
1  ¯2  ¯3  4  4
```

So $p(\lambda) = \lambda^4 - 2\lambda^3 - 3\lambda^2 + 4\lambda + 4$. The roots of a polynomial divide the constant term, so we should try $\lambda = \pm 1, \pm 2, \pm 4$.

$$p(-1) = 1 + 2 - 3 - 4 + 4 = 0$$
$$p(2) = 16 - 16 - 12 + 8 + 4 = 0$$

and these are the only two roots among the possibilities. Long division produces

$$p(\lambda) = (\lambda + 1)(\lambda^3 - 3\lambda^2 + 4) = (\lambda + 1)(\lambda - 2)(\lambda^2 - \lambda - 2)$$

But $\lambda^2 - \lambda - 2 = (\lambda + 1)(\lambda - 2)$, so $p(\lambda)$ factors completely as

$$p(\lambda) = (\lambda + 1)^2(\lambda - 2)^2$$

The eigenvalues are thus $\lambda = -1, -1, 2, 2$. Both have a multiplicity of 2.

```
      ECHELON A-¯1×ID 4
 1.00E0      1.00E0      3.25E¯19    1.00E0
¯3.47E¯18   ¯3.47E¯18    1.00E0     ¯1.00E0
 0.00E0      0.00E0      0.00E0      6.94E¯18
 0.00E0      0.00E0      0.00E0      6.94E¯18
      ECHELON A-2×ID 4
 1.00E0     ¯1.73E¯18    1.00E0      5.00E¯1
 0.00E0      1.00E0     ¯1.00E0      0.00E0
 0.00E0      0.00E0      0.00E0      0.00E0
 0.00E0      0.00E0      0.00E0      0.00E0
```

So the eigenspace belonging to $\lambda = -1$ is the column space of

$$\begin{bmatrix} -1 & -1 \\ 1 & 0 \\ 0 & 1 \\ 0 & 1 \end{bmatrix}$$

and the eigenspace belonging to $\lambda = 2$ is the column space of

$$\begin{bmatrix} -1 & -\frac{1}{2} \\ 1 & 0 \\ 1 & 0 \\ 0 & 1 \end{bmatrix} \text{ or } \begin{bmatrix} -1 & -1 \\ 1 & 0 \\ 1 & 0 \\ 0 & 2 \end{bmatrix}$$

Thus putting

$$P = \begin{bmatrix} -1 & -1 & -1 & -1 \\ 1 & 0 & 1 & 0 \\ 0 & 1 & 1 & 0 \\ 0 & 1 & 0 & 2 \end{bmatrix}$$

we have

$$P^{-1}AP = \begin{bmatrix} -1 & 0 & 0 & 0 \\ 0 & -1 & 0 & 0 \\ 0 & 0 & 2 & 0 \\ 0 & 0 & 0 & 2 \end{bmatrix}$$

The matrix A is diagonalizable. ■

If the eigenspace of λ has too low a dimension, then A will not be diagonalizable. There is another way in which an n-by-n matrix A may fail to be diagonalizable. The characteristic polynomial may fail to have n roots. There is a way out of the latter problem, however. The fundamental theorem of algebra states that a polynomial always factors completely into linear factors if complex numbers are allowed as roots. This means that a matrix may become diagonalizable if we allow our scalars to be complex numbers as well as real numbers. This is a very useful device for some applications, particularly for the theory of linear differential equations.

DEFINITION 7.5 The *eigenvalues* of a matrix A are all the roots, real and complex, of the characteristic polynomial.

The definitions of eigenspace and eigenvector remain unchanged, with the understanding that vectors and matrices may have complex as well as real entries. The space of n-tuples of complex numbers is denoted by $\mathbf{C}^n$.

EXAMPLE 7.11 Is the matrix $R_{\pi/2}$ diagonalizable?

Solution A rotation matrix R_θ will not be diagonalizable ($\theta \neq k\pi$) if we restrict our attention to real scalars, because the only vector v parallel to $R_\theta v$ is the vector $v = 0$. Using complex scalars, however, we have, for $\theta = \pi/2$,

$$p(\lambda) = |R_{\pi/2} - \lambda I| = \begin{bmatrix} -\lambda & -1 \\ 1 & -\lambda \end{bmatrix}$$

$$= \lambda^2 + 1 = (\lambda - i)(\lambda + i)$$

where $i = \sqrt{-1}$. Thus the eigenvalues are $\lambda = \pm i$.

The corresponding eigenspaces can be found by row reduction.

$\lambda = i$: We row reduce the matrix

$$A_1 = \begin{bmatrix} -i & -1 \\ 1 & -i \end{bmatrix}$$

to find its null space. Interchanging the two rows, we have

$$\begin{bmatrix} 0 & 1 \\ 1 & 0 \end{bmatrix}\begin{bmatrix} -i & -1 \\ 1 & -i \end{bmatrix} = \begin{bmatrix} 1 & -i \\ -i & -1 \end{bmatrix}$$

Pivoting on the 1;1 entry gives

$$\begin{bmatrix} 1 & 0 \\ i & 1 \end{bmatrix}\begin{bmatrix} 1 & -i \\ -i & -1 \end{bmatrix} = \begin{bmatrix} 1 & -i \\ 0 & -(i^2+1) \end{bmatrix} = \begin{bmatrix} 1 & -i \\ 0 & -((-1)+1) \end{bmatrix} = \begin{bmatrix} 1 & -i \\ 0 & 0 \end{bmatrix}$$

Thus the null space consists of all vectors of the form $t[i \;\; 1]^T$ where t is now any *complex* number.

$\lambda = -i$: We row reduce

$$A_2 = \begin{bmatrix} i & -1 \\ 1 & i \end{bmatrix}$$

to find the eigenspace belonging to $-i$.

$$\begin{bmatrix} 0 & 1 \\ 1 & 0 \end{bmatrix}\begin{bmatrix} i & -1 \\ 1 & i \end{bmatrix} = \begin{bmatrix} 1 & i \\ i & -1 \end{bmatrix}$$

$$\begin{bmatrix} 1 & 0 \\ -i & 1 \end{bmatrix}\begin{bmatrix} 1 & i \\ i & -1 \end{bmatrix} = \begin{bmatrix} 1 & i \\ 0 & -i^2-1 \end{bmatrix} = \begin{bmatrix} 1 & i \\ 0 & -((-1)-1) \end{bmatrix} = \begin{bmatrix} 1 & i \\ 0 & 0 \end{bmatrix}$$

and the eigenspace consists of all vectors of the form $t[-i \;\; 1]$ where t is any complex scalar.

Notice that

$$\begin{bmatrix} 0 & -1 \\ 1 & 0 \end{bmatrix}\begin{bmatrix} i \\ 1 \end{bmatrix} = \begin{bmatrix} -1 \\ i \end{bmatrix} = i\begin{bmatrix} i \\ 1 \end{bmatrix}$$

$$\begin{bmatrix} 0 & -1 \\ 1 & 0 \end{bmatrix}\begin{bmatrix} -i \\ 1 \end{bmatrix} = \begin{bmatrix} -1 \\ -i \end{bmatrix} = -i\begin{bmatrix} -i \\ 1 \end{bmatrix}$$

So $(i, 1)$ and $(-i, 1)$ are eigenvectors belonging to i and $-i$, respectively. Taking

$$p = \begin{bmatrix} i & -i \\ 1 & 1 \end{bmatrix}$$

and using the coordinate change $X = PX'$ in $\mathbf{C}^2$, we find

$$P^{-1}R_{\pi/2}P = \frac{1}{2i}\begin{bmatrix} 1 & i \\ -1 & i \end{bmatrix}\begin{bmatrix} 0 & -1 \\ 1 & 0 \end{bmatrix}\begin{bmatrix} i & -i \\ 1 & 1 \end{bmatrix} = \frac{1}{2i}\begin{bmatrix} 1 & i \\ -1 & i \end{bmatrix}\begin{bmatrix} -1 & -1 \\ i & -i \end{bmatrix}$$

$$= \frac{1}{2i}\begin{bmatrix} -1+i^2 & -1-i^2 \\ 1+i^2 & 1-i^2 \end{bmatrix} = \frac{1}{2i}\begin{bmatrix} -1+(-1) & -1-(-1) \\ 1+(-1) & 1-(-1) \end{bmatrix}$$

$$= \frac{1}{2i}\begin{bmatrix} -2 & 0 \\ 0 & 2 \end{bmatrix} = \begin{bmatrix} -\frac{1}{i} & 0 \\ 0 & \frac{1}{i} \end{bmatrix}$$

$$= \begin{bmatrix} i & 0 \\ 0 & -i \end{bmatrix}$$

Thus $R_{\pi/2}$ is diagonalizable when complex scalars are allowed. ∎

As the computations in the above example imply, many of the techniques developed in previous chapters for vectors and matrices of real numbers hold for complex numbers as well. The techniques that remain unchanged include row reduction and the related theory of flats and subspaces, including dimension, intersections, and so on. Modifications are required for techniques involving distances, angles, and any formula using transposes.

PROPOSITION 7.16 Allowing complex scalars, the matrix A is diagonalizable if and only if the dimension of the eigenspace of λ is equal to the multiplicity of λ. In particular, A is diagonalizable if all the eigenvalues have multiplicity equal to 1. ∎

A method of handling matrices with complex entries with an APL processor that does not recognize complex numbers is sketched in exercise 21.

JORDAN BLOCKS

The matrix of Example 7.8 is typical of those which are not diagonalizable. The basic nondiagonalizable pattern is

$$J_\lambda = \begin{bmatrix} \lambda & 1 & 0 & \dots & 0 \\ 0 & \lambda & 1 & \dots & 0 \\ \cdot & \cdot & \cdot & \dots & \cdot \\ & & & & 1 \\ 0 & 0 & 0 & \dots & \lambda \end{bmatrix} \tag{7.1}$$

The matrix J_λ has λ's on the diagonal and 1's on the first superdiagonal. It is not difficult to see (exercise 24) that λ is the only eigenvalue of J_λ and that the corresponding eigenspace is one-dimensional.

DEFINITION 7.6 The matrix of equation (7.1) is a *Jordan block* matrix.

The matrix of Example 7.8 is a 2-by-2 Jordan block with $\lambda = 1$. A Jordan block may be regarded as a "generalized eigenvalue" with an ordinary eigenvalue being simply a 1-by-1 Jordan block. It may be shown that there is a unique set of Jordan blocks associated to each square matrix and that, if the definition of "diagonal matrix" is loosened up to include matrices with Jordan blocks on the diagonal and zeros elsewhere, then every square matrix is "diagonalizable." For the sake of completeness we record the following proposition without proof.

PROPOSITION 7.17 Allowing complex scalars, given any square matrix A there is an invertible matrix P such that

$$P^{-1}AP = \begin{bmatrix} J_{\lambda_1} & & 0 \\ & \ddots & \\ 0 & & J_{\lambda_k} \end{bmatrix}$$

where $J_{\lambda_1}, \ldots, J_{\lambda_k}$ are Jordan blocks. ∎

EXERCISES 7.2

In exercises 1 through 10 decide if the given matrix is diagonalizable. (The characteristic polynomials have been designed to factor.)

1. $\begin{bmatrix} -1 & -2 \\ 0 & 1 \end{bmatrix}$

2. $\begin{bmatrix} 1 & 1 \\ -1 & 3 \end{bmatrix}$

3. $\begin{bmatrix} -2 & 2 \\ -5 & 4 \end{bmatrix}$

4. $\begin{bmatrix} 1 & 3 & -2 \\ 0 & 2 & 0 \\ 0 & 3 & -1 \end{bmatrix}$

5. $\begin{bmatrix} 0 & -1 & 0 \\ 2 & 3 & 0 \\ -2 & -2 & 1 \end{bmatrix}$

6. $\begin{bmatrix} 4 & 2 & 1 \\ -3 & -1 & -1 \\ -1 & -1 & 1 \end{bmatrix}$

7. $\begin{bmatrix} 3 & -2 & -1 \\ 4 & -3 & -2 \\ -2 & 2 & 1 \end{bmatrix}$

8. $\begin{bmatrix} 1 & 0 & 0 & 0 \\ 0 & -3 & 2 & 0 \\ 0 & -4 & 3 & 0 \\ -2 & -2 & 2 & -1 \end{bmatrix}$

9. $\begin{bmatrix} 1 & 4 & -3 & -1 \\ -1 & -3 & 3 & 0 \\ -2 & -4 & 4 & -1 \\ 3 & -6 & 5 & 5 \end{bmatrix}$

10. $\begin{bmatrix} 2 & -2 & -2 & -2 \\ -3 & 5 & 5 & 4 \\ -1 & -3 & -1 & -2 \\ 5 & -2 & -4 & -2 \end{bmatrix}$

11. Show that the eigenvalues of $A + \alpha I$ are $\lambda_i + \alpha$, where the λ_i are the eigenvalues of A.

12. Show that the eigenvalues of αA are $\alpha\lambda_i$, where the λ_i are the eigenvalues of A.

13. Show that if λ_i is an eigenvalue of A with eigenvector v, then λ_i^2 is an eigenvalue of A^2 with eigenvalue v.

14. Show that if the λ_i is an eigenvalue of A with the eigenvector v, then $a_n\lambda_i^n + a_{n-1}\lambda_i^{n-1} + \cdots + a_0$ is an eigenvalue of $a_nA^n + a_{n-1}A^{n-1} + \cdots + a_0I$ with eigenvector v.

15. Show that A and A^T have the same eigenvalues.

16. Show that

(a) A is singular if and only if 0 is an eigenvalue of A.

(b) If A is nonsingular, then the eigenvalues of A^{-1} are $1/\lambda_i$, where the λ_i are the eigenvalues of A.

17. If A has zeros below the main diagonal, show that the eigenvalues of A are the diagonal entries.

Hint: Expand det $(A - \lambda I)$ by minors.

18. Given any polynomial $p(\lambda) = (-1)^n(a_0 + a_1\lambda + \cdots + a_{n-1}\lambda^{n-1} + \lambda^n)$, show that it is the characteristic polynomial of the matrix

$$\begin{bmatrix} -a_{n-1} & -a_{n-2} & \cdots & -a_1 & -a_0 \\ 1 & 0 & \cdots & 0 & 0 \\ 0 & 1 & \cdots & 0 & 0 \\ & & \cdots & & \\ & & \cdots & & \\ 0 & 0 & \cdots & 1 & 0 \end{bmatrix}$$

Hint: Expand by minors.

19. Let P_S be the perpendicular projection matrix for the subspace S.

(a) Show that every $v \neq 0$ in S is an eigenvector of P_S for the eigenvalue $\lambda = 1$.

(b) Show that every $v \neq 0$ in $S^\perp$ is an eigenvector of P_S for the eigenvalue $\lambda = 0$.

(c) Show that $\lambda = 0$, $\lambda = 1$ are the only eigenvalues of P_S.

Hint: Let $P_S v = \lambda v$ and write $v = v^\parallel + v^\perp$.

(d) Let the columns of U form an orthonormal basis of S and the columns of V an orthonormal basis of $S^\perp$. Let $Q = [U \mid V]$. Show $Q^{-1} = Q^T$ and use matrix algebra to compute $Q^{-1}P_S Q$.

Hint: $P_S = UU^T$.

(e) From (d) show that det $(P_S - \lambda I) = (-1)^n \lambda^{h-k}(\lambda - 1)^k$, where k is the dimension of S.

20. Let R_S be the reflection matrix for the subspace S.

(a) Show that every $v \neq 0$ in S is an eigenvector of R_S for the eigenvalue $\lambda = 1$.

(b) Show that every $v \neq 0$ in $S^\perp$ is an eigenvector of R_S for the eigenvalue $\lambda = -1$.

(c) Show that $\lambda = 1$ and $\lambda = -1$ are the only eigenvalues of R_S.

Hint: Let $R_S v = \lambda v$ and write $v = v^\parallel + v^\perp$.

(d) Let the columns of U form an orthonormal basis of S and the columns of V an orthonormal basis of $S^\perp$. Let $Q = [U \mid V]$. Show $Q^{-1} = Q^T$ and use matrix algebra to compute $Q^{-1}R_S Q$.

Hint: $R_S = 2UU^T - I$.

(e) From (d) show that det $(R_S - \lambda I) = (-1)^n(\lambda - 1)^k(\lambda + 1)^{n-k}$, where k is the dimension of S.

21. (Inverting complex matrices in APL) Let the complex number $Z = \alpha + \beta i$ correspond to the matrix $A = \begin{bmatrix} \alpha & -\beta \\ \beta & \alpha \end{bmatrix}$

(a) Show that if Z_1 corresponds to A_1 and Z_2 corresponds to A_2, then Z_1Z_2 corresponds to A_1A_2.

(b) Show that if Z_1 corresponds to A_1 and Z_2 corresponds to A_2, then $Z_1 \div Z_2$ corresponds to $A_1 ⌹ A_2$.

(c) Show that the complex conjugate $\bar{Z}$ corresponds to A^T.

Let the complex matrix

$$\mathcal{Z} = \begin{bmatrix} Z_{11} & Z_{12} & Z_{13} & \cdots \\ Z_{21} & Z_{22} & Z_{23} & \cdots \\ \cdot & \cdot & \cdot & \\ \cdot & \cdot & \cdot & \\ \cdot & \cdot & \cdot & \end{bmatrix}$$

correspond to the partitioned matrix

$$\mathcal{A} = \left[\begin{array}{c|c|c|c} A_{11} & A_{12} & A_{13} & \cdots \\ \hline A_{21} & A_{22} & A_{23} & \cdots \\ \hline \vdots & \vdots & \vdots & \end{array}\right]$$

(d) Show that if $\mathcal{Z}_1$ corresponds to $\mathcal{A}_1$ and $\mathcal{Z}_2$ corresponds to $\mathcal{A}_2$, then $\mathcal{Z}_1\mathcal{Z}_2$ corresponds to $\mathcal{A}_1\mathcal{A}_2$.

(e) Show that if $\mathcal{Z}_1$ corresponds to $\mathcal{A}_1$ and $\mathcal{Z}_2$ corresponds to $\mathcal{A}_2$, then $\mathcal{Z}_2^{-1}\mathcal{Z}_1$ corresponds to $\mathcal{A}_1 ⌹ \mathcal{A}_2$. In particular $⌹ \mathcal{A}_2$ corresponds to $\mathcal{Z}_2^{-1}$.

(f) Let $\mathcal{Z}^*$ denote the coordinatewise conjugate of the transpose of $\mathcal{Z}$. Show that $\mathcal{Z}^*$ corresponds to $\mathcal{A}^T$.

(Computer assignment) Use the result of exercise 21(e) to do exercises 22 through 25 using domino.

22. Solve

$$\begin{aligned} x + iy + (1 - i)z &= 2 \\ 2x - iy + (1 + i)z &= i \\ x - y + Z &= 2i \end{aligned}$$

23. Solve

$$\begin{aligned} x + y + z &= i \\ x - y - z &= 2i \\ y + 3z &= 1 + i \end{aligned}$$

24. Invert

$$\begin{bmatrix} 1 & i & -1 \\ -1 & 0 & i \\ 2i & 1 & 0 \end{bmatrix}$$

25. Invert

$$\begin{bmatrix} i & 1 & -1 & i \\ -i & 0 & 1 & 1 \\ i & 0 & i & -1 \\ -i & 1 & 1 & -i \end{bmatrix}$$

26. (a) Show that the n-by-n Jordan block J_α has characteristic polynomial $p(\lambda) = (-1)^n(\lambda - \alpha)^n$.

    Hint: Expand by minors.

    (b) Show that the eigenspace of J_α is one-dimensional.

    Hint: What is the echelon form of $J_\alpha - \alpha I$?

27. Let J_α be an n-by-n Jordan block and let e_i be the vector with ith component equal to 1 and all other components zero. Show

$$\begin{aligned} J_\alpha e_1 &= \alpha e_1 \\ J_\alpha e_k &= e_{k-1} + \alpha e_k, \qquad k > 1 \end{aligned}$$

28. Suppose that $P^{-1}AP = J_\alpha$, a Jordan block. Let $P = [v_1|v_2|\ldots|v_k]$. Show that the vectors $v_1, v_2, \ldots, v_n$ satisfy the recursive equations

$$\begin{aligned} (A - \alpha I)v_1 &= 0 \\ (A - \alpha I)v_k &= v_{k-1}, \qquad k > 1 \end{aligned}$$

Hint: Exercise 27.

29. The matrices A below have characteristic polynomials of the form $p(\lambda) = (-1)^n(\lambda - \alpha)^n$ and one-dimensional eigenspaces. Find α and use the equations of exercise 28 to find matrices P such that $P^{-1}AP = J_\alpha$.

(a) $\begin{bmatrix} 4 & -1 \\ 4 & 0 \end{bmatrix}$ (b) $\begin{bmatrix} 2 & 1 \\ -1 & 4 \end{bmatrix}$ (c) $\begin{bmatrix} 2 & 1 & 0 \\ -1 & 4 & 1 \\ 0 & 0 & 3 \end{bmatrix}$ (d) $\begin{bmatrix} 2 & -1 & 2 \\ -1 & 1 & 3 \\ -1 & 2 & 6 \end{bmatrix}$

7.3* Powers of Matrices Revisited

Powers of matrices were discussed in optional Section 3.6*, where difference equations and stochastic matrices were discussed. If A can be diagonalized, then there is another approach to analyzing the powers of A.

First notice that if D is a diagonal matrix, then

$$D^k = \begin{bmatrix} d_1 & & & \\ & d_2 & 0 & \\ & 0 & \ddots & \\ & & & d_n \end{bmatrix}^k = \begin{bmatrix} d_1^k & & & \\ & d_i^k & 0 & \\ & 0 & \ddots & \\ & & & d_n^k \end{bmatrix}$$

So we need not carry out a full-blown matrix multiplication; we just take the powers of the diagonal. Now suppose that A can be diagonalized

$$P^{-1}AP = D$$

Then $A = PDP^{-1}$, $A^2 = PDP^{-1}PDP^{-1} = PD^2P^{-1}, \ldots, A^k = PD^kP^{-1}$. This gives us a formula for the nth power of A.

EXAMPLE 7.12. In Example 3.23 the first twenty Fibionacci numbers were computed by recursion. Find a formula for the nth Fibionacci number.

Solution The Fibionacci numbers are defined by

$$x_1 = x_2 = 1, \qquad x_n = x_{n-1} + x_{n-2}$$

Following the prescription of Section 3.6*, we write this as

$$v_n = Av_{n-1}$$

where

$$v_n = \begin{bmatrix} x_n \\ x_{n-1} \end{bmatrix} \quad \text{and} \quad A = \begin{bmatrix} 1 & 1 \\ 1 & 0 \end{bmatrix}$$

Then

$$v_n = A^{n-2}\begin{bmatrix} 1 \\ 1 \end{bmatrix}$$

To compute A^{n-2} we diagonalize A. The characteristic polynomial is

$$p(\lambda) = \begin{vmatrix} 1-\lambda & 1 \\ 1 & -\lambda \end{vmatrix} = \lambda(\lambda - 1) - 1 = \lambda^2 - \lambda - 1$$

so the eigenvalues of A are

$$\lambda = \frac{1 \pm \sqrt{1 - 4(-1)}}{2} = \frac{1 \pm \sqrt{5}}{2}$$

To find the eigenvectors we compute the null space of $A - \lambda I$.

$$\begin{bmatrix} 1 - \dfrac{1+\sqrt{5}}{2} & 1 \\ 1 & -\dfrac{1+\sqrt{5}}{2} \end{bmatrix} \text{ row-reduces to } \begin{bmatrix} 1 & -\dfrac{1+\sqrt{5}}{2} \\ 0 & 0 \end{bmatrix}$$

and

$$\begin{bmatrix} 1 - \dfrac{1-\sqrt{5}}{2} & 1 \\ 1 & -\dfrac{1-\sqrt{5}}{2} \end{bmatrix} \text{ row-reduces to } \begin{bmatrix} 1 & -\dfrac{1-\sqrt{5}}{2} \\ 0 & 0 \end{bmatrix}$$

The corresponding eigenspaces are

$$t\begin{bmatrix} \dfrac{1+\sqrt{5}}{2} \\ 1 \end{bmatrix} \quad \text{and} \quad t\begin{bmatrix} \dfrac{1-\sqrt{5}}{2} \\ 1 \end{bmatrix}$$

Taking

$$P = \begin{bmatrix} \dfrac{1+\sqrt{5}}{2} & \dfrac{1-\sqrt{5}}{2} \\ 1 & 1 \end{bmatrix}$$

we have

$$P^{-1} = \frac{2}{\sqrt{5}}\begin{bmatrix} 1 & -\dfrac{1-\sqrt{5}}{2} \\ -1 & \dfrac{1+\sqrt{5}}{2} \end{bmatrix}$$

and

$$\begin{bmatrix} x_n \\ x_{n-1} \end{bmatrix} = P\begin{bmatrix} \dfrac{1+\sqrt{5}}{2} & 0 \\ 0 & \dfrac{1-\sqrt{5}}{2} \end{bmatrix}^{n-2} P^{-1}$$

$$= P\begin{bmatrix} \left(\dfrac{1+\sqrt{5}}{2}\right)^{n-2} & 0 \\ 0 & \left(\dfrac{1-\sqrt{5}}{2}\right)^{n-2} \end{bmatrix}\begin{bmatrix} 1 & -\dfrac{1-\sqrt{5}}{2} \\ -1 & \dfrac{1+\sqrt{5}}{2} \end{bmatrix}\begin{bmatrix} 1 \\ 1 \end{bmatrix}$$

$$= \frac{2}{\sqrt{5}}\begin{bmatrix} \left(\dfrac{1+\sqrt{5}}{2}\right)^{n} - \left(\dfrac{1-\sqrt{5}}{2}\right)^{n} \\ \left(\dfrac{1+\sqrt{5}}{2}\right)^{n-1} - \left(\dfrac{1-\sqrt{5}}{2}\right)^{n-1} \end{bmatrix}$$

or

$$x_n = \frac{2}{\sqrt{5}}\left(\left(\frac{1+\sqrt{5}}{2}\right)^{n} - \left(\frac{1-\sqrt{5}}{2}\right)^{n}\right) \quad ■$$

Notice that although the nth Fibionacci number is obviously an integer, irrational numbers appear in the formula for the nth term. In fact complex numbers may even be necessary to answer a question that seemingly involves only whole numbers. This is why complex scalars must be introduced.

EXAMPLE 7.13 Find a formula for x_n if

$$x_1 = 1, \quad x_2 = -1, \quad x_3 = 1, \quad x_n = x_{n-1} - x_{n-2} + x_{n-3}$$

Solution

$$\begin{bmatrix} x_n \\ x_{n-1} \\ x_{n-2} \end{bmatrix} = \begin{bmatrix} 1 & -1 & 1 \\ 1 & 0 & 0 \\ 0 & 1 & 0 \end{bmatrix} \begin{bmatrix} x_{n-1} \\ x_{n-2} \\ x_{n-3} \end{bmatrix}$$

$$\begin{aligned} p(\lambda) &= \begin{vmatrix} 1-\lambda & -1 & 1 \\ 1 & -\lambda & 0 \\ 0 & 1 & -\lambda \end{vmatrix} \\ &= (1-\lambda)\begin{vmatrix} -\lambda & 0 \\ 1 & -\lambda \end{vmatrix} - \begin{vmatrix} -1 & 1 \\ 1 & -\lambda \end{vmatrix} \\ &= -\lambda^3 + \lambda^2 - \lambda + 1 \\ &= -(\lambda - 1)(\lambda^2 + 1) \\ &= -(\lambda - 1)(\lambda - i)(\lambda + i) \end{aligned}$$

The eigenvalues $\lambda = 1, i, -i$ are distinct, so A is diagonalizable.

The eigenspaces are the null spaces of $A - \lambda I$ for $\lambda = 1, \pm i$.

$\lambda = 1$:

$$\begin{bmatrix} 0 & -1 & 1 \\ 1 & -1 & 0 \\ 0 & 1 & -1 \end{bmatrix} \text{ row-reduces to } \begin{bmatrix} 1 & 0 & -1 \\ 0 & 1 & -1 \\ 0 & 0 & 0 \end{bmatrix}$$

The eigenspace is

$$t\begin{bmatrix} 1 \\ 1 \\ 1 \end{bmatrix}$$

$\lambda = i$:

$$\begin{bmatrix} 1-i & -1 & 1 \\ 1 & -i & 0 \\ 0 & 1 & -i \end{bmatrix} \text{ row-reduces to } \begin{bmatrix} 1 & 0 & 1 \\ 0 & 1 & -i \\ 0 & 0 & 0 \end{bmatrix}$$

The eigenspace is

$$t\begin{bmatrix} -1 \\ i \\ 1 \end{bmatrix}$$

$\lambda = -i$:

$$\begin{bmatrix} 1+i & -1 & 1 \\ 1 & i & 0 \\ 0 & 1 & i \end{bmatrix} \text{ row-reduces to } \begin{bmatrix} 1 & 0 & 1 \\ 0 & 1 & i \\ 0 & 0 & 0 \end{bmatrix}$$

The eigenspace is

$$t\begin{bmatrix}-1\\-i\\1\end{bmatrix}$$

Let

$$P = \begin{bmatrix}1 & -1 & -1\\1 & i & -i\\1 & 1 & 1\end{bmatrix}$$

Row-reducing $[P\,|\,I]$ to $[I\,|\,P^{-1}]$ gives

$$P^{-1} = \frac{1}{2}\begin{bmatrix}1 & 0 & 1\\ \dfrac{i-1}{2} & -i & \dfrac{i+1}{2}\\ -\dfrac{i+1}{2} & i & \dfrac{1-i}{2}\end{bmatrix}$$

Since

$$v_n = \begin{bmatrix}x_n\\x_{n-1}\\x_{n-2}\end{bmatrix} = A^{n-3}\begin{bmatrix}1\\1\\1\end{bmatrix}$$

we have

$$\begin{bmatrix}x_n\\x_{n-1}\\x_{n-2}\end{bmatrix} = \frac{1}{2}\begin{bmatrix}1 & -1 & -1\\1 & i & -i\\1 & 1 & 1\end{bmatrix}\begin{bmatrix}1 & 0 & 0\\0 & i & 0\\0 & 0 & -i\end{bmatrix}^{n-3}\begin{bmatrix}1 & 0 & 1\\ \dfrac{i-1}{2} & -i & \dfrac{i+1}{2}\\ -\dfrac{i+1}{2} & i & \dfrac{1-i}{2}\end{bmatrix}\begin{bmatrix}1\\-1\\1\end{bmatrix}$$

$$= \frac{1}{2}\begin{bmatrix}1 & -1 & -1\\1 & i & -i\\1 & 1 & 1\end{bmatrix}\begin{bmatrix}1 & 0 & 0\\0 & i^{n-3} & 0\\0 & 0 & (-i)^{n-3}\end{bmatrix}\begin{bmatrix}2\\2i\\-2i\end{bmatrix}$$

$$= \begin{bmatrix}1 & -1 & -1\\1 & i & -i\\1 & 1 & 1\end{bmatrix}\begin{bmatrix}1\\i^{n-2}\\(-i)^{n-2}\end{bmatrix}$$

$$= \begin{bmatrix}1-(i^{n-2}+(-i)^{n-2})\\1+i^{n-1}+(-i)^{n-1}\\1+i^{n-2}+(-i)^{n-2}\end{bmatrix}$$

or

$$x_n = 1 - i^{n-2} - (-i)^{n-2} \qquad \blacksquare$$

In the last two examples it was not necessary to compute P^{-1}. If we think of the coordinate change given by $X = PX'$, we see that P^{-1} was used only to get the coordinates of the starting vector v_1 in the X' coordinate system. This can be done by solving

$$PX' = v_1$$

which is usually less work than inverting P when P is larger than 2 by 2. With this procedure one first obtains W, the solution of $PX' = v_1$, and then obtains an expression for v_n as

$$v_n = PD^m W$$

where $m = n - k$ for v_n in $\mathbf{R}^k$.

The second application of the powers of a matrix discussed in Section 3.6* involved stochastic matrices. In Section 3.6* a matrix was called stochastic if it had nonnegative entries and if the sum of the entries in each row was 1. We prefer to work with columns rather than rows below, so we will call a matrix *stochastic* if it has nonnegative entries and the sum of the entries in each *column* is one.

In Example 3.24 a stochastic matrix A was exhibited for which the rows of A^n all approached the same vector as n became large. Since we are working with columns in this section, we replace A by A^T and have a matrix for which the *columns* of A^n approach a fixed vector.

We are now in a position to analyze this phenomenon in much more detail. The basic fact is given in the next proposition.

PROPOSITION 7.18 If the entries of the columns of the square matrix A sum to 1, then $\lambda = 1$ is an eigenvalue of A.

Proof First notice that A and A^T have the same characteristic polynomial and hence the same eigenvalues. In fact, since $\det(B) = \det(B^T)$ by Proposition 7.4,

$$p(\lambda) = |A - \lambda I| = |(A - \lambda I)^T| = |A^T - \lambda I|$$

Now let $v = (1, 1, \ldots, 1)$. For any matrix B, Bv is $+/B$. Thus if the entries of the columns of A sum to 1, then

$$A^T v = (1, 1, \ldots, 1) = v$$

which shows that v is an eigenvector for A^T belonging to the eigenvalue $\lambda = 1$. It follows that A also has an eigenvalue $\lambda = 1$ (v is not, however, an eigenvector for A). ■

We will now investigate the existence of

$$\lim_{k \to \infty} A^k$$

by diagonalizing A. The interested reader may check that, if A is not diagonalizable, the same conclusions may be reached using Jordan blocks (see exercise 17).

Since $\lambda = 1$ is an eigenvalue of A, there is, assuming A diagonalizable, a matrix P with

$$P^{-1}AP = \Lambda = \begin{bmatrix} 1 & & & \\ & \lambda_2 & & 0 \\ & & \ddots & \\ & 0 & & \lambda_n \end{bmatrix}$$

Now A and Λ are representations of the same transformation of $\mathbf{R}^n$ in different coordinates (the coordinate change is $X = PX'$, which is continuous), hence $\lim_{k\to\infty} A^k$ exists if and only if $\lim_{k\to\infty} \Lambda^k$ exists. Using the formula

$$\Lambda^k = \begin{bmatrix} 1^k & & & \\ & \lambda_2^k & & 0 \\ & & \ddots & \\ & 0 & & \lambda_n^k \end{bmatrix}$$

we see that $\lim_{k\to\infty} \Lambda^k$ exists if and only if $\lim_{k\to\infty} \lambda_i^k$ exists for all $i = 2, 3, \ldots, n$. Now $|\lambda_i^k| = |\lambda_i|^k \to \infty$ if $|\lambda_i| > 1$, so if the limit $\lim_{k\to\infty} A^k$ exists, one must have $|\lambda_i| \leq 1$ for all i. Further, if $|\lambda_i| = 1$, then $\lim_{k\to\infty} \lambda_i$ exists only if $\lambda_i = 1$. For example, $\lim_{k\to\infty} (-1)^k$ does not exist. (We leave the general case as exercise 16.) We conclude, then, that $\lim_{k\to\infty} A^k$ exists if and only if $|\lambda_i| \leq 1$ for all i and $|\lambda_i| = 1$ only if $\lambda_i = 1$.

Now consider the case where $\lambda_1 = 1$ and $|\lambda_i| < 1$ for $i = 2, 3, \ldots, n$. Then

$$\lim_{k\to\infty} A^k = P(\lim_{k\to\infty} \Lambda^k)P^{-1} = P \left[\begin{array}{c|ccc} 1 & 0 & \ldots & 0 \\ \hline 0 & & & \\ \vdots & & 0 & \\ 0 & & & \end{array}\right] P^{-1}$$

$$= [P[;1]|0]\left[\frac{P^{-1}[1;]}{0}\right]$$

So the columns of A^k approach multiples of $P[;1]$ as k grows without bound.

Now one can show (exercise 15) that if the entries of the columns of A sum to 1, then the same is true of A^k for all k and hence, by continuity, the entries of the columns of $\lim_{k\to\infty} A^k$ sum to 1. In the case under consideration this means that *all* the columns of A^k approach that multiple of $P[;1]$ whose components sum to 1. Notice that $P[;1]$ can be chosen to be any eigenvector belonging to $\lambda_1 = 1$.

To summarize: If $\lambda_1 = 1$ and $|\lambda_i| < 1$ for $i > 1$ and, further, the entries of the columns of A sum to 1, then as k goes to infinity the columns of k approach that eigenvector belonging to $\lambda = 1$ whose components sum to 1.

It may be shown that a stochastic matrix with no entries equal to zero satisfies the condition $\lambda_1 = 1$ and $|\lambda_i| < 1$ for $i \geq 1$ (see Proposition 7.25). This gives the following basic result on stochastic matrices.

PROPOSITION 7.19. Let A be a stochastic matrix and suppose that there is an integer l such that A^l has only positive entries. Then as k goes to infinity, the columns of A^k all approach that eigenvector of $\lambda = 1$ whose components sum to 1. ■

Note: If v is any vector, then the multiple of v whose columns sum to 1 is $V \div + /V$.

EXAMPLE 7.14

$$A = \begin{bmatrix} \frac{1}{2} & \frac{1}{3} & \frac{1}{5} \\ 0 & \frac{1}{3} & \frac{1}{5} \\ \frac{1}{2} & \frac{1}{3} & \frac{3}{5} \end{bmatrix}$$

is the transpose of the matrix of Example 3.24, where it was discovered that A^{400} is fuzzily equal to A^{800}. The matrix A^2 has no zero cntries, so Proposition 7.19 applies.

The matrix $(A - I)$ row-reduces to

$$\begin{bmatrix} 1 & 0 & -\frac{3}{5} \\ 0 & 1 & -\frac{3}{10} \\ 0 & 0 & 0 \end{bmatrix}$$

Hence the null space of $A - I$ is

$$t\begin{bmatrix} \frac{3}{5} \\ \frac{3}{10} \\ 1 \end{bmatrix} \quad \text{or} \quad t\begin{bmatrix} 6 \\ 3 \\ 10 \end{bmatrix}$$

Thus the steady state approximated by the columns of A^{400} is $(\frac{6}{19}, \frac{3}{19}, \frac{10}{19})$. ■

EXAMPLE 7.15

$$A = \begin{bmatrix} 0 & 1 \\ 1 & 0 \end{bmatrix}$$

is clearly a stochastic matrix. The limit $\lim_{k\to\infty} A^k$ does not exist, however. In fact

$$A^k = \begin{cases} A & \text{if } k \text{ is odd} \\ I & \text{if } k \text{ is even} \end{cases}$$

The characteristic polynomial of A is

$$p(\lambda) = \begin{vmatrix} -\lambda & 1 \\ 1 & -\lambda \end{vmatrix} = \lambda^2 - 1$$

and hence the eigenvalues are $\lambda_1 = 1, \lambda_2 = -1$. Thus not every stochastic matrix satisfies the hypotheses of Proposition 7.18. ■

EXERCISES 7.3*

In exercises 1 through 6 a sequence of vectors of scalars is defined inductively. Find an expression for the nth term.

1. $x_1 = 1,\ x_2 = 1,\ x_n = -x_{n-2}$
2. $x_1 = x_2 = 1,\ x_n = 2x_{n-1} + x_{n-2}$
3. $x_1 = x_2 = 1,\ x_n = 3x_{n-2}$
4. $x_1 = x_2 = 1,\ x_n = 2x_{n-1}$
5. $x_1 = 1,\ x_2 = 0,\ x_3 = -1,\ x_n = 2x_{n-1} - x_{n-2} + 2x_{n-3}$
6. $x_1 = 0,\ x_2 = x_3 = -1,\ x_n = x_{n-1} - 2x_{n-3}$
7. Suppose that a linear difference equation is defined by

$$x_n = a_0x_{n-k} + a_1x_{n-k+1} + \cdots + a_{k-1}x_{n-1}$$

when $x_1, x_2, \ldots, x_k$ are given. Show that the associated eigenvalues are the roots of the polynomial

$$a_0 + a_1\lambda + \cdots + a_{k-1}\lambda^{k-1} + \lambda^k$$

Hint: Exercise 18 of Exercises 7.2.

In exercises 8 through 14 decide if the limit $\lim_{k\to\infty} A^k$ exists by diagonalizing A. If the limit exists, what is it? Is the matrix stochastic? Does the matrix satisfy the hypotheses of Proposition 7.19?

8. $\begin{bmatrix} .7 & .7 \\ .3 & .3 \end{bmatrix}$
9. $\begin{bmatrix} .8 & .5 \\ .2 & .5 \end{bmatrix}$
10. $\frac{1}{2}\begin{bmatrix} 1 & 0 & 1 \\ 1 & 1 & 0 \\ 0 & 1 & 1 \end{bmatrix}$
11. $\begin{bmatrix} 1 & \frac{1}{2} & \frac{1}{3} \\ 0 & \frac{1}{2} & \frac{1}{3} \\ 0 & 0 & \frac{1}{3} \end{bmatrix}$
12. $\begin{bmatrix} 2 & 3 \\ -\frac{1}{2} & -\frac{1}{2} \end{bmatrix}$
13. $\begin{bmatrix} .9996 & .0002 \\ -.0006 & 1.0003 \end{bmatrix}$
14. $\frac{1}{2}\begin{bmatrix} 2 & -1 & 1 \\ 0 & 2 & 0 \\ 0 & 1 & 1 \end{bmatrix}$

15. Let v be a vector whose components sum to 1(1 = +/v) and let A be a matrix each of whose columns sums to 1(1∧.=+⌿A)

    (a) Show that the components of the vector $w = Av$ sum to 1

    Hint: $w = v_1 A[;1] + v_2 A[;2] + \cdots + v_n A[;n]$.

    (b) If B is a second matrix with 1∧.=+⌿B, show that the same is true of AB. Hence, by induction, the same is true of A^k, $k \geq 0$.

    (c) Show that (a) and (b) above may be derived from these APL identities: (+⌿A+.×B) = (+⌿A)+.×B, A a matrix, B a vector or matrix, and (+⌿A) = 1+.×A, A a matrix.

16. Use the fact that any complex number z may be written

$$z = |z|(\cos\theta + i\sin\theta)$$

for suitable $\theta(= \mathrm{Arg}(z))$ and the fact that

$$z^k = |z|^k(\cos k\theta + i\sin k\theta)$$

to show that if $\lim_{k\to\infty} z^k$ exists, then $|z| < 1$ or $z = 1$.

17. Let J_λ be a Jordan block. Show that if $\lim_{k\to\infty} J_\lambda^k$ exists, then $|\lambda| < 1$.

7.4* Congruences and Affine Transformations in Space

A working knowledge of the congruences (isometries) of three-dimensional space is useful in such diverse applications as crystallography, control of artificial satellites, computer graphics, and theoretical physics. In this section we will use the theory of eigenvalues and eigenvectors to prove a fundamental result about congruences of three-dimensional space: every such congruence may be factored into a rotation, a reflection, and a translation (Proposition 7.21).

We also give a geometric description of an arbitrary, but nonsingular, affine transformation of three-dimensional space.

Recall (Section 5.1) that a congruence $T: \mathbf{R}^3 \to \mathbf{R}^3$ is defined to be a mapping that preserves distances: $d(p, q) = d(T(p), T(q))$. In Section 5.1 it was shown (Proposition 5.7) that any congruence of $\mathbf{R}^2$ can be factored as reflection in a line, a rotation, and a translation. We will reduce the present case to this one after a preliminary transformation. It is first necessary, however, to verify that an isometry of $\mathbf{R}^3$ is an affine function.

Proposition 7.20 An isometry $T: \mathbf{R}^3 \to \mathbf{R}^3$ is affine.

Proof First notice that T must carry lines to lines. This is because three points p, q, r are collinear, with q between p and r say, if and only if

$$d(p, r) = d(p, q) + d(q, r)$$

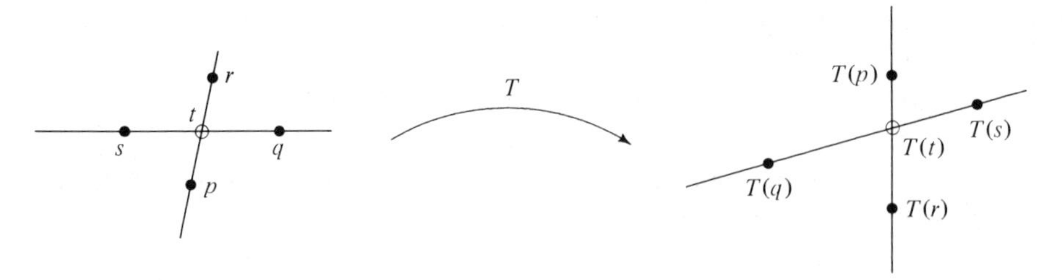

FIGURE 7.8

Since T is a congruence, we have

$$d(T(p), T(r)) = d(T(p), T(q)) + d(T(q), T(r))$$

and so $T(p)$, $T(q)$, $T(r)$ are collinear with $T(q)$ between $T(p)$ and $T(r)$.

Now we can use the fact that T takes lines to lines to show that it also takes planes to planes. The precise statement we need is that if the four points p, q, r, s are coplanar (but no three of them are collinear), then the four points $T(p)$, $T(q)$, $T(r)$, and $T(s)$ are also coplanar (Figure 7.8).

The four points may always be labeled so that the line through s and q cuts the line through p and r in a fifth point t (you should convince yourself of this).

Now since p, q, and r are not collinear, the above reasoning shows that $T(p)$, $T(q)$, and $T(r)$ are not collinear and hence determine a plane π that contains the line through $T(p)$ and $T(r)$. Hence π contains $T(t)$. It follows that π contains the line through $T(t)$ and $T(q)$ and hence π contains $T(s)$ as well.

In particular T must carry parallelograms to parallelograms, and so the proof of Proposition 5.7 applies to the current case as well. ∎

Now we know that our congruence $T: \mathbf{R}^3 \to \mathbf{R}^3$ can be written $T(X) = b + AX$ for some vector b and matrix A. Since adding b to AX is a translation, we will restrict our attention to the linear part of T, $L(X) = AX$, and show that A can be factored as a reflection and a rotation.

First notice that since the characteristic polynomial

$$p(\lambda) = \det (A - \lambda I) = -\lambda^3 + \cdots$$

is a cubic, it must have a real root. This is because

$$\lim_{\lambda \to +\infty} p(\lambda) = -\infty \quad \text{and} \quad \lim_{\lambda \to -\infty} p(\lambda) = +\infty$$

so the graph of $p(\lambda)$ must somewhere cross the axis.

If λ is a real eigenvalue of A and v is an eigenvector belonging to λ, then v and $Av = \lambda v$ must have the same length, so $\lambda = \pm 1$.

Let u_1 be a unit eigenvector belonging to a real eigenvalue of A. Let $\{u_2, u_3\}$ be an orthonormal basis of $S^\perp$, where S is the subspace generated by u_1. Then

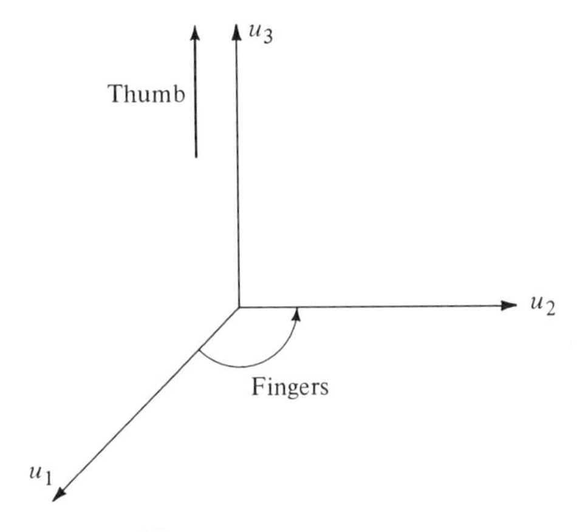

FIGURE 7.9

$U = [u_1 | u_2 | u_3]$ is orthogonal. By changing the direction of u_3 if necessary, we can assure that the new coordinate system given by $X = UX'$ is right-handed; that is, curling the fingers of the right hand from u_1 to u_2 causes the thumb to point in the direction of u_3 (Figure 7.9).

Isometries preserve angles, and linear transformations carry subspaces to subspaces. It follows that L carries $S^\perp$ to $S^\perp$, since it carries S to S. In particular, Au_2 and Au_3 are perpendicular to u_1. Thus in the new coordinate system we have $L(X') = A'X'$, where

$$A' = U^T A U = \begin{bmatrix} u_1^T \\ u_2^T \\ u_3^T \end{bmatrix} [Au_1 | Au_2 | Au_3]$$

$$= \begin{bmatrix} u_1^T \\ u_2^T \\ u_3^T \end{bmatrix} [\pm u_1 | Au_2 | Au_3]$$

$$= \left[\begin{array}{c|cc} \pm 1 & 0 & 0 \\ \hline 0 & \multicolumn{2}{c}{\multirow{2}{*}{B}} \\ 0 & & \end{array}\right]$$

Now $A' = U^T A U$ is orthogonal, so $B^T = B^{-1}$. Thus the 2-by-2 matrix B can be thought of as a congruence of $\mathbf{R}^2$ and by Proposition 5.7 it can be factored as a reflection and a rotation. More precisely, Example 5.6 shows that

$$\text{either } B = R_\theta \quad \text{or} \quad B = R_\theta \begin{bmatrix} 1 & 0 \\ 0 & -1 \end{bmatrix}$$

for some angle of rotation θ. Thus A' has one of the following forms:

(1)
$$A' = \left[\begin{array}{c|cc} 1 & 0 & 0 \\ \hline 0 & \multicolumn{2}{c}{\multirow{2}{*}{R_θ}} \\ 0 & & \end{array}\right]$$

This is a rotation through the angle θ about the x'-axis — that is, a rotation through the angle θ about the line $\ell(t) = tu_1$.

(2)
$$A' = \left[\begin{array}{c|cc} -1 & 0 & 0 \\ \hline 0 & \multicolumn{2}{c}{\multirow{2}{*}{R_θ}} \\ 0 & & \end{array}\right] = \begin{bmatrix} -1 & 0 & 0 \\ 0 & 1 & 0 \\ 0 & 0 & 1 \end{bmatrix} \left[\begin{array}{c|cc} 1 & 0 & 0 \\ \hline 0 & \multicolumn{2}{c}{\multirow{2}{*}{R_θ}} \\ 0 & & \end{array}\right]$$
$$= \left[\begin{array}{c|cc} 1 & 0 & 0 \\ \hline 0 & \multicolumn{2}{c}{\multirow{2}{*}{R_θ}} \\ 0 & & \end{array}\right] \begin{bmatrix} -1 & 0 & 0 \\ 0 & 1 & 0 \\ 0 & 0 & 1 \end{bmatrix}$$

This A' is rotation about the x' axis either followed or preceded by a reflection in the $y'z'$ plane — that is, rotation through θ about $\ell(t) = tu_1$ either followed or preceded by reflection in $S^\perp$.

In both cases above the sense of the rotation is determined by a right-hand rule. Point the thumb in the direction of u_1; then the fingers curl in the direction of positive θ.

At first glance it appears that there are two more cases to consider:

(3)
$$A' = \left[\begin{array}{c|c} 1 & 0 \quad 0 \\ \hline 0 & \multirow{2}{*}{$R_\theta \begin{bmatrix} 1 & 0 \\ 0 & -1 \end{bmatrix}$} \\ 0 & \end{array}\right]$$

(4)
$$A' = \left[\begin{array}{c|c} -1 & 0 \quad 0 \\ \hline 0 & \multirow{2}{*}{$R_\theta \begin{bmatrix} 1 & 0 \\ 0 & -1 \end{bmatrix}$} \\ 0 & \end{array}\right]$$

These fall under the two cases above, however. In fact by rotating axes in the plane of u_2 and u_3 the symmetric matrix

$$R_\theta \begin{bmatrix} 1 & 0 \\ 0 & -1 \end{bmatrix}$$

may be diagonalized (exercise 10). The eigenvalues are ± 1, so by properly choosing u_2, u_3 we have

$$A' = \begin{bmatrix} \pm 1 & 0 & 0 \\ 0 & 1 & 0 \\ 0 & 0 & -1 \end{bmatrix}$$

Taking $\lambda_1 = 1$, we have reflection in the $x'y'$ plane (case 2 with $\theta = 0$ and the axes interchanged), and taking $\lambda_1 = -1$ we have a $180°$ rotation about the y' axis (case 1 with $\theta = 180$ and axes interchanged).

Summarizing the information above gives the next proposition.

PROPOSITION 7.21 A congruence $T: \mathbf{R}^3 \to \mathbf{R}^3$ can be factored into at most three transformations: a reflection in a plane, a rotation (about an axis perpendicular the plane), and a translation. The rèflection is present only when the determinant of the linear part of T is -1. ■

EXAMPLE 7.16 Proposition 7.20 states that the reflection can be taken with respect to a plane perpendicular to the axis of rotation. How is this possible? Suppose that we reflect in a plane and then rotate about a line in that plane — to be specific, suppose we reflect in the xy plane and then rotate $45°$ about the x axis. How does the decomposition of Proposition 7.20 go in this case?

Solution Since

$$R_{\pi/4} = \frac{1}{\sqrt{2}}\begin{bmatrix} 1 & -1 \\ 1 & 1 \end{bmatrix}$$

we have

$$\begin{aligned} T(X) &= \frac{1}{\sqrt{2}}\begin{bmatrix} \sqrt{2} & 0 & 0 \\ 0 & 1 & -1 \\ 0 & 1 & 1 \end{bmatrix}\begin{bmatrix} 1 & 0 & 0 \\ 0 & 1 & 0 \\ 0 & 0 & -1 \end{bmatrix} X \\ &= \frac{1}{\sqrt{2}}\begin{bmatrix} \sqrt{2} & 0 & 0 \\ 0 & 1 & 1 \\ 0 & 1 & -1 \end{bmatrix} X = AX \end{aligned}$$

Notice that A falls under case 3 discussed above. The characteristic polynomial of A is

$$\begin{aligned} p(\lambda) &= \begin{vmatrix} 1-\lambda & 0 & 0 \\ 0 & \dfrac{1}{\sqrt{2}} - \lambda & \dfrac{1}{\sqrt{2}} \\ 0 & \dfrac{1}{1\sqrt{2}} & -\dfrac{1}{\sqrt{2}} - \lambda \end{vmatrix} \\ &= (1-\lambda)\begin{vmatrix} \dfrac{1}{\sqrt{2}} - \lambda & \dfrac{1}{\sqrt{2}} \\ \dfrac{1}{\sqrt{2}} & -\dfrac{1}{\sqrt{2}} - \lambda \end{vmatrix} \\ &= (1-\lambda)(\lambda^2 - 1) \\ &= -(\lambda - 1)^2(\lambda + 1) \end{aligned}$$

Thus the eigenvalues of A are 1, 1, -1. If we take $\lambda_1 = -1$, then the eigenspace is the null space of

$$A + I = \begin{bmatrix} 2 & 0 & 0 \\ 0 & 1 + \frac{1}{\sqrt{2}} & \frac{1}{\sqrt{2}} \\ 0 & \frac{1}{\sqrt{2}} & 1 - \frac{1}{\sqrt{2}} \end{bmatrix}$$

which has echelon form

$$\begin{bmatrix} 1 & 0 & 0 \\ 0 & 1 & \sqrt{2} - 1 \\ 0 & 0 & 0 \end{bmatrix}$$

Hence the eigenspace is $t(0, 1 - \sqrt{2}, 1)$ and a unit vector in this direction is obtained by taking $t = 1/\sqrt{4 - 2\sqrt{2}}$. To compute u_2 and u_3 we could apply the Gram-Schmidt process to u, $(1, 0, 0)$, $(0, 1, 0)$, and $(0, 0, 1)$ or simply observe that $u_2 = (1, 0, 0)$ and $u_3 = (0, -1, 1 - \sqrt{2}) \div \sqrt{4 - 2\sqrt{2}}$ will do. Then, setting $U = [u_1 | u_2 | u_3]$, we have

$$\begin{aligned} A' &= U^T A U \\ &= \frac{1}{\sqrt{2}} \frac{1}{4 - 2\sqrt{2}} \begin{bmatrix} 0 & 1 - \sqrt{2} & 1 \\ \sqrt{4 - 2\sqrt{2}} & 0 & 0 \\ 0 & -1 & 1 - \sqrt{2} \end{bmatrix} \begin{bmatrix} \sqrt{2} & 0 & 0 \\ 0 & 1 & 1 \\ 0 & 1 & -1 \end{bmatrix} \\ &\qquad \cdot \begin{bmatrix} 0 & \sqrt{4 - 2\sqrt{2}} & 0 \\ 1 - \sqrt{2} & 0 & -1 \\ 1 & 0 & 1 - \sqrt{2} \end{bmatrix} \\ &= \frac{1}{4\sqrt{2} - 4} \begin{bmatrix} 4 - 4\sqrt{2} & 0 & 0 \\ 0 & 4\sqrt{2} - 4 & 0 \\ 0 & 0 & 4\sqrt{2} - 4 \end{bmatrix} = \begin{bmatrix} -1 & 0 & 0 \\ 0 & 1 & 0 \\ 0 & 0 & 1 \end{bmatrix} \end{aligned}$$

Thus $T(X)$ is a reflection in the plane spanned by u_2 and u_3, and the rotation does not appear ($\theta = 0$). A sketch shows that T is a reflection in the plane containing the x axis and cutting the yz plane in a line inclined 22.5° to the y axis (Figure 7.10). ■

If $T_1(X) = b_1 + A_1 X$ and $T_2(X) = b_2 + A_2 X$ are two congruences, then the composition is given by

$$T_2 \circ T_1(X) = T_2(b_1) + (A_2 A_1) X$$

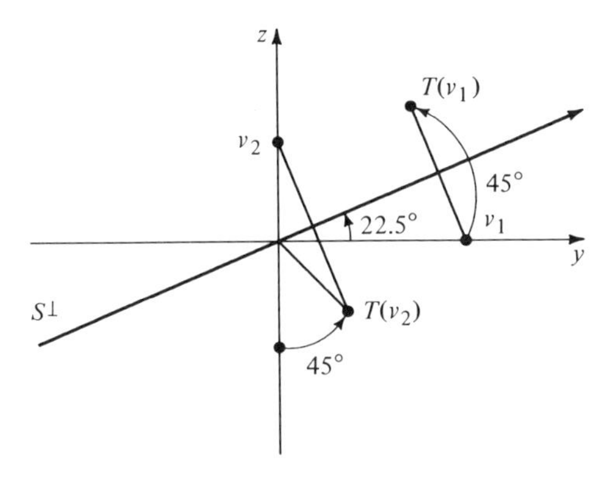

FIGURE 7.10 T is a reflection in the xy plane followed by a rotation of 45° about the x axis (the x axis points out of the page)

If A_1 and A_2 both involve reflections, then

$$\det(A_1A_2) = \det(A_1)\det(A_2) = (-1)(-1) = 1$$

and so $T_2 \circ T_1$ is a rotation followed by a translation — no reflection is involved. Similarly if T_1, say, involves a reflection but T_2 does not, then $T_2 \circ T_1$ involves a reflection, whereas if neither T_1 nor T_2 involves a reflection, then $T_2 \circ T_1$ does not involve a reflection. In particular, if T_1 and T_2 are pure rotations $[(b_i = 0, \det(A_i) = 1;\ i = 1, 2)]$, then $T_2 \circ T_1$ is also a pure rotation. Thus any composition of pure rotations is a (single) pure rotation.

EXAMPLE 7.17 Let T_1 be a rotation of 30° about the x axis, T_2 a rotation of 45° about the y axis, and T_3 a rotation of $-60°$ about the z axis. Find the axis and angle of rotation of $T_3 \circ T_2 \circ T_1$.

Solution In radians the rotation angles are $\pi/6$, $\pi/4$, and $-\pi/3$, so we need the matrices $R_{\pi/6}$, $R_{\pi/4}$, and $R_{-\pi/3}$. These can be produced by the function *ROT* of Section 4.2. The matrices of T_1, T_2, T_3 are given by

```
A1←A2←A3←ID 3
A1[2 3;2 3]←ROT ○30÷180
A2[1 3;1 3]←ROT ○45÷180
A3[1 2;1 2]←ROT ○¯60÷180
```

```
      A1                      A2                         A3
1.000  0.000  0.000    0.707  0.000  ¯0.707     0.500  0.866  0.000
0.000  0.866 ¯0.500    0.000  1.000   0.000    ¯0.866  0.500  0.000
0.000  0.500  0.866    0.707  0.000   0.707     0.000  0.000  1.000
```

and the matrix of $T_3 \circ T_2 \circ T_1$ is

```
      +A←A3+.×A2+.×A1
0.354 0.573 ¯0.739
¯0.612 0.739 0.280
0.707 0.354 0.612
```

The axis of a rotation is the eigenspace belonging to $\lambda = 1$ — that is, the null space of $A - I$. Row-reducing $A - I$ gives

```
      +E←ECHELON A-ID 3
1.00E0         4.34E¯19    6.18E¯2
0.00E0         1.00E0     ¯1.22E0
1.63E¯19       5.15E¯19    1.41E¯18
```

The last column gives the null space, but it is in an inconvenient form. We will use the function *BACKSUB* from Section 5.5 to get a pair of points on this line. Since the null space is the solution set of $AX = 0$ or, equivalently, $EX = 0$, two points (one of which will be the origin) are given by

```
      +B←1 BACKSUB E,0
¯0.061    0.000
 1.220    0.000
 1.000    0.000
```

and so the vector $B[;1]$ points along the axis of rotation. To find the angle of rotation we can shift to an orthonormal coordinate system with $B[;1]$ pointing along the x' axis. Then the matrix of $T_3 \circ T_2 \circ T_1$ will be of the form

$$A = \left[\begin{array}{c|cc} 1 & 0 & 0 \\ \hline 0 & \multicolumn{2}{c}{\multirow{2}{*}{R_θ}} \\ 0 & & \end{array}\right]$$

where $\theta = \mathrm{Cos}^{-1} A[2;2]$. Recall that the function *HSHLDR* from Section 5.5 returns an orthonormal basis of $\mathbf{R}^n$ extending an orthonormal basis of the column space of its argument.

```
      +U←HSHLDR B
0.039 ¯0.773 ¯0.633
¯0.773 ¯0.425 0.471
¯0.633 0.471 ¯0.614
```

```
      +A←(⍉U)+.×A+.×U
 1.00E0       3.09E¯18    ¯2.06E¯18
¯5.42E¯18     3.53E¯1     ¯9.36E¯1
 2.20E¯18     9.36E¯1      3.53E¯1
```

```
      ⍞THETA←÷○÷180×¯2○A[2;2]
69.4
```

Thus $T_3 \circ T_2 \circ T_1$ is a rotation of about 70° in the direction that the fingers of the right hand curl when the thumb points along $B[;1]$. ■

Proposition 7.21 immediately leads to a geometric description of an arbitrary affine transformation of space. For simplicity we assume that the affine transformation has an invertible linear part, although the general case may be included by taking a zero stretch factor to indicate a perpendicular projection. The proof of Proposition 5.12 easily adapts to give the next result.

PROPOSITION 7.22 Let $f(X) = b + AX$ be an affine transformation of space with A invertible. Then f may be viewed as successive applications of

1. Three stretches with mutually perpendicular stretch lines
2. A rotation
3. (Possibly) a reflection
4. A translation

The stretch factors are the singular values of A. The stretch lines are eigenspaces of A^TA. The reflection is present if and only if $\det(A) < 0$. If the rotation is nontrivial, the reflection may be taken in the subspace perpendicular to the axis of rotation. ■

EXERCISES 7.4*

In exercises 1 through 7 the matrix A of a congruence $f(X) = b + AX$ is given. Decide if the reflection and rotation of Proposition 7.21 are present.

Hint: The reflection is present if $\det(A) = -1$. The rotation must be present if the matrix has complex eigenvalues (but this is not necessary).

1. $\begin{bmatrix} \frac{1}{\sqrt{3}} & -\frac{1}{\sqrt{2}} & \frac{1}{\sqrt{6}} \\ \frac{1}{\sqrt{3}} & 0 & -\frac{2}{\sqrt{6}} \\ \frac{1}{\sqrt{3}} & \frac{1}{\sqrt{2}} & \frac{1}{\sqrt{6}} \end{bmatrix}$ 2. $\begin{bmatrix} \frac{1}{\sqrt{3}} & \frac{1}{\sqrt{3}} & \frac{1}{\sqrt{3}} \\ \frac{1}{\sqrt{2}} & 0 & -\frac{1}{\sqrt{2}} \\ \frac{1}{\sqrt{6}} & -\frac{2}{\sqrt{6}} & \frac{1}{\sqrt{6}} \end{bmatrix}$ 3. $\frac{1}{3}\begin{bmatrix} 2 & 2 & -1 \\ 2 & -1 & 2 \\ -1 & 2 & 2 \end{bmatrix}$

4. $\frac{1}{6}\begin{bmatrix} 2\sqrt{3} & 2\sqrt{3} & 2\sqrt{3} \\ -3-\sqrt{3} & 2\sqrt{3} & 3-\sqrt{3} \\ -3+\sqrt{3} & -2\sqrt{3} & 3+\sqrt{3} \end{bmatrix}$ 5. $\frac{1}{6}\begin{bmatrix} 2\sqrt{3} & -3+\sqrt{3} & 3+\sqrt{3} \\ 2\sqrt{3} & -2\sqrt{3} & -2\sqrt{3} \\ 2\sqrt{3} & 3+\sqrt{3} & -3+\sqrt{3} \end{bmatrix}$

6. $\frac{1}{3}\begin{bmatrix} 1 & -2 & -2 \\ -2 & 1 & -2 \\ -2 & -2 & 1 \end{bmatrix}$ 7. $\frac{1}{3}\begin{bmatrix} -1 & 2 & 2 \\ 2 & -1 & 2 \\ 2 & 2 & -1 \end{bmatrix}$

8. (a) Let T_i be a reflection in the plane π_i. Suppose π_1 and π_2 are parallel. Show that $T_2 \circ T_1$ is a pure translation.

   Hint: Choose a coordinate system such that π_1 is the xy plane, write formulas for T_1 and T_2, and compute $T_2 \circ T_1$.

   (b) Let T_i be a reflection in the plane π_i. Suppose that π_1 and π_2 are not parallel. Show that $T_2 \circ T_1$ is a rotation about the line of intersection of π_1 and π_2.

   Hint: Choose a coordinate system such that the line of intersection is the z axis, π_1 is the xz plane, and π_2 cuts the xy plane in a line making an angle of $\theta/2$ with the x axis. Compute $T_2 \circ T_1$.

9. Show that the determinant of the n-by-n matrix A is the product of the eigenvalues.

   Hint: If $p(\lambda) = \det(A - \lambda I)$, then $\det(A) = p(0)$, the constant term of $p(\lambda)$. Now factor $p(\lambda)$ into linear factors.

10. By diagonalizing the symmetric matrix $R_\theta \begin{bmatrix} 1 & 0 \\ 0 & -1 \end{bmatrix}$ show that it represents reflection in a line in $\mathbf{R}^2$.

11. Write an APL function:

Name:	*AXNANG*
Input:	The matrix of a pure rotation
Output:	*A* vector V with four components. The vector $3\uparrow V$ is a unit vector pointing along the axis of the rotation and $V[4]$ is the signed angle of rotation, the positive direction being the direction that the fingers of the right hand will curl when the thumb points in the direction of $3\uparrow V$.

12. (Computer assignment) Use the function of exercise 11 to compute the axes and angles of rotation of the rotation matrices among the matrices of exercises 1 through 7. (Use the function *DET* from Section 7.1 to locate the pure rotations.)

7.5* Estimating Eigenvalues (Gerschgorin's Theorem)

If a matrix is diagonal, then the diagonal entries are the eigenvalues. But suppose that the matrix is not diagonal. How close might the diagonal entries be to the eigenvalues?

Each diagonal entry is the center of a disc, the Gerschgorin disc, in the complex plane. The eigenvalues lie within the Gerschgorin discs. If the radii are reasonably small, we may have a useful estimate of the eigenvalues.

DEFINITION 7.7 Let A be an n-by-n matrix with (possibly) complex entries. The ith *Gerschgorin disc*, D_i, for A is the disc in the complex plane with center $A[i;i]$ and radius

$$r_i = \sum_{j \neq i} |A[i;j]|$$

EXAMPLE 7.18 Let

$$A = \begin{bmatrix} -1 & \frac{1}{2} & -\frac{1}{2} \\ 0 & 3 & 1 \\ -\frac{3}{2} & 0 & 6 \end{bmatrix}$$

The Gerschgorin discs for A are

D_1: center -1, radius 1
D_2: center 3, radius 1
D_3: center 6, radius $\frac{3}{2}$

These discs are sketched in Figure 7.11.

Gerschgorin's theorem (see below) states that the eigenvalues lie within the discs. For the matrix A it also states that there is exactly one eigenvalue in each disc. This has two immediate consequences. The three eigenvalues of A are distinct, so A is diagonalizable. Further, since A has real entries, the characteristic polynomial of A has real coefficients and hence complex eigenvalues of A would occur in complex conjugate pairs. If $z = x + iy$ lies in one of the discs of Figure 7.11, then so does the complex conjugate $\bar{z} = x - iy$. It follows that the eigenvalues of A are real. ■

Although the fact that the Gerschgorin discs contain the eigenvalues is not difficult to prove, a proof of the sharper statement concerning the distribution of eigenvalues among the discs is more difficult and will be omitted.

PROPOSITION 7.23 (*Gerschgorin's theorem*) The eigenvalues of a matrix A lie in the portion of the complex plane enclosed by the Gerschgorin discs. Further, if k of the discs enclose a portion of the plane disjoint from the other discs, then the portion of the plane enclosed by the k discs contains exactly k eigenvalues.

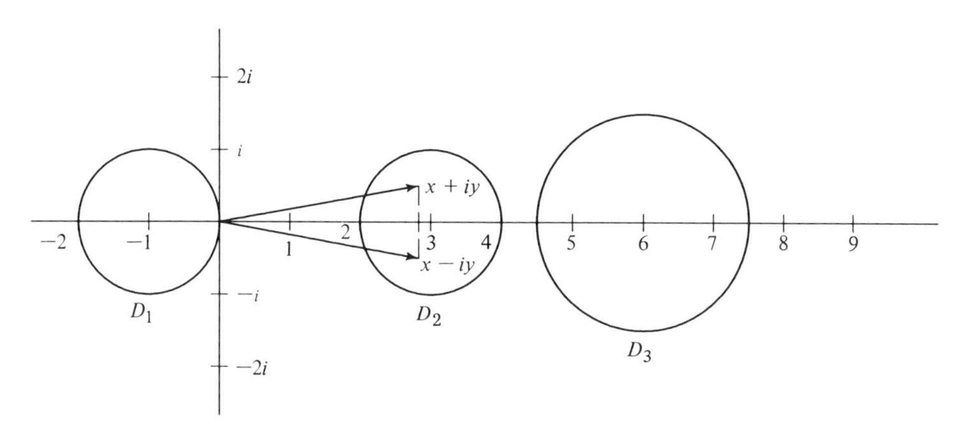

FIGURE 7.11

Proof We prove only the first statement. Let λ be an eigenvalue of A and v an eigenvector. Suppose that the largest absolute value among the components of v is $v[i]$. Now

$$\lambda v[i] = (\lambda v)[i] = (Av)[i] = A[i;] +.\times v = \sum_j A[i;j]v[j]$$

Hence

$$(\lambda - A[i;i])v[i] = \sum_{j \neq i} A[i;j]v[j]$$

Now if $v[i] = 0$, then $v = 0$ and cannot be an eigenvector, hence $v[i] \neq 0$. Dividing through by $v[i]$ and taking absolute values gives

$$|\lambda - A[i;i]| = \left| \sum_{j \neq i} \frac{A[i;j]v[j]}{v[i]} \right| \leq \sum_{j \neq i} |A[i;j]| \left| \frac{v[j]}{v[i]} \right| \leq r_i$$

since $|v[j]/v[i]| \leq 1$. Thus λ lies in D_i, where $|v[i]|$ is the largest absolute value in v, any eigenvector belonging to λ. ■

We will now discuss three applications of Gerschgorin's theorem, all of which touch on previous topics.

ACCURACY OF THE JACOBI ALGORITHM

In Section 5.2 the Jacobi algorithm for the diagonalization of a symmetric matrix was coded into a function *JACOBI*. If A is a symmetric matrix and `Q←JACOBI A`, then Q is approximately orthogonal and

$$Q^T A Q = \begin{bmatrix} a_{11} & & & \\ * & a_{22} & & * \\ & & \ddots & \\ & & & a_{nn} \end{bmatrix} = A'$$

where the off-diagonal entries are negligible compared to b, the largest absolute value in the original matrix A. From Gerschgorin's theorem we see that if the largest off-diagonal absolute value in A' is c, then the eigenvalues of A' may be taken to be the a_{ii} with an error of $c' = (n - 1)c$ at most. The error c' is close to negligible compared to b (unless n is larger than most APL workspaces can handle). The accuracy with which the numbers a_{ii} approximate eigenvalues of A depends upon how closely the machine calculation `(⍉Q)+.×A+.×Q` approximates an exact coordinate-change calculation. With the coding of Section 5.2 the difference between a_{ii} and an eigenvalue of A is usually close to negligible compared to b.

BOUNDING THE ROOTS OF POLYNOMIALS

In Section 3.7* the sectioning method of approximating roots of $y = f(x)$ was described and coded into a function *CENTSECT*. Before using *CENTSECT*, however, one must first locate an interval that contains a solution of $f(x) = 0$.

If f is a polynomial, Gerschgorin's theorem gives an interval that contains all real roots of $f(x)$.

We begin with the observation that any polynomial

$$p(x) = (a_0 + a_1x + a_2x^2 + \cdots + a_nx^n)(-1)^n$$

with $a_n = 1$ is $\det(A - xI)$ for some matrix A. In fact

$$A = \begin{bmatrix} -a_{n-1} & -a_{n-2} & -a_{n-3} & \ldots & -a_1 & -a_0 \\ 1 & 0 & 0 & \ldots & 0 & 0 \\ 0 & 1 & 0 & \ldots & 0 & 0 \\ & & & \ldots & & \\ & & & \ldots & & \\ & & & \ldots & & \\ 0 & 0 & 0 & \ldots & 1 & 0 \end{bmatrix} \tag{7.4}$$

This fact is exercise 18 in Exercises 7.2. The matrix A is sometimes called a *companion matrix* for $p(x)$.

The Gerschgorin discs for A are

$$D_1\text{:} \quad \text{center } -a_{n-1}, \quad \text{radius } \tau_1 = \sum_{i=0}^{n-2} |a_i|$$

and, for $i > 1$,

$$D_i\text{:} \quad \text{center } 0, \quad \text{radius } r_i = 1$$

The radius r_1 is often overlarge. Since A and A^T have the same characteristic polynomial, we can use A^T as well. The Gerschgorin discs for A^T are

$$D_1\text{:} \quad \text{center } -a_{n-1}, \quad \text{radius } r_1 = 1$$

for $1 < i < n$,

$$D_i\text{:} \quad \text{center } 0, \quad \text{radius } r_i = 1 + |a_{n-i}|$$

and

$$D_n\text{:} \quad \text{center } 0, \quad \text{radius } r_n = |a_0|$$

since the disc D_1 lies within the disc with center 0 and radius $1 + |a_{n-1}|$. This gives us Cauchy's theorem.

PROPOSITION 7.24 (*Cauchy's theorem*) A (possibly complex) root λ of the polynomial

$$p(x) = a_0 + a_1x + \cdots + a_{n-1}x^{n-1} + x^n$$

must satisfy

$$|\lambda| \leq r = \max\{|a_0|, 1 + |a_1|, \ldots, 1 + |a_{n-1}|\}$$

In particular, if λ is real, then $-r \leq \lambda \leq r$. ■

EXAMPLE 7.19 Let $f(x) = x^5 + x + 1$. Then Cauchy's theorem states that real roots of $f(x)$ must lie in the interval $[-2, 2]$. Since the degree of f is odd, it must have at least one real root. Using the function *CENTSECT* from Section 3.7* and the function *AT* from Example 3.5, we have

```
    ∇ Z←CFCN X
[1]     Z←1 1 0 0 0 1 AT X
    ∇
        1↑1E¯10 CENTSECT ¯2 2
¯0.755
```

So f has a root at approximately $-.755$. ■

STOCHASTIC MATRICES

A matrix A is *stochastic* if

1. The entries of A are nonnegative, and
2. The sum of the entries of each column of A is 1 (1∧.=+⌿*A*).

Stochastic matrices were discussed in Sections 3.6* and 7.3*. In Section 7.3* it was shown that $\lambda = 1$ is always an eigenvalue of A, and the question of the existence of

$$\lim_{n\to\infty} A^n$$

was discussed under the general assumption that the eigenvalues of A are $\lambda_1 = 1$, $\lambda_2, \ldots, \lambda_n$ with $|\lambda_i| < 1$ for $i > 1$. Here we will use Gerschgorin's theorem to prove a slightly different result on the existence of the above limit.

First notice that if there is an invertible matrix P such that

$$P^{-1}AP = \left[\begin{array}{c|c} I & 0 \\ \hline 0 & J \end{array}\right] \tag{7.5}$$

where I is an identity matrix and $J^n \to 0$ as $n \to \infty$, then

$$\lim_{n\to\infty} A^n$$

exists. In fact

$$\begin{aligned} A^n &= \left(P \left[\begin{array}{c|c} I & 0 \\ \hline 0 & J \end{array}\right] P^{-1} \right)^n \\ &= P \left[\begin{array}{c|c} I & 0 \\ \hline 0 & J \end{array}\right]^n P^{-1} \\ &= P \left[\begin{array}{c|c} I & 0 \\ \hline 0 & J^n \end{array}\right] P^{-1} \\ &\to P \left[\begin{array}{c|c} I & 0 \\ \hline 0 & 0 \end{array}\right] P^{-1} \qquad \text{as } n \to \infty \end{aligned}$$

In Section 7.3* the case where I is 1 by 1 was discussed. It follows from the discussion in Section 7.3* that $J^n \to 0$ as $h \to \infty$ if the eigenvalues of J all have absolute value less than 1. It can be shown that this is true even if J is not diagonalizable.

We have from Section 7.3* (Proposition 7.18) that $\lambda = 1$ is always an eigenvalue of a stochastic matrix. If for a given stochastic matrix with eigenvalue λ we can show that either $\lambda = 1$ or $|\lambda| < 1$, then the decomposition of Equation (7.5) will exist and the sequence $\{A^n\}$ will converge. The proposition we will prove is

Proposition 7.25 *Let A be a stochastic matrix with $A[i; i] > 0$ for $i = 1, 2, \ldots, n$. Then*

$$B = \lim_{n\to\infty} A^n$$

exists. The columns of B are vectors in the eigenspace of A belonging to the eigenvalue $\lambda = 1$. Further, the entries of the columns of B sum to 1.

Proof First we look at the Gerschgorin disc for A^T which has the same eigenvalues as A. Since $1\wedge.=+/A$, the ith disc has center $A[i; i]$ and radius $r_i = 1 - A[i; i]$. Thus if $A[i; i] > 0$, the only complex number in the ith disc with absolute value equal to 1 is 1 itself (Figure 7.12). Since the eigenvalues lie within the area of the complex plane enclosed by the discs, we have $|\lambda| \le 1$ and $|\lambda| = 1$ if and only if $\lambda = 1$.

This does not quite finish the proof, since we should verify that the eigenvalue $\lambda = 1$ cannot appear as part of a Jordan block before we can assert that Equation (7.5) holds. We leave this fact as exercise 5.

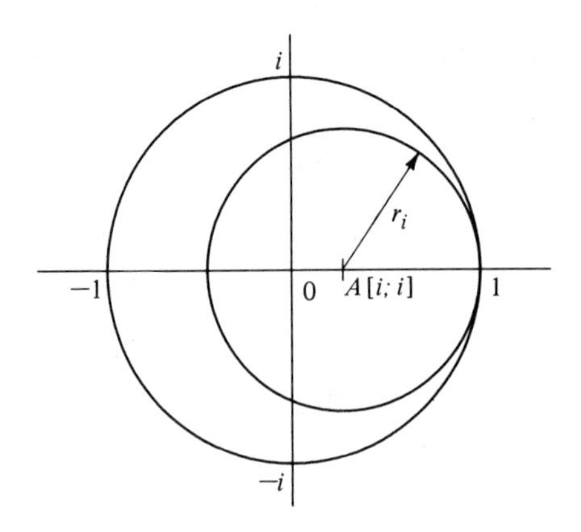

FIGURE 7.12

The equation $1\wedge.=+/B$ is true, since $1\wedge.=+/A^n$ for every n. The columns of B are eigenvectors, since

$$AB = A \lim_{n\to\infty} A^n = \lim_{n\to\infty} A^{n+1} = B. \qquad \blacksquare$$

EXAMPLE 7.20 The matrix

$$A = \begin{bmatrix} 0 & 0 & 1 \\ 1 & 0 & 0 \\ 0 & 1 & 0 \end{bmatrix}$$

is stochastic but $A[i; i] = 0$ for all i. Taking powers, we find

$$A^2 = \begin{bmatrix} 0 & 1 & 0 \\ 0 & 0 & 1 \\ 1 & 0 & 0 \end{bmatrix}, \quad A^3 = I, \quad A^4 = A, \quad A^5 = A^2, \quad A^6 = I, \ldots$$

So the limit does not exist. The characteristic polynomial of A is

$$\det(A - \lambda I) = \begin{vmatrix} -\lambda & 1 & 0 \\ 0 & -\lambda & 1 \\ 1 & 0 & -\lambda \end{vmatrix} = -(\lambda^3 - 1)$$

which has roots $\lambda = 1$, $(1 + \sqrt{3}i)/2$, $(1 - \sqrt{3}i)/2$, all of which have absolute value equal to 1. $\blacksquare$

EXAMPLE 7.21 The matrix

$$A = \begin{bmatrix} .5 & 0 & 0 & .7 \\ 0 & .2 & .6 & 0 \\ 0 & .8 & .4 & 0 \\ .5 & 0 & 0 & .3 \end{bmatrix}$$

satisfies the hypotheses of Proposition 7.25 and thus

$$\lim_{n\to\infty} A^n = B$$

exists. We can approximate B using the function *TOTHE* from Section 3.4 to compute $A^{100,000}$:

```
      + B←((A TOTHE 20)TOTHE 20)TOTHE 25
0.583 0.000 0.000 0.583
0.000 0.429 0.429 0.000
0.000 0.571 0.571 0.000
0.417 0.000 0.000 0.417
```

Checking to see if the columns of B are in the null space of $A - I$, we have

```
     B-A+.×B
0.00E0      0.00E0      0.00E0      0.00E0
0.00E0      8.67E¯19    1.30E¯18    0.00E0
0.00E0      8.67E¯19    8.67E¯19    0.00E0
1.30E¯18    0.00E0      0.00E0      1.30E¯18
```

which is quite good (⎕CT is about 1E¯15 here). The matrix B should be a stochastic matrix, but round-off error has accumulated during the 65 multiplications to the extent that the columns sums of B do not fuzz to 1.

```
      1-+⌿B
8.9E¯15   1.94E¯14   1.94E¯14   8.9E¯15
```

In this case the eigenspace of A for the eigenvalue $\lambda = 1$ is two-dimensional:

```
     ECHELON A-ID 4
1.00E0      0.00E0      0.00E0      ¯1.40E0
0.00E0      1.00E0     ¯7.50E¯1      0.00E0
0.00E0      8.67E¯19   ¯2.60E¯18     0.00E0
0.00E0      0.00E0      0.00E0      ¯8.67E¯19
```

Proposition 7.19 states that if all components of some power of A are nonzero, then this eigenspace will be one-dimensional and the columns of B will all be equal. The verification is left as exercise 6. ■

EXERCISES 7.5*

1. Give the centers and radii of the Gerschgorin discs for the following matrices. Can you deduce that the eigenvalues are distinct?

(a) $\begin{bmatrix} 1 & 3 & 2 \\ 7 & 0 & 9 \\ 0 & 1 & 2 \end{bmatrix}$ (b) $\begin{bmatrix} 1 & 0 & 1 \\ 1 & -3 & 1 \\ 2 & 0 & 5 \end{bmatrix}$ (c) $\begin{bmatrix} 1 & 0 & \frac{1}{2} \\ 0 & 2 & \frac{1}{2} \\ \frac{1}{2} & \frac{1}{3} & 4 \end{bmatrix}$

2. For the matrix A below sketch the Gerschgorin discs and the eigenvalues in the complex plane for the given values of a.

$$A = \begin{bmatrix} 0 & -a \\ a & 10 \end{bmatrix}$$

(a) $a = 3$ (b) $a = 4$ (c) $a = 5$ (d) $a = \sqrt{34}$

3. Use Cauchy's theorem to find discs in the complex plane that must contain the roots of the polynomial.

(a) $f(x) = x^3 - 1$ (b) $g(x) = \dfrac{x^3 - 1}{x - 1} = x^2 + x + 1$ (c) $10x^3 + x - 1$

4. A matrix A is called *diagonally dominant* if

$$|A[i;\, i]| > \sum_{j \neq i} |A[i;\, j]|$$

Show that a diagonally dominant matrix is invertible.

Hint: Show that zero cannot be an eigenvalue.

5. In the proof of Proposition 7.25 it was stated that a stochastic matrix cannot have an associated Jordan block with eigenvalue 1 unless the block is 1 by 1. This exercise indicates why this is true. The general fact is sufficiently indicated by a special case.

(a) Show, by induction, that

$$J_1^n = \begin{bmatrix} 1 & 1 & 0 \\ 0 & 1 & 1 \\ 0 & 0 & 1 \end{bmatrix}^n = \begin{bmatrix} 1 & n & \dfrac{n(n-1)}{2} \\ 0 & 1 & n \\ 0 & 0 & 0 \end{bmatrix}$$

and hence the 1;3 component goes to infinity at the same rate as n^2.

(b) Let J_1 be the matrix of part (a) and consider

$$A = PJ_1^nQ$$

as n goes to infinity. Show that once n is so large that only the $n^2/2$ term of $J_1[1;\, 3]$ needs to be considered, the matrix A cannot have both nonnegative entries and columns summing to 1 unless the first column of P or the last row of Q is zero. Hence we cannot have both $Q = P^{-1}$ and A stochastic.

6. Let A be a stochastic matrix with $A[i; j] > 0$ for all i and j. Let $v = (v_1, v_2, \ldots, v_n)$ be an eigenvector of $B = A^T$ for $\lambda = 1$ adjusted so that $1 = \max_i |v_i|$. Suppose $|v_{i_0}| = 1$.

   Show $v_{i_0} = v_{i_0} + \sum_{j \neq i_0} B[i_0; j](v_j - v_{i_0})$ and deduce $v_j = v_{i_0}$ for all j.

   Hint: $v_{i_0} = (Bv)[i_0]$.

   Note: This shows that the only eigenvectors of B for $\lambda = 1$ are multiples of $(1, 1, \ldots, 1)$. Thus if B has no Jordan blocks with eigenvalue 1, then $\lambda = 1$ is a simple eigenvalue. An argument similar to that of exercise 5 will show that B has no such Jordan blocks.

7.6* Linear Differential Equations

In Section 2.6* the linearization of a nonlinear differential equation at a critical point was discussed. The nonlinear equation was approximated at the critical point by a linear differential equation. In this section we show how eigenvalue-eigenvector calculations may be used to solve a linear differential equation. No knowledge of Section 2.6* is assumed.

We begin with the simplest linear differential equation

$$\frac{dy}{dt} = \lambda y \tag{7.6}$$

The general solution of Equation (7.6) is

$$y = f(t) = ce^{\lambda t} \tag{7.7}$$

where c is a constant. This can be checked by substituting $f(t)$ into Equation (7.6). The variable t is real, but (7.7) gives the solution of (7.6) even when λ is a complex number. If $\lambda = a + bi$, where a and b are real numbers, then $e^{\lambda t}$ may be computed using Euler's formula

$$e^{(a+bi)t} = e^{at}(\cos bt + i \sin bt) \tag{7.8}$$

A *linear autonomous system* of differential equations is a system of the form

$$\frac{dy_1}{dt} = a_{11}y_1 + a_{12}y_2 + \cdots + a_{1n}y_n$$

$$\frac{dy_2}{dt} = a_{21}y_1 + a_{22}y_2 + \cdots + a_{2n}y_n$$

$$\frac{dy_n}{dt} = a_{n1}y_1 + a_{n2}y_2 + \cdots + a_{nn}y_n$$

where the a_{ij} are constants.

If we introduce the notation

$$\frac{d}{dt}\begin{bmatrix} y_1 \\ \vdots \\ y_n \end{bmatrix} = \begin{bmatrix} \frac{dy_1}{dt} \\ \vdots \\ \frac{dy_n}{dt} \end{bmatrix}$$

and set $Y[i] = y_i$, $A[i, j] = a_{ij}$; then a linear autonomous system is a system that can be written

$$\frac{d}{dt} Y = AY \tag{7.9}$$

where A is a matrix of constants.

The next proposition is quite easy to check and is left as exercise 9. For more general statements, see Section 2.5*, particularly Proposition 2.21.

PROPOSITION 7.26 Let P be a constant matrix.

$$\frac{d}{dt} PY = P \frac{d}{dt} Y \quad \blacksquare$$

Suppose now that the matrix A of Equation (7.9) is diagonalizable, say $P^{-1}AP = \Lambda$. Then, introducing the change of variable

$$Y = PZ$$

and using Proposition 7.26, Equation (7.9) becomes

$$\frac{d}{dt} PZ = APZ$$

$$P \frac{d}{dt} Z = APZ$$

$$\frac{d}{dt} Z = \Lambda Z$$

Equation (7.10) is quite easy to solve. It is the scalar system of equations

$$\frac{dz_1}{dt} = \lambda_1 z_1$$

$$\frac{dz_2}{dt} = \lambda_2 z_2$$

$$\vdots$$

$$\frac{dz_n}{dt} = \lambda_n z_n$$

where the λ_i are the eigenvalues of A. These equations are all of the form (7.6) and have the solutions

$$z_i(t) = c_i e^{\lambda_i t}, \qquad 1 \leq i \leq n$$

where the c_i are constants. The equation

$$Y(t) = PZ(t)$$

now gives the solution in terms of the original variables $y_1, y_2, \ldots, y_n$.

The procedure may be carried out even when A is not diagonalizable if one knows the solutions of equations of the form (7.9), when A is a Jordan block (see exercise 10).

EXAMPLE 7.22 Solve

$$\frac{dy_1}{dt} = -40y_1 + 22y_2$$
$$\frac{dy_2}{dt} = -66y_1 + 37y_2$$

Solution In matrix form this is

$$\frac{d}{dt}\begin{bmatrix} y_1 \\ y_2 \end{bmatrix} = \begin{bmatrix} -40 & 22 \\ -66 & 37 \end{bmatrix}\begin{bmatrix} y_1 \\ y_2 \end{bmatrix}$$

So we wish to diagonalize $\begin{bmatrix} -40 & 22 \\ -66 & 37 \end{bmatrix}$. The characteristic polynomial is

$$\begin{vmatrix} -40-\lambda & 22 \\ -66 & 37-\lambda \end{vmatrix} = \lambda^2 + 3\lambda - 28 = (\lambda - 4)(\lambda + 7)$$

The eigenspaces corresponding to these eigenvalues are

$\lambda = 4$:

$$A - 4I = \begin{bmatrix} -44 & 22 \\ -66 & 33 \end{bmatrix}$$

which has echelon form

$$\begin{bmatrix} 1 & -\frac{1}{2} \\ 0 & 0 \end{bmatrix}$$

and hence the eigenspace is generated by $\begin{bmatrix} \frac{1}{2} \\ 1 \end{bmatrix}$.

$\lambda = -7$:

$$A + 7I \quad \text{row-reduces to} \quad \begin{bmatrix} 1 & -\frac{2}{3} \\ 0 & 0 \end{bmatrix}$$

and so the eigenspace is generated by $\begin{bmatrix} \frac{2}{3} \\ 1 \end{bmatrix}$.

Any nonzero vector from the appropriate eigenspace will do for a column of the coordinate-change matrix, so to avoid fractions we use the variable change

$$Y = \begin{bmatrix} y_1 \\ y_2 \end{bmatrix} = \begin{bmatrix} 1 & 2 \\ 2 & 3 \end{bmatrix} \begin{bmatrix} z_1 \\ z_2 \end{bmatrix} = PZ$$

Then

$$P^{-1}AP = \begin{bmatrix} 4 & 0 \\ 0 & -7 \end{bmatrix}$$

Notice that we need not carry out this computation; we *know* that $P^{-1}AP$ will be diagonal and we know the diagonal entries. In particular P^{-1} need not be computed — at least not at this stage. The solution of

$$\frac{d}{dt} Z = \begin{bmatrix} 4 & 0 \\ 0 & -7 \end{bmatrix} Z$$

is

$$Z(t) = \begin{bmatrix} c_1 e^{4t} \\ c_2 e^{-7t} \end{bmatrix}$$

and the solution of the original system is

$$Y(t) = PZ(t) = \begin{bmatrix} c_1 e^{4t} + 2c_2 e^{-7t} \\ 2c_1 e^{4t} + 3c_2 e^{-7t} \end{bmatrix} \qquad \blacksquare$$

In the above example the matrix P^{-1} was not needed, but that is because only the general solution of the system of equations was obtained. If one wishes, for example, the particular solution for which, say,

$$y_1(0) = 1$$
$$y_2(0) = 1$$

then one must solve the system

$$\begin{bmatrix} 1 \\ 1 \end{bmatrix} = Y(0) = PZ(0) = P\begin{bmatrix} c_1 \\ c_2 \end{bmatrix}$$

If P is larger than 2 by 2, then solving such a system by row reduction will still be less work than computing P^{-1}.

Linear autonomous systems are more general than they appear. Any homogeneous system of differential equations with constant coefficients may be rewritten as such a system.

For example, consider the differential equation of the form

$$a\frac{d^2y}{dt^2} + b\frac{dy}{dt} + cy = 0$$

where a, b, c are constants. If we set $y_1 = y$ and $y_2 = dy/dt$, we have

$$a\frac{d^2y}{dt^2} + b\frac{dy}{dt} + cy = a\frac{dy_2}{dt} + by_2 + cy_1 = 0$$

and hence the original second-order equation can be rewritten as the 2-by-2 linear system

$$\frac{dy_1}{dt} = y_2$$
$$\frac{dy_2}{dt} = -\left(\frac{c}{a}\right)y_1 - \left(\frac{b}{a}\right)y_2$$

or

$$\frac{d}{dt}Y = \begin{bmatrix} 0 & 1 \\ -\dfrac{c}{a} & -\dfrac{b}{a} \end{bmatrix} Y$$

Similarly, a third-order equation could be rewritten as a 3-by-3 system, a system of two second-order equations could be rewritten as a 4-by-4 system, and so on.

This trick is used for the next example, which shows how complex numbers arise and are disposed of in systems with real coefficients.

EXAMPLE 7.23 Solve the differential equation of simple harmonic motion

$$\frac{d^2y}{dt^2} = -\omega^2 y$$

with initial position $y(0) = 1$ and initial velocity $(dy/dt)(0) = 0$ by the methods of this section.

Solution Set $y_1 = y$ and $y_2 = dy/dt$. Then we have the system

$$\frac{dy_1}{dt} = y_2$$
$$\frac{dy_2}{dt} = -\omega^2 y_1$$

which is a linear autonomous system with matrix

$$A = \begin{bmatrix} 0 & 1 \\ -\omega^2 & 0 \end{bmatrix}$$

The characteristic polynomial of A is

$$\begin{vmatrix} -\lambda & 1 \\ -\omega^2 & -\lambda \end{vmatrix} = \lambda^2 + \omega^2 = (\lambda + i\omega)(\lambda - i\omega)$$

The matrix

$$\begin{bmatrix} -i\omega & 1 \\ -\omega^2 & -i\omega \end{bmatrix} \text{ row-reduces to } \begin{bmatrix} \omega & i \\ 0 & 0 \end{bmatrix}$$

and the matrix

$$\begin{bmatrix} i\omega & 1 \\ -\omega^2 & i\omega \end{bmatrix} \text{ row-reduces to } \begin{bmatrix} \omega & -i \\ 0 & 0 \end{bmatrix}$$

so the eigenspace of $\lambda_1 = -i\omega$ is generated by

$$\begin{bmatrix} -i \\ \omega \end{bmatrix}$$

the eigenspace of $\lambda_2 = -i\omega$ is generated by

$$\begin{bmatrix} -i \\ \omega \end{bmatrix}$$

and setting

$$P = \begin{bmatrix} -i & i \\ \omega & \omega \end{bmatrix}$$

gives

$$P^{-1}AP = \begin{bmatrix} i\omega & 0 \\ 0 & -i\omega \end{bmatrix} = \Lambda$$

The general solution of the system $(d/dt)Z = \Lambda Z$ is

$$Z(t) = \begin{bmatrix} c_1 e^{i\omega t} \\ c_2 e^{-i\omega t} \end{bmatrix}$$

and so the general solution of the original system is

$$Y(t) = PZ(t) = \begin{bmatrix} -ic_1 e^{i\omega t} + ic_2 e^{-i\omega t} \\ \omega c_1 e^{i\omega t} + \omega c_2 e^{-i\omega t} \end{bmatrix}$$

The general solution of our differential equation involves complex numbers, although the original statement of the problem could be stated in a physical context where complex numbers are meaningless. How does this come about?

By allowing complex numbers into the calculation we have answered a more general mathematical question than the original. We have solved the differential equation for y under the assumption that $y(t)$ may be a complex number as well as a real number. The connection of this general answer with the more special original question is this: real initial conditions produce real solutions $y(t)$.

To see how this works in a specific instance, we finish the solution of the problem.

We must determine c_1, c_2 so that $y_1(0) = 1$ and $y_2(0) = 0$. That is,

$$\begin{bmatrix} 1 \\ 0 \end{bmatrix} = Y(0) = PZ(0) = P\begin{bmatrix} c_1 \\ c_2 \end{bmatrix}$$

and hence

$$\begin{bmatrix} c_1 \\ c_2 \end{bmatrix} = P^{-1}\begin{bmatrix} 1 \\ 0 \end{bmatrix} = \frac{i}{2\omega}\begin{bmatrix} \omega & -i \\ -\omega & -i \end{bmatrix}\begin{bmatrix} 1 \\ 0 \end{bmatrix} = \frac{i}{2}\begin{bmatrix} 1 \\ -1 \end{bmatrix}$$

and hence

$$y(t) = y_1(t) = i\omega\frac{i}{2}(-ie^{i\omega t} - ie^{-i\omega t})$$

or

$$y(t) = \frac{e^{i\omega t} + e^{-i\omega t}}{2}$$

Using Euler's formula (7.8), we have

$$y(t) = \tfrac{1}{2}(\cos\omega t + i\sin\omega t + \cos\omega t - i\sin\omega t)$$
$$y(t) = \cos\omega t$$

■

EXERCISES 7.6*

In exercises 1 through 4 find the solutions of the linear autonomous system with the given matrix.

1. $\begin{bmatrix} 1 & 1 \\ 1 & 3 \end{bmatrix}$

2. $\begin{bmatrix} 1 & 1 \\ -1 & 1 \end{bmatrix}$

3. $\begin{bmatrix} 1 & 3 & -2 \\ 0 & 2 & 0 \\ 0 & 3 & -1 \end{bmatrix}$

4. $\begin{bmatrix} 1 & 0 & 0 & 0 \\ 0 & -3 & 2 & 0 \\ 0 & -4 & 3 & 0 \\ -2 & -2 & 2 & -1 \end{bmatrix}$

Find the solution that satisfies the given initial conditions in exercises 5 through 8.

5. $\dfrac{d^2y}{dt^2} - y = 0;\ y(0) = 0,\ \dfrac{dy}{dt}(0) = 1$

6. $\dfrac{d^2y}{dt^2} + 2\dfrac{dy}{dt} + 6y = 0;\ y(0) = 1,\ \dfrac{dy}{dt}(0) = 0$

7. $\dfrac{d^2y}{dt^2} - 2\dfrac{dy}{dt} + 2y = 0;\ y(0) = 1,\ \dfrac{dy}{dt}(0) = 0$

8. $\dfrac{d^3y}{dt^3} - \dfrac{dy}{dt} = 0;\ y(0) = 2,\ \dfrac{dy}{dt}(0) = -1,\ \dfrac{d^2y}{dt^2}(0) = 1$

   Hint: Set $y_1 = y$, $y_2 = dy/dt$, and $y_3 = d^2y/dt^2$.

9. Prove Proposition 7.26.

10. Find the solution of the linear autonomous system

$$\frac{d}{dt}\begin{bmatrix} y_1 \\ y_2 \\ y_3 \end{bmatrix} = \begin{bmatrix} \lambda & 1 & 0 \\ 0 & \lambda & 1 \\ 0 & 0 & \lambda \end{bmatrix}\begin{bmatrix} y_1 \\ y_2 \\ y_3 \end{bmatrix}$$

by writing out the individual equations, solving for y_3, and then "back-substituting."

Hint: To solve

$$\frac{dy}{dt} - \lambda y = f(t)$$

first multiply through by $e^{-\lambda t}$, obtaining

$$e^{-\lambda t}\frac{dy}{dt} - \lambda e^{-\lambda t}y = e^{-\lambda t}f(t)$$

or

$$\frac{d}{dt}(e^{-\lambda t}y) = e^{-\lambda t}f(t)$$

or

$$y(t) = e^{\lambda t}\int e^{-\lambda s}f(s)\,ds$$

7.7 The *QR* Algorithm

The basic *QR* algorithm is as mysterious as it is simple to state. Given a real square matrix A, define a sequence of matrices $\{A_n\}$ as follows:

1. $A_0 = A$.
2. Given A_{n-1}, perform a *QR* factorization (Definition 5.13) with Q square:

$$A_{n-1} = QR$$

3. $A_n = RQ$.

If A has real eigenvalues, then the sequence $\{A_n\}$ usually converges to an upper triangular matrix

$$\begin{bmatrix} \lambda_1 & & * \\ & \lambda_2 & \\ 0 & & \ddots \\ & & & \lambda_3 \end{bmatrix}$$

where the eigenvalues of A are $\lambda_1, \lambda_2, \ldots, \lambda_n$.

What if A has complex eigenvalues? Since A is assumed real, the eigenvalues of A come in complex conjugate pairs: $\lambda = \alpha + i\beta$ and $\bar{\lambda} = \alpha - i\beta$, where α and β are real. In the case of complex eigenvalues the sequence $\{A_n\}$ usually converges to a matrix of the form

$$\begin{bmatrix} B_1 & & * \\ & B_2 & \\ 0 & & \ddots \\ & & & B_k \end{bmatrix} \tag{7.11}$$

where either $B_i = \lambda_i$, a real eigenvalue of A, or B_i is a 2-by-2 matrix with two complex conjugate eigenvalues λ_i, $\bar{\lambda}_i$ that are also eigenvalues of A.

Before explaining how this comes about we must develop some facts about invariant subspaces.

DEFINITION 7.8 Let T: $\mathbf{R}^n \to \mathbf{R}^n$ be a linear transformation. A subspace S of $\mathbf{R}^n$ is an *invariant subspace* for T if $T(S)$ is contained in S.

EXAMPLE 7.24 Let T: $\mathbf{R}^n \to \mathbf{R}^n$ be given by $T(X) = AX$.

1. Let S be the null space of A. Then $T(v) = 0$ for every v in S. Thus $T(S) = \{0\}$, which is a subset of *any* subspace. In particular $T(S)$ is contained in S.

2. Let λ be an eigenvalue of A and let S be the corresponding eigenspace. If v is in S, then $T(v) = Av = \lambda v$ is also in S (Proposition 4.25) and hence $T(S)$ is contained in S.

3. Let λ_1, λ_2 be two eigenvalues of A and let S be the subspace generated by the eigenvectors of λ_1 and λ_2.

If v is in S, then $v = \alpha_1 v_1 + \alpha_2 v_2$, where v_i is an eigenvector of λ_i and hence $T(v) = A(\alpha_1 v_1 + \alpha_2 v_2) = (\alpha_1\lambda_1)v_1 + (\alpha_2\lambda_2)v_2$, which lies in S. ■

The third example in Examples 7.23 generalizes. The subspace generated by a collection of eigenspaces is an invariant subspace. The next proposition gives a matrix algebra test for invariant subspaces.

PROPOSITION 7.27 Let S be the column space of V. S is an invariant subspace for $T(X) = AX$ if and only if there is a matrix C such that $AV = VC$.

Proof Recall that a linear transformation carries subspaces to subspaces (Proposition 4.23). The subspace $T(S)$ is the column space of AV, hence the proposition follows from Proposition 4.12. ■

The next proposition gives the basic mechanism by which the blocks B_i of matrix (7.11) are obtained.

PROPOSITION 7.28 Let $T: \mathbf{R}^n \to \mathbf{R}^n$ be given by $T(X) = AX$, and introduce the coordinate change $X = PX'$. Suppose that $P = [P_1 \mid P_2]$, where the column space of P_1 is an invariant subspace of T. Then $T(X') = A'X'$, where

$$A' = P^{-1}AP = \left[\begin{array}{c|c} B_1 & * \\ \hline 0 & B_2 \end{array}\right]$$

Here $AP_1 = P_1B_1$. If, in addition, the column space of P_2 is also an invariant subspace for T, then

$$A' = \left[\begin{array}{c|c} B_1 & 0 \\ \hline 0 & B_2 \end{array}\right]$$

and $AP_2 = P_2B_2$.

Proof If $T(X) = AX$, then $T(X') = (P^{-1}AP)X'$. Now suppose that

$$P^{-1} = \left[\begin{array}{c} Q_1 \\ \hline Q_2 \end{array}\right]$$

where Q_1 has k rows. Then

$$I = P^{-1}P = \left[\begin{array}{c} Q_1 \\ \hline Q_2 \end{array}\right] [P_1 \mid P_2] = \left[\left[\begin{array}{c} Q_1 \\ \hline Q_2 \end{array}\right] P_1 \;\middle|\; \left[\begin{array}{c} Q_1 \\ \hline Q_2 \end{array}\right] P_2 \right] = \left[\begin{array}{c|c} Q_1P_1 & Q_1P_2 \\ \hline Q_2P_1 & Q_2P_2 \end{array}\right]$$

So $Q_1P_1 = I$, $Q_2P_1 = 0$, $Q_1P_2 = 0$, $Q_2P_2 = I$.

Further, by Proposition 7.27, there is a matrix C such that $AP_1 = P_1C$. Hence

$$P^{-1}AP = \left[\begin{array}{c} Q_1 \\ \hline Q_2 \end{array}\right] A[P_1 \mid P_2] = \left[\begin{array}{c} Q_1 \\ \hline Q_2 \end{array}\right] [AP_1 \mid AP_2] = \left[\begin{array}{c|c} Q_1AP_1 & Q_1AP_2 \\ \hline Q_2AP_1 & Q_2AP_2 \end{array}\right]$$

Now $Q_2AP_1 = Q_2P_1C = 0C = 0$. Similarly, if the column space of P_2 can be chosen invariant, we can show $Q_1AP_2 = 0$. ■

The next task is to indicate how the complex eigenvalues of a real matrix produce invariant subspaces of $\mathbf{R}^n$. We begin with the next proposition, the proof of which is left as exercise 27.

PROPOSITION 7.29 Let $f(x) = x^n + a_{n-1}x^{n-1} + \cdots + a_0$ be a polynomial and let A be an m-by-m matrix. Let $f(A)$ denote the matrix $A^n + a_{n-1}A^{n-1} + \cdots + a_0I$. If λ is an eigenvalue of A with eigenvector v, then $f(\lambda)$ is an eigenvalue of $f(A)$ with eigenvector v. ■

It follows from Proposition 7.29 that, for diagonalizable matrices at least, $f(A)$ and A have exactly the same eigenspaces. A related statement can be proved for general matrices.

Now suppose that A is a diagonalizable real matrix with a pair of complex conjugate eigenvalues λ, $\bar{\lambda}$. If $f(x) = (x - \lambda)(x - \bar{\lambda}) = x^2 - (\lambda + \bar{\lambda})x + \lambda\bar{\lambda}$, then $f(\lambda) = f(\bar{\lambda}) = 0$ and the subspace generated by the eigenvectors of A belonging to λ and $\bar{\lambda}$ is the subspace generated by the eigenvectors of $f(A)$ belonging to the eigenvalue 0 — that is, the null space of $f(A)$. But $\lambda + \bar{\lambda}$ and $\lambda\bar{\lambda}$ are real numbers, so $f(A)$ is a real matrix and its null space can be computed in real arithmetic. This is how the 2-by-2 blocks of (7.11) will arise — via Proposition 7.28, where the invariant subspace is the null space of $(A - \lambda I)(A - \bar{\lambda}I)$.

EXAMPLE 7.25 Let

$$A = \begin{bmatrix} -1 & -1 & -1 \\ 1 & 0 & 0 \\ 0 & 1 & 0 \end{bmatrix}$$

The characteristic polynomial of A is $\det(A - \lambda I) = -(\lambda + 1)(\lambda^2 + 1)$ and the matrix has two complex eigenvalues, the roots of $\lambda^2 + 1 = 0$. Thus the subspace generated by the corresponding eigenvectors is contained in the null space of $A^2 + I$.

The null space of a matrix is the orthogonal complement of the row space. A convenient way to compute the desired null spaces here is to apply the function *PERP* of Section 5.5 to the transposes of the matrices involved. (PERP A computes an orthonormal basis of the orthogonal complement of the column space of A).

```
      +P1←PERP ⍉A+ID 3
0.577
¯0.577
0.577
      +P2←PERP ⍉(ID 3)+A+.×A
 0.000  0.707
¯1.000  0.000
 0.000 ¯0.707
      P←P1,P2
      +B←(A+.×P)⌹P
¯1.00E0      ¯1.76E¯18   ¯1.50E¯18
¯2.60E¯18     1.02E¯18   ¯7.07E¯1
 4.34E¯18     1.41E0     ¯1.23E¯18
```

Up to fuzz (⎕CT is about 1E¯15 here) this last matrix is

$$\left[\begin{array}{c|cc} 1 & 0 & 0 \\ \hline 0 & 0 & \frac{\sqrt{2}}{2} \\ 0 & \sqrt{2} & 0 \end{array}\right]$$

and the 2-by-2 block has eigenvalues $\lambda = \pm i$. ■

We now return to the consideration of the QR algorithm. Our approach is somewhat oblique. We begin with a particular example that illustrates the main properties of the algorithm.

The particular example is a geometric one (Figure 7.13). Given an ellipse and its center, find the axes of the ellipse. We assume that the center is at the origin of the coordinate system. Then the ellipse has an equation of the form, taking $X = \begin{bmatrix} x \\ y \end{bmatrix}$,

$$X^T A X = 1$$

where A is two by two and symmetric. In Figure 7.13 the axes of the ellipse lie along the $x'y'$ axes, the x and y axes are not shown. The problem is to find the x', y' axes.

Starting with an initial guess at an axis, S_0, we construct a sequence of lines (subspaces) as follows. Given S_n we find t_n, the tangent to the ellipse at the point

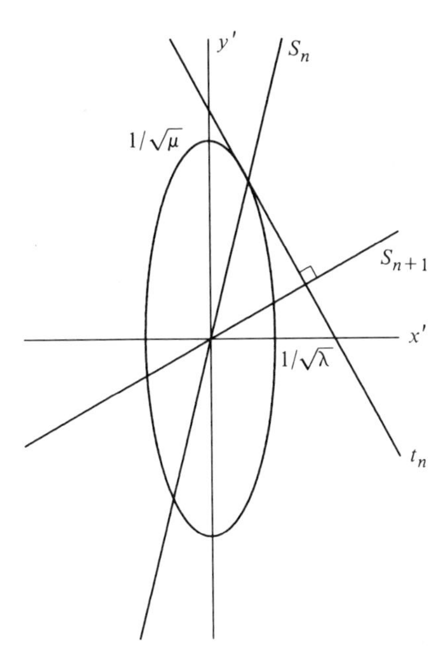

FIGURE 7.13

at which the ellipse meets S_n (there are two such points, but the resulting tangent lines are parallel). The next term in the sequence, S_{n+1}, is then obtained by dropping a perpendicular from the origin to t_n.

We will see below that this geometric algorithm is equivalent to the QR algorithm.

There are two things to notice about the sequence $\{S_n\}$. The first is that if S_n is the x' or y' axis, then $S_n = S_{n+1}$. The second is that if S_0 is not the y' axis, then S_n converges to the x' axis, the shorter axis of the ellipse.

Suppose that the eigenvalues of A are $\lambda > 0$ and $\mu > 0$. Then in the $x'y'$ coordinate system the equation of the ellipse is

$$\lambda(x')^2 + \mu(y')^2 = 1$$

The intersection of the ellipse with the x' axis occurs at $x' = \pm 1/\sqrt{\lambda}$. Thus $\{S_n\}$ converges to the eigenspace of the *largest* eigenvalue.

Differentiating this equation gives

$$2\lambda x' + 2\mu y' \frac{dy'}{dx'} = 0$$

and so if S_n meets the ellipse at (x_0', y_0'), then the slope of t_n is $-\lambda x_0'/\mu y_0'$. Now, if the slope of a line is m, then the slope of a perpendicular line is $-1/m$ and so the slope of S_{n+1} is $\mu y_0'/\lambda x_0'$. Thus S_n is generated by (x_0', y_0') and S_{n+1} is generated by the vector

$$\begin{bmatrix}\lambda x'_0\\ \mu y'_0\end{bmatrix}=\begin{bmatrix}\lambda & 0\\ 0 & \mu\end{bmatrix}\begin{bmatrix}x'_0\\ y'_0\end{bmatrix} \tag{7.13}$$

Now let T be the linear transformation with matrix

$$\begin{bmatrix}\lambda & 0\\ 0 & \mu\end{bmatrix}$$

in $x'y'$ coordinates. Since the coordinate change $X = PX'$ is orthonormal (Section 4.3), $P^T = P^{-1}$ and in the xy coordinate system T is given by $T(X) = AX$. In terms of the transformation T the description of the sequence of subspaces $\{S_n\}$ becomes

$$\begin{aligned}&(1)\quad \text{Guess at } S_0\\ &(2)\quad S_n = T(S_{n-1})\end{aligned} \tag{7.14}$$

If we take S_0 to be generated by a unit vector u_0, a convenient computational algorithm defining a sequence of unit vectors $\{u_n\}$ is

1. Guess at u_0.
2. $v_n = Au_{n-1}$.
3. $u_n = v_n/\|v_n\|$.

Step 3 is necessary to keep the components of v_n from growing too large or too small for accurate numerical work. In this form the algorithm applies to any n-by-n matrix A and is one version of the *Power method* of approximating an eigenvector belonging to the largest (or *dominant*) eigenvalue of A.

We are interested in a different generalization of Figure 7.13, however. If S_0 is a k-dimensional subspace of $\mathbf{R}^n$, then the sequence $\{S_n\}$ of Equations (7.14) usually approaches a k-dimensional subspace of $\mathbf{R}^n$. We will refer to this as the *k-dimensional power method.*

This is somewhat shaky ground. In what sense can a sequence of subspaces approach a subspace? Look again at Figure 7.13. The angle between S_n and the x' axis is becoming small. The tangent of this angle is the slope of the line S_n and Equation (7.13) shows us that the slope of S_n goes to zero as n becomes large. In fact, if m_n is the slope of S_n, then the slope of S_{n+1} is

$$m_{n+1}=\frac{\mu y'_0}{\lambda x'_0}=\left(\frac{\mu}{\lambda}\right)m_n$$

Thus, beginning with m_0, the slope of S_0, we have

$$m_1=\left(\frac{\mu}{\lambda}\right)m_0,\quad m_2=\left(\frac{\mu}{\lambda}\right)m_1=\left(\frac{\mu}{\lambda}\right)^2 m_0,\quad \ldots,\quad m_n=\left(\frac{\mu}{\lambda}\right)^n m_0$$

and since $\mu/\lambda < 1$, $m_n \to 0$ as $n \to \infty$. Notice that the rate of convergence is controlled by the size of the ratio μ/λ.

This shows that the sequence of lines S_n approaches the x' axis in the sense that the angle between them goes to zero. It works as long as m_0, the slope of S_0, is defined — that is, as long as S_0 is not the y' axis.

The arguments above extend to the case of an m-by-m matrix A. Arrange the eigenvalues of A in order of descending magnitude,

$$|\lambda_1| \geq |\lambda_2| \geq \cdots \geq |\lambda_m|$$

If $|\lambda_k| > |\lambda_{k+1}|$, then the sequence of subspaces (7.14) approaches a k-dimensional subspace containing the eigenspaces of $\lambda_1, \lambda_2, \ldots, \lambda_k$ for almost all choices of S_0. This is true even if the matrix A is not diagonalizable.

To implement Equations (7.14) we replace the subspace S_n with an m-by-k matrix V_n whose columns generate S_n.

Then (7.14) gives a sequence of n-by-k matrices $\{V_n\}$ defined by

$$\begin{aligned} &(1) \quad \text{Guess at } V_0 \\ &(2) \quad V_n = AV_{n-1} \end{aligned} \tag{7.15}$$

If, arranging the eigenvalues $\lambda_1, \ldots, \lambda_n$ of A in order of descending magnitude, we have $|\lambda_k| > |\lambda_{k+1}|$, then for almost all choices of V_0 the column space of V_n approaches an invariant subspace of A.

Since we do not know the eigenvalues of A, we do not know how to pick k. However, looking at V column by column, we have

$$A[v_1 | v_2 | \cdots | v_k] = [Av_1 | Av_2 | \cdots | Av_k]$$

This means that (7.15) is actually doing many different power methods at once. Looking at individual columns, we have k different one-dimensional power-method sequences. Looking at pairs of columns, we have several $\left(\binom{k}{2} \text{ in fact}\right)$ two-dimensional power-method sequences. In fact, we have l-dimensional power methods going for all $l \leq k$. Since we do not know what k to pick, we may as well take $k = m - 1$. Then we will be doing all dimensions at once! In what follows we will only try to keep track of one method for each l: the l-dimensional power method obtained by using the first l columns.

Further, since we have no good method for choosing V_0, we will always make the same choice: the first $m - 1$ columns of the identity matrix. With this choice of V_0, Equations (7.15) collapse to

$$V_n = \text{first } l \text{ columns of } A^n \tag{7.16}$$

Those values of l for which $|\lambda_l| = |\lambda_{l+1}|$ or V_0 happens to be a bad choice will not give convergence. All other values of l will give convergence to an l-dimensional invariant subspace of A. Suppose we fix an l that gives convergence. Proposition 7.28 suggests the following procedure. First compute A^n for a large n. Next introduce a coordinate change $X = PX'$, where $P = [P_1 | P_2]$ and P_1 has the same column space as V_n in Equation (7.16). Then, since V_n is close to an invariant subspace of A, we might expect that

$$P^{-1}AP = \left[\begin{array}{c|c} * & * \\ \hline \epsilon & * \end{array}\right]$$

where ϵ is $(m - l)$ by l with entries near zero. Unfortunately, this does not work. The problem is that an arbitrary coordinate change $X = PX'$ does not preserve dot products. Two points that are "near" in one coordinate system need not be "near" in the other. In fact, the most obvious choice for P, $P = A^n$, does not work at all, since

$$P^{-1}AP = (A^n)^{-1}AA^n = A,$$

putting us exactly where we started.

It can be shown that the problem is avoided by using orthonormal coordinate changes, which do preserve dot products.

Notice that we can get, for each choice of l, an orthonormal basis of the column space of the matrix V_n of Equation (7.16) by performing a QR factorization of A^n:

$$A^n = QR$$

since the triangular form of R shows that the first l columns of Q generate the same subspace as the first l columns of A^n.

Here is the form of the QR algorithm that we will implement.

QR algorithm Let A be m by m. Fix an integer p and define a sequence of matrices $\{A_n\}$ as follows:

$$\begin{aligned} A_0 &= A \\ \alpha_n A_{n-1}^p &= QR, \text{ a } QR \text{ factorization with } \alpha_n \text{ a scalar} \\ A_n &= Q^T A_{n-1} Q \end{aligned} \tag{7.17}$$

The scalar α_n is a scaling factor used to adjust the size of the components of A_{n-1}^p. For $p = \alpha_n = 1$ we obtain the basic form of the QR algorithm given at the beginning of the section. In fact, $A_n = Q^T A_{n-1} Q = Q^T(QR)Q = RQ$.

PROPOSITION 7.30 Let A be an m-by-m real matrix and suppose that the eigenvalues $\lambda_1, \lambda_2, \ldots, \lambda_m$ of A have been arranged so that $|\lambda_i| \geq |\lambda_{i+1}|$. Let $\{A_n\}$ be a sequence of matrices defined by the QR algorithm. Suppose that $|\lambda_k| > |\lambda_{k+1}|$.

Then for almost every A the sequence converges to a matrix of the form

$$\left[\begin{array}{c|c} B & * \\ \hline 0 & C \end{array}\right]$$

where B is k by k. The eigenvalues of B are $\lambda_1, \ldots, \lambda_k$ and the eigenvalues of C are $\lambda_{k+1}, \ldots, \lambda_m$. ■

The phrase "almost every A" may be made precise in a probabilistic sense. If one chooses matrices A "at random," then the probability of getting a "bad" matrix A is zero. People, however, do not choose matrices "at random." In fact, the probability that a real matrix chosen "at random" will have an integer entry is also zero.

The important point about the set of such "bad" matrices is that it is unstable. Consider Figure 7.13. In this context a 2-by-2 matrix A will be "bad" if its first column is parallel to the y' axis. If A is perturbed ever so slightly, so that the first column is no longer precisely parallel to the y' axis, then the perturbed matrix is no longer bad and S_n will converge to the x' axis. Most of the time, as the computations proceed, round-off error will provide such perturbations and, after a slow start, the algorithm converges anyway. Here we have a case where the algorithm is actually helped by round-off error!

APL FUNCTIONS

The idea is to iterate the QR algorithm until one has

$$A_n = \left[\begin{array}{c|c} B & * \\ \hline \epsilon & C \end{array}\right] \tag{7.18}$$

where ϵ is negligible compared to, say, the input data. Then the problem deflates to that of finding the eigenvalues of the smaller matrices B and C. This suggests that the algorithm proceed by induction on the size of A. Since 2-by-2 matrices can be handled by the quadratic formula, it is not hard to start the induction.

The QR algorithm as given in (7.17) often converges slowly. If $|\lambda_k| > |\lambda_{k+1}|$, then, as in the 2-by-2 case of Figure 7.15, the rate of convergence of the k-dimensional power method is governed by the ratio $|\lambda_{k+1}/\lambda_k|$. The smaller this ratio is, the faster the algorithm converges. Typically the algorithm given in (7.17) proceeds rapidly at first and then slows down as the problem deflates to small matrices with close eigenvalues. In practice the convergence of the algorithm is accelerated by the technique of *shifting*, which will now be described. The technique is based on Proposition 7.29.

As the algorithm (7.17) proceeds, the lower right-hand entry $a_n = A_n[m;m]$ approaches λ_m, the smallest eigenvalue of A. By Proposition 7.29 the eigenvalues of $A_n - a_n I$ are $\lambda_i - a_n$, and, if a_n is close to λ_m, then $\lambda_m - a_n$ is small and hence the ratio $|(\lambda_m - a_n)/(\lambda_{m-1} - a_n)|$ is also small. Replacing A_n by $A_n - a_n I$ is called a *shift*.

If λ_m is complex, however, and we are using real arithmetic, then clearly a_n, being real, cannot become close to λ_m. In this case the lower-right 2-by-2 block $B_n = {}^{-}2\uparrow A_n$ should have eigenvalues μ_1, μ_2 approaching λ_m and $\lambda_{m-1} = \bar{\lambda}_m$.

Now the characteristic polynomial of B_n is

$$p_n(x) = (x - \mu_1)(x - \mu_2) = x^2 + a_1 x + a_0$$

and by Proposition 7.29 the eigenvalues of $p_n(A)$ are

$$p_n(\lambda_i) = \lambda_i^n + a_1\lambda_i + a_0$$

If μ_1 is sufficiently close to λ_n, then, by continuity, $p_n(\lambda_m)$ is close to $p_n(\mu_1) = 0$. Replacing A_n by $p_n(A_n)$ is called an *implicit double shift* because it can be thought of as simultaneously performing two complex conjugate single shifts using only real arithmetic. It follows from Proposition 7.29 that, for diagonalizable matrices at least, $p_n(A_n)$ and A_n have exactly the same eigenspaces as A. A related statement can be proved for general matrices.

The shifted QR algorithm is implemented in the function QR listed below. We will first briefly describe the purpose of each line, then explain the details.

```
    ∇ Z←S QR A ;N;B;Q;K;C
[1]    Z←STRT A
[2]    →(2≥N←1↑ρZ)/0
[3]    →(S∧.=S+,A)/0
[4]    C←1
[5]   L:A←(⍉Q)+.×A+.×Q←S HHLDR 4 SCPOW(¯2 ¯2↑A)DBLSHFT A
[6]    Z←Z+.×Q
[7]    →(10=C←C+1)/0
[8]    →(N=K←(1 1⍉⊖∧⍀⊖∧\S=S+1 ¯1↓A)⍳1)/L
[9]    Z←Z+.×((N,K)↑S QR(K,K)↑A),(-N,K-N)↑S QR(K,K)↓A
    ∇
```

The right argument A of QR is the matrix to be diagonalized. The left argument S is a scalar that sets the scale for fuzzy comparisons. Normally we take S to be of the same order of magnitude as the lengths of the columns of A.

The function *STRT* on line 1 starts the induction if A is 1 by 1 or 2 by 2. It returns an orthogonal matrix Q such that $Q^T AQ$ is triangular — that is, has a zero in the 2;1 position, if A has real eigenvalues, and returns an identity matrix if A has complex eigenvalues. *STRT* also returns an identity matrix if A is not 2 by 2.

Line [2] stops the function if A is 2 by 2 and line [3] stops the function if A is a zero matrix (compared to S). Line [5] does all the work. The expression `B DBLSHFT A` computes $p(A)$, where $p(\lambda)$ is the characteristic polynomial of the 2-by-2 matrix B. That is, line [5] begins (on the right) with an implicit double shift.

The function *SCPOW* computes a power of A, scaled so that the entries do not become too large or too small. The powers are doubled up for efficiency. The expression n *SCPOW* A computes a rescaling of A^{2^n}. We are using $n = 4$ and hence $p = 16$ in (7.17). The function *HHLDR* is discussed in Section 5.5. It uses the Householder algorithm to produce the orthogonal matrix Q of (7.17). Finally, A_{n-1} is replaced by A_n.

In APL, matrix multiplications are cheap compared to Householder reductions. The value $p = 16$ gives four matrix multiplications per Householder reduction. Too high a value of p wastes computation in too many multiplications, and too low a value wastes computation in Householder reductions. In *FORTRAN*, for example, matrix multiplications are relatively expensive, and $p = 1$ is the usual choice.

Line [6] accumulates the coordinate-change matrices Q. The function QR returns an orthogonal matrix Q such that Q^TAQ is in the form (7.11). If QR iterates for l steps, computing the successive coordinate-change matrices $Q_1, Q_2, \ldots, Q_l$, then Z will be the orthogonal matrix $Q_1Q_2 \cdots Q_l$.

Line [7] limits the iterations to ten, a rather arbitrary choice. There are matrices for which this coding of QR will not converge (see below).

Line [8] detects the presence of the matrix ϵ in (7.18). Here "small" is taken to mean "negligible compared to S."

Line [9] is the induction step. The results of applying the QR algorithm to the matrices B and C of (7.18) are combined with the results of lines [5] and [6] to give the final result.

Now for the details.

The most puzzling line is probably [8]. To begin with, it contains some primitive APL functions that we have not had occasion to use previously.

The functions ⊖ and ⌽ (which will be used in coding *STRT*) are typographically and conceptually similar to the transpose ⍉.

The symbol ⊖ (theta) is ○, backspace, - and the symbol ⌽ (phi) is ○, backspace, |. Just as ⍉A rotates A about a diagonal axis, ⊖A rotates A about a horizontal axis and ⌽A rotates A about a vertical axis.

```
      A
1 2 3
4 5 6
7 8 9
      ⊖ A
7 8 9
4 5 6        The order of the rows is reversed
1 2 3
      ⌽ A
3 2 1
6 5 4        The order of the columns is reversed
9 8 7
```

The *scan* operator, \, is similar to the reduction operator /. If V is a vector and

f is a scalar dyadic function, then $f\backslash v$ is the vector of partial reductions. Suppose that v has n components.

```
(f\V)[1]  is  V[1]
(f\V)[2]  is  f/V[1 2]
(f\V)[3]  is  f/V[1 2 3]
   ⋮
(f\V)[n]  is  f/V
```

In particular, if v is a vector of zeros and ones as in line $QR[8]$, then ∧\V consists of ones up to the component of V that contains the first zero. Thereafter ∧\V consists of zeros:

```
      V
1 1 1 0 1 0 1 1
      ∧\V
1 1 1 0 0 0 0 0
```

For matrices A one has `(f\A)[;k] = f/A[;⍳k]` and `(f⍀A)[k;] = f⌿A[⍳k;]`.

The action of line [8] is now easily understood by working out just how the indicated block of zeros in

$$\left[\begin{array}{cc|cc} * & 0 & 0 & * \\ * & 0 & * & * \\ \hline 0 & 0 & 0 & * \\ 0 & 0 & * & 0 \end{array}\right]$$

is flagged.

Line [9] takes the orthogonal matrix Z computed by the loop $QR[5\ 6\ 7\ 8]$ together with Z_B and Z_C computed by $S\ QR\ B$ and $S\ QR\ C$ [B and C here refer to (7.18)] and forms

$$Q = Z\begin{bmatrix} Z_B & 0 \\ 0 & Z_C \end{bmatrix}$$

Then

$$\begin{aligned} Q^T A Q &= \begin{bmatrix} Z_B^T & 0 \\ 0 & Z_C^T \end{bmatrix} (Z^T A Z) \begin{bmatrix} Z_B & 0 \\ 0 & Z_C \end{bmatrix} \\ &= \begin{bmatrix} Z_B^T & 0 \\ 0 & Z_C^T \end{bmatrix} \left[\begin{array}{c|c} B & * \\ \hline 0 & C \end{array}\right] \begin{bmatrix} Z_B & 0 \\ 0 & Z_C \end{bmatrix} \\ &= \left[\begin{array}{c|c} Z_B^T B Z_B & * \\ \hline 0 & Z_C^T C Z_C \end{array}\right] \end{aligned}$$

which should be of the form (7.11).

To write the function *STRT* we use the following formula, which is left as exercise 28.

PROPOSITION 7.31 Let

$$A = \begin{bmatrix} a & b \\ c & d \end{bmatrix}$$

have real eigenvalues λ_1, λ_2. If $b = 0$, set

$$Q = \begin{bmatrix} 0 & 1 \\ 1 & 0 \end{bmatrix}$$

If $b \neq 0$, set $Q = R_\theta$, where

$$\theta = \operatorname{Tan}^{-1} \frac{(a - d) \pm \sqrt{(a - d)^2 + 4bc}}{2b}$$

Then Q is orthogonal and

$$Q^T A Q = \begin{bmatrix} \lambda_1 & * \\ 0 & \lambda_2 \end{bmatrix} \quad \blacksquare$$

In the formula for θ given in Proposition 7.31 we choose the sign of the radical to be the sign of $a - d$. This prevents a loss of accuracy through cancellation.

```
    ∇ Z←STRT A ;T;D;S
[1]   Z←ID 1↑ρA
[2]   →(4≠ρA←,A)/0
[3]   →(0>D←(((T←-/A[1 4])*2)+4××/A[2 3]))/0
[4]   Z←⌽Z
[5]   →(S=A[2]+S←⌈/,|A)/0
[6]   Z←ROT ¯3○(-T+((T≥0)-T<0)×D*÷2)÷2×A[2]
    ∇
```

Let

$$B = \begin{bmatrix} a & b \\ c & d \end{bmatrix}$$

Then the characteristic polynomial of B is

$$\det(B - \lambda I) = \lambda^2 - (a + d)\lambda + ad - bc$$

This formula is used in coding *DBLSHFT*.

```
    ∇ Z←B DBLSHFT A ;I
[1]     Z←((B[1;]-.×⌽B[2;])×I)+A+.×A-(+/1 1⍉B)×I←ID 1↑ρA
    ∇
```

To write the function *SCPOW* notice that A^{2^p} could be defined inductively as

$$A^{2^p} = \begin{cases} A & \text{if } p = 0 \\ (A^{2^{p-1}})^2 & \text{if } p \geq 1 \end{cases}$$

```
    ∇ Z←P SCPOW A
[1]     Z←A
[2]     →(P=0)/0
[3]     A←A+.×A←(P-1)SCPOW A
[4]     Z←A÷(|,A)+.÷ρ,A
    ∇
```

The last line rescales the matrix by dividing by the average absolute value of its entries.

EXAMPLE 7.26 Consider the matrix

$$A = \begin{bmatrix} 2 & -26 & 10 & -18 & 11 \\ -14 & 13 & -23 & 17 & -5 \\ 7 & 8 & 8 & 1 & -2 \\ 7 & -18 & 16 & -15 & 9 \\ -20 & -20 & -20 & 0 & 11 \end{bmatrix}$$

If we apply the function *QR*, we obtain

```
      +Q←10 QR A
 1.33E¯1    ¯6.61E¯1    3.66E¯1    5.83E¯1    2.67E¯1
 3.85E¯1     3.52E¯1   ¯1.46E¯1    6.48E¯1   ¯5.35E¯1
¯1.33E¯1     6.61E¯1    4.39E¯1    2.59E¯1    5.35E¯1
¯1.13E¯17   ¯8.67E¯19   8.04E¯1   ¯2.59E¯1   ¯5.35E¯1
 9.03E¯1     4.37E¯2    7.31E¯2   ¯3.24E¯1    2.67E¯1

      +T←(⍉Q)+.×A+.×Q
 2.36E0     ¯7.44E0    ¯1.40E1    ¯3.66E1    ¯1.09E1
 2.20E0      3.64E0     4.94E0     1.12E1    ¯3.27E1
¯2.21E¯17    2.95E¯17   1.00E0    ¯2.42E0     4.14E1
¯1.21E¯17    3.47E¯18  ¯1.11E¯16   2.00E0    ¯9.70E0
¯1.17E¯17    8.67E¯18   8.67E¯18   1.73E¯17   1.00E1
```

and *A* appears to have three real and two complex eigenvalues. To “diagonalize”

A we proceed as in Example 7.25 to find bases of the appropriate invariant subspaces.

```
P←(PERP⍉(2 2↑T)DBLSHFT A),(PERP⍉A−T[3;3]×ID 5),PERP⍉A−T[4;4]×ID 5
+P←P,PERP⍉A − T[5;5]×ID 5

¯6.72E¯1    ¯6.70E¯2    ¯7.56E¯1    ¯3.89E¯16   ¯5.77E¯1
 3.13E¯1    ¯4.18E¯1     3.78E¯1    ¯4.08E¯1     5.77E¯1
 6.71E¯1     6.70E¯2     3.78E¯1     4.08E¯1    ¯1.24E¯17
¯1.04E¯17    2.92E¯17   ¯3.78E¯1     8.16E¯1    ¯5.77E¯1
¯4.53E¯2    ¯9.03E¯1    ¯1.03E¯16    4.81E¯17    4.88E¯17

      +D←(A+.×P)⌹P
 4.14E0     ¯2.38E0     ¯1.88E¯16    1.26E¯16    6.08E¯16
 7.27E0      1.86E0      2.18E¯16    1.56E¯17   ¯1.98E¯16
 2.48E¯16    3.28E¯17    1.00E0     ¯4.78E¯16   ¯6.93E¯16
¯1.26E¯16   ¯7.70E¯17    4.72E¯16    2.00E0     ¯3.47E¯16
¯1.52E¯16    2.81E¯17    8.05E¯16    4.29E¯16    1.00E1
```

To two significant figures, A has eigenvalues 1, 2, and 10 and the columns of P[;3 4 5] are associated eigenvectors. In addition, A has two more eigenvalues, which are the same as the eigenvalues of the 2-by-2 matrix D[1 2;1 2] or the 2-by-2 matrix T[1 2;1 2]. The corresponding invariant subspace is generated by P[;1 2]. ■

In the above example it is awkward to actually obtain the complex eigenvalues. It is possible to choose a basis of the invariant subspaces that puts the 2-by-2 blocks in a more convenient form. In the example we used the function *DBLSHFT* to compute the matrix $A^2 + bA + cI$, where $\lambda^2 + b\lambda + cI = \det(B - \lambda I)$. Suppose that S is the null space of this matrix and $v \neq 0$ is a vector in S. In the case of interest S is two-dimensional and contains no eigenvectors of A. Thus v and $w = Av$ form a basis of S. Further, since $A^2v + bAv + cv = 0$, we have

$$A[v\,|\,w] = [v\,|\,w]\begin{bmatrix} 0 & -c \\ 1 & -b \end{bmatrix}$$

It follows that if we use bases of the form v, Av, then the 2-by-2 blocks will have the form

$$\begin{bmatrix} 0 & -c \\ 1 & -b \end{bmatrix}$$

and the complex eigenvalues belonging to such a block are the roots of $\lambda^2 + b\lambda + c = 0$.

Example 7.27 Let A and P be the matrices defined in Example 7.26. Let us replace the first two columns of P by v, Av, where $v \neq 0$ is in the column space of $P[;1\ 2]$. Say $v = P[;1]$.

```
      +P←P[;1],(A+.×P[;1]),0 2↓P
¯6.71E¯1     ¯3.26E0      ¯7.56E¯1     ¯3.89E¯16    ¯5.77E¯1
 3.13E¯1     ¯1.75E0       3.78E¯1     ¯4.08E¯1      5.77E¯1
 6.71E¯1      3.26E0       3.78E¯1      4.08E¯1     ¯1.24E¯17
¯1.04E¯17     6.25E¯17    ¯3.78E¯1      8.16E¯1     ¯5.77E¯1
¯4.53E¯2     ¯6.75E0      ¯1.03¯16      4.81E¯17     4.88E¯17

      +D←(A+.×P)⌹P
 0.00E0      ¯2.50E1      ¯3.13E¯17     1.32E¯16     6.95E¯16
 1.00E0       6.00E0       3.00E¯17     1.81E¯18    ¯2.80E¯17
 0.00E0       9.48E¯16     1.00E0      ¯4.91E¯16    ¯6.50E¯16
 0.00E0      ¯2.47E¯16     4.78E¯16     2.00E0      ¯3.21E¯16
 0.00E0      ¯1.02E¯15     8.09E¯16     4.29E¯16     1.00E1
```

Thus, to two significant figures, the complex eigenvalues of A are roots of $\lambda^2 - 6\lambda + 25 = 0$ or $\lambda = 3 \pm 4i$. ■

The function QR is faster than $JACOBI$ on symmetric matrices. For the system used to compute the examples in this book the function QR works three times faster than $JACOBI$ on the 6-by-6 Hilbert matrix and six times faster than $JACOBI$ on a randomly chosen 10-by-10 symmetric matrix.

Example 7.28 In Example 5.8 the action of $JACOBI$ on the matrix

$$A = \begin{bmatrix} 1 & 2 & 3 \\ 2 & 4 & 5 \\ 3 & 5 & 6 \end{bmatrix}$$

was followed by setting a trace flag. The matrix A was diagonalized in eight iterations. Let us repeat the experiment for QR.

```
      TΔQR←5
      Q←1 QR A
QR[5]
 1.13E1      ¯1.96E¯7      6.78E¯11
¯1.96E¯7     ¯5.16E¯1      4.08E¯4
 6.78E¯11     4.08E¯4      1.71E¯1
QR[5]
 1.13E1       6.94E¯18     2.71E¯18
¯1.03E¯24    ¯5.16E¯1      4.08E¯4
 4.04E¯28     4.08E¯4      1.71E¯1
```

```
      +D←(⍉Q)+.×A+.×Q
1.13E1        ¯2.55E¯18    ¯7.37E¯18
¯6.94E¯18     1.71E¯1      0.00E0
 0.00E0       ¯2.06E¯18    ¯5.16E¯1
```

Two iterations locate the eigenvalue near 11.3, and *STRT* takes care of the rest. Of course one iteration of *QR*, involving as it does a Householder reduction, involves more computation than one iteration of *JACOBI*. ∎

Notice that for symmetric matrices A the columns of `S QR A` are eigenvectors and functions such as *PERP* are not needed.

POLYNOMIALS

The *QR* algorithm can be used to find the roots of polynomials. This follows from the next proposition, which is exercise 18 of Exercises 7.2.

PROPOSITION 7.32 The polynomial $p(\lambda) = (-1)^n(\lambda^n + a_{n-1}\lambda^{n-1} + \cdots + a_1\lambda + a_0)$ is the characteristic polynomial of the matrix

$$\begin{bmatrix} -a_{n-1} & -a_{n-2} & \cdots & -a_1 & -a_0 \\ 1 & 0 & \cdots & 0 & 0 \\ 0 & 1 & \cdots & 0 & 0 \\ & & \cdots & & \\ & & \cdots & & \\ & & \cdots & & \\ 0 & 0 & \cdots & 1 & 0 \end{bmatrix}$$ ∎

The multiplier $(-1)^n$ in Proposition 7.32 does not affect the eigenvalues and will be ignored below. The matrix of Proposition 7.32 [without the multiplier $(-1)^n$] will be referred to as the *companion matrix* of $p(\lambda)$.

EXAMPLE 7.29 Find the roots of $p(x) = x^3 - 6x^2 + 11x - 6$.

Solution The companion matrix is

$$A = \begin{bmatrix} 6 & -11 & 6 \\ 1 & 0 & 0 \\ 0 & 1 & 0 \end{bmatrix}$$

Taking $S = 10$, which is roughly the size of the columns of A, we have

```
      Q←10 QR A

      +D←(⍉Q)+.×A+.×Q
 3.00E0       ¯2.55E0      ¯1.26E1
 0.00E0        1.00E0       4.09E0
¯8.67E¯19     ¯1.73E¯18     2.00E0
```

Thus, to two digits anyway, the roots of $p(\lambda)$ appear to be $\lambda = 1, 2, 3$. We can check this result using the function *AT* from Example 3.5:

```
      ¯6 11 ¯6  1  AT  1  1 ⍉ D
¯5.55E¯17  ¯1.39E¯17  ¯2.78E¯17
```

which is as close as can be expected for this computer system. The exact roots of $p(\lambda)$ are indeed 1, 2, 3, as can be easily checked. ∎

EXAMPLE 7.30 Factor $p(x) = x^5 - 1$.

Solution The companion matrix is

$$A = \begin{bmatrix} 0 & 0 & 0 & 0 & 1 \\ 1 & 0 & 0 & 0 & 0 \\ 0 & 1 & 0 & 0 & 0 \\ 0 & 0 & 1 & 0 & 0 \\ 0 & 0 & 0 & 1 & 0 \end{bmatrix}$$

```
      +T←(⍉Q)+.×A+.×Q←10 QR A
 1.00E0      1.63E¯19    6.51E¯18    8.67E¯19   ¯1.95E¯18
¯1.73E¯18    3.09E¯1    ¯9.51E¯1     3.90E¯18   ¯1.52E¯18
¯3.52E¯18    9.51E¯1     3.09E¯1     1.73E¯18   ¯2.69E¯18
¯2.17E¯19    3.20E¯18    1.30E¯18   ¯8.09E¯1     5.88E¯1
¯1.52E¯18    3.79E¯19    1.30E¯18   ¯5.88E¯1    ¯8.09E¯1
```

We have the obvious root $\lambda = 1$ in the 1;1 position and then two 2-by-2 blocks, indicating complex roots. Let us "diagonalize" *A*.

```
      P1←PERP⍉A−ID 5
      P2←PERP⍉ T[2 3;2 3] DBLSHFT A
      P3←PERP⍉ (¯2 ¯2↑T)DBLSHFT A

      P←P1,(P2[;1],A+.×P2[;,1]),P3[;1],A+.×P3[;,1]
      +D←(A+.×P)⌹P
 1.00E0      0.00E0      3.80E¯17    2.41E¯35    2.04E¯17
¯3.67E¯20    0.00E0     ¯1.00E0      0.00E0     ¯9.17E¯18
¯5.99E¯19    1.00E0      6.18E¯1    ¯1.01E¯34   ¯2.03E¯17
¯2.51E¯18    0.00E0      3.63E¯17   ¯1.73E¯18   ¯1.00E0
¯4.43E¯18    0.00E0      1.54E¯16    1.00E0     ¯1.62E0
```

The blocked-out submatrices (which are variants of companion matrices) show that, to three significant figures,

$$x^5 - 1 = (x - 1)(x^2 - .618x + 1)(x^2 + 1.62x + 1) \quad ∎$$

Line [7] of the function QR prevents infinite loops. When could these arise? From the discussion of the convergence of the k-dimensional power method it appears that whenever a matrix has an eigenvalue of multiplicity 3 or more, an infinite loop should result. This is because QR stops only after the problem has deflated entirely to 2-by-2 matrices (or zero matrices). In fact, QR will work on almost any matrix that is diagonalizable with real eigenvalues. If such a matrix has an eigenvalue λ of multiplicity 3, say, then QR will produce a 3-by-3 block that has eigenvalues λ, λ, and λ and is diagonalizable. But such a block is diagonal in any coordinate system $[P^{-1}(\lambda I)P = \lambda I]$, and the problem deflates to three 1-by-1 problems.

Complex eigenvalues of multiplicity two or greater will pose difficulties for QR, since it cannot separate a complex eigenvalue from its conjugate.

If QR stops on line [7] at some level, then the result will be a "triangular" matrix $Q^T A Q$ with a block larger than 2 by 2. Such a block is not too difficult to handle, since it is not a complete mystery but is known to have repeated eigenvalues. We will not pursue the subject, however (see exercise 29).

There are less obvious ways of tricking QR as well (see exercises 24 and 25). Convergence problems with QR usually become transparent when trace flags are set on lines [5] and [6].

We close the section with an example of a matrix whose eigenvalues are ill conditioned (i.e., unstable). The behavior exhibited is rather typical of non-diagonalizable matrices.

EXAMPLE 7.31 The companion matrix of $(\lambda - 1)^5 = \lambda^5 - 5\lambda^4 + 10\lambda^3 - 10\lambda^2 + 5\lambda - 1$ is

$$A = \begin{bmatrix} 5 & -10 & 10 & -5 & 1 \\ 1 & 0 & 0 & 0 & 0 \\ 0 & 1 & 0 & 0 & 0 \\ 0 & 0 & 1 & 0 & 0 \\ 0 & 0 & 0 & 1 & 0 \end{bmatrix}$$

Since the eigenvalues of A are all of the same magnitude, the QR algorithm, if done in exact arithmetic, cannot find an invariant subspace. Setting a trace flag on line [5] of QR, we have

```
      TΔQR←5
      Q←1 QR A
QR[5]
 1.13E0     ¯7.85E¯1    ¯9.55E¯1    ¯2.67E0     ¯8.24E0
 5.61E¯3     1.06E0      1.71E0      2.51E0      9.11E0
¯1.46E¯13   ¯3.41E¯3     9.92E¯1     2.85E0      7.05E0
¯1.05E¯13   ¯3.15E¯13   ¯1.76E¯3     9.36E¯1     4.84E0
 6.72E¯15   ¯4.18E¯13    1.31E¯10   ¯6.00E¯4     8.87E¯1
```

From the first iteration a zero block appears in the lower left

```
QR[5]
 1.01E0       7.14E¯1     ¯6.35E¯1      2.18E0      ¯6.86E0
¯6.74E¯5      1.01E0      ¯1.69E0       2.50E0      ¯9.59E0
 5.34E¯15     3.79E¯5      9.99E¯1     ¯2.95E0       7.74E0
¯1.52E¯18    ¯6.14E¯15     1.96E¯5      9.93E¯1     ¯5.24E0
 1.28E¯22     3.66E¯19     1.81E¯14     6.72E¯6      9.88E¯1
QR[5]
 1.00E0      ¯7.08E¯1     ¯6.02E¯1     ¯2.13E0      ¯6.70E0
 7.90E¯7      1.00E0       1.69E0       2.50E0       9.63E0
 9.15E¯19    ¯4.40E¯7      1.00E0       2.96E0       7.82E0
¯1.90E¯23    ¯8.83E¯19    ¯2.27E¯7      9.99E¯1      5.29E0
¯5.76E¯26    ¯2.21E¯23     2.66E¯18    ¯7.74E¯8      9.99E¯1
QR[5]
 1.00E0       7.07E¯1     ¯5.99E¯1      2.12E0      ¯6.69E0
¯8.30E¯8      1.00E0      ¯1.69E0       2.50E0      ¯9.64E0
¯2.80E¯22    ¯2.30E¯11     1.00E0      ¯2.96E0       7.82E0
 1.10E¯25    ¯9.37E¯16     5.53E¯8      1.00E0      ¯5.29E0
¯1.30E¯29    ¯1.77E¯23     7.44E¯16     1.56E¯11     1.00E0
QR[5]
 1.00E0      ¯7.07E¯1     ¯5.98E¯1     ¯2.12E0      ¯6.69E0
 7.64E¯8      1.00E0       1.69E0       2.50E0       9.64E0
 2.75E¯24     6.03E¯14     9.99E¯1      2.96E0       7.82E0
 2.73E¯28     1.02E¯17    ¯1.29E¯7      1.00E0       5.29E0
¯1.43E¯37     4.50E¯34    ¯1.07E¯26     1.52E¯23     1.00E0
```

The 5-by-5 problem deflates to a 4-by-4 problem

```
QR[5]
 1.00E0       7.07E¯1     ¯5.98E¯1     ¯2.12E0
¯1.04E¯7      1.00E0      ¯1.69E0      ¯2.50E0
¯6.19E¯30    ¯6.10E¯26     9.99E¯1      2.96E0
¯3.14E¯33     1.23E¯30    ¯1.29E¯7      1.00E0
```

The 4-by-4 deflates to two 2-by-2 problems, which are handled by *STRT*

The resulting block-triangular form is

```
      +T←(⍉Q)+.×A+.×Q
 1.00E0       7.07E¯1     ¯5.98E¯1     ¯2.12E0      ¯6.68E0
¯1.04E¯7      1.00E0      ¯1.69E0      ¯2.50E0      ¯9.64E0
 1.04E¯17     1.73E¯18     9.99E¯1      2.96E0       7.82E0
 7.59E¯18     2.60E¯18    ¯1.29E¯7      1.00E0       5.29E0
 1.41E¯18     4.34E¯19    ¯8.67E¯19     0.00E0       1.00E0
```

We have two blocks, indicating complex eigenvalues and a real eigenvalue near 1.

The matrix A has the eigenvalue 1 with multiplicity 5. The matrix Q is orthogonal to about seventeen digits.

```
      (ID 5)-(⍉Q)+.×Q
 0.00E0       2.17E 18    ¯3.04E¯18    ¯6.29E¯18    ¯1.95E¯18
 2.17E¯18     0.00E0      ¯2.60E¯18    ¯1.52E¯18    ¯9.76E¯19
```

```
¯3.04E¯18   ¯2.60E¯18   1.73E¯18   ¯1.73E¯18   1.08E¯18
¯6.29E¯18   ¯1.52E¯18   ¯1.73E¯18   ¯3.47E¯18   ¯1.14E¯18
¯1.95E¯18   ¯9.76E¯19   1.08E¯18   ¯1.14E¯18   ¯8.67E¯18
```

How near to 1 is the real eigenvalue of T?

```
      )DIGITS 18
WAS 3
      T[5;5]
0.9996068055611783
```

So $T[5;5]$ differs from 1 in the fourth digit. Similarly, computation shows that the complex eigenvalues of T are of the form $1 + z$, where z is a complex number with $|z| \sim 10^{-3}$. ■

The function *QR* produced a result so quickly because inexact arithmetic quickly perturbed A to a diagonalizable matrix. The eigenvalues of A are *ill conditioned* because a perturbation in the eighteenth digit of Q produces a change in the fourth digit of the eigenvalues.

EXERCISES 7.7

(Computer assignment) In exercises 1 through 12 use the function *QR* to find the eigenvalues of the given matrix. (You may wish to use the function *CXEIG* of exercise 23 to compute complex eigenvalues.)

1. $\begin{bmatrix} 2 & 1 & 0 & -1 \\ -1 & 2 & 2 & 3 \\ 1 & -1 & 2 & 0 \\ -1 & 1 & 2 & 4 \end{bmatrix}$

2. $\begin{bmatrix} -8 & -2 & 6 & 3 \\ -1 & 1 & 1 & 0 \\ -12 & -4 & 10 & 3 \\ -12 & 0 & 6 & 7 \end{bmatrix}$

3. $\begin{bmatrix} -5 & 1 & -5 & -3 \\ 2 & 1 & 2 & 2 \\ 10 & -2 & 10 & 6 \\ -2 & 1 & -2 & 0 \end{bmatrix}$

4. $\begin{bmatrix} 4 & -4 & 0 & -3 \\ 4 & -2 & 2 & -4 \\ -5 & 4 & -1 & 5 \\ -4 & 0 & -4 & 5 \end{bmatrix}$

5. $\begin{bmatrix} 3 & 6 & -2 & -2 \\ -4 & -9 & 4 & 2 \\ -4 & -10 & 5 & 2 \\ -4 & -8 & 4 & 1 \end{bmatrix}$

6. $\begin{bmatrix} 1 & -2 & 0 & -2 \\ -2 & -3 & -2 & -4 \\ -2 & 2 & -1 & 2 \\ 2 & 2 & 2 & 3 \end{bmatrix}$

7. $\begin{bmatrix} 3 & 2 & 2 & 2 \\ -4 & -3 & -2 & -2 \\ -4 & -2 & -3 & -2 \\ 4 & 2 & 2 & 1 \end{bmatrix}$

8. $\begin{bmatrix} 3 & 8 & 4 & -4 \\ -6 & -13 & -6 & 6 \\ 2 & 4 & 1 & -2 \\ -8 & -16 & -8 & 7 \end{bmatrix}$

9. $\begin{bmatrix} 2 & 5 & -7 & 6 \\ 3 & -2 & 1 & -4 \\ 3 & 1 & -4 & 2 \\ -8 & 10 & -4 & 13 \end{bmatrix}$

10. $\begin{bmatrix} 14 & -28 & -40 & -27 \\ -7 & 15 & 20 & 13 \\ 15 & -36 & -49 & -35 \\ -12 & 28 & 40 & 29 \end{bmatrix}$

11. $\begin{bmatrix} -1 & -2 & -2 & 2 \\ 4 & 3 & 4 & 0 \\ 12 & 4 & 11 & -8 \\ 12 & 4 & 12 & -5 \end{bmatrix}$

12. $\begin{bmatrix} 1 & 0 & 4 & -2 \\ 2 & 3 & 0 & -2 \\ 0 & 8 & 7 & -8 \\ 2 & 4 & 4 & -3 \end{bmatrix}$

(Computer assignment) In exercises 13 through 20 factor the given polynomial into linear and quadratic factors with real coefficients. (You may wish to use the function *CXCOEF* of exercise 22 to compute the coefficients of the quadratic factors.)

13. $x^4 - x^3 - 7x^2 + x + 6$
14. $x^4 - x^3 - 3x^2 + x + 2$
15. $x^5 - 15x^4 + 85x^3 - 225x^2 + 274x - 120$
16. $x^4 - 9x^3 + 45x^2 - 87x + 50$
17. $x^4 - 8x^3 + 42x^2 - 80x + 125$
18. $x^4 - 2x^3 + x^2 + 2x - 2$
19. $x^6 - 4x^5 - x^4 + 30x^3 - 66x^2 + 64x - 24$
20. $x^5 - 6x^4 + 23x^3 + 8x^2 - 26x - 100$
21. Let B be a 2-by-2 block with complex eigenvalues. Show that if $P[;1] = [1 \;\; 0]^T$ and $P[;2] = BP[;1]$ then $P^{-1}BP = \begin{bmatrix} 0 & -c \\ 1 & -b \end{bmatrix}$ where $x^2 + bx + c$ is the characteristic polynomial of B.

    Hint: See the discussion preceding Example 7.26.
22. Use the result of exercise 21 to write an APL function:

    Name: *CXCOEF*
    Input: A 2-by-2 matrix B with complex eigenvalues
    Result: The vector (b, c) where $x^2 + bx + c = \det(B - xI)$

23. Use the result of exercise 21 to write an APL function:

    Name: *CXEIG*
    Input: A 2-by-2 matrix B with complex eigenvalues
    Result: The vector (α, β) where the eigenvalues of B are $\alpha \pm \beta i$

24. (Computer assignment)

$$A = \begin{bmatrix} -1 & -2 & -2 & 0 \\ -4 & 1 & -2 & -2 \\ 4 & 0 & 3 & 2 \\ -4 & 4 & 0 & -3 \end{bmatrix}$$

    (a) Show that 10 *QR A* returns an identity matrix.
    (b) Define a "random" orthogonal matrix *Q* by

```
Q←100 HHLDR? 4 4ρ100
```

Show that `10 QR (⍉Q)+.A+.×Q` is upper triangular.

(c) Explain why the function QR works on Q^TAQ but not on A.

Hint: Compute `4 SCPOW(¯2↑A)DBLSHFT A` "by hand" (i.e., in calculator mode).

25. Let $A = \left[\begin{array}{c|c} 0 & I \\ \hline I & 0 \end{array}\right]$ where I is an identity matrix of size 2 by 2 or greater. Show that the matrix Q computed on line [5] of QR is an identity matrix and hence line [5] does not change A at all.

26. Let V, W be n-by-k matrices of rank k. Show that V and W have the same column space if and only if $V = WC$, where C is k by k and invertible.

Hint: Propositions 4.12, 2.9, and 2.5.

27. Prove Proposition 7.29.

Hint: Exercises 11, 12, 13, 14 of Exercises 7.2.

28. Prove Proposition 7.31.

Hint: Compute the roots of $\det(A - I)$ obtaining λ_1, λ_2. Then row-reduce $A - \lambda_i I$.

29. Let A be a $2m$-by-$2m$ matrix with eigenvalues $\alpha \pm \beta i$, each repeated m times. Let t be the sum of the diagonal elements of A and let $\delta = \det(A)$.

(a) Show that $\det(A - \lambda I) = \lambda^{2m} - t\lambda^{2m-1} + \cdots + \delta$.

Hint: Elaborate on the proof of Proposition 7.12.

(b) Show that, since $\det(A - \lambda I) = [\lambda - (\alpha + \beta i)]^m[\lambda - (\alpha - \beta i)]^m$,

$$\alpha = -\frac{t}{2m} \quad \text{and} \quad \alpha^2 + \beta^2 = \delta^{1/m}.$$

30. Sketch the geometric algorithm (Figure 7.16) in the case where the ellipse is replaced by a hyperbola.

Note: The asymptotes must be assumed tangent to the hyperbola (tangents at infinity) and both the hyperbolas $\lambda(x')^2 - \mu(y')^2 = \pm 1$ must be drawn in.

What happens with the rectangular hyperbola $(x')^2 - (y')^2 = \pm 1$? Explain.

APPENDIX A

Answers to Selected Exercises

EXERCISES 1.1

1. `2*÷2`   3. `3*2+1`   5. `7*-3÷2`   7. `*4`   9. `(*2)*2`
11. `7+3*÷2`   13. `○÷180` (Notice that this is *monadic* ÷.)   15. `1○○3÷2`
17. `1+÷(2○3)*2` (Notice that $\sec\theta = 1/\cos\theta$.)   19. `((5○4)*2)+(6○5)*3`
21. `10⍟⍟3`   23. `(*○1)-(○1)**1`   25. `*X×Y`   27. `((X*2)+Y*2)*÷2`
29. `B×H÷2`   31. `(○H×R*2)÷3`   33. `÷(1+X*2)*÷2`
35. `(1+X*÷2)÷1-X*2`   37. $\frac{5}{3}$ 39. -22 41. -9 43. $\frac{1}{9}$ 45. 6
47. π 49. 0   51. 1   53. $\frac{2}{3}$ 55. 1   57. 6   59. 1   61. ac/b
63. $z - b + c + d$ 64. $a(b - c)$ 65. x^{y^2} 67. $\tan(\mathrm{Tan}^{-1} x) = x$
69. $1 + e^x$ (The leftmost '+' is monadic.)   71. Hint: Which gives the largest answer on the computer: `(*○1)` or `(○1)**1`?   73. 4   75. 0
77. 1   79. 1

EXERCISES 1.2

1. `¯2 ¯2`   3. $(-\frac{1}{2}, -1)$ 5. `1 0 3 0 5 0`

7. `0 0 0`   9. `5 ¯3 2 ¯2`   11. `0 0 1 2 0 0 4 3`

13. $(-49.832, -49.832, -49.832, -49.832)$ Recall that $\log_b(b^x) = x$.

15. `+/3 ¯2 1×X* 2 1 0`   *Ans.* $= 2.9$

17. `+/1 ¯1 ¯1 3 ¯2×X* 11 10 2 1 0`   *Ans.* $= -.0367, 0, .0210$

19.
```
      (6×1 0 1)+3×0 1 0
6 3 6
      (6×1,(2÷3),1)+1 2 3÷2
```
21. `6.5 5 7.5`

23.
```
      +/⍳30
465
```
25.
```
      K←⍳100
      +/1+K*2
338500
```
27.
```
      I←1+2×0,⍳20
      -/(3*I)÷ I
0.1411
```

29. For Univac 1100 series machines, $n = 49$, $n = 50$ is slightly worse.

31. $\int_0^1 x^2\,dx = \frac{1}{3}$

```
      K←(⍳100)÷100
      (+/K*2)÷100
.338
```

33. $\int_0^\pi \sin x\,dx = 2$

```
      K←○(⍳100)÷100
      (+/1○K)×○÷100
1.99984
```

35. $\int_0^{2\pi} \cos 3x \sin 5x\,dx = 0$

```
      K←○2×(⍳100)÷100
(+/(2○3×K)×1○5×K)×○÷50
```

Answer machine dependent — very close to zero.

37. `2 2 4 4 5 6 6 8 9`

39. `¯3 5 ¯3 5 ¯3 5`

41. `¯6 2 ¯6 4 5 6 ¯6`

43. `⍳3`

45. `V,W` is $(\alpha_1, \alpha_2, \ldots, \alpha_n, \beta_1, \beta_2, \ldots, \beta_m)$, which has $n + m$ or `(⍴V)+⍴W` components.

47. `+/V,W` is $\alpha_1 + \alpha_2 + \cdots + \alpha_n + \beta_1 + \beta_2 + \cdots + \beta_m$ or $\left(\sum_1^n \alpha_i\right) + \left(\sum_1^n \beta_i\right)$.

49. `⍴⍴V` $= \begin{cases} 1 & \text{if } V \text{ is a vector} \\ 0 & \text{if } V \text{ is a scalar} \end{cases}$ $\begin{pmatrix}\text{these are vectors} \\ \text{with one component}\end{pmatrix}$

51. 48 53. 0 55. 6 57. $2^{2^2} = 16$ 59. $\cos(\sin 2\pi) = 1$

60. (a) Years = 1926 + .2534 · steel

61. The result is given in Equations (1.4).

63.

```
      +/(100×C÷G)÷8
78.5
```

65.

```
      K ← ⍳7
      +ΔM← M[K+1]-M[K]
203   202   199   236   175   160   146
```

67. (a)

```
      K ← ⍳7
      +MILAGE ← ΔM÷G[1+K]
13.9   15.5   14.9   15.8   12.7   11   10.7
```

where `ΔM` is the answer to exercise 65

EXERCISES 1.3

1.
```
1 2
3 4
```

3.
```
1 2
```

5.
```
1
0
1
```

7.
```
1  2  ¯1  4
5  1  ¯1  3
0  0   0  0
3  4  ¯1  1
2  3  ¯1  5
```

9.
```
1 2 3 4 5
6 1 2 3 4
3 2 1 6 5
4 5 6 1 2
3 4 5 6 1
```

11.
```
1 1 1
2 2 2
3 3 3
```

13.
```
9 2 7
4 5 6
3 8 1
```

15.
```
1 2 3
4 5 6
1 1 1
```

16. (a) Size is `2 2`. (c) Size is `, 2`. (e) Size is `1 2`. (g) Size is `10 1`.
(i) Size is `1 1`. (k) Size is `, 1`. (m) Size is `2 2`.

17.
```
2 4 5
6 2 4
```

19.
```
 2  2 12
12  2  2
```

21.
```
2 2 4
4 2 2
```

23.
```
1 2 3 2 1 4
4 1 2 3 2 1
```

25.
```
1 2 3
4 1 2
2 1 4
3 2 1
```

27.
```
1 2 3 ¯1
4 1 2 ¯1
```

29.
```
2 1 1 2 3
2 1 4 1 2
```

31. (c) The *z*-scores for *CON* are

−1.40	−1.30	−1.00	−.92	−.40	−.62	.66	.91
1.00	.90	.70	.64	1.00	.14	.45	−.92
.90	.90	.39	−.01	1.30	.64	1.10	.53
.15	.25	.96	1.10	−1.60	.23	−2.70	−1.40
.05	.02	−.46	−.51	.16	.46	.74	1.30
−.42	−.51	−.16	−.11	.41	1.40	.67	1.50
−.42	−.30	−.38	−.13	−.71	−.62	−.60	−.57
.43	.57	−.07	−.08	−.74	−1.30	−.95	−1.30
−1.70	−1.60	−1.20	−1.20	−.46	−1.30	.94	.61
.71	.65	−.07	−.28	−.46	−.26	−.58	.05
.26	1.00	−.84	−.85	−.15	−.89	.82	.42
1.70	2.00	1.00	.98	1.40	1.20	.52	−.47
−1.10	−.88	−.77	−.80	−1.30	−.13	−1.00	1.20
1.10	−.41	2.70	2.80	2.00	2.20	−.37	−1.60
−1.30	−1.30	−.83	−.70	−.43	−1.10	.37	−.15

33.
```
      I←ι10
      J←ι20
      +/+/I∘.−J
−1000
```

35.
```
      I←ι10
      J←ι20
      +/+/(I*2)∘.+J*2
36400
```

37.
```
      I←J←○(ι10)÷6
      +/+/(1○I)∘.×2○J
−.933013
```

39.
```
      I←J←¯1+ι11
      +/+/I∘.÷J+1
166.093
```

41.
```
      I←J←○(ι30)÷30
      ΔA←(○÷30)*2
      A×+/+/I∘.×1○J
10.2
```

43.
```
      I←J←○(ι30)÷30
      ΔA←(○÷30)*2
      ΔA×+/+(2○I)∘.×1○J
¯0.209
```

45.
```
      I←J←○(ι25)÷25×2
      ΔA←(○÷50)*2
      ΔA×+/+/3○I∘.×J
¯0.521
```

46. (b) $A[k; i] = 0$ if and only if $k \mid i$ is zero — that is, if and only if k divides i.

EXERCISES 2.1

1. (a) (0, 4) (c) (0, 18) (e) (b, a)
   (g) $(-2, -6, -4)$ Notice that `3  1÷ρA`, `1=ρV` and `3=ρW`.

2. (b) $\begin{bmatrix} -5 & 5 \\ -1 & 7 \end{bmatrix}$ (d) $\begin{bmatrix} 1 & 0 & 0 \\ 0 & 1 & 0 \\ 0 & 0 & 1 \end{bmatrix}$ (f) Not defined

   (h) $\begin{bmatrix} 2 & 3 & 3 & -2 \\ -2 & -7 & -8 & 6 \\ 0 & 4 & 5 & -4 \\ -2 & -3 & -3 & -2 \end{bmatrix}$ (j) $\begin{bmatrix} 2 \\ 8 \\ 6 \\ -2 \end{bmatrix}$ (l) $\begin{bmatrix} 0 & 0 \\ 0 & 0 \end{bmatrix}$

   (n) $\begin{bmatrix} 2 & 1 \\ 4 & 3 \end{bmatrix}$ (p) $\begin{bmatrix} 4 & 3 \\ 2 & 1 \end{bmatrix}$ (r) $\begin{bmatrix} 1 & 2 & 3 \\ 0 & 0 & 0 \\ 7 & 8 & 9 \end{bmatrix}$

5. Suppose that $A = \begin{bmatrix} a & b \\ c & d \end{bmatrix}$ and $B = \begin{bmatrix} 0 & 1 \\ 0 & 0 \end{bmatrix}$. If $AB = BA$, then $\begin{bmatrix} 0 & a \\ 0 & c \end{bmatrix} = \begin{bmatrix} c & d \\ 0 & 0 \end{bmatrix}$ so $a = d$ and $c = 0$. Another properly chosen B will show $b = 0$.

7. (a) $\begin{bmatrix} 3 & 4 \\ 7 & -6 \end{bmatrix}\begin{bmatrix} x \\ y \end{bmatrix} = \begin{bmatrix} 2 \\ 1 \end{bmatrix}$ (c) $\begin{bmatrix} 3 & 1 & -4 \\ 0 & 1 & 1 \\ 4 & 0 & 0 \end{bmatrix}\begin{bmatrix} x_1 \\ x_2 \\ x_3 \end{bmatrix} = \begin{bmatrix} 7 \\ 1 \\ 0 \end{bmatrix}$

   (e) $\begin{bmatrix} x \\ y \end{bmatrix} = \begin{bmatrix} 4 \\ 3 \end{bmatrix}[t]$ (g) $\begin{bmatrix} 3 & 12 \\ 12 & -1 \\ 0 & 1 \end{bmatrix}\begin{bmatrix} x \\ y \end{bmatrix} = \begin{bmatrix} 0 \\ 0 \\ 1 \end{bmatrix} + \begin{bmatrix} 7 & -1 & 0 \\ 1 & 6 & 0 \\ 0 & 0 & 2 \end{bmatrix}\begin{bmatrix} u \\ v \\ t \end{bmatrix}$

9. (a) 12 (c) 14 (e) $2^{4^6} = 2^{4096}$ (g) 3 (i) 1
   (k) 1 (m) 1 (o) 0 1 0 (q) 2 (s) 2

10. (a) `+/A∧.=0` (c) `+/0∧.=A`

11. For example, $(A^TA)[1; 1] = \Sigma_k A^T[1; k]A[k; 1] = \Sigma_k A[k; 1]^2 = 0$ shows $A[k; 1] = 0$ for all k.

EXERCISES 2.2

1. $I = \begin{bmatrix} 1 & 0 \\ 0 & 0 \end{bmatrix}\begin{bmatrix} a & b \\ c & d \end{bmatrix}$ gives $0 = 1$.

3. $(\alpha L_1 + \beta L_2)A = \alpha L_1 A + \beta L_2 A = \alpha I + \beta I = (\alpha + \beta)I = I$

5. (a) $(B^{-1}A^{-1})(AB) = B^{-1}A^{-1}AB = B^{-1}IB = B^{-1}B = I$ shows $(AB)^{-1} = B^{-1}A^{-1}$ by remarks following Proposition 2.5.

   (b) Choose 2-by-2 matrices at random until you find a pair that works. Use Proposition 2.7 to compute the answers.

7. $AA^{-1} = I$ shows $A = (A^{-1})^{-1}$ by remarks following Proposition 2.5.

9. $A(R_1 - R_2) = AR_1 - AR_2 = I - I = 0$

11. (b) If $d_k = 0$, then `X←K=⍳N` gives a nonzero solution of $AX = 0$.

12. (b) (1, .4) (d) (0, 0)
13. (a) ι3 (c) ι4 (e) ι2
15. (a) No left inverse (c)
```
0.007  0.019  0.086  0.024
0.076  0.084  0.431  0.130
0.022  0.018  0.102  0.027
```
(e) No left inverse
16. (b) No solution (d) (.919, −1.14, 2.52, −1.77) (f) (.661, .188, .248)
17. (a)
```
6.49E¯3   ¯8.91E¯3
5.05E¯3   ¯7.76E¯3
5.25E¯4   ¯9.49E¯6
```
(c)
```
¯0.006 ¯0.006 ¯0.011
 0.003 ¯0.009 ¯0.003
 0.001 ¯0.010 ¯0.008
 0.002  0.000 ¯0.004
```
(e)
```
¯0.007  0.006 ¯0.011  0.001
¯0.001 ¯0.007  0.003 ¯0.003
¯0.002  0.008 ¯0.015  0.006
 0.004  0.006 ¯0.006 ¯0.003
 0.000  0.006  0.000 ¯0.003
```
18. (2)
$$\begin{aligned}((B + C)D)[i;j] &= \sum_k (B + C)[i;k]D[k;j] \\ &= \sum_k (B[i;k] + C[i;k])D[k;j] \\ &= \sum_k B[i;k]D[k;j] + \sum_k C[i;k]D[k;j] \\ &= (BD)[i;j] + (CD)[i;j]\end{aligned}$$
(4)
$$\begin{aligned}((\alpha + \beta)A)[i;j] &= (\alpha + \beta)A[i;j] = \alpha A[i;j] + \beta A[i,j] \\ &= (\alpha A)[i;j] + (\beta A)[i;j] \\ &= (\alpha A + \beta A)[i;j]\end{aligned}$$

EXERCISES 2.3

2. (b) $X = (C - B)U = \begin{bmatrix} 0 & 1 & 0 \\ 2 & -1 & 1 \\ 3 & 1 & 2 \end{bmatrix}$ (d) No solution

   (f) $X = 0$ (h) $X = -HJ = [-3 \ \ -10]$

3. (a) $ZA = B(L_1 - L_2)A = BL_1A - BL_2A = B - B = 0$

   (c) Take the matrix constructed in (b) and either delete rows or add rows of zeros until it is square.
5. (a) (2, 3) is a solution
   (c) 1 1÷4 is not a solution.
   (e) (1, −2, 1) is a solution.
6. (b) (1.2, .313, .702) is not a solution. (d) (−1, 1, −1, 1) is a solution.
9. 3942

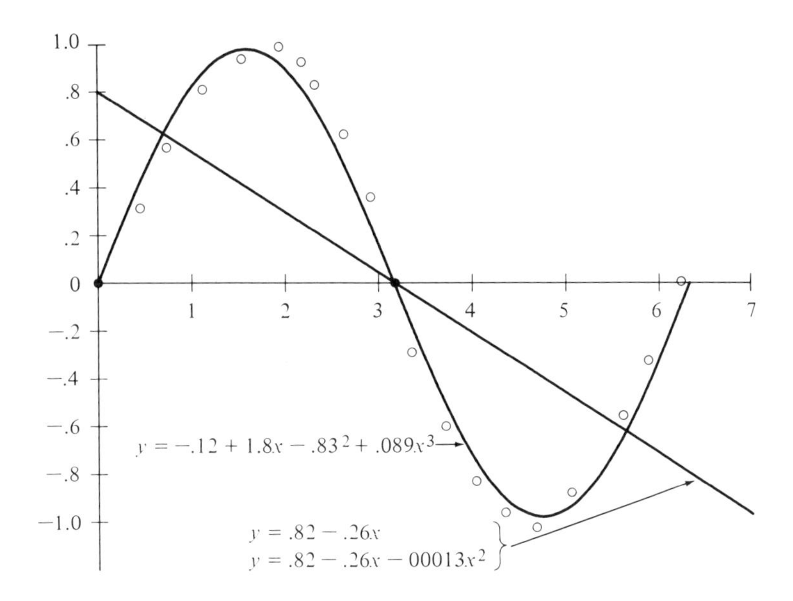

11. The line $y = .82 - .26x$ and the parabola $y = .82 - .26x - .00013x^2$ cannot be distinguished on the plot above.

13. Age = −4.21 + .218 height + .0224 weight
    Height = 15 + 1.6 age + .387 weight
    Weight = 2.52 − .447 age + 1.06 height

15. $a = 1.047$, $b = -1.091$ (see plot on next page)

EXERCISES 2.4

1. Linear, $A = \begin{bmatrix} 2 & -3 \\ 2 & 3 \end{bmatrix}$

3. Affine, $B = \begin{bmatrix} 1 \\ 2 \end{bmatrix}$, $A = \begin{bmatrix} 4 & -12 \\ 6 & 7 \end{bmatrix}$

5. Affine, $B = [1]$, $A = [2 \quad 3 \quad 4]$

7. Quadratic form, $A = \frac{1}{2}\begin{bmatrix} 0 & 1 & 0 & 0 \\ 1 & 0 & 0 & 0 \\ 0 & 0 & 0 & 1 \\ 0 & 0 & 1 & 0 \end{bmatrix}$

9. Quadratic, $c = 4$, $B = [9 \quad 12 \quad -1]$, $A = \begin{bmatrix} 2 & \frac{3}{2} & -1 \\ \frac{3}{2} & 0 & -\frac{3}{2} \\ -1 & -\frac{3}{2} & 0 \end{bmatrix}$

11. Affine, $B = \begin{bmatrix} 1 \\ 1 \\ 1 \end{bmatrix}$, $A = I$

13. Linear, $A = -I$

15. Otherwise

17. Linear, $A =$ 3 6ρ 1 1 , 6ρ0

19. Linear, $A =$ ⍉ 3 12ρ 1 0 0 2 0 0 3 0 0 4 0 0 0

21. Otherwise

23. Quadratic, $c = 1$, $B = (1 = \iota n)$, $A = I$

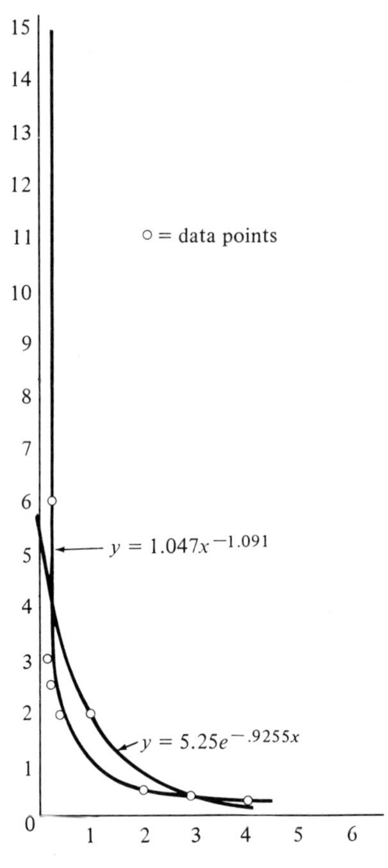

(graph for exercise 15 of Exercises 2.3)

25. Linear, $A = \begin{bmatrix} a & c \\ b & d \end{bmatrix}$

27. Linear, $A = M^T + N$

29. $f(X) = A_1X$, $g(X) = A_2X$ implies $g \circ f(X) = (A_2A_1)X$

31. $f(X) = B + AX$, $g(X) = c + DX + X^TEX$ with $E = E^T$ gives

$$g \circ f(X) = g(B) + (DA + B^TEA)X + X^T(A^TEA)X$$

(Notice that X^TA^TEB is symmetric.)

33. (a) $X = LY - LB$

(c) $g(f(X)) = D + CB + CAX = X$. Taking $X = 0$ gives $D + CB = 0$. So $g \circ f(X) = CAX = X = IX$, which shows $CA = I$, since the matrix of a linear transformation is uniquely determined.

35. (a) $(\alpha A)^T[i;j] = (\alpha A)[j;i] = \alpha A[j;i] = \alpha A^T[i;j]$

(b) $(\frac{1}{2}(A + A^T))^T = \frac{1}{2}(A + A^T)^T = \frac{1}{2}(A^T + A)$

EXERCISES 2.5*

1. (1, 1)

3. $\begin{bmatrix} 1 & 1 & 0 \\ 1 & 0 & -1 \end{bmatrix}$

5. $(-\sin t, \cos t, 1)$ (using $(\rho\ Df) = (\rho Y), \rho X$, we see that Df is a vector).

7. $Df(X) = 3 \times X * 2$

9. $(1\ \ 1\ ⍉\ Df) = 2 \circ X$; $Df[i;j] = 0$ if $i \neq j$

11. $\dfrac{\partial}{\partial x_j} \sum_i x_i^{\alpha} = \alpha x_j^{\alpha-1}$

13. $(D(*X))[i;j] = \dfrac{\partial e^{x_i}}{\partial x_j} = \begin{cases} e^{x_i} & \text{if } i = j \\ 0 & \text{otherwise} \end{cases}$

15. $(2 \div X + . * 2) \times X$

17. $(0, 0)$, neither max nor min

19. $(0, 0)$, second-derivative test fails. Neither max nor min, since $f(x, 0)$ changes sign at $(0, 0)$.

21. $X = 0$ is critical point, second-derivative test fails. Neither max nor min, since a component of $f(X)$ changes sign near 0.

23. Neither max nor min

25. $D(\alpha f)[i;j] = \dfrac{\partial \alpha f_i}{\partial x_j} = \alpha \dfrac{\partial f_i}{\partial x_j}$

27. $g : \mathbf{R}^n \to \mathbf{R}^n$ by $g(X) = \left(\dfrac{\partial f}{\partial x_1}(X), \ldots, \dfrac{\partial f}{\partial x_n}(X)\right)$, so

$$(D^2 f(X))[i;j] = (Dg)[i;j] = \frac{\partial^2 f}{\partial x_j\, \partial x_i}(X).$$

EXERCISES 2.6*

1. At $(0, 0)$, $a(X) = 0$. At $(1, 1)$, $a(X) = \begin{bmatrix} -2 \\ 0 \end{bmatrix} + \begin{bmatrix} 2 & 2 \\ 2 & -2 \end{bmatrix} \begin{bmatrix} x \\ y \end{bmatrix}$

3. $a(X) = B + AX$

5. Using column vectors, $p = 0$ gives $a(t) = \begin{bmatrix} 1 \\ 0 \\ 0 \end{bmatrix} + \begin{bmatrix} 0 \\ 1 \\ 1 \end{bmatrix} [t]$

$$p = 2\pi \text{ gives } a(t) = \begin{bmatrix} 1 & & 0 \\ -2\pi & + & 1 \\ 0 & & 1 \end{bmatrix} [t]$$

7. Curve is unit circle. Tangent is $x + y = \sqrt{2}$.

9. Curve is helix winding counterclockwise about z axis. Tangent is intersection of planes $x = 1$ and $y = z$.

11. $X_1 = (20, 15)$

13. $f(x) = \tan x - x$, $Df(x) = \tan^2(x)$, $X_1 = .7702$ [$\tan(1) = 1.544$].

15. $f(x, y, z, w) = (x^2 - y^2 + z^2 - w^2 - 4, \sin(xyzw), x + y + z + w,$ $x + y \ln zw - 4)$,

$$Df = \begin{bmatrix} 2x & -2y & 2z & -2w \\ yzw \cos(xyzs) & xzw \cos(xyzw) & xyw \cos(xyzw) & xyz \cos(xyzw) \\ 1 & 1 & 1 & 1 \\ 1 & \ln zw & y/z & y/w \end{bmatrix}$$

$x_1 = (5.65, .991, {}^{-}4.75, {}^{-}1.89)$ (use ⌹).

17. Let $w = dz/dt$. Critical point is $(z, w) = (1, 0)$. At $(1, 0)$

$$\begin{bmatrix} \dot{h}_1 \\ \dot{h}_2 \end{bmatrix} = \begin{bmatrix} 0 & 1 \\ 0 & 0 \end{bmatrix} \begin{bmatrix} h_1 \\ h_2 \end{bmatrix} \quad \text{or} \quad \frac{d^2z}{dt^2} = 0$$

19. $\frac{d}{dt} \begin{bmatrix} y_1 \\ y_2 \\ y_3 \end{bmatrix} = \begin{bmatrix} 1 \\ y_3 \\ y_1 y_2 \end{bmatrix} \neq \begin{bmatrix} 0 \\ 0 \\ 0 \end{bmatrix}$ since $0 \neq 1$.

EXERCISES 3.1

1.
```
    ∇Y←F1 X
[1]Y←1+X
    ∇
```

3.
```
    ∇Y←F3 X
[1]Y←+⌿1 2∘.○X
    ∇
```

5.
```
    ∇Y←F5 X
[1]Y←*-X*2
    ∇
```

7.
```
    ∇Y←F7 X
[1]Y←(3⊛÷/4 ¯4∘.○X)÷2
    ∇
```

9.
```
    ∇Z←X F9 Y
[1]Z←10+(X*2)+2○Y
    ∇
```

11.
```
    ∇Z←X F11 Y
[1]Z←(X*Y)-Y*X
    ∇
```

12. (b)
```
    ∇A←B TABLE H
[1]A←B∘.×H÷2
    ∇
```

13.
```
    ∇T←TRIGTABLE TH;A
[1]A←⍉1 2 3∘.○ TH
[2]T←A,÷A
    ∇
```

14. (b)
```
    ∇A←AREA3 M
[1]S←M+.÷2
[2]A←(×/(⍉(4,ρS)ρS)-0,M)*÷2
    ∇
```

15.
```
    ∇P←N DEGFIT M
[1]P←M[;2]⌹M[;1]∘.*0,ιN
    ∇
```

17.
```
∇Z←SIN X;Y
[1] Y←2○500 CHOP 0,X
[2] Z←(X×(1,(499ρ 4 2),1)+.×Y)÷1500
∇
```

The worst approximation is for $x = \pi/2$. On a UNIVAC machine carrying about 18 digits the value of `SIN ○÷2` is correct to 13 digits.

EXERCISES 3.2

1. $\begin{bmatrix} 1 & 0 & 0 & 0 \\ -4 & 1 & 0 & 0 \\ -3 & 0 & 1 & 0 \\ 0 & 0 & 0 & 1 \end{bmatrix}$

3. $\begin{bmatrix} 1 & -\frac{1}{4} & 0 & 0 \\ 0 & 1 & 0 & 0 \\ 0 & -\frac{3}{4} & 1 & 0 \\ 0 & 0 & 0 & 1 \end{bmatrix}$

5. $\begin{bmatrix} 1 & 0 & -4 & 0 \\ 0 & 1 & 2 & 0 \\ 0 & 0 & 1 & 0 \\ 0 & 0 & -4 & 1 \end{bmatrix}$

7. Yes, E[;1 3]

9. No

11. Yes, E[;3]

13. Yes, E[;ι0] (empty set)

15. No

17. $\begin{bmatrix} 1 & 0 & 0 \\ 0 & 1 & 0 \\ 0 & 0 & \frac{1}{2} \end{bmatrix}$

19. $\begin{bmatrix} 1 & 0 & 0 \\ 0 & -1 & 0 \\ 0 & 0 & 1 \end{bmatrix}$

21. $\begin{bmatrix} 1 & 0 & 0 & 0 \\ 0 & 0 & 0 & 1 \\ 0 & 0 & 1 & 0 \\ 0 & 1 & 0 & 0 \end{bmatrix}$

23. $\begin{bmatrix} 1 & 0 & -2 \\ 0 & 1 & -3 \\ 0 & 0 & 1 \end{bmatrix}$

25. $\begin{bmatrix} 1 & -1 & 0 & 0 \\ 0 & 1 & 0 & 0 \\ 0 & -1 & 1 & 0 \\ 0 & -1 & 0 & 1 \end{bmatrix}$

In exercises 26 through 35 different row-reductions may give different augmentation columns.

27. $[E|F] = \left[\begin{array}{ccc|c} 1 & 0 & 0 & 2 \\ 0 & 1 & 1 & 3 \\ 0 & 0 & 0 & 4 \end{array}\right]$, no solution

29. $[E|F] = \left[\begin{array}{cccc|c} 1 & 0 & 0 & 2 & 5 \\ 0 & 1 & 0 & 3 & 6 \\ 0 & 0 & 1 & 4 & 7 \\ 0 & 0 & 0 & 0 & 8 \end{array}\right]$, no solution

31. $[E|F] = \left[\begin{array}{cccc|c} 1 & 2 & 0 & 2 & 5 \\ 0 & 0 & 1 & 4 & 6 \\ 0 & 0 & 0 & 0 & 7 \\ 0 & 0 & 0 & 0 & 0 \end{array}\right]$, no solution

33. $[E|F] = \left[\begin{array}{ccc|c} 1 & 0 & 0 & 2 \\ 0 & 1 & 0 & 3 \\ 0 & 0 & 1 & 4 \\ 0 & 0 & 0 & 5 \\ 0 & 0 & 0 & 6 \end{array}\right]$, no solution

35. $[E|F] = \left[\begin{array}{cccccc|c} 1 & 0 & 2 & 4 & 6 & 0 & 8 \\ 0 & 1 & 3 & 5 & 7 & 0 & 9 \\ 0 & 0 & 0 & 0 & 0 & 1 & 10 \\ 0 & 0 & 0 & 0 & 0 & 0 & 0 \\ 0 & 0 & 0 & 0 & 0 & 0 & 0 \\ 0 & 0 & 0 & 0 & 0 & 0 & 0 \end{array}\right]$, $X = \begin{bmatrix} 8 \\ 9 \\ 0 \\ 0 \\ 0 \\ 10 \end{bmatrix} + t_1 \begin{bmatrix} -2 \\ -3 \\ 1 \\ 0 \\ 0 \\ 0 \end{bmatrix} + t_2 \begin{bmatrix} -4 \\ -5 \\ 0 \\ 1 \\ 0 \\ 0 \end{bmatrix} + t_3 \begin{bmatrix} -6 \\ -7 \\ 0 \\ 0 \\ 1 \\ 0 \end{bmatrix}$

37. No inverse

39. One left inverse is

$\begin{bmatrix} 5 & 2 & 1 & 1 & -1 \\ 7 & 2 & 3 & 0 & -1 \\ 8 & 3 & 3 & 1 & -1 \end{bmatrix}$

41. One left inverse is

$\begin{bmatrix} 2 & 0 & -1 & -1 & 0 & 1 \\ 0 & 1 & 0 & -1 & 0 & 1 \end{bmatrix}$

43. $A^{-1} = \begin{bmatrix} -7 & -2 & 3 & -1 \\ 0 & -1 & 0 & 2 \\ -5 & -1 & 2 & -1 \\ 2 & 0 & -1 & 1 \end{bmatrix}$

45. $\begin{bmatrix} 1 & 0 & 2 & 0 \\ 0 & 1 & 3 & 0 \\ 0 & 0 & 0 & 1 \end{bmatrix}$

47. $\begin{bmatrix} 1 & 0 & .586 & 0 & -.767 & 1.21 \\ 0 & 1 & 2.95 & 0 & -4.39 & 6.92 \\ 0 & 0 & 0 & 1 & -1.65 & 2.24 \end{bmatrix}$

49. $\begin{bmatrix} 1 & 0 & .0612 & 0 & 0 & 0 \\ 0 & 1 & 1.24 & 0 & 0 & 0 \\ 0 & 0 & 0 & 1 & 0 & -1.14 \\ 0 & 0 & 0 & 0 & 1 & .901 \end{bmatrix}$

51. 3

53. −2 4 5

55. 5+ι9

57. 5 6
8 9

59. 3
6
9

61. 1 1ρ 7

63. 0 0 ρ 0

65. 3 3ρ 29

EXERCISES 3.3

1. The echelon form of $[v_1 | v_2 | w_1 | w_2 | w_3 | w_4]$ is
$$\begin{bmatrix} 1 & 0 & 1 & 2 & 3 & 4 \\ 0 & 1 & 5 & 6 & 7 & 8 \end{bmatrix}$$

3. The echelon form of $[v_1 | v_2 | v_3 | w_1 | w_2 | w_3]$ is
$$\begin{bmatrix} 1 & 0 & 0 & 1 & -1 & -1 \\ 0 & 1 & 0 & -1 & 1 & -1 \\ 0 & 0 & 1 & 1 & -1 & 1 \end{bmatrix}$$

5. The echelon form of $[v_1 | v_2 | v_3 | v_4 | w_1 | w_2]$ is
$$\begin{bmatrix} 1 & 0 & 0 & 0 & -2 & 0 & -2 \\ 0 & 1 & 0 & 0 & 3 & -1 & 4 \\ 0 & 0 & 1 & 0 & -4 & 2 & -5 \\ 0 & 0 & 0 & 1 & 7 & 0 & 7 \end{bmatrix}$$

7. rank $= 2$

9. rank $= 3$

For exercises 11 through 15 the echelon forms of $[v_1 | v_2 | \cdots | v_n]$ are

11. $\begin{bmatrix} 1 & 2 & 0 & 3 \\ 0 & 0 & 1 & 4 \end{bmatrix}$

13. $\begin{bmatrix} 1 & 0 & -1 & -2 & 0 & 2 \\ 0 & 1 & 1 & 1 & 0 & 3 \\ 0 & 0 & 0 & 0 & 1 & 4 \end{bmatrix}$

15. $\begin{bmatrix} 1 & 0 & 2 & -1 \\ 0 & 1 & -2 & 0 \\ 0 & 0 & 0 & 0 \\ 0 & 0 & 0 & 0 \end{bmatrix}$

17. (a) Equations 1, 3, and 5 are irredundant.
(b) $n = 2$, $k = 3$, so no solution.

19. (a) Equations 1 and 2 are irredundant.
(b) $n = 3$, $k = 2$, one arbitrary parameter.

22. If $G_1A = E = G_2B$, then $B = G_1^{-1}G_2A$, so $F = G_1^{-1}G_2$. If $B = FA$, F invertible, then F is a product of elementary matrices. So $E = GB = GFA$. Show A and B both reduce to E.

EXERCISES 3.4

1.
```
     ∇Z←F X
[1]Z←X
[2]→(X≤1)/0
[3]Z←X*2
     ∇
```

3.
```
     ∇Z←F X
[1]Z←−1+X
[2]→(X≤−1/0
[3]Z←X−1
[4]→(X≥1)/0
[5]Z←−0○X
     ∇
```

5.
```
     ∇Z←F X
[1]Z←X
[2]→(0=2|⌈X)/0
[3]Z←−X
     ∇
```

7.
```
     ∇Z←X F Y
[1]Z←X×Y
[2]→(3≥(X,Y)+.×(X,Y)+.× 2 2ρ3 −3 −3 4)/0
[3]Z←0
     ∇
```

9. Yes 11. Yes 13. Yes 15. Yes

17. $a = 9999$, $b = 10{,}001 + 1/9999$; no

19.
```
     ∇ Z←F N
[1]    Z←,2
[2]    →(1=N)/0
[3]    Z←F N−1
[4]    Z←Z,1 2+.×Z[N−1]*1 2
     ∇
       F 9
2  10  210  88410  15630000000  4.888E20  4.778E41  4.566E83
   4.169E167
```

21.
```
     ∇ Z←F N
[1]    Z←,1
[2]    →(1=N)/0
[3]    Z←F N−1
[4]    Z←Z,Z[N−1]+2○Z[N−1]
     ∇
       F 5
1  1.54  1.571  1.571  1.571
```

23.
```
     ∇ Z←F N
[1]    Z←1 1
[2]    →(2≥N)/0
[3]    Z←F N−1
[4]    Z←Z,4 1+.×¯2↑Z
     ∇
       F 20
1  1  5  9  29  65  181  441  1165  2929  7589  19310  49660
     126900  325500  833000  2135000  5467000  14010000
       35880000
```

25.
```
     ∇ Z←F N
[1]    Z←1 2 3
[2]    →(3≥N)/0
[3]    Z←F N−1
[4]    Z←Z,1 0 ¯1+.×(¯3↑Z)*3 1 2
     ∇

       F 12
1  2  3  ¯8  ¯56  ¯3109  ¯9666000  ¯9.344E13  ¯8.731E27
    ¯7.623E55  ¯5.811E111   ¯3.376E223
```

EXERCISES 3.5

1. $\begin{bmatrix} x \\ y \\ z \end{bmatrix} = \begin{bmatrix} -2 \\ 3 \\ 0 \end{bmatrix} + t\begin{bmatrix} 1 \\ -2 \\ 1 \end{bmatrix}$

3. No solution

5. $\begin{bmatrix} x_1 \\ x_2 \\ x_3 \\ x_4 \end{bmatrix} = \begin{bmatrix} -.08642 \\ .8898 \\ .2234 \\ .8169 \end{bmatrix}$

7. $\begin{bmatrix} -1.333 & 0 & .333 \\ 1.1670 & 0 & -.1667 \end{bmatrix}$

9. No left inverse

EXERCISES 3.6*

1. $v_1 = (1, 1)$, $a = 0$, $A = \begin{bmatrix} 1 & -1 \\ 1 & 0 \end{bmatrix}$

3. $v_1 = (2, 1)$, $a = (-5, 0)$, $A = \begin{bmatrix} -1 & 1 \\ 1 & 0 \end{bmatrix}$

5. $a = \begin{bmatrix} \sqrt{2} & -1 \\ 1 & 0 \end{bmatrix}$, hence $A^8 = ((A^2)^2)^2 = I$ and $v_{n+8} = A^8 v_n = v_n$

7. $v_1 = (1, 2, 3)$, $a = (2, 0, 0)$, $A = \begin{bmatrix} 0 & -1 & 1 \\ 1 & 0 & 0 \\ 0 & 1 & 0 \end{bmatrix}$, hence

$$(I - A)^{-1} = \begin{bmatrix} 1 & -1 & 1 \\ 1 & 1 & 1 \\ 1 & 1 & 2 \end{bmatrix}, \quad A^8 = \begin{bmatrix} -2 & -3 & 3 \\ 3 & -2 & 0 \\ 0 & 3 & -2 \end{bmatrix}, \quad x_9 = 6$$

9. $A = \begin{bmatrix} .9 & .05 & .05 \\ .05 & .75 & .20 \\ .10 & .20 & .70 \end{bmatrix}$ and $A^{1600} = (A^{40})^{40} = \begin{bmatrix} .424 & .303 & .273 \\ .424 & .303 & .273 \\ .424 & .303 & .273 \end{bmatrix}$

and so the long-term distribution of fleas is 42 percent on color 1, 30 percent on color 2, and 27 percent on color 3.

EXERCISES 3.7*

1.
```
     ∇Z←E CENTSECT I
[1]L:Y←FCN X← 100 CHOP I
[2]Z←I←X[−1 0 + ((×Y)≠×Y[1])⍳1]
[3]→(E≤|−/I)/L
```

3. 2.331

5. 1.230

7. $(-.9684, \pm.1228)$

9. $(.1435, \pm 1.549, \pm 1.849)$

EXERCISES 3.8*

1. $x_1 = 0, \quad x_2 = 1, \quad x_3 = 2; \quad y_1 = 0, \quad y_2 = 1, \quad y_3 = 8; \quad s_1(x) = \frac{3}{2}x^3 - \frac{1}{2}x$, $s_2(x) = \frac{3}{2}(x-2)^3 + 8(x-1) + \frac{1}{2}(x-2)$. The four conditions thus become

$$\begin{aligned} &(1) \quad s_1(0) = 1, \quad s_2(1) = 1 \\ &(2) \quad s_1(1) = 1, \quad s_2(2) = 8 \\ &(3) \quad s_1'(1) = s_2'(1) \\ &(4) \quad s_1''(1) = s_2''(1) \end{aligned}$$

which are easily checked.

3. The output of *SPLINE* is

```
CF
 ¯2.31130    0.00000    87.21100   ¯66.60000
 ¯5.35480    2.31130    93.95500   ¯87.21100
  9.43050    5.35480    68.56900   ¯93.95500
 ¯2.96730   ¯9.43050    99.76700   ¯68.56900
 ¯7.96130    2.96730   113.16000   ¯99.76700
 14.41300    7.96130    78.78700  ¯113.16000
¯19.28900  ¯14.41300   130.89000   ¯78.78700
 21.04400   19.28900    67.25600  ¯130.89000
¯12.88600  ¯21.04400   129.89000   ¯67.25600
  0.00000   12.88600   115.20000  ¯129.89000
```

EXERCISES 4.1

1. The point is *not* on the line.
3. The point is on the line.
5. The lines intersect at the point $(-6, 2)$
7. Intersection for $t = 3$.
9. Parallel and distinct.
11. The point is *not* on the plane.
13. Intersection for $(t_1, t_2, t) = (38, -15, 22)$.
15. No intersection, line parallel to plane.
17. Line of intersection given by $s_1 = 7 - 6s_2$ or $t_1 = 3 - 2\alpha$, $t_2 = 5 - 4\alpha$.
19. Line of intersection given by $s_2 = \frac{7}{6}$ or $t_1 = 3 - 2\alpha$, $t_2 = 5 - 4\alpha$.
21. Distinct parallel planes.
23. Write $s = 1 - t$; then $r = (1-t)p + tq = p + t(q-p) = l(t)$. Now use Proposition 4.1.
25. Since the figure is a parallelogram: $(b-a) + (d-a) = c - a$ or $b + d = c + a$.
27. Let the plane be $p(t_1, t_2) = p_0 + t_1 v_1 + t_2 v_2$ and let the line be $l(t) = q + w$.

If there is no intersection, then the echelon form of $[v_1|v_2|-w|q-p_0]$ is

$$\begin{bmatrix} 1 & 0 & * & 0 \\ 0 & 1 & * & 0 \\ 0 & 0 & 0 & 1 \end{bmatrix}$$

Now apply Proposition 4.5.

EXERCISES 4.2

1.

x	1	0	−2	−9
y	1	4	8	25

x'	−7	0	−10	11
y'	−3	0	−4	5

3.

x	1	2	2	−1	−19
y	−2	−5	−6	5	68

x'	1	0	−2	−3	5
y'	−2	−1	2	3	−7

5.

x	−2	−2	−2
y	3	4	−2
z	1	2	0

x'	5	9	10	13
y'	7	11	11	13
z'	−3	−4	−3	−3

7.

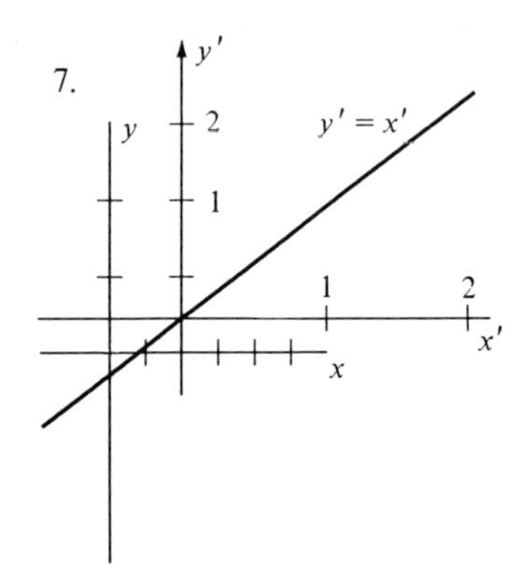

9.

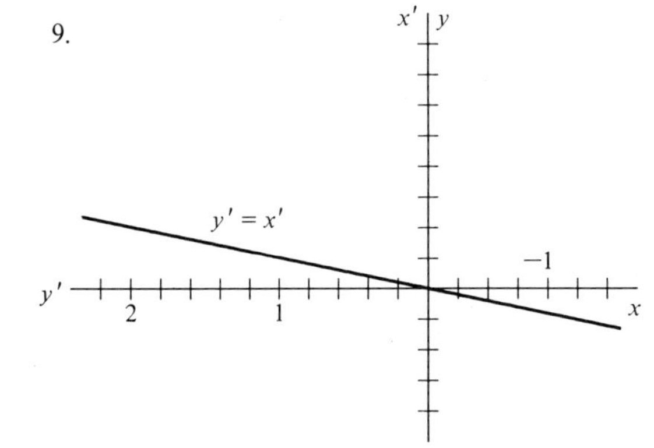

11.

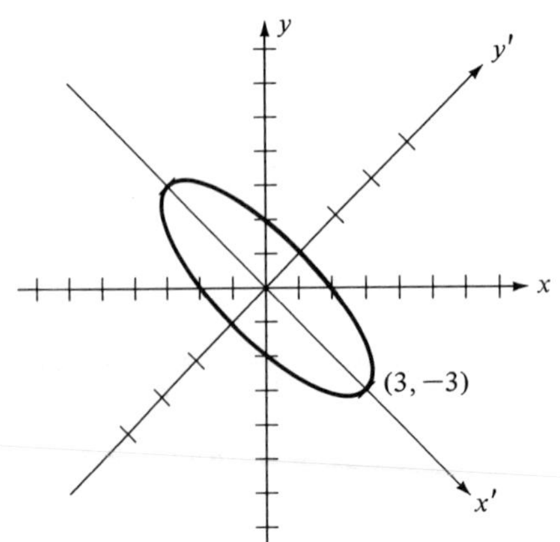

```
      C←2 1∘.○ 10 CHOP 0,○2
      (⍉C),⍉(2 2⍴3 1 ¯3 1)+.×C
 1.00E0       0.00E0       3.00E0      ¯3.00E0
 8.09E¯1      5.88E¯1      3.01E0      ¯1.84E0
 3.09E¯1      9.51E¯1      1.88E0       2.40E¯2
¯3.09E¯1      9.51E¯1      2.40E¯2      1.88E0
¯8.09E¯1      5.88E¯1     ¯1.84E0       3.01E0
¯1.00E0       1.03E¯18    ¯3.00E0       3.00E0
¯8.09E¯1     ¯5.88E¯1     ¯3.01E0       1.84E0
¯3.09E¯1     ¯9.51E¯1     ¯1.88E0      ¯2.40E¯2
 3.09E¯1     ¯9.51E¯1     ¯2.40E¯2     ¯1.88E0
 8.09E¯1     ¯5.88E¯1      1.84E0      ¯3.01E0
 1.00E0      ¯2.07E¯18     3.00E0      ¯3.00E0
```

13. $p_0 = (0, 0)$, $p_1 = (2, 0)$, $p_2 = (0, \frac{1}{2})$, $X = \begin{bmatrix} 2 & 0 \\ 0 & .5 \end{bmatrix} X'$.

15. $p_0 = (0, 0)$, $p_1 = (0, -1)$, $p_2 = (1, 0)$, $X = \begin{bmatrix} 0 & -1 \\ -1 & 0 \end{bmatrix} X'$.

17. $2x + 3y = 8$ and $2x - y = 0$ intersect at $p_0 = (1, 2)$. $2x + 3y = 8$ has parametric equation

$$l(t) = \begin{bmatrix} 4 \\ 0 \end{bmatrix} + t \begin{bmatrix} -\frac{3}{2} \\ 1 \end{bmatrix}$$

so $\begin{bmatrix} -\frac{3}{2} \\ 1 \end{bmatrix}$ points along the x' axis and

$$p_1 - p_0 = a \begin{bmatrix} -\frac{3}{2} \\ 1 \end{bmatrix} \qquad \text{for some } a$$

Similarly

$$p_2 - p_0 = b \begin{bmatrix} \frac{1}{2} \\ 1 \end{bmatrix} \qquad \text{for some } b$$

$$X = \begin{bmatrix} 1 \\ 2 \end{bmatrix} + \begin{bmatrix} -\dfrac{3a}{2} & \dfrac{b}{2} \\ a & b \end{bmatrix} X'$$

19. (a) Coplanar (c) Coplanar

21. $Y' = \begin{bmatrix} -2 & 0 \\ 0 & 2 \end{bmatrix} X'$ 23. $Y = \begin{bmatrix} 0 & 0 \\ 0 & -1 \end{bmatrix} X'$ 25. $Y' = \begin{bmatrix} 1 & 0 \\ 0 & -1 \end{bmatrix} X'$

27. $Y' = \begin{bmatrix} 1 & 0 & 0 \\ 0 & 0 & 0 \\ 0 & 0 & 3 \end{bmatrix} X'$ 29. $Y' = \begin{bmatrix} 1 & 0 & 0 \\ 0 & 0 & 1 \\ 0 & 1 & 0 \end{bmatrix} X'$

31. $Y = \dfrac{1}{\sqrt{2}} \begin{bmatrix} \sqrt{2} - 1 \\ 1 \end{bmatrix} + \dfrac{1}{\sqrt{2}} \begin{bmatrix} 1 & -1 \\ 1 & 1 \end{bmatrix} X$; use $X = \begin{bmatrix} 1 \\ 0 \end{bmatrix} + X'$ and $Y' = R_{\pi/4} X'$

33. $Y = \begin{bmatrix} -1 \\ -2 \end{bmatrix} + \begin{bmatrix} 2 & -1 \\ 2 & -1 \end{bmatrix} X$; use $X = \begin{bmatrix} -1 \\ -2 \end{bmatrix} + \begin{bmatrix} 1 & 1 \\ 1 & 2 \end{bmatrix} X'$ and

$Y' = \begin{bmatrix} 1 & 0 \\ 0 & 0 \end{bmatrix} X'$.

35. $Y = \begin{bmatrix} 0 & 0 & 1 \\ 4 & 1 & -4 \\ 0 & 0 & 1 \end{bmatrix} X$; use $X = \begin{bmatrix} -2 \\ 0 \end{bmatrix} + \begin{bmatrix} 1 & -1 & -1 \\ -3 & 4 & 4 \\ 1 & -1 & 0 \end{bmatrix} X'$ and

$$Y' = \begin{bmatrix} 1 & 0 & 0 \\ 0 & 1 & 0 \\ 0 & 0 & 0 \end{bmatrix} X'$$

36. (b) $X = \begin{bmatrix} -2 \\ 0 \end{bmatrix} + t\begin{bmatrix} 1 \\ 1 \end{bmatrix}$, the line through $(-2, 0)$ and $(1, 1)$

(d) $Y'' = \begin{bmatrix} 1 & 0 \\ 0 & 1 \end{bmatrix} X''$

37. (a) $Y = \begin{bmatrix} -\frac{\sqrt{3}}{2} & \frac{1}{2} & 0 \\ -\frac{1}{2} & -\frac{\sqrt{3}}{2} & 0 \\ 0 & 2 & 1 \end{bmatrix} \begin{bmatrix} 0 & 0 & 1 \\ 0 & 1 & 0 \\ -1 & 0 & 0 \end{bmatrix} \begin{bmatrix} 1 & 0 & 0 \\ 0 & \frac{\sqrt{3}}{2} & -\frac{1}{2} \\ 0 & \frac{1}{2} & \frac{\sqrt{3}}{2} \end{bmatrix}$

$$X = \begin{bmatrix} 0 & 0 & -1 \\ 0 & -1 & 0 \\ -1 & 0 & 0 \end{bmatrix} X$$

39. (a) Let $\underline{A}$ and $\underline{B}$ be the partitioned matrices. Let $\underline{C} = \underline{A}\underline{B}$. Then (Proposition 2.1)

$$\underline{C}[i; j] = \sum_{k=1}^{n+1} \underline{A}[i; k]\underline{B}[k; j]$$

Analyze the cases $i = 1, j = 1$; $i = 1, j \neq 1$; $i \neq 1, j = 1$; and $i \neq 1, j \neq 1$ separately.

(b) Use (a) to check that the given A^{-1} is in fact an inverse.

41. The hint is sufficient.

43.
```
     ∇ X←P NEWTOOLD XPRIME
[1]    X←1 0↓(PREP P)+.×1, [1]XPRIME
     ∇
```

45.
```
     ∇ Z←P NEWAFFN BA ; P
[1]    P←PREP P
[2]    Z←1 0↓(((1=⍳1↓⍴BA),[1]BA)+.×P)⌹P
     ∇
```

EXERCISES 4.3

1. Ellipse
3. Imaginary — by inspection
5. Imaginary ellipse
7. Degenerate hyperbola

9. A line
11. Parabola
13. Minimum at $(.5, -1)$
15. Line of minima through $(.5, 0)$ and $(1, 1)$
17. Saddle point at $(-.5, .5)$
19. Parabola
21. Multiply out $\underline{X}^T\underline{QX} = [1 \quad X^T]\left[\begin{array}{c|c} c & \frac{1}{2}B \\ \hline \frac{1}{2}B & A \end{array}\right]\left[\begin{array}{c} 1 \\ \hline X \end{array}\right]$
23.
```
     ∇ Z←P NEWQUAD Q ;P̲
[1]    P̲←PREP P
[2]    Z←(⍉P̲)+.×(0.5×Q+⍉Q)+.×P̲
[3]    Z←(Z[1;1],2×1↓Z[1;]),[1]0,1 1↓Z
```

EXERCISES 4.4

1. $p_3 - p_0 = 2(p_1 - p_0) + 3(p_2 - p_0)$. A plane in $\mathbf{R}^4$.
3. $p_3 - p_0 = 2(p_1 - p_0) + 3(p_2 - p_0)$. A plane in $\mathbf{R}^4$.
5. $p_3 - p_0 = 2(p_1 - p_0) + 3(p_2 - p_0)$. $p_5 - p_0 = 4(p_1 - p_0) + 5(p_2 - p_0)$. A 3-flat in $\mathbf{R}^5$.
7. The line (1-flat) in $\mathbf{R}^4$ through $(4, 5, 0, 6)$ and $(2, 2, 1, 6)$.
9. The point (0-flat) $(2, 3, 4, 5)$ in $\mathbf{R}^4$.
11. The point $(-15, -20, 47, 25)$.
13. The line and the 3-flat do not intersect.
15. The plane is contained in the 3-flat.
17. Linear with matrix $\begin{bmatrix} 1 & 0 & 0 & 0 & 0 \\ 0 & 2 & 0 & 0 & 0 \\ 0 & 0 & 3 & 0 & 0 \\ 0 & 0 & 0 & 4 & 0 \\ 0 & 0 & 0 & 0 & 5 \end{bmatrix}$
19. $f(x_1', x_2', x_3', x_4') = x_1'^2 - x_2'^2 + x_3'^2$.
21. $f(x_1', x_2', x_3', x_4') = x_1'^2 - x_2'^2 + x_3'^2 + x_4'^2$.
23. `(nρ0)=((n,0)ρ0)+.×ι0`
25. The hint is sufficient.

No change necessary for exercises 27, 29, 31, 33, 35.

EXERCISES 4.5

1. Basis $\{p_i - p_0 \mid i = 1, 2\}$
3. Basis $\{p_i - p_0 \mid i = 1, 2, 3\}$
5. Basis $\{p_i - p_0 \mid i = 1, 2, 3, 4\}$
7. Condition 3 fails.
9. Conditions 1 and 3 fail.

The echelon forms of the matrices of exercises 13 through 19 are

13. $\begin{bmatrix} 1 & 0 & 2 & 0 \\ 0 & 1 & 3 & 0 \\ 0 & 0 & 0 & 1 \end{bmatrix}$
15. $\begin{bmatrix} 1 & 0 & 2 & 4 \\ 0 & 1 & 3 & 5 \\ 0 & 0 & 0 & 0 \end{bmatrix}$

17. $\begin{bmatrix} 1 & 2 & 0 & 1 \\ 0 & 0 & 1 & -1 \\ 0 & 0 & 0 & 0 \\ 0 & 0 & 0 & 0 \end{bmatrix}$

19. $\begin{bmatrix} 1 & 2 & 0 & 1 & -1 \\ 0 & 0 & 1 & -1 & 1 \\ 0 & 0 & 0 & 0 & 0 \\ 0 & 0 & 0 & 0 & 0 \end{bmatrix}$

The answers for exercise 13 are

Column space: $(-2, 3, -1)$, $(-2, 2, 1)$, $(1, -1, 1)$
Null space: $(-2, -3, 1, 0)$
Row space: $(1, 0, 2, 0)$, $(0, 1, 3, 0)$, $(0, 0, 0, 1)$

$$\dim(\text{null space } A^T) = 3 - \dim(\text{column space } A^T) = 0$$

21. Intersecting in a line, the planes cannot be parallel.
23. The planes are skew: nonintersecting and not parallel.
25. The hint is sufficient.

In exercises 27 through 33 the null space is the column space of the following full-rank matrices, which are displayed to two significant digits.

27.
```
 6.7E¯1
 1.0E0
¯3.3E¯1
¯3.5E¯18
```

29.
```
 0.40   0.80
 0.00   1.00
 1.00   0.00
¯0.60  ¯0.20
```

31.
```
 1.00   0.00
¯0.50  ¯0.50
 0.00   1.00
 0.00   1.00
```

33.
```
 1.00   0.00   0.00
¯0.50  ¯0.50   0.50
 0.00   1.00  ¯1.00
 0.00   1.00   0.00
 0.00   0.00   1.00
```

35. $F^{-1}(S)$ spanned by $(-2. -3, 1)$

37. $F^{-1}(S)$ spanned by $(-2, -3, 1)$

39.
```
    ∇ Z←M PIVS E ;I
[1]   Z←⍳0
[2]   I←(,M≠M+(1,1↓⍴E)↑E)⍳1
[3]   →(I>1↓⍴E)/0
[4]   Z←I,M PIVS 1 0↓E
    ∇
```

By induction on the number of nonzero rows of E. The result is ⍳0 if $E = 0$. Otherwise find the index of the first nonzero entry in the first row, call it I. Then the answer is I catenated with the answer for E with the first row discarded.

41.
```
    ∇ Z←COLSPACE A ;E
[1]   E←ECHELON A
[2]   Z←A[;(⌈/,|A)PIVS E]
    ∇
```

43.
```
    ∇ Z←NULLSPACE2 A ;G;E;M
[1]   E←(G←GAUSS A)+.×A←⍉A
[2]   M←⌈/,|A
[3]   Z←⍉((+/(E+M)∨.≠M),0)↓G
    ∇
```

The expression `+/(E+M)∨.≠M` counts the number of rows of E that are not negligible compared to the input data.

EXERCISES 5.1

1. $(\sqrt{2}, \pi/2)$
3. $(\sqrt{2}, \text{Cos}^{-1} 1/\sqrt{3})$
5. $(\sqrt{n-1}, \text{Cos}^{-1} 1/\sqrt{n}) \to (\infty, \pi/2)$ as $n \to \infty$

(*Note:* In exercises 7 through 11, $(A^T A)$ [1; 2] refers to the *exterior* angle at p, whereas A[1; 3] and A[2; 3] refer to interior angles.)

7. (b), (e)
9. (a)
11. (d) — notice that $\sqrt{2} + \sqrt{18} = \sqrt{32}$.
13.
```
    ∇Z←V DIST W
[1]Z←NORM V−W
    ∇
```
15.
```
    ∇Z←V DSPREAD W
[1]Z←(V SPREAD W)÷○÷180
    ∇
```
17.
```
    ∇Z←SIDES A
[1]A←(⍉A)+.×A←A[;1 3 3]−A[;2 1 2]
[2]Z←(1 1⍉A)*÷2
    ∇
```
19. If $u^T X = 0$, then

$$X = \begin{bmatrix} 1 & -1 \\ 1 & 0 \\ 0 & 1 \end{bmatrix} \begin{bmatrix} t_1 \\ t_2 \end{bmatrix} \text{ so take } (1, 1, 0) \text{ and } (-1, 0, 1).$$

21. If $u^T X = 0$, then

$$X = \begin{bmatrix} 1 & -1 \\ 1 & 0 \\ 0 & 1 \end{bmatrix} T,$$

so take $v = (1, 1, 0)$, say. Then $[u|v]^T X = 0$ implies

$$X = t \begin{bmatrix} -\frac{1}{2} \\ \frac{1}{2} \\ 1 \end{bmatrix} \text{ so take } w = (-1, 1, 2), \text{ say.}$$

23. Yes
25. Yes, $f(x, y) = \begin{bmatrix} 3 \\ 4 \end{bmatrix} + \begin{bmatrix} -1 & 0 \\ 0 & 1 \end{bmatrix} \begin{bmatrix} x \\ y \end{bmatrix}$
27. Yes
29. Yes (cf. exercise 23)
31. Yes (cf. exercise 25)
33. No (cf. exercise 28)

35. If the triangle has sides u, v and $u + v$ as in Examples 5.3 and 5.4, then $\|u + v\|^2 = (u + v) \cdot (u + v) = \|u\|^2 + \|v\|^2 + 2u \cdot v$, but $u \cdot v = \|u\| \, \|v\| \cos\theta$ by the definition of θ.

37. The columns of $A = [u|v]$ are linearly dependent if and only if $A^T A$ is not invertible.

$$A^T A = \begin{bmatrix} u \cdot u & u \cdot v \\ v \cdot u & v \cdot v \end{bmatrix}$$

is invertible if and only if $(u \cdot u)(v \cdot v) - (u \cdot v)^2 \neq 0$.

39. $d(f(p), f(q))^2 = \|f(p) - f(q)\|^2 = \|A(p-q)\|^2 = (A(p-q))^T A(p-q)$
$= (p-q)^T A^T A(p-q) = \|p-q\|^2 = d(p,q)$
Similarly for angles.

41. $d(h(p), h(q)) = d(f(g(p)), f(g(q))) = d(g(p), g(q)) = d(p,q)$

43. Reflection in x axis of $\mathbf{R}^2$ is

$$\begin{bmatrix} y_1 \\ y_2 \end{bmatrix} = \begin{bmatrix} 1 & 0 \\ 0 & -1 \end{bmatrix} \begin{bmatrix} x_1 \\ x_2 \end{bmatrix}$$

Rotation of 180° about x axis in $\mathbf{R}^3$ is

$$\begin{bmatrix} y_1 \\ y_2 \\ y_3 \end{bmatrix} = \begin{bmatrix} 1 & 0 & 0 \\ 0 & -1 & 0 \\ 0 & 0 & -1 \end{bmatrix} \begin{bmatrix} x_1 \\ x_2 \\ x_3 \end{bmatrix}$$

If we identify (x_1, x_2) with $(x_1, x_2, 0)$, the effect is the same for points in the xy plane.

45. f is linear,

$$\frac{f(v) \cdot f(w)}{\|f(v)\| \, \|f(w)\|} = \frac{\alpha^2 v \cdot w}{\alpha^2 \|v\| \, \|w\|} = \frac{v \cdot w}{\|v\| \, \|w\|}$$

$d(f(p), f(q)) = \|f(p) - f(q)\| = \|\alpha p - \alpha q\| = |\alpha| \, \|p - q\| = |\alpha| d(p,q).$

47. $y = q + R_\theta X$, where $q = p - R_\theta p$.

49. Let $g(X) = (1/\alpha)X$. Then see the answer to exercise 33, $d(g(p), g(q)) = (1/\alpha) d(p,q)$. Then

$$d\left(\frac{1}{\alpha} f(p), \frac{1}{\alpha} f(q)\right) = \frac{1}{\alpha} d(f(p), f(q)) = \frac{\alpha}{\alpha} d(p,q) = d(p,q)$$

so g is a congruence and, for example,

$$f(X) = \alpha g(X) = \alpha\left(b + R_\theta \begin{bmatrix} 1 & 0 \\ 0 & -1 \end{bmatrix}\right)$$
$$= b' + \alpha R_\theta \begin{bmatrix} 1 & 0 \\ 0 & -1 \end{bmatrix}$$

where $b' = \alpha b$.

51. $\sqrt{v \cdot v} = \sqrt{v_1^2 + v_2^2 + \cdots + v_n^2} \geq 0$. Further $v \cdot v = 0$ must mean that $v_i^2 = 0$ or $v_i = 0$ for all i, since there can be no cancellation among the v_i^2.

53. (c) No

EXERCISES 5.2

1.
```
      ∇ Z←ANG A  ;S
[1]     Z←0
[2]     →(S=A[1;2]+S←⌈/1 1⍉|A)/0
[3]     Z←((○÷2)-¯3○(A[1;1]-A[2;2])÷2×A[1;2])÷2
      ∇
```

3. $Q = \frac{1}{\sqrt{5}}\begin{bmatrix} 1 & 2 \\ -2 & 1 \end{bmatrix}$; eigenvalues: 5, 0; positive $B = \begin{bmatrix} 1 & -2 \\ 0 & 0 \end{bmatrix}\sqrt{2}$

5. $Q = \frac{1}{\sqrt{2}}\begin{bmatrix} 1 & -1 \\ 1 & 1 \end{bmatrix}$; eigenvalues ± 1, not positive

7. $$Q = \begin{cases} \begin{bmatrix} \cos\theta & -\sin\theta \\ \sin\theta & \cos\theta \end{bmatrix}, & (\sin 2\theta) > 0, \\ \begin{bmatrix} \cos\theta & \sin\theta \\ -\sin\theta & \cos\theta \end{bmatrix}, & (\sin 2\theta) \le 0, \end{cases} \qquad \text{eigenvalues } \pm 1$$

The matrix represents reflection in the line through the origin at an angle θ to the x axis. The matrix Q obtained from `ROT ANG A` would just be

$$\begin{bmatrix} \cos\theta & -\sin\theta \\ \sin\theta & \cos\theta \end{bmatrix}$$

9. $\frac{1}{2}\sqrt{30 \pm 2\sqrt{221}}$

11. $\sqrt{3}$, $\sqrt{2}$

13.
```
      +Q←JACOBI A
·0.914697 ¯0.393316 ¯0.092911
 0.390436 0.800646 0.454451
¯0.104353 ¯0.451960 0.885913
```
Eigenvalues are .459, 4.18, −3.64.
A is not positive.

15.
```
      +Q←JACOBI A
 0.657192 ¯0.260956 ¯0.707107
 0.369048 0.929410 0.000000
 0.657192 ¯0.260956 0.707107

      B
  1.403620   0.788205  ¯1.403620
 ¯0.172793   0.615412  ¯0.172793
  0.000000   0.000000   0.000000
```
Eigenvalues are 4.56, .439, 0.
A is positive.

17.
```
      +Q←JACOBI A←HILB 6
¯0.062 ¯0.615 ¯0.749 0.011  0.001  0.240
 0.491  0.211 ¯0.441 ¯0.180 ¯0.035 ¯0.698
¯0.535 0.366 ¯0.321 0.604 0.241 ¯0.231
¯0.417 0.395 ¯0.254 ¯0.444 ¯0.625 0.133
 0.047 0.388 ¯0.212 ¯0.442 0.690 0.363
 0.541 0.371 ¯0.181 0.459 ¯0.272 0.503
```
Eigenvalues are 1.62, .242, .0163, .000616, .0000126.
A is positive definite.

```
    B
¯1.54E¯3    1.22E¯2    ¯1.33E¯2    ¯1.03E¯2    1.17E¯3    1.34E¯2
¯3.03E¯1    1.04E¯1     1.80E¯1     1.94E¯1    1.91E¯1    1.82E¯1
¯9.53E¯1   ¯5.61E¯1    ¯4.08E¯1    ¯3.24E¯1   ¯2.69E¯1   ¯2.31E¯1
 3.95E¯5   ¯6.37E¯4     2.14E¯3    ¯1.57E¯3   ¯1.57E¯3    1.63E¯3
 4.11E¯7   ¯1.17E¯5     7.92E¯5    ¯2.06E¯4    2.27E¯4   ¯8.94E¯5
 3.07E¯2   ¯8.91E¯2    ¯2.96E¯2     1.70E¯2    4.63E¯2    6.42E¯2
```

19. Let $p_0 = (-1, -1.5, 1.5)$, $Q = JACOBI\ A$, and $X = p_0 + QX'$. Then $f(x', y', z') = 9.25 - 1.555(x')^2 - .1981(y')^2 - 3.247(z')^2$, so p_0 is a maximum point.

21. Writing $f(X) = c + BX + X^TAX$, one has

```
        ECHELON A,⍉-B÷2
1.00000E0      ¯2.60209E¯18    2.00000E0      ¯8.67362E¯19
1.73472E¯18     1.00000E0      3.00000E0       4.33681E¯19
0.00000E0      ¯2.60209E¯18   ¯2.60209E¯18     1.00000E0
```

Thus the linear term cannot be eliminated and there are no critical points.

23. If $f(X) = c + B \cdot X + X^TAX$, B considered a vector, then the critical points are the solutions of $(Df)(X) = B + 2AX = 0$. On the other hand the coordinate change $X = p_0 + X'$ will eliminate the linear term if and only if $B + 2Ap_0 = 0$.

25. The hint is sufficient.

27. $(Q^TQ)[K;K] = Q^T[K;K]Q[K;K] + Q^T[K;L]Q[L;K] = R_\theta^T R_\theta + 0 = I$;
$(Q^TQ)[K;L] = Q^T[K;K]Q[K;L] + Q^T[K;L]Q[L;L] = 0 + 0 = 0$, and so on.

29. (a) $D[K;K]^2 = (Q^T(A^TA)Q)[K;K] = Q^T[K;]((A^TA)Q)[;K]$
$= Q[;K]^T(A^TA)Q[;K]$

(b) $U^TU = (AQ[;K]D[K;K]^{-1})^TAQ[;K]D[K;K]^{-1}$
$= D[K;K]^{-1}Q[;K]^TA^TAQ[;K]D[K;K]^{-1}$
$= D[K,K]^{-1}D(K,K]^2D[K,K]^{-1}$ by part (a)

(c) $0 = (Q^TA^TAQ)[;L] = Q^TA^TAQ[;L]$ implies $A^TAQ[;L] = 0$, since Q^T has the left inverse Q. Proposition 2.14 now shows that $AQ[;L] = 0$.

(d) $QQ^T = (QQ^T)[\iota N, \iota N] = Q[\iota N;K]Q^T[K;\iota N] + Q[\iota N;L]Q^T[L;\iota N]$
$= Q[;K]Q[;K]^T + Q[;L]Q[;L]^T$

(e) $U\Lambda V = AQ[;K]D[K;K]^{-1}D[K;K]Q[;K]^T$
$= AQ[;K]Q[;K]^T$
$= AQQ^T - AQ[;L]Q[;L]^T$ by part (d)
$= A$, since $Q^T = Q^{-1}$

31. The hint is sufficient.

33. The hint is sufficient.

EXERCISES 5.3*

1. (a) Obvious (b) The eigenvalues are $-1, 2$.
3. (a) Eigenvalues: $-6.16, -6, -4.24, -4, .162, .236$; saddle
   (b) Eigenvalues: $4.24, 4, 6.16, 6, -.236, -.162$; saddle
   (c) Eigenvalues: $-4.24, -4, -2.41, -2, .236, .414$; saddle

EXERCISES 5.4

1. $(-2, 1)$

3. $S^\perp = \{0\}$, basis is empty set $((n, 0)\rho 0)$.

5. $(-2, -3, 1)$

7. $(1, -1, 1)$

9. $\begin{bmatrix} 0 & -3 \\ -2 & -4 \\ 1 & 0 \\ 0 & -5 \\ 0 & 1 \end{bmatrix}$

11. (a) Since s in S implies $s \perp v$ for every v on $S^\perp$ one has S contained in $(S^\perp)^\perp$. but $\dim S = \dim (S^\perp)^\perp$.

    (b)Null space $(A) =$ (row space $A)^\perp$, so null space $(A)^\perp =$ row space (A). The rows of

$$\begin{bmatrix} 1 & 1 & -1 \\ 2 & -1 & -1 \end{bmatrix}$$

are independent.

13. $v^\parallel = (1, 0)$, $v^\perp = (0, 1)$, dist $= 1$

15. $v^\parallel = (1, .5, .5)$, $v^\perp = (0, -.5, .5)$, dist $= .707$

17. $v^\parallel = (3.96, -.515, 3.97, .099)$, $v^\perp = (.039, .515, .029, -.099)$, dist $= .526$

19. 54.8

21. $U = \begin{bmatrix} 1 & 0 \\ 0 & 1 \end{bmatrix}$

23. $U \quad \begin{bmatrix} 1/\sqrt{2} & 1/\sqrt{6} & -1/\sqrt{3} \\ 1/\sqrt{2} & -1/\sqrt{6} & 1/\sqrt{3} \\ 0 & 2/\sqrt{6} & 1/\sqrt{3} \end{bmatrix}$

25. $U = \frac{1}{2}\begin{bmatrix} 1 & 1 \\ -1 & 1 \\ 1 & 1 \\ -1 & 1 \end{bmatrix}$

27. $P_S = \frac{1}{2}\begin{bmatrix} 1 & -1 \\ -1 & 1 \end{bmatrix}, \quad P_{S^\perp} = \frac{1}{2}\begin{bmatrix} 1 & 1 \\ 1 & 1 \end{bmatrix}, \quad R_S = \begin{bmatrix} 0 & -1 \\ -1 & 0 \end{bmatrix}$

    $v^\parallel = (0, 0), \quad v^\perp = (1, 1) = v, \quad \text{dist} = \sqrt{2}$

29. $P_S = \frac{1}{2}\begin{bmatrix} 2 & 0 & 0 \\ 0 & 1 & 1 \\ 0 & 1 & 1 \end{bmatrix}, \quad P_{S^\perp} = \frac{1}{2}\begin{bmatrix} 0 & 0 & 0 \\ 0 & 1 & -1 \\ 0 & -1 & 1 \end{bmatrix}, \quad R_S = \begin{bmatrix} 1 & 0 & 0 \\ 0 & 0 & 1 \\ 0 & 1 & 0 \end{bmatrix}$

    $v^\parallel = \frac{1}{2}(2, 5, 5), \quad v^\perp = \frac{1}{2}(0, -1, 1), \quad \text{dist} = 1/\sqrt{2}$

31. $P_S = \frac{1}{4}\begin{bmatrix} 3 & -1 & 1 & 1 \\ -1 & 3 & 1 & 1 \\ 1 & 1 & 3 & -1 \\ 1 & 1 & -1 & 3 \end{bmatrix}, \quad P_S^\perp = \frac{1}{4}\begin{bmatrix} 1 & 1 & -1 & -1 \\ 1 & 1 & -1 & -1 \\ -1 & -1 & 1 & 1 \\ -1 & -1 & 1 & 1 \end{bmatrix}$

$$R_S = \frac{1}{2}\begin{bmatrix} 1 & -1 & 1 & 1 \\ -1 & 1 & 1 & 1 \\ 1 & 1 & 1 & -1 \\ 1 & 1 & -1 & 1 \end{bmatrix}$$

    $v^\parallel = (1, 0, 1, 0), \quad v^\perp = (0, 0, 0, 0), \quad \text{dist} = 0$

33. The hint is sufficient.

35. $Y = \frac{1}{2}\begin{bmatrix} 1 \\ 1 \end{bmatrix} + \frac{1}{2}\begin{bmatrix} 1 & -1 \\ -1 & 1 \end{bmatrix}X; \quad Y = \begin{bmatrix} 1 \\ 1 \end{bmatrix} + \begin{bmatrix} 0 & -1 \\ -1 & 0 \end{bmatrix}X; \quad \frac{1}{2}\begin{bmatrix} 1 \\ 1 \end{bmatrix}$

37. $P_S = UU^T = \begin{bmatrix} 0 & 0 \\ 0 & 0 \end{bmatrix}$, hence: $Y = \begin{bmatrix} 1 \\ 1 \end{bmatrix}$ (constant function);

    $Y = \begin{bmatrix} 2 \\ 2 \end{bmatrix} - X; \qquad \begin{bmatrix} 1 \\ 1 \end{bmatrix}$

39. The plane is parallel to the xy plane. Hence, by inspection,

$$Y = \begin{bmatrix} 0 \\ 0 \\ 1 \end{bmatrix} + \begin{bmatrix} 1 & 0 & 0 \\ 0 & 1 & 0 \\ 0 & 0 & 0 \end{bmatrix} X; \quad Y = \begin{bmatrix} 0 \\ 0 \\ 2 \end{bmatrix} + \begin{bmatrix} 1 & 0 & 0 \\ 0 & 1 & 0 \\ 0 & 0 & -1 \end{bmatrix} X; \quad \begin{bmatrix} 0 \\ 0 \\ 1 \end{bmatrix}$$

41. $$Y = \tfrac{1}{2}\begin{bmatrix} 1 \\ -1 \\ 2 \end{bmatrix} + \tfrac{1}{2}\begin{bmatrix} 1 & 1 & 0 \\ 1 & 1 & 0 \\ 0 & 0 & 0 \end{bmatrix} X; \quad Y = \begin{bmatrix} 1 \\ -1 \\ 2 \end{bmatrix} + \begin{bmatrix} 0 & 1 & 0 \\ 1 & 0 & 0 \\ 0 & 0 & -1 \end{bmatrix} X; \quad \tfrac{1}{2}\begin{bmatrix} 1 \\ -1 \\ 2 \end{bmatrix}$$

43. $$Y = \tfrac{1}{3}\begin{bmatrix} 1 \\ -1 \\ 3 \\ 1 \end{bmatrix} + \tfrac{1}{3}\begin{bmatrix} 2 & 1 & 0 & -1 \\ 1 & 2 & 0 & 1 \\ 0 & 0 & 0 & 0 \\ -1 & 1 & 0 & 2 \end{bmatrix}$$

$$Y = \tfrac{2}{3}\begin{bmatrix} 1 \\ -1 \\ 3 \\ 1 \end{bmatrix} + \tfrac{1}{3}\begin{bmatrix} 1 & 2 & 0 & -2 \\ 2 & 1 & 0 & 2 \\ 0 & 0 & -3 & 0 \\ -2 & 2 & 0 & 1 \end{bmatrix} X; \quad \tfrac{1}{3}\begin{bmatrix} 1 \\ -1 \\ 3 \\ 1 \end{bmatrix}$$

45. $$Y = \tfrac{1}{2}\begin{bmatrix} 2 \\ -1 \\ 0 \\ 1 \end{bmatrix} + \tfrac{1}{6}\begin{bmatrix} 0 & 0 & 0 & 0 \\ 0 & 1 & -2 & 1 \\ 0 & -2 & 4 & -2 \\ 0 & 1 & -2 & 1 \end{bmatrix} X$$

$$Y = \begin{bmatrix} 2 \\ -1 \\ 0 \\ 1 \end{bmatrix} + \tfrac{1}{3}\begin{bmatrix} -3 & 0 & 0 & 0 \\ 0 & -2 & -2 & 1 \\ 0 & -2 & 1 & -2 \\ 0 & 1 & -2 & -2 \end{bmatrix} X; \quad \tfrac{1}{2}\begin{bmatrix} 2 \\ -1 \\ 0 \\ 1 \end{bmatrix}$$

47. $$Y = \tfrac{1}{4}\begin{bmatrix} 1 \\ 1 \\ 1 \\ 1 \end{bmatrix} + \tfrac{1}{4}\begin{bmatrix} 3 & -1 & -1 & -1 \\ -1 & 3 & -1 & -1 \\ -1 & -1 & 3 & -1 \\ -1 & -1 & -1 & 3 \end{bmatrix} X$$

$$Y = \tfrac{1}{2}\begin{bmatrix} 1 \\ 1 \\ 1 \\ 1 \end{bmatrix} + \tfrac{1}{2}\begin{bmatrix} 1 & -1 & -1 & -1 \\ -1 & 1 & -1 & -1 \\ -1 & -1 & 1 & -1 \\ -1 & -1 & -1 & 1 \end{bmatrix} X; \quad \tfrac{1}{4}\begin{bmatrix} 1 \\ 1 \\ 1 \\ 1 \end{bmatrix}$$

49. $$Y = f(X) = (1, 7, 6, -10); \quad Y = \begin{bmatrix} 2 \\ 14 \\ 12 \\ -20 \end{bmatrix} - X; \quad (1, 7, 6, -10)$$

EXERCISES 5.5

1. (b) *SPFLCTR* $A[;1]$ is

$$Q_1 = \begin{bmatrix} 1 & \epsilon & 0 & 0 \\ \epsilon & -1 & 0 & 0 \\ 0 & 0 & -1 & 0 \\ 0 & 0 & 0 & -1 \end{bmatrix}$$

and $Q_1^T A$ is

$$\begin{bmatrix} 1 & 1 & 1 \\ 0 & 0 & \epsilon \\ 0 & -\epsilon & 0 \\ 0 & 0 & \epsilon \end{bmatrix}$$

`1 1↓`$Q_1 A$ is

$$A_2 = \begin{bmatrix} 0 & \epsilon \\ -\epsilon & 0 \\ 0 & \epsilon \end{bmatrix}$$

and `SPRFLCTR` A_2[;2] is

$$Q_2 = \begin{bmatrix} 1/\sqrt{2} & -1/\sqrt{2} & 0 \\ -1/\sqrt{2} & -1/\sqrt{2} & 0 \\ 0 & 0 & 1 \end{bmatrix}$$

Set $A_3 =$ `1 1↓`$Q_2 A_2$ and $Q_3 =$ `SPRFLCTR` A_3. Then

$$Q = Q_1 \left[\begin{array}{c|c} 1 & 0 \\ \hline 0 & Q_2 \end{array}\right] \left[\begin{array}{c|c} I & 0 \\ \hline 0 & Q_3 \end{array}\right]$$

3. $A = \left(\frac{1}{\sqrt{2}} \begin{bmatrix} 1 & -1 \\ 1 & 1 \end{bmatrix}\right) \left(\frac{1}{\sqrt{2}} \begin{bmatrix} 2 & 4 & 5 \\ 0 & 0 & 1 \end{bmatrix}\right)$

5. $A = \begin{bmatrix} 1/\sqrt{2} & 1/\sqrt{6} & -1/\sqrt{3} \\ 0 & 2/\sqrt{6} & 1/\sqrt{3} \\ 1/\sqrt{2} & -1/\sqrt{6} & 1/\sqrt{3} \end{bmatrix} \begin{bmatrix} \sqrt{2} & 1/\sqrt{2} & 1/\sqrt{2} \\ 0 & 3/\sqrt{6} & 1/\sqrt{6} \\ 0 & 0 & 2/\sqrt{3} \end{bmatrix}$

7.
```
    ∇Z←PRJCTR A;U
[1] Z←U+.×⍉U←ORTHO A
    ∇
```

9.
```
    ∇Z←RFLCTR A
[1] Z←-(ID 1↑ρA)-2×PRJCTR A
    ∇
```

11.
```
    ∇Z←ORDIST A
[1] Z←((A[;1]-(PRJCTR SBSP A)+.×A[;1])+.*2)*÷2
    ∇
```

or

```
    ∇Z←ORDIST A
[1] Z←((A[;1]COMPS SBSP A)[;2]+.*2)*÷2
    ∇
```

13.
```
    ∇Z←REFL A;P
[1] Z←-(ID 1↑ρA)-2×P←PRJCTR SBSP A
[2] Z←(2×A[;1]-P+.×A[;1]),Z
    ∇
```

14. Check answers by comparing `A+.×U` to zero. The dimension can be checked from the answers to exercises 40 through 47 of Exercises 5.4.

16. No solution. There is a plane of least-squares solutions. $p_0 = (-5.580, -8.060, 1,1)$, $p_1 = (-3.580, -5.060, 0, 1)$, $p_2 = (.424, -.060, 0, 0)$

18. $(-.08641, .88981, .22340, .81690)$

EXERCISES 5.6*

1. (a) $\bar{p} = (0, 0)$, $AA^T = 4\begin{bmatrix} a^2 & 0 \\ 0 & 1 \end{bmatrix}$, so the principal 1-flat is the x axis if $a > 1$ and the y axis if $a < 1$.

   (b) This is part (a) for $a = 1$. Since $\bar{p} = 0$, $p_i^{\perp} = p$ is $p_i \cdot u$, where $u = (-\sin\theta, \cos\theta)$ and $\Sigma\,(p_i, u)^2 = 4$, independent of θ.

3. The stable axes are parallel to the longest and shortest sides; the unstable axis is parallel to the side of intermediate length.

5. The stable axes are parallel to (.927, .376, 0) and (0, 0, 1). The unstable axis is parallel to (−.376, .927, 0).

6. (a) $Q^TQ = I$, hence $A = (Q^T)^T = (Q^T)^{-1}$.

   (b) Obvious. (c) $1 = \|(U, V)[i;]\| \geq \|U[i;]\|$.

8. The hint is sufficient.

9. The hint is sufficient.

EXERCISES 6.1

1. Maximize $z = 2x + 3y$
   subject to $-5x + 6y \leq -7$
   $7x + 8y \leq 9$
   $x, y \geq 0$

3. Minimize $z = -2x - 3y$
   subject to $5x - 6y \geq 7$
   $-7x - 8y \geq -9$
   $x, y \geq 0$

5. Maximize $z = 2x + 3y$
   subject to $5x - 6y \leq 7$
   $-5x + 6y \leq -7$
   $7x + 8y \leq 9$
   $x, y, \geq 0$

7. Maximize $z = 2x + 3(y' - z)$
   subject to $5x - 6(y' - z) \leq 7$
   $7x + 8(y' - z) \leq 9$
   $x, y', z \geq 0$

9. Maximize $z = 2x + 3y$
   subject to $5(x' - z) - 6(y' - z) \leq 7$
   $-5(x' - z) + 6(y' - z) \leq -7$
   $7(x' - z) + 8(y' - z) \leq 9$
   $-7(x' - z) - 8(y' - z) \leq -9$
   $x', y', z \geq 0$

11. Maximize $z = -(d^+ + d^-)$
    subject to $2(x^+ - x^-) + 3(y^+ - y^-) - (d^+ - d^-) \leq 4$
    $-2(x^+ - x^-) - 3(y^+ - y^-) + (d^+ - d^-) \leq -4$
    $x^+, x^-, y^+, y^-, d^+, d^- \geq 0$

13. Maximize $z = -(d_1^+ + d_1^- + d_2^+ + d_2^-)$
    subject to $2(x^+ - x^-) + 3(y^+ - y^-) - (d_1^+ - d_1^-) \leq 4$
    $-2(x^+ - x^-) - 3(y^+ - y^-) + (d_1^+ - d_1^-) \leq -4$
    $2(x^+ - x^-) + 3(y^+ - y^-) - (d_2^+ - d_2^-) \leq 5$
    $-2(x^+ - x^-) - 3(x^+ - x^-) + (d_2^+ - d_2^-) \leq -5$
    $x^+, x^-, y^+, y^-, d_1^+, d_1^-, d_2^+, d_2^- \geq 0$

15. Maximize $z = 2x' + 3y^+ - 3y^-$
    subject to $x' \leq 9 \qquad (x' = x + 10)$
    $x', y^+, y^- \geq 0$

17. Maximize $z = d^+ + d^-$
subject to $4x' + 5y' + d^+ - d^- = 23$
$x', y', d^+, d^- \geq 0 \quad (x' = x + 2, y' = y + 3)$

19. Maximize $z = 2x^+ - 3x^-$
subject to $2(x^+ - x^-) + 3(y^+ - y^-) \leq 4$

21. Maximize $z = 2x + 5y$
subject to $10x + 70y \leq 490$
$2x + 3y \leq 32$
$20x + 10y \leq 240$
$x, y \geq 0$

23. Minimize $w = 4x_{11} + 6x_{12} + 8x_{13} + 5x_{21} + 7x_{22} + 12x_{23}$
subject to $x_{11} + x_{12} + x_{13} \leq 10$
$x_{21} + x_{22} + x_{23} \leq 5$
$x_{11} + x_{21} \geq 2$
$x_{12} + x_{22} \geq 5$
$x_{13} + x_{23} \geq 3$
$x_{ij} \geq 0$, all i, j

25. $A = \begin{bmatrix} 1 & 0 \\ 1 & 1 \\ 1 & 2 \\ 1 & 4 \end{bmatrix}, \quad B = \begin{bmatrix} 1 \\ 0 \\ 3 \\ 5 \end{bmatrix},$ 12 unknowns, 8 constraints

27. $A = \begin{bmatrix} 1 & 1 \\ 1 & 2 \\ 1 & 4 \\ 1 & -5 \\ 1 & 6 \\ 1 & 9 \end{bmatrix}, \quad B = \begin{bmatrix} 1 \\ 2 \\ 5 \\ 4 \\ 7 \\ 8 \end{bmatrix},$ 21 unknowns, 18 constraints

29. $A = \begin{bmatrix} 0 & 0 & 0 \\ 1 & 2 & 4 \\ 9 & 6 & 4 \\ 1 & 1 & 16 \\ 4 & 2 & 1 \\ 1 & 1 & 1 \end{bmatrix}, \quad B = \begin{bmatrix} 0 \\ 3 \\ 1 \\ 1 \\ -2 \\ 1 \end{bmatrix},$ 19 variables, 18 constraints

31. (a), (b) Minimize $w = \|1 - x - 2y\|_1$ (or $w = \|1 - x - 2y\|_\infty$)
subject to $x \geq 0, y \geq 0$

The minimum value of w is 0 if and only if the answer to (a) is "Yes." Part (b) is the same problem as part (a).

33. Minimize $w = \|2x + 3y - 6\|_1$ (or $w = \|2x + 3y - 6\|_\infty$)
subject to $-1 \leq x \leq -2, 2 \leq y \leq 3$

35. Minimize $w = \left\| \begin{bmatrix} 1 \\ 0 \end{bmatrix} - \begin{bmatrix} 1 & 2 & -3 \\ 2 & 4 & -7 \end{bmatrix} \begin{bmatrix} x \\ y \\ z \end{bmatrix} \right\|_1$

subject to $3 \leq x \leq 4, 3 \leq y \leq 4, 2 \leq z \leq 3$

EXERCISES 6.2

1. Vertices: $(-2, 1)$, $(1, 2)$, $(0, -1)$; max at $(1, 2)$, min at $(0, -1)$
3. Vertices: $(-2, 2)$, $(1, 3)$, $(3, 1)$; max at $(1, 3)$, no min
5. Vertices: $(-1, 1)$, $(2, 2)$, $(1, -1)$; max from $(-1, 1)$ to $(1, -1)$, min at $(2, 2)$
7. $v = (-1, 2) + t(1, -1),\ 0 \le t \le 3;\ w = (0, 0, 1)$
9. $v = (-2, -2),\ w = \begin{bmatrix} \frac{3}{2} & & -\frac{3}{2} & -\frac{11}{6} \\ -\frac{3}{2} & + & \frac{5}{2} & \frac{13}{6} \\ 0 & & 1 & 0 \\ 0 & & 0 & 1 \end{bmatrix} \begin{bmatrix} t \\ s \end{bmatrix}$

   for $\begin{bmatrix} 9 & 11 \\ -15 & -13 \end{bmatrix} \begin{bmatrix} t \\ s \end{bmatrix} \le \begin{bmatrix} 9 \\ -9 \end{bmatrix}, \quad t, s \ge 0$
11. $X = (3, 4),\ X' = (0, 0, 3),\ Y = (1, 1, 0),\ Y' = (0, 0)$
13. $X = (4, 4),\ X' = (0, 0, 0),\ Y = (-\frac{1}{3}, \frac{10}{3}, 0) + t(\frac{1}{3}, -\frac{7}{3}, 1),\ 1 \le t \le \frac{10}{7},\ Y' = (0, 0)$
15. $X = (0, 1) + t(1, 1),\ t \ge 0;\ X' = (0, 3, 2),\ Y = (2, 0, 0),\ Y' = (0, 0)$
17. (a) Minimize $w = d^+ + d^-$
   subject to
   $$\begin{aligned} 2x + d^+ - d^- &\ge -1 \\ -2x - d^+ + d^- &\ge 1 \\ x, d^+, d^- &\ge 0 \end{aligned}$$

   (b) The solution occurs all along the ray $l(t) = (0, 1) + t(1, 1),\ t \ge 0$.

   (c) $X = (0, 1) + t(1, 1),\ t \ge 0;\ X' = (2, 2, 0);\ Y' = (0, 0),\ Y = (0, 0, 1)$; that is, $x = 0,\ d^+ = 0,\ d^- = 1$, and hence $w = 1$.

EXERCISES 6.3

1. $X = (4, 1),\ X' = (0, 0, 1),\ Y = (6, 4, 0),\ Y' = (0, 0)$
3. $X = (8, 0),\ X' = (0, 4, 10),\ Y = (2, 0, 0),\ Y' = (0, 10)$
5. $X = (5, 3),\ X' = (0, 3, 0),\ Y = (1, 0, 4),\ Y' = (0, 0)$
7. $X = (6, 5),\ X' = (3, 0, 0),\ Y = (0, 10, 1),\ Y' = (0, 0)$
9. $X = (4, 2),\ X' = (0, 6, 8, 0),\ Y = (10, 0, 0, 10),\ Y' = (0, 0, 110)$
11. $X = (7, 7, 4),\ X' = (0, 0, 8, 0, 1),\ Y = (6, 4, 10, 0, 0),\ Y' = 0$
13. $X = (1, 10, 10),\ X' = (8, 0, 0, 9, 0),\ Y = (0, 9, 8, 0, 8),\ Y' = 0$
15. $X = (4, 10, 0),\ X' = (0, 8, 9, 4, 0),\ Y = (9, 0, 0, 0, 8),\ Y' = (0, 0, 90)$
17. $(2, 2)$, $(5, 4)$, $(4, 1)$
19. $(0, 4)$, $(1, 2)$, $(3, 1)$, $(7, 0)$
21. $(2, 2)$. Set is a single point (maximum is unique).
23. The hint is sufficient.
25. (a) The maximum is attained at a unique vertex in $\mathbf{R}^5$. The intersection of the line and the triangle is the point $(2, 1)$.

   (b) The max is attained along an edge in $\mathbf{R}^5$ but it projects to the single point $(2, 3)$ in $\mathbf{R}^2$.

   (c) The max is attained along a face in $\mathbf{R}^5$ but it projects to the segment from $(2, 1.9)$ to $(4, 1)$ in $\mathbf{R}^2$.

EXERCISES 6.4*

Answers to exercises 1 through 6 can be found in reference 1 or 2.

7. The hint is sufficient.

9. The hint is sufficient.

11. (a) The payoff matrix is skew-symmetric.
    (b) $p = q = (\frac{1}{3}, \frac{1}{6}, \frac{1}{2})$

13. $p = (\frac{1}{2}, 0, \frac{1}{2})$, $q = (0, \frac{1}{2}, \frac{1}{2})$, payoff $= 0$

16. $p = (0, 1)$ is the pure dove strategy. $q = (1, 0)$ is pure hawk. $E(p, p) = (a/2) - c$, $E(q, p) = a > E(p, p)$, the first condition of Definition 6.8 is violated.

18. The hint is sufficient.

20. Adding $\alpha = -a$, $A = \begin{bmatrix} 0 & b \\ 0 & b \end{bmatrix}$, and $E(p, q) = bq[2] \leq b$. So $p = (0, 1)$ is evolutionarily stable.

EXERCISES 7.1

For exercises 1, 3, 5, and 7, the determinant is 6.
For exercises 9, 11, 13, and 15 the determinant is -6.
For exercises 17, 19, 21, and 23 the determinant is 1.
For exercises 25 and 27 the determinant is -6.

29. -5.5
31. 4
33. 8
35. 2
37. $(-1)^{n-k}$
39. 1.414
41. 4
43. 1.414
45. 19.60

47. The matrix is not invertible, since its columns are dependent. Hence $\det(A) = 0$.

49. The hint is sufficient.

51. (b) $\det(A^TA) = (\det A)^2$. Since A^TA is symmetric, apply part (a).

EXERCISES 7.2

1. $\lambda = \pm 1$
3. $\lambda = 1 \pm i$
5. $\lambda = 1, 1, 2$; diagonalizable
7. $\lambda = 1, \pm i$
9. $\lambda = 1, 1, 2, 3$; not diagonalizable

11. If $Av = \lambda_i v$, then $(A + \alpha I)v = Av + \alpha v = (\lambda_i + \alpha)v$.

13. If $Av = \lambda_i v$, then $A^2v = A(Av) = \lambda_i Av = \lambda_i^2 v$.

15. $\det(A^T - \lambda I) = \det((A - \lambda I)^T) = \det(A - \lambda I)$, since $\det(B) = \det(B^T)$.

17. The hint is sufficient.

19. (a) $P_s v = v^{\|} = v = 1 \cdot v$
    (b) $P_s v = P_s v^{\perp} = 0 = 0 \cdot v$
    (c) $\lambda_v = P_s v = P_s(v^{\|} + v^{\perp}) = v^{\|}$, so $\lambda v^{\|} + \lambda v^{\perp} = v^{\|}$, and this equation, which implies $(\lambda - 1)v^{\|} = \lambda v^{\perp}$, so $v^{\|} \cdot (\lambda - 1)v^{\|} = v^{\|} \cdot \lambda v^{\perp}$ or $(\lambda - 1)\|v^{\|} D2\|^2 = 0$, so $\lambda - 1 = 0$ if $v^{\|} \neq 0$. If $v^{\|} = 0$, $v^{\perp} \cdot (\lambda - 1)v^{\|} = v^{\perp} \cdot \lambda v$ or $\lambda \|v^{\perp}\|^2 = 0$.

(d) $[U\,|\,V]^T UU^T[U\,|\,V] = \left[\frac{U^T}{V^T}\right] UU^T[U\,|\,V] = \left[\frac{U^T}{0}\right](U\,|\,V]$

$$= \left[\left[\frac{U^T}{0}\right]U \,\middle|\, \left[\frac{U^T}{0}\right]V\right] = \begin{bmatrix} I & 0 \\ 0 & 0 \end{bmatrix}$$

(e) $\det(A - \lambda I) = \det(Q^{-1}AQ - \lambda I)$, so apply part (d).

21. (a) $z_1z_2 = (\alpha_1 + \beta_1 i)(\alpha_2 + \beta_2 i) = (\alpha_1\alpha_2 - \beta_1\beta_2) + (\alpha_1\beta_2 + \beta_1\alpha_2)i$

$$\begin{pmatrix} \alpha_1 & -\beta_1 \\ \beta_1 & \alpha_1 \end{pmatrix}\begin{pmatrix} \alpha_2 & -\beta_2 \\ \beta_2 & \alpha_2 \end{pmatrix} = \begin{pmatrix} \alpha_1\alpha_2 - \beta_1\beta_2 & -(\alpha_1\beta_2 + \beta_1\alpha_2) \\ \alpha_1\beta_2 + \beta_1\alpha_2 & -\beta_1\beta_2 + \alpha_1\alpha_2 \end{pmatrix}$$

Parts (b), (c), (d), (e), (f) are similar.

23. $\frac{1}{4}(6i, -2 - 5i, 2 + 3i)$

25. $\frac{1}{4}\begin{bmatrix} -1-i & -4 & -2-2i & 1+i \\ 2 & 0 & 0 & 2 \\ 0 & 2-2i & 2-2i & 0 \\ 1-i & 2-2i & 0 & -1+i \end{bmatrix}$

27. Simply perform the multiplications.

29. (a) $\alpha = 2$, a choice of P is $\begin{bmatrix} 1 & 2 \\ 2 & 3 \end{bmatrix}$

(b) $\alpha = 3$, a choice of P is $\begin{bmatrix} 1 & 1 \\ 1 & 2 \end{bmatrix}$

(c) $\alpha = 3$, a choice of P is $\begin{bmatrix} 1 & -1 & 1 \\ 1 & 0 & 0 \\ 0 & 0 & 1 \end{bmatrix}$. Then $P^{-1} = \begin{bmatrix} 0 & 1 & 0 \\ -1 & 1 & 1 \\ 0 & 0 & 1 \end{bmatrix}$

(d) $\alpha = 3$, a choice of P is $\begin{bmatrix} 1 & 1 & -1 \\ 1 & 2 & -2 \\ 1 & 2 & -1 \end{bmatrix}$. Then $P^{-1} = \begin{bmatrix} 2 & -1 & 0 \\ -1 & 0 & 1 \\ 0 & -1 & 1 \end{bmatrix}$

EXERCISES 7.3*

1. $x_n = \dfrac{i^{n-2}}{2}[1 + i + (-1)^n(1 - i)]$

3. $x_n = \dfrac{(\sqrt{3})^{n-2}}{2}[1 + \sqrt{3} + (-1)^n(1 - \sqrt{3})]$

5. $x_n = \dfrac{i^{n-3}}{2}[1 - (-1)^{n-2}]$, eigenvalues are 2, i, $-i$.

7. The hint is sufficient.

9. $\begin{bmatrix} .8 & .5 \\ .2 & .5 \end{bmatrix}^k \to \frac{1}{7}\begin{bmatrix} 5 & 5 \\ 2 & 2 \end{bmatrix}$

The matrix is stochastic and satisfies the hypothesis of Proposition 7.19.

11. $\begin{bmatrix} 1 & \frac{1}{2} & \frac{1}{3} \\ 0 & \frac{1}{2} & \frac{1}{3} \\ 0 & 0 & \frac{1}{3} \end{bmatrix}^k \to \begin{bmatrix} 1 & 1 & 1 \\ 0 & 0 & 0 \\ 0 & 0 & 0 \end{bmatrix}$

The matrix is stochastic but does not satisfy the hypothesis of Proposition 7.19 (although the conclusion of Proposition 7.19 is true).

13. $\begin{bmatrix} .9996 & .0002 \\ -.0006 & 1.003 \end{bmatrix}^k \to \begin{bmatrix} -3 & 2 \\ -6 & 4 \end{bmatrix}$ The matrix is not stochastic.

15. (a) Adding the columns first (see hint) gives $w = v_1 \cdot 1 + v_2 \cdot 1 + \cdots v_n \cdot 1 = 1$

(b) $(AB)[;i] = A(B[;i])$, so apply part (a).

(c)
```
(+⌿A+.×B) = (+⌿A)+.×B
          = 1 +.×B        since +⌿A is (1↓ρA)ρ1
          = +⌿B
```

17. $\begin{bmatrix} \lambda & & * \\ & \lambda & \\ 0 & \ddots & \lambda \end{bmatrix}^k = \begin{bmatrix} \lambda^k & & * \\ & \ddots & \\ 0 & & \lambda^k \end{bmatrix}$, so apply exercise 16

EXERCISES 7.4*

1. Rotation present, reflection not present
3. Rotation not present, reflection present
5. Rotation present, reflection not present
7. Rotation present, reflection not present
9. Notice that $p(\lambda) = (-1)^n(\lambda - \lambda_1)(\lambda - \lambda_2) \ldots (\lambda - \lambda_n)$, which has constant term $(-1)^n(-1)^n\lambda_1\lambda_2 \ldots \lambda_n = \lambda_1\lambda_2 \ldots \lambda_n$.

11.
```
     ∇ Z←AXNANG A ; Q
[1]    Q←HSHLDR 1 BACKSUB(ECHELON A-ID 3), 0
[2]    Z←Q[;1], ÷○÷180×¯2○((⍉Q)+.×A+.×Q)[2;2]
     ∇
```

EXERCISES 7.5*

1. (a) D_1: center 1, radius 5
D_2: center 0, radius 16
D_3: center 2, radius 1
No, the discs overlap.

(b) D_1: center 1, radius 1
D_2: center -3, radius 2
D_3: center 5, radius 2
Yes, the three discs do not meet. Notice that D_2 for A^T has center -3, radius 0; that is, -3 is an eigenvalue.

(c) D_1: center 1, radius $\frac{1}{2}$
D_2: center 2, radius $\frac{1}{2}$
D_3: center 4, radius $\frac{5}{6}$.
No, since D_1 and D_2 meet. The answer is yes for A^T, however.

3. (a) Center 0, radius 1

(b) Center 0, radius 2 — but notice that roots of $g(x)$ are also roots of $f(x)$ of part (a).

(c) Center 0, radius 1.1

5. (b) For $P = \left[\begin{array}{c|c} a & \\ b & * \\ c & \end{array}\right]$, $\quad Q = \left[\begin{array}{ccc} & * & \\ \hline e & f & g \end{array}\right]$ one has

$$PJ_1^nQ = n^2 \begin{bmatrix} a \\ b \\ c \end{bmatrix} [e \;\; f \;\; g] + B$$

where the entries of B will be very much smaller than n^2 for n sufficiently large.

EXERCISES 7.6*

1. $y(t) = \begin{bmatrix} \sqrt{2} - 1 & -\sqrt{2} - 1 \\ 1 & 1 \end{bmatrix} \begin{bmatrix} c_1 e^{(2+\sqrt{2})t} \\ c_2 e^{(2-\sqrt{2})t} \end{bmatrix}$

3. $y(t) = \begin{bmatrix} 1 & 1 & 1 \\ 0 & 1 & 0 \\ 0 & 1 & 1 \end{bmatrix} \begin{bmatrix} c_1 e^{t} \\ c_2 e^{2t} \\ c_3 e^{-t} \end{bmatrix}$

5. $y(t) = \frac{1}{2}(e^t + e^{-t}) = \cosh(t)$

7. $y(t) = e^t(\cos t - \sin t)$

EXERCISES 7.7

1. $\lambda = 1, 2, 3, 4$
3. $\lambda = 0, 1, 2, 3$
5. $\lambda = 1, 1, -1, -1$
7. $\lambda = 1, -1, -1, -1$
9. $\lambda = 1, 2, 3 \pm 4i$
11. $\lambda = 1 \pm 2i, 3 \pm 4i$

13. $x^4 - x^3 - 7x^2 + x + 6 = (x - 1)(x + 1)(x - 3)(x + 2)$

15. $x^5 - 15x^4 + 85x^3 - 225x^2 + 274x - 120 =$
$(x - 1)(x - 2)(x - 3)(x - 4)(x - 5)$

17. $x^4 - 8x^3 + 42x^2 - 80x + 125 = (x^2 - 6x + 25)(x^2 - 2x + 5)$

19. $x^6 - 4x^5 - x^4 + 30x^3 - 66x^2 + 64x - 24 =$
$(x - 1)(x - 2)^2(x + 3)(x^2 - x + 1)$

Note: Although the double eigenvalue 2 is computed somewhat inaccurately, the polynomial $(x - 2)^2 = x^2 - 4x + 4$ is computed quite accurately by *CXCOEF*.

21. The hint is sufficient.

23.
```
    ∇ Z←CXEIG A ;P
[1]   A←(A+.×P)⌹P←1 0,[1.5]A[;1]
[2]   Z←(-(A[2;2]*2)+4×A[1;2])*÷2
[3]   Z←(A[2;2],Z)÷2
    ∇
```

25. `(-2↑A)DBLSHFT A` or just A, since `-2↑A` is a zero matrix. It follows that $A^{16} = I$ and so Q is the result of `HHLDR` working on a matrix of the form αI. In this case `HHLDR` returns an identity matrix.

27. The hint is sufficient.

29. The hint is sufficient.

APPENDIX B

A Short List of APL Functions

Primitive Scalar Functions

Symbol	*Monadic function*	*Dyadic function*
+	Identity (The graph is $y = x$)	Addition
−	Negation −1 ¯2 is ¯1 2	Subtraction
×	$\times x = \begin{cases} 1 & \text{if } x > 0 \\ 0 & \text{if } x = 0 \\ ¯1 & \text{if } x < 0 \end{cases}$	Multiplication
÷	Reciprocal	Division
*	Exponential function $*x$ is e^x	Exponentiation $a*b$ is a^b
⍟ (○,*)	Natural logarithm ⍟x is $\ln x$	Logarithm a⍟b is $\log_a b$
⌈	Ceiling ⌈ 1.3 ¯.7 is 2 0	Maximum a⌈b is max (a, b)
⌊	Floor (greatest integer function) ⌊ 1.3 ¯.7 is 1 ¯1	Minimum a⌊b is min (a, b)
∣	Absolute value ∣¯3 is 3	Residue (remainder) 3∣14 is 2
! (',.)	Factorial !x is $\Gamma(x + 1)$	Binomial coefficient x!y is related to $B(x, y)$
○	π Times ○2 is 2π	Trigonometric and hyperbolic functions. See below
?	Roll ?5 is a random digit between 1 and 5	Deal 5?52 picks 5 random digits from 1 to 52 without replacement

Trigonometric and Hyperbolic Functions

Symbol	*Function*	*Symbol*	*Function*
0○x	$\sqrt{1 - x^2}$		
1○x	$\sin x$	¯1○x	$\mathrm{Sin}^{-1}\, x$
2○x	$\cos x$	¯2○x	$\mathrm{Cos}^{-1}\, x$
3○x	$\tan x$	¯3○x	$\mathrm{Tan}^{-1}\, x$
4○x	$\sqrt{1 + x^2}$	¯4○x	$\sqrt{x^2 - 1}$
5○x	$\sinh x$	¯5○x	$\mathrm{Sinh}^{-1}\, x$
6○x	$\cosh x$	¯6○x	$\mathrm{Cosh}^{-1}\, x$
7○x	$\tanh x$	¯7○x	$\mathrm{Tanh}^{-1}\, x$

Logical functions — all dyadic except ∼

∧	and	=	equal
∨	or	≠	not equal
$<$	less	⍲(∧,∼)	nand
$>$	greater	⍱(∨,∼)	nor
≤	not greater	∼	not
≥	not less		

Miscellaneous Functions

Symbol	*Name*	*Monadic*	*Dyadic*
⌹ (⎕,÷)	domino	⌹A is a left inverse for A	B⌹A is the least-squares solution of $AX = B$
⍳	index, index of	⍳n is the vector of integers 1 to n	V⍳A gives the index of the first occurrence of each component of A in the vector V
⍴	shape reshape	⍴A is a vector giving the shape of A	V⍴A reshapes A to an array of shape V
⍉(○,\)	transpose	⍉A is A^T	1 1⍉A is the main diagonal of A
↑	take	—	V↑A extracts components from A in the manner specified by V
↓	drop	—	V↓A deletes components from A in the manner specified by V
∊	member	—	A∊B flags the components of A that are components of B

Miscellaneous Functions (*Continued*)

Symbol	*Name*	*Monadic*	*Dyadic*
/	compress	—	*L*/*A* deletes components from *A* corresponding to the zeros in the logical vector *L*. Deletes columns from matrices
⌿(/,−)	compress	—	Same as / but deletes rows from matrices
\	expand	—	*L**A* expands *A* by putting zeros in the places corresponding to the zeros in the logical vector *L*. Adds columns to matrices
⍀(\\,−)	expand	—	Same as \\ but adds rows to matrices
⍒ (∣,⍒)	down-grade	*W*←⍒ *V* defines *W* to be the vector of indices such that *V*[*W*] has components arranged in descending order	—
⍋ (∣,⍋)	upgrade	*W*←⍋ *V* defines *W* to be the vector of indices such that *V*[*W*] has components arranged in ascending order	—
→	goto	→3 means *GOTO* line 3	—
,	ravel catenate	, *A* strings *A* out into a vector	*A* , *B* sticks arrays *A* and *B* together
⌽ (○,∣)	reverse rotate	⌽ *A* reverses the order of the components or columns of *A*	*V*⌽*Z* shifts row entries in manner specified by *V*
⊖(○,−)	reverse rotate	⊖ *A* reverses the order of the components or rows of *A*	*V*⊖*A* shifts column entries in manner specified by *V*

Operators

Symbol	*Name*	*Result*
/	reduction	$f/(v_1, v_2, v_3, \ldots)$ is $v_1 f v_2 f v_3 f \ldots$, where f is any primitive dyadic function. $+/$ is $\Sigma\, v_i$ and $\times/$ is Πv_i. Reduces the rows of matrices
⌿(/,–)	reduction	Same as / but reduces columns of matrices
.	inner product	Takes two primitive dyadic functions as argument $(v_1, v_2, \ldots) f.g (w_1, w_2, \ldots)$ is $(v_1 g w_1) f (v_2 g w) f \ldots$ $+ . \times$ is matrix multiplication
∘	jot	Turns inner-product operator into outer-product operator. $(v_1, v_2, \ldots) \circ .\ f(w_1, w_2 \ldots)$ is $v_1 fw,\ v_1 fw_2\ v_1 fw_3 \ldots$ $v_2 fw_1,\ v_2 fw_2\ v_2 fw_3 \ldots$ $v_3 fw_1,\ v_3 fw_2\ v_3 fw_3 \ldots$

APPENDIX C

Some Miscellaneous APL

The mechanics of writing, editing, and debugging APL functions vary with the computer system used. For IBM systems any commercially published APL manual will do. For other systems the computer-center personnel should be consulted. A few facts may hold generally, however.

SUSPENDED FUNCTIONS

When an error occurs in the execution of a user-defined function, the function is *suspended.* Suspended functions often cause confusion for beginners because the workspace is in a subtly altered state.

When a function, for example *GAUSS* of Section 3.5, suspends an error message accompanied by the line on which the error occurs prints out. For example,

```
SYNTAX ERROR
GAUSS [4]   A←A,IS 1↑T
              ∧
```

(The problem here is that a nonexistent function, IS, has been called.) At this point all the local variables in *GAUSS* that have already been assigned values (*A*, *S*, *P*, *L*, *T*) are alive in the workspace. They "mask" any other variables with these names that may have been defined before the function was called. Thus, for example, a matrix *P* may be mysteriously changed to the scalar 0 (see line [2] of GAUSS). When the suspension is removed, however, the original *P* will be available again. The confusion is compounded by the fact that the suspended function may be executed again and again, resulting in additional suspensions. If a function has been suspended more than once, however, it cannot be edited or displayed! At this point there is a real danger that you may hurt yourself while destroying the keyboard.

When an error occurs and a function suspends,

1. Do not call the function again.
2. Display the local variables (type their names) if you think that knowing their values may help to find the trouble.

3. Remove the suspension. This is usually done by typing →0 or just → (called the *niladic branch*). In some systems special action may be necessary when a recursive function suspends.

To get a list of suspended functions type)SI. (SI stands for *state indicator* — an SI DAMAGE message usually means that you must remove some function suspensions before the machine will let you proceed.)

TRACE VECTOR, STOP VECTOR

The trace vector and stop vector are debugging aids. The trace vector for the function GAUSS is denoted T∆GAUSS and the stop vector for GAUSS is denoted S∆GAUSS. These so-called vectors are lists of line numbers. Typing

```
T∆GAUSS←2  3  8
```

sets trace flags on lines 2, 3, and 8 of *GAUSS*. Whenever these lines of *GAUSS* are executed, a message to that effect is printed and the last quantity computed on the line is printed. For example,

```
GAUSS[2]  0
```

To turn off tracing, use the expression T∆GAUSS←⍳0. The stop vector is similar to the trace vector. It is used to suspend a function at a specific line in order to inspect the values of the local variables. The expression S∆GAUSS←5 will cause GAUSS to suspend just before line [5]. After the local variables have been inspected, the function may be restarted on line [5] by typing →5.

Some systems do not have trace and stop vectors.

SOME SYSTEM COMMANDS

)CONTINUE
: Signs off APL and saves contents of current workspace for automatic reload at next signon.

)COPY 240 ROWREDUCTION GAUSS PROBLEM1
: Goes to Library 240, workspace ROWREDUCTION, finds the function GAUSS and the matrix PROBLEM1, and adds them to the active workspace.

)DIGITS 3
: Causes all numbers to be displayed to three significant digits. Does not affect the number of digits used for computation or storage. An alternate command is ⎕PP←3.

)ERASE GAUSS PROBLEM1
: Erases the function GAUSS and the variable PROBLEM1 from the active workspace.

)LIB
: Lists your saved workspaces.)LIB 6 lists the workspaces in public library 6.

)LOAD SAM
: Loads your saved workspace SAM into the active workspace. Destroys current contents.)LOAD 6 SAM gets WS SAM from public library 6.

)OFF

Signs off APL. Current contents of active workspace discarded.

)SAVE SAM

Saves current workspace (which may be called SAM, CLEAR, or CONTINUE) under the name SAM.

)SI

List of pendent and suspended functions. Suspensions denoted by *. (*Pendent* functions are waiting for results from another function.)

Index